T3-BSE-389

0800-2

16-02-89

IN.

Elementary
Mathematics
for Computing

Elementary Mathematics for Computing

Larry R. Lance
Columbus Technical Institute

John R. Hinton
Otterbein College

RETIRÉ DE LA COLLECTION · UQO

UNIVERSITÉ DU QUÉBEC À HULL
BIBLIOTHÈQUE

Addison-Wesley Publishing Company
Reading, Massachusetts • Menlo Park, California
Don Mills, Ontario • Wokingham, England • Amsterdam
Sydney • Singapore • Tokyo • Mexico City
Bogotá • Santiago • San Juan

QA
76
.9
M 35
L 35
1986

06-10216663

Sponsoring Editor: Jeffrey Pepper
Production Supervisor: Herbert Nolan
Editorial, Design, and Production Services: Quadrata, Inc.
Illustrator: Scientific Illustrators
Manufacturing Supervisor: Ann DeLacey
Cover Designer: Marshall Henrichs

CREDITS

p. 120, The Granger Collection. p. 245, The Bettman Archive. Fig. 13.1, parts (d), (e), (i), (k), (l), and Fig. 13.5(b), reprinted with permission of *USA Today*. Fig. 13.10(a), copyright 1984 *USA Today*; reprinted with permission. Fig. 13.12(a), U.S. Bureau of the Census: Statistical Abstract of the United States: 1985 (105th ed.), Washington, D.C., 1984.

Library of Congress Cataloging in Publication Data

Lance, Larry R., 1942–
    Elementary mathematics for computing.

    Includes index.
    1. Electronic data processing—Mathematics.
I. Hinton, John R. 1943–   II. Title.
QA76.9.M35L36 1986   519.4   85-1258
ISBN 0-201-05123-0

Copyright © 1986 by Addison-Wesley Publishing Company, Inc.
All rights reserved. No part of this publication may be reproduced, stored in a retrieval system, or transmitted, in any form or by any means, electronic, mechanical, photocopying, recording, or otherwise, without the prior written permission of the publisher. Printed in the United States of America. Published simultaneously in Canada.

ABCDEFGHIJ-MA-89876

Dedication

To my two lovely daughters, Jackie and Jennifer, in appreciation for their encouragement and inspiration through the completion of this project.

L.R.L.

My grateful appreciation and sincere thanks to my loving, encouraging, and sacrificing family: my parents LeRoy and Wilma Hinton, my wife Barbara, and my son Andy; and also to an educator who has had a profound effect on my professional growth, Dr. Harold Trimble.

J.R.H.

Preface

"I know what you're thinking about, but it isn't so, nohow." . . . "if it was so, it might be; and if it were so, it would be; but as it isn't, it ain't. That's logic." So goes the conversation between Tweedledum and Tweedledee, overheard by Alice in *Through the Looking Glass* written by the English mathematician, Charles Lutwidge Dodgson, better known in literary circles as Lewis Carroll.

This is logic to Tweedledum and Tweedledee. But what is logic to the modern student beginning to prepare for a life of work with computers? To answer this and many other similar questions, logic is just one of the topics examined in this text *designed to present mathematical skills and concepts that are important to beginning students in the world of computing*. Although intended for students who have completed at least one year of high school algebra, this text begins with a brief review of algebra. To be successful with the concepts that follow in succeeding chapters, a student must be familiar with the elementary algebraic concepts presented in Chapter 0. For a brief overview of those topics in mathematics essential to computing, Chapter 1 presents an introduction with references. The instructor is encouraged to use the information in this chapter to discuss the content and relevancy of the mathematics to computing.

The book was designed for students studying data processing and/or computer programming in two-year colleges. The text is also appropriate for use in a four-year college or university as a precalculus course in a computer science program. The content and organization make the text suitable for use in a two-academic-quarter sequence or in a one-semester course. The following are the book's special pedagogical features.

- **Intuitive Approach.** Each topic is presented from an intuitive standpoint. The mathematics is correct and accurately stated but the emphasis is on understanding rather than on the rigor of mathematical theory. Each topic is carefully explained and illustrated with appropriate examples.

- **Flexibility.** The layout and order of the chapters and the sequencing of topics will allow individual schools and/or faculty to pick and choose the specific topics best suited to their course(s). A list of chapter objectives is included with each chapter.

- **Examples.** Hundreds of carefully selected examples with their solutions are included to illustrate both the technique and the application of various concepts.

- **Illustrations.** A more than ample supply of flowcharts, graphs, tables, charts, simplex tableaus, sample programs, and statistics articles are contained throughout as visual aids to accompany the narrative.

- **Chapter Exercises.** Enough exercises are included in each chapter to provide the student with a more than adequate opportunity to practice and master each concept.

- **Summaries.** A chapter summary listing important terms, concepts, and formulas is presented at the end of each chapter.

- **Review Exercises.** At the end of each chapter are review exercises that provide additional student practice on the topics presented in the chapter. These problems can also be used by the instructor for testing.

- **Solutions.** Answers to the odd-numbered problems are included at the back of the book.

- **Glossary.** A glossary of all important terms appears at the end of the text. This should provide a quick source of explanation and definitions for the student. Terms given in boldface type throughout the text will be found in the glossary.

- **Appendixes.** Tables for conversion between decimal and hexadecimal and decimal and octal numeration systems are provided in two appendixes.

- **Instructor's Manual.** A complete Instructor's Manual is also available, and includes not only the answers but also the solutions to all even-numbered exercises, as well as two tests for each chapter. Also included are an introduction to the computer language BASIC and an assortment of BASIC programs.

Acknowledgments

We would like to thank the following reviewers for their helpful comments and suggestions: Ken Balts, Western Wisconsin Technical Institute; Harvey Braverman, NYC Technical College, City University of New York; Dawson Carr, Sandhills Community College; Louis Edwards, Valencia Community College; Lister Horn, Pensacola Junior College; Byron V. Johnson, Southern Illinois University at Carbondale; Charles McSurdy, Nashville State Technical Institute; Nelson G. Rich, Monroe Community College; Al Schroeder, Richland College; Elaine Shillito, Clark Technical College; Paul Szuch, Niceland, Florida; and Ralph M. Whitmore, Southwestern University. We are grateful also for the assistance of Amy Stearns, John Horvath, and J. R. Titko in preparing the answers and solutions.

L.R.L.
J.R.H.

Contents

3
Algorithms, Flowcharts, Pseudocode, and Decision Tables   64

4
Decimal and Nondecimal Numeration Systems   80

5
Sets   118

6
Intermediate Algebra   144

Basic Elements of Algebra: A Review

CHAPTER 0 OBJECTIVES

After completing this chapter, you should be able to:

1. Find the value of an algebraic expression, given numeric values for the literals.
2. Perform computations involving integral exponents.
3. Add, subtract, multiply, and divide algebraic expressions.
4. Simplify algebraic expressions.
5. Solve first-degree equations and inequalities.
6. Factor expressions containing common factors, differences between perfect squares, perfect-square trinomials, and quadratic trinomials.
7. Add, subtract, multiply, and divide algebraic fractions.
8. Solve literal equations and formulas for any one of the literals.
9. Translate a verbal problem into an algebraic equation and solve the equation.

0

0.1

The Real-Number System

An important basis on which to build a good working knowledge of the mathematics of science, engineering, business, and digital computers is to understand how the concepts of our numeration system have developed. The first human attempts to count involved setting up a one-to-one correspondence between a set of objects, such as fingers or stones, and a set of possessions. For example, in order to count the sheep in a flock, a stone would be set aside for each sheep that passed through a gate. Later, when the flock returned, a stone would be replaced for each sheep that passed back through the gate. If any stones remained after the sheep returned, it was clear that some sheep had been lost.

1

As human intelligence and ability to symbolize relationships improved, the Hindu-Arabic numerals 0, 1, 2, 3, 4, 5, 6, 7, 8, 9 evolved as symbols representing quantity relationships. The principle of place value gave this system a versatility some earlier systems such as those developed by the Chinese, Egyptians, Babylonians, Greeks, and Romans lacked. Its evolution took thousands of years. The Egyptian numeration system dates back to at least 3400 B.C., while the Babylonian system has been traced to between 2000 and 3000 B.C. The Hindu-Arabic system, however, was not completed until around A.D. 800 and is still the system in use today. Its numerals became known as the *counting numbers,* although we formally refer to this set as the **natural numbers.**

■

NATURAL NUMBERS = {1, 2, 3, 4, 5, 6, 7, 8, 9, 10, 11, ...}

This set of numerals proved sufficient as problems of everyday life led to the development of the operations of addition and multiplication. Further dealings with other people led to the development of the operations of subtraction and division. However, whereas the natural numbers had been sufficient to handle all addition and multiplication problems, deficiencies within the set became apparent when certain subtraction and division problems were attempted. For example, what natural numbers answer the problems 7 − 7 and 3 − 7 or the problem 4/8? Since no natural number satisfied any of these problems, it was clear that additional numerals were required. The ancient Arabs are given credit for inventing a symbol for zero as the symbol representing the solution to problems like 7 − 7. The negatives of the natural numbers were formed as solutions to problems like 3 − 7. Thus, 7 − 7 = 0 and 3 − 7 = −4. The resulting number system is called the system of **integers.**

■

INTEGERS = {..., −5, −4, −3, −2, −1, 0, 1, 2, 3, 4, 5 ...}

People were able to visualize an *order* to this system on a graph called a *number line,* as shown in Fig. 0.1. Once again this evolution took hundreds of years, and the negative integers did not appear on the number line until the seventeenth century.

On this line, any number to the left of another number is less than that number, and any number to the right is greater than that num-

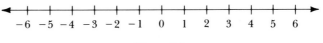

Figure 0.1

ber. We express this relationship using the less than symbol , <, or the greater than symbol, >.

■ **EXAMPLE 1**

a) $-4 < -1$ and conversely $-1 > -4$
b) $-1 < 1$ and conversely $1 > -1$
c) $3 < 7$ and conversely $7 > 3$ □

Within the system of integers it then became possible to perform any addition, subtraction, or multiplication problem, although division remained a problem. Division was discovered through its relationship to multiplication, as, for example, $12/3 = 4$ only because $3 \times 4 = 12$. Integers were adequate for a division of this type, but people could not find an integer to satisfy a problem like $4/8$. The resulting multiplication, $8 \times ? = 4$, obviously could not be an integer since multiplication of positive integers always resulted in a product greater than either factor. Clearly this was different since the product, 4, was smaller than the factor 8. Thus a new number, called a fraction, was introduced into the system, and it represented the quotient of two integers. It provided the solution to the division problem $4/8$, since it could now be expressed as $8 \times 1/2 = 4$.

The addition of fractions to the system of integers completes the *rational number system*. It is not possible to list rational numbers as members of a set as we did with the natural numbers and the integers, because between any two rational numbers there is an infinite number of other rational numbers. For example, between $1/2$ and $3/4$ are the rational numbers $3/5$, $2/3$, $7/10$, and $18/25$, to name just a few. We can use the number line to show the rationals, but remember that although every rational number is on the line, not every point on the line represents a rational number.

A **rational number** is defined as any number that can be put in the form of a/b, where a and b are integers, but $b \neq 0$. Note that such integers as 5 and -3 satisfy this definition since $5 = 5/1$ and $-3 = -3/1$.

For a period of time it was thought that the invention of the rationals completely filled the number line. However, because of the early Greek mathematicians' preoccupation with associating numbers with line-segment lengths (approximately 600 B.C.), a number was found that did not satisfy the definition of a rational number. Up until that time the popular belief was that any length that could be found by a geometric construction should represent a number. However, the Pythagorean Theorem showed that a line segment could be found with length $\sqrt{2}$. This number did not fit into any of the known number groups. The radical was not the problem: Man could deal with radicals like $\sqrt{4}$ and $\sqrt{9}$ and knew they were rational. But $\sqrt{2}$ was a puzzle since no rational number could be found whose square was 2. This meant that $\sqrt{2}$ could not belong to the group of rationals. So a new group—the **irrational numbers**—was named, and $\sqrt{2}$ was the first known irrational number. Other examples from this number

---

THE REAL-NUMBER PROPERTIES

1. CLOSURE FOR ADDITION: Whenever two real numbers are added, the result is also a real number.

2. COMMUTATIVE PROPERTY FOR ADDITION: The order property

$$a + b = b + a$$

3. ASSOCIATIVE PROPERTY FOR ADDITION: The grouping property

$$(a + b) + c = a + (b + c)$$

4. IDENTITY ELEMENT FOR ADDITION

$$a + 0 = a$$

5. ADDITIVE INVERSE

$$a + (-a) = 0$$

6. CLOSURE PROPERTY FOR MULTIPLICATION: When any two real numbers are multiplied, the product is a real number.

7. COMMUTATIVE PROPERTY FOR MULTIPLICATION

$$a \times b = b \times a$$

8. ASSOCIATIVE PROPERTY FOR MULTIPLICATION

$$(a \times b) \times c = a \times (b \times c)$$

9. IDENTITY ELEMENT FOR MULTIPLICATION

$$a \times 1 = a$$

10. MULTIPLICATIVE INVERSE

$$a \times 1/a = 1, \qquad a \neq 0$$

11. DISTRIBUTIVE PROPERTY

$$a(b + c) = ab + ac$$

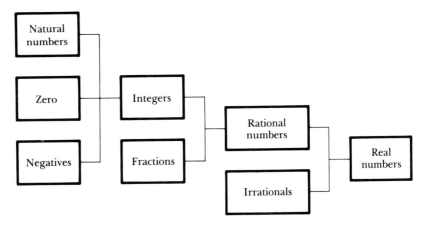

Figure 0.2 The real-number system

group are $\sqrt{3}$, $\sqrt{5}$, $\sqrt{7}$, $\sqrt{11}$, and π. With the discovery of the irrationals, all remaining gaps on the number line were filled. The conclusion that can be drawn from this is that a number may be either rational or irrational, but not both. When rationals and irrationals are combined, the result is called the *real-number system,* a diagram of which is shown in Fig. 0.2.

A set of rules was necessary in order to have a system of numbers together with operations on those numbers. These rules, called the *real-number properties,* control the operations with the real numbers.

Note that no properties are stated for subtraction and division. In the algebra of real numbers, subtraction is defined as the addition of two numbers with unlike signs.

■ **DEFINITION**

SUBTRACTION

The addition of numbers with unlike signs, that is,

$$a - b = a + (-b).$$

■ **DEFINITION**

DIVISION

The multiplication of the numerator by the reciprocal of the divisor, that is,

$$a/b = a \times 1/b, \qquad b \neq 0.$$

When dealing with real numbers, we are occasionally concerned only with the magnitude of a number, and not its direction from zero. The magnitude of a number is called the **absolute value** and refers to that number's distance from zero. Vertical bars are used as the symbol for absolute value.

■ **DEFINITION**

ABSOLUTE VALUE

The distance between the number and zero on the number line is its *absolute value*.

We can determine the absolute value of a number by using the following rule.

■

Rule for determining absolute value:

$$|a| = \begin{cases} a & \text{if } a > 0, \\ 0 & \text{if } a = 0, \\ -a & \text{if } a < 0. \end{cases}$$

■ **EXAMPLE 2**

Find the absolute value of (a) 5, (b) 0, and (c) −5.

a) $|5| = 5$
b) $|0| = 0$
c) $|-5| = -(-5) = 5$ □

Exercise Set 0.1

For Exercises 1–5, be able to explain your answer. Whenever your answer is no, give an example to illustrate.

1. Is the set of odd natural numbers, that is, {1, 3, 5, 7, 9, 11, ...}, closed with respect to addition? multiplication?

2. Is the set of even natural numbers closed with respect to addition? multiplication?

3. Is the set of integral powers of 3 > 0, that is, {1, 3, 9, 27, 81, ...}, closed with respect to addition? multiplication?

4. Is the set of negative numbers closed with respect to addition? multiplication?

5. Is the set {0, 1} closed with respect to addition? multiplication? (*Hint:* This is a finite set. Consider all possible addition and multiplication combinations.)

State which real-number property is illustrated in each of the following.

6. $3.2(4.7x + 2.1y) = 15.04x + 6.72y$

7. $(4x * y) * 3 = 4x * (y * 3)$ [The symbol $*$ is used to indicate multiplication in some computer languages.]

8. $17 + (-17) = 0$

9. $(4 + 2x) + y = (2x + 4) + y$

10. $23 * (1/23) = 1$

In Exercises 11 and 12, be prepared to explain your answer.

11. Does $|a - b| = |b - a|$ for any real numbers a and b?

12. Does $|a| + |b| = |a + b|$? Are there exceptions? What could we say in general?

13. If a and b are real numbers, consider the implications of a and b in each of the following.

    a) $a * b < 0$ **b)** $a * b = 0$ **c)** $a * b > 0$

0.2
Operations with Zero

Zero is the only number that requires special attention when reviewing algebra. The operations with zero are defined as follows, with a being any real number.

■ **DEFINITION**

a) $a + 0 = a$

b) $a \cdot 0 = 0$

c) $\dfrac{0}{a} = 0$, when $a \neq 0$

Note that division by zero is not defined. It is easy to understand why when we look at division, as follows.

■ **DEFINITION**

$$\frac{a}{b} = q \text{ only if } a = b \cdot q$$

Consider 5/0: If we say that $5/0 = 0$, then the corresponding multiplication is not true, since $5 \neq 0 \times 0 = 0$. If we say that $5/0 = 5$, we get another false result, since $5 \neq 0 \times 5 = 0$. In fact, the same happens for any q. The conclusion is that division by zero is not defined. The only exception might occur if you consider $0/0 = q$. Actually any q satisfies the corresponding multiplication since $0 = 0 \times 1$ and $0 = 0 \times 17$, and so on. Since no one value of q can be determined, we assign no value to 0/0. Again the conclusion is that division by zero is not defined.

0.3
Properties of Integral Exponents

A number of the form a^n is called an exponential number and is an abbreviation for the repeated multiplication of a, n times. In other words, a is taken as a factor n times:

$$\underbrace{a \cdot a \cdot a \cdot a \cdot \ldots \cdot a = a^n}_{n}$$

a is called the base and

n is the exponent or power

$a^1 = a,$

$a^2 = a \cdot a,$ **a squared**

$a^3 = a \cdot a \cdot a,$ **a cubed**

$a^4 = a \cdot a \cdot a \cdot a,$ **a to the fourth power**

etc.

In algebra we are constantly encountering exponential numbers and therefore need rules with which to operate. If a, b, m, and n are any integers, we have the following.

LAWS OF EXPONENTS	DEFINITIONS
1. $a^m \cdot a^n = a^{m+n}$	1. $a^0 = 1$, if $a \neq 0$
2. $\dfrac{a^m}{a^n} = a^{m-n}$, if $a \neq 0$	2. $a^{-n} = \dfrac{1}{a^n}$, if $a \neq 0$
3. $(a^m)^n = a^{mn}$	
4. $(ab)^n = a^n b^n$	
5. $\left(\dfrac{a}{b}\right)^n = \dfrac{a^n}{b^n}$, if $b \neq 0$	

■ **EXAMPLE 3**

a) $2^3 \cdot 2^4 = 2 \cdot 2 \cdot 2 \cdot 2 \cdot 2 \cdot 2 \cdot 2 = 2^{3+4} = 2^7$ **By Law 1**

b) $\dfrac{2^5}{2^2} = 2^{5-2} = 2^3$ **By Law 2**

c) $\dfrac{2^2}{2^5} = 2^{2-5} = 2^{-3} = \dfrac{1}{2^3} = \dfrac{1}{8}$ **By Law 2 and Definition 2**

d) $\dfrac{2^3}{2^3} = 2^{3-3} = 2^0 = 1$ **By Law 2 and Definition 1**

e) $(2^3)^2 = 2^6$ **By Law 3**

f) $(2x^3)^2 = 2^2 \cdot x^6 = 4x^6$ **By Law 3**

g) $\left(\dfrac{x}{2}\right)^3 = \dfrac{x^3}{2^3} = \dfrac{x^3}{8}$ **By Law 5**

h) $2^{-3} \cdot 2^4 = 2^{-3+4} = 2^1$ **By Law 1**

i) $\dfrac{2^{-3}}{2^4} = 2^{-3-4} = 2^{-7} = \dfrac{1}{2^7}$ **By Law 2 and Definition 2**

j) $\dfrac{2^3}{2^{-4}} = 2^{3-(-4)} = 2^{3+4} = 2^7$ **By Law 2** □

Exercise Set 0.3

Perform the indicated operations, if possible.

1. 7×0 **2.** $0 \times (-3)$ **3.** $\dfrac{0}{17}$ **4.** $\dfrac{-16}{0}$ **5.** $\dfrac{0}{0}$

Perform the indicated operations using the laws and definitions of exponents.

6. $x^4 \cdot x^2$ **7.** $y^{-2} \cdot y^6$ **8.** $x^0 \cdot x^5$

9. $3^2 \cdot 3^3$ **10.** $(a^2)^5$ **11.** $(-3b^2)^2$

12. $\dfrac{x^7}{x^2}$ **13.** $\dfrac{x^3}{x^5}$ **14.** $\dfrac{y^{-3}}{y^4}$

15. $\dfrac{x^2}{x^{-6}}$ **16.** $\dfrac{a^4}{a^0}$ **17.** $\dfrac{b^0}{b^3}$

18. $\dfrac{(6x^3)^2}{(4x^2)^3}$ **19.** $(-2x^3)(-2x)^3$ **20.** $x^{m-2} \cdot x^2$

21. $\dfrac{x^{a+2}}{x^{a-2}}$

Perform the indicated operations and simplify. Do not express your answers with zero or negative exponents.

22. $(2x^2y)^{-2}$ **23.** $(3x^{-2}y^2)^{-3}$ **24.** $[(-2a^2b)^2]^3$

25. $\left(\dfrac{a^2}{b^3}\right)^{-2}$ **26.** $\dfrac{-2x}{y^{-2}}$ **27.** $(-3a^2b^{-1})^{-2}$

28. $\dfrac{3x^2}{2y^3} \cdot \dfrac{y^3}{x^4}$ **29.** $\left(\dfrac{2x^3}{3y^2}\right)^{-2}\left(\dfrac{-2x^2}{3y}\right)^3$

0.4

Evaluating Algebraic Expressions

The most basic unit in algebra is the algebraic term. An *algebraic term* is the product of numerical and literal factors. **Literals** are nonnumerical symbols used to represent sets of numbers, and they may be either *constants* or *variables*. Typically, letters at the beginning of the alphabet (*a*, *b*, *c*, etc.) represent constants whereas letters at the end of the alphabet (*x*, *y*, *z*, etc.) represent variables. Some examples of algebraic terms are $3xy$, $-2x^2$, ab, $-a^2cd^3$, and 10. The numerical factors of algebraic terms are called *numerical coefficients*. In the term $3xy$, 3 is the numerical coefficient. In ab, the numerical coefficient is understood to be 1 and in $-a^2cd^3$, it is understood to be -1.

Algebraic terms with exactly the same literals raised to identical powers are called *like* terms (for example, $3x^2y$ and $-7x^2y$). Note that like terms need not possess the same numerical coefficients. However, $3x^2y$ and $3xy^2$ are *not* like terms even though they have the same numerical coefficients. All constant terms are considered like terms.

When algebraic terms are combined by addition and/or subtraction, the result is called an *algebraic expression*. An algebraic expression consisting of only one term is a *monomial;* an expression consisting of two terms is a *binomial;* three terms, a *trinomial*. Although specific names can be given to expressions of more than three terms, we generally refer to these as *polynomials* (also *multinomials*). A polynomial is actually any expression containing two or more terms, although we use the more specific names binomial and trinomial for expressions containing two and three terms, respectively.

■ **EXAMPLE 4**

a) Monomial: $-3xy^2z$
b) Binomial: $-3xy^2z + 2ab$
c) Trinomial: $-3xy^2z + 2ab + 5c$
d) Polynomial: $-3xy^2z + 2ab + 5c - 10$ □

Any algebraic expression, such as those in Example 4, is nothing more than a general symbol and does not represent any particular number until we substitute values for specific literals. The value of an algebraic expression is thus dependent on the actual number values substituted into the expression.

■ **EXAMPLE 5**   Evaluate the expression $-3xy^2z$ if $x = 1$, $y = -2$, and $z = 3$.

Solution   Making the substitutions, we have

$$-3xy^2z = -3(1)(-2)^2(3) = -3(1)(4)(3) = -36. \quad □$$

■ **EXAMPLE 6**   Evaluate the expression $-3xy^2z$ if $x = -1$, $y = 2$, and $z = 3$.

Solution   This time we have

$$-3xy^2z = -3(-1)(2)^2(3)$$
$$= -3(-1)(4)(3)$$
$$= 36. \quad \square$$

We must be concerned with the order in which we perform the arithmetic operations when finding the value of an algebraic expression. This order, called the *hierarchy of operations,* is as follows. The first operation has the highest priority.

■

> HIERARCHY OF OPERATIONS
> 1. Quantities: Operations within the grouping symbols
> 2. Exponentiations
> 3. Multiplication and/or division
> 4. Addition and/or subtraction

When we encounter two or more operations of the same level, we perform the operations from left to right.

Exercise Set 0.4

Find the value of each of the following expressions if $x = 1$, $y = -2$, and $z = 3$.

1. $xy - z$

2. $x - 2y$

3. $x^2 - y^2$

4. $x^2 + y^2$

5. $(x - y)^2$

6. $(x + y)^2$

7. $x - 2(y - z)$

8. $x^2 + 4(z - y)$

9. $xy^2 - z$

10. $3(x - y) + z$

11. $xy - (y^2 - z)$

12. $2xy - z^2$

13. $xz(2 - y)$

14. $2x[y - 2(x - 3)]$

15. $2y[3 - x(z + 2)]$

16. $-2[(x - y) - z]$

17. $3x[y - 2(xz)]$

18. $-\{x - 2[x - (y + z)]\}$

0.5

Operations with Polynomials

Addition and Subtraction

In an algebraic expression, only like terms can be combined. We do so by adding and/or subtracting their numerical coefficients. Two or more algebraic expressions can be added by removing the parentheses and combining the like terms.

■ **EXAMPLE 7**

a) Simplify by combining like terms: $2xy + 5xy - 3xy$.

$$2xy + 5xy - 3xy = (2 + 5 - 3)xy$$
$$= 4xy$$

b) Simplify by removing the parentheses and combining like terms: $(3ab + 7ac) + (2ab - 3ac)$.

$$(3ab + 7ac) + (2ab - 3ac) = 3ab + 7ac + 2ab - 3ac$$
$$= 3ab + 2ab + 7ac - 3ac$$
$$= 5ab + 4ac \quad \square$$

One algebraic expression can be subtracted from another by removing the parentheses, changing the signs on *each* term of the expression, and then combining like terms.

■ **EXAMPLE 8**

a) Subtract $(2ab - 3ac)$ from $(3ab + 7ac)$.

$$(3ab + 7ac) - (2ab - 3ac) = 3ab + 7ac - 2ab + 3ac$$
$$= 3ab - 2ab + 7ac - 3ac$$
$$= ab + 10ac$$

b) Simplify by removing parentheses and combining like terms: $(3ab + 7ac) - (2ab - 3ac) - (-5ab + 2ac - 3)$.

$$(3ab + 7ac) - (2ab - 3ac) - (-5ab + 2ac - 3)$$
$$= 3ab + 7ac - 2ab + 3ac + 5ab - 2ac + 3$$
$$= 3ab - 2ab + 5ab + 7ac + 3ac - 2ac + 3$$
$$= 6ab + 8ac + 3 \quad \square$$

Simplifying Nested Algebraic Expressions

Sometimes one expression may be contained as part of another expression. If this is the case, we use the grouping symbols, brackets [] and braces {}, along with parentheses. Such an expression can be simplified by working from the inside out by first removing the innermost grouping symbols and combining like terms, and then removing the next set of grouping symbols and combining like terms, and continuing this process until all grouping symbols have been removed and all like terms have been combined.

■ **EXAMPLE 9**

From each of the following, remove the grouping symbols, starting with the innermost quantity, and combine like terms.

a)

$2\{x - 7[5 - 2(x + 4)]\}$	**First remove parentheses**
$2\{x - 7[5 - 2x - 8]\}$	**Combine like terms**
$2\{x - 7[-2x - 3]\}$	**Remove brackets**
$2\{x + 14x + 21\}$	**Combine like terms**
$2\{15x + 21\}$	**Remove braces**
$30x + 42$	

b)

$2 - \{5 - 3[x - 2(3 - x) + 4]\}$	**Remove parentheses**
$2 - \{5 - 3[x - 6 + 2x + 4]\}$	**Combine like terms**
$2 - \{5 - 3[3x - 2]\}$	**Remove brackets**
$2 - \{5 - 9x + 6\}$	**Combine like terms**
$2 - \{11 - 9x\}$	**Remove braces**
$2 - 11 + 9x$	**Combine like terms**
$9x - 9$ □	

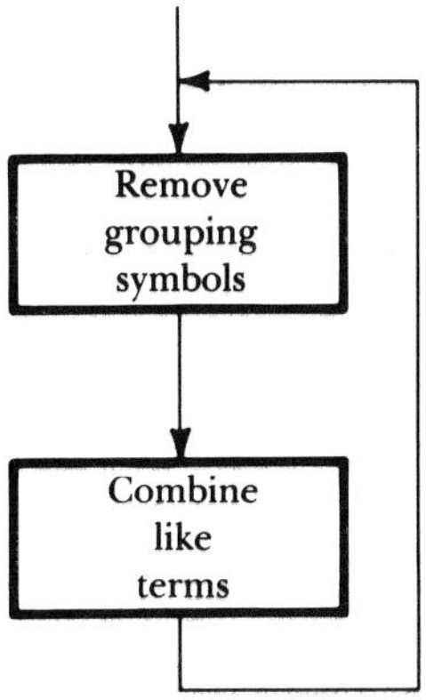

Consider the flowchart in Fig. 0.3, which describes the repetitive nature of the steps required to simplify this type of expression.

Continue in this manner, removing grouping symbols and combining like terms until all grouping symbols have been removed and all like terms have been combined.

Figure 0.3

Multiplication and Division

There are several different forms of multiplication involving algebraic expressions. The most basic form involves *multiplying monomials*. To multiply two or more monomials:

1. Multiply the coefficients, paying close attention to the law of multiplication of signed numbers.
2. Multiply the *like* literals together, using the proper law of exponents.
3. Express the product of these different factors.

The associative and commutative laws are used in these steps.

■ **EXAMPLE 10**

a) Multiply $-3x^2yz$ and $2xy^3z$.

$$(-3x^2yz)(2xy^3z) = (-3)(2)(x^2)(x)(y)(y^3)(z)(z) = -6x^3y^4z^2$$

b) Find the following product: $(-2abc)(3a^2b^3c)(-4ab^2)$.

$$(-2abc)(3a^2b^3c)(-4ab^2) = (-2)(3)(-4)(a)(a^2)(a)(b)(b^3)(b^2)(c)(c)$$
$$= 24a^4b^6c^2 \quad □$$

Another form of multiplication involves multiplying a monomial and a polynomial. To find a product of this type multiply the monomial times *each* term of the polynomial using the *distributive law*. The laws for multiplying signed numbers and multiplying exponents must be followed.

■ **EXAMPLE 11**

Find each of the following products.

a) $2x(3xy - 5xy^2 + 4) = 2x(3xy) + 2x(-5xy^2) + 2x(4)$
$$= 6x^2y - 10x^2y^2 + 8x$$

b) $-3x^2y(4xz - 3x^2y^2 + 2y)$
$$= -3x^2y(4xz) + (-3x^2y)(-3x^2y^2) + (-3x^2y)(2y)$$
$$= -12x^3yz + 9x^4y^3 - 6x^2y^2 \quad \square$$

We can also multiply two polynomials. To do so, we multiply *each* term of the first polynomial by each term of the second polynomial, using the distributive property. Again, we must remember to follow the laws of signed numbers and multiplication of exponents. If any *like* terms result from the multiplication, they must be combined so that the final product is expressed in simplified form.

■ **EXAMPLE 12**

Find each of the following products.

a) $(x + 2)(2x - 5) = x(2x - 5) + 2(2x - 5)$
$$= 2x^2 - 5x + 4x - 10$$
$$= 2x^2 - x - 10$$

b) $(3x - 4)(2x - 3) = 3x(2x - 3) - 4(2x - 3)$
$$= 6x^2 - 9x - 8x + 12$$
$$= 6x^2 - 17x + 12$$

c) $(2x - 3)(x^2 - 4x + 5) = 2x(x^2 - 4x + 5) - 3(x^2 - 4x + 5)$
$$= 2x^3 - 8x + 10x - 3x^2 + 12x - 15$$
$$= 2x^3 - 11x^2 + 22x - 15$$

d) $(3x - 4)^2 = (3x - 4)(3x - 4)$
$$= 3x(3x - 4) - 4(3x - 4)$$
$$= 9x^2 - 12x - 12x + 16$$
$$= 9x^2 - 24x + 16$$

e) $(x - 5)(2x + 3)(3x - 4) = [x(2x + 3) - 5(2x + 3)](3x - 4)$
$$= [2x^2 + 3x - 10x - 15](3x - 4)$$
$$= (2x^2 - 7x - 15)(3x - 4)$$
$$= (3x - 4)(2x^2 - 7x - 15)$$
$$= 3x(2x^2 - 7x - 15) - 4(2x - 7x - 15)$$
$$= 6x^3 - 21x^2 - 45x - 8x^2 + 28x + 60$$
$$= 6x^3 - 29x^2 - 17x + 60 \quad \square$$

When dividing a monomial by another monomial, we must pay close attention to the signs of the coefficients. We divide the *like* literals using the laws and definitions for dividing exponents.

■ **EXAMPLE 13**   **a)** Divide $-18x^3y^2z$ by $3xyz$.

$$\frac{-18x^3y^2z}{3xyz} = \frac{-18}{3}x^{3-1} \cdot y^{2-1} \cdot z^{1-1}$$

$$= -6x^2y$$

b) Divide $-32a^2bc^3$ by $-8ab^2c$.

$$\frac{-32a^2bc^3}{-8ab^2c} = 4ab^{-1}c^2 \quad \text{or} \quad \frac{4ac^2}{b} \quad \square$$

If a polynomial is to be divided by a monomial, *each* term of the polynomial must be divided by the monomial, following the procedure described above.

■ **EXAMPLE 14**   Perform each of the following divisions.

a) $\dfrac{16x^2y^3 - 8x^2y + 4xy}{2xy} = \dfrac{16x^2y^3}{2xy} + \dfrac{-8x^2y}{2xy} + \dfrac{4xy}{2xy}$

$$= 8xy^2 - 4x + 2$$

b) $\dfrac{-24x^3y^2z + 16x^2z^2 - 8xz}{-8xz} = \dfrac{-24x^3y^2z}{-8xz} + \dfrac{16x^2z^2}{-8xz} + \dfrac{-8xz}{-8xz}$

$$= 3x^2y^2 - 2xz + 1$$

c) $\dfrac{12x^2y - 4xy^2 - 8xz}{4xyz} = \dfrac{12x^2y}{4xyz} + \dfrac{-4xy^2}{4xyz} + \dfrac{-8xz}{4xyz}$

$$= \frac{3x}{z} - \frac{y}{z} - \frac{2}{y} \quad \square$$

Exercise Set 0.5

Combine all like terms in each of the following algebraic expressions. Remove parentheses when necessary.

1. $5xy - 7xz + 2xz - 4xy$

2. $2x^2y + 4xy^2 - 7x^2y - xy^2$

3. $2x - 2(x + 3x^2)$

4. $(7x^2y - 3xz) + (2x^2y + 4xz)$

5. $(3ab^2 + 5abc) - (2abc - 4ab^2)$

6. $(18wx - 7bx) - (4bx + 3wx)$

7. $(4x^3y^2 - xy^2) - (2xy^2 - 3x^3y^2) - (2x^3y^2 - 5xy^2)$

8. $(12a^2b - 4ab^2 - 3) - (4a^2b - ab^2 - 5) - (7 + 2ab^2 - 5a^2b)$

Simplify by removing the grouping symbols and combining like terms.

9. $8 - [4 - (-3x - 2) + 7]$

10. $8x - [4y - (-3x + 3y) + x]$

11. $5 - \{7 - 3[2 - (x + 4)]\}$

12. $-2\{8 - 2[7x - (3 - x)]\}$

13. $-\{8x - [5x - (3 - 2x) + 4]\}$

14. $5 - \{2x + [3 - 5x - (7 + x)]\}$

15. $2 - \{9 - 3[x - (4 - 2x) - 3(x + 2)]\}$

16. $-\{x - 4[5 - (x + 2) + 2(x - 3) + 3x]\}$

Perform the following multiplications.

17. $(-2abc)(4ac)$

18. $(-3a^2bx)(5ax^3)$

19. $(-8x^2y)(-3xy^3)$

20. $(-3xy)(-4x^2yz)(-2y^2z)$

21. $(-x^3z)(a^2xy)(-yz)$

22. $(3ac^2y)(-4cy^3)(-2a^4cy^2)$

23. $2ab(-3a^2x - 5abx)$

24. $4x^2y(3xy^2 - 5x^2y^2)$

25. $-3xc(4xy - 7xac^2)$

26. $-4ay(5ax + 9ay)$

27. $6a^2bc(5a^2 - 7ab^2 - 9b^4c)$

28. $-2xy^2z(4 - 2xz^2 + 3yz^2 - 6xz^2)$

29. $(x + 2)(x + 5)$

30. $(x - 3)(x + 7)$

31. $(2x - 3)(x + 3)$

32. $(3x - 5y)(2x + 3y)$

33. $(x + y)(x - y)$

34. $(x - y)(2x + z)$

35. $(a + b)^2$

36. $(2a - 3b)^2$

37. $(a - b)^2$

38. $(4x + 2y)(4x - 2y)$

39. $(x^2 + y^2)(x^2 - y^2)$

40. $(3x + 2y)^2$

41. $(x - 3)(x^2 + 2x - 3)$

42. $(x - 1)(x^2 - 2x + 3)$

43. $(2x + 3)(x^2 - 5x + 6)$

44. $(3x - 2)(2x^2 + 4x - 5)$

45. $(x + y - z)(x - y + z)$

46. $(2x - y + 3z)(x - 3y + 4z)$

Perform the following divisions.

47. $\dfrac{27x^3}{3x}$

48. $\dfrac{-16a^3}{8a^2}$

49. $\dfrac{40x^2y^3z}{5axz}$

50. $\dfrac{20a^2bc}{4a^3b^2c}$

51. $\dfrac{-32xyz}{-4x^2y}$

52. $\dfrac{-24x^2y}{-8x^3y^2}$

53. $\dfrac{(16xy^2 - 24x^2y)}{-8xy}$

54. $\dfrac{(-12a^2xy + 2ay)}{-4y}$

55. $\dfrac{(35xy^4 - 14x^2y + 21x^2y^2)}{7xy^2}$

56. $\dfrac{(24x^2y - 18xy^2 - 12xy)}{-6xy}$

57. $\dfrac{(20a^2b^3c + 30ab^2c^2 - 10a^3bc^2)}{12a^2bc}$

58. $\dfrac{(12xyz - 18x^2y^2z + 24x^2y^2z^2)}{8x^2y}$

0.6

First-Degree Equations and Inequalities

Equations

Probably the most important use of algebra is in the area of problem solving, when an everyday situation from business or industry is translated into the symbolism of algebra and expressed as an **equation.** We will see specific examples of this in Section 0.11. Our intent at this point is to review the process of solving an equation. We will concern ourselves with equations of the form $ax + b = c$, where $a \neq 0$. Any equation that can be written as $ax + b = c$ is called a *linear equation* or *first-degree equation* in one variable. The term "first degree" refers to the fact that the highest exponent on the variable is 1.

■ **DEFINITION**

EQUATION
A statement that two algebraic expressions are equal is an *equation.*

■

There are three types of equation.

1. *Identity:* An equation that is true for any value substituted for the variable.
2. *Conditional:* An equation that is true for only certain values.
3. An equation that is not true for any value.

The values that make an equation true form its *solution set.*

■ **EXAMPLE 15**

a) The equation $2(x + 3) = 2x + 6$ is an identity since it is true for any value of x. For example, if $x = 4$,

$$2(4 + 3) = 2(4) + 2(3)$$
$$2(7) = 8 + 6$$
$$14 = 14.$$

If $x = -5$,

$$2(-5 + 3) = 2(-5) + 2(3)$$
$$2(-2) = -10 + 6$$
$$-4 = -4.$$

The given equation is true regardless of the value substituted for x. The numbers 4 and -5 were chosen only to illustrate.

b) The equation $x + 3 = 7$ is a conditional equation since it is true if and only if $x = 4$.

c) The equation $x + 1 = x$ is *not* true for any real number. Try substituting some values for x and see if you can get a true result. □

Any number that makes an equation a true statement of equality is called a *root* or *solution* to the equation. The process of finding the solution(s) to an equation is called *solving the equation*. We now turn our attention to the steps involved in the process of solving equations. One important point to remember is that we start with a basic statement of equality and each step must maintain that equality. There are two properties of equality that help illustrate the process.

■

---

1. ADDITION PROPERTY

For any real numbers a, b, and c:

$$\text{If } a = b, \text{ then } a + c = b + c.$$

2. MULTIPLICATION PROPERTY

For any real numbers a, b, and c:

$$\text{If } a = b, \text{ then } a \cdot c = b \cdot c.$$

The addition property indicates that we can add the same number to both sides of an equation without destroying the original equality, as shown in the following examples.

■ **EXAMPLE 16**   Using the addition property, we have

$$17 = 17$$
$$17 + 3 = 17 + 3$$
$$20 = 20. \quad \square$$

■ **EXAMPLE 17**   Solve each of the following for x.

a)
$$x - 4 = 8$$
$$x - 4 + (4) = 8 + (4)$$
$$x + 0 = 12$$
$$x = 12$$

Twelve is the solution to this equation as well as the original equation. We have changed the form of the original equation but not its solution.

Check:
$$12 - 4 = 8$$
$$8 = 8$$

b)
$$x + 3 = 7$$
$$x + 3 + (-3) = 7 + (-3)$$
$$x + 0 = 4$$
$$x = 4$$

Check:
$$4 + 3 = 7$$
$$7 = 7 \quad \square$$

The multiplication property suggests that we can multiply both sides of an equation by the same number without destroying the original equality, as shown in the following examples.

■ **EXAMPLE 18**   Using the multiplication property, we have

$$17 = 17$$
$$3(17) = 3(17)$$
$$51 = 51. \quad \square$$

■ **EXAMPLE 19**   Solve each of the following for x.

a)
$$\frac{x}{4} = 8$$
$$4\left(\frac{x}{4}\right) = 4 \cdot 8$$
$$x = 32$$

b)
$$3x = 24$$
$$\frac{1}{3} \cdot 3x = \frac{1}{3} \cdot 24$$
$$x = 8 \quad \square$$

Examples 17 and 19 illustrate the application of either the addition or the multiplication property of equality in solving simple linear algebraic equations. Often, however, both properties must be applied in solving a single equation.

■ **EXAMPLE 20**   Solve each of the following for x.

a)
$$2x - 3 = 9$$
$$2x - 3 + (3) = 9 + (3)$$ **We need to combine like terms using the addition property as our first step.**
$$2x + 0 = 12$$
$$2x = 12$$
$$\left(\frac{1}{2}\right) \cdot 2x = \left(\frac{1}{2}\right) \cdot 12$$ **Multiplication property**
$$x = 6$$

b)
$$3x + 4 = 13$$
$$3x + 4 + (-4) = 13 + (-4)$$ **Addition property**
$$3x + 0 = 9$$
$$3x = 9$$
$$\left(\frac{1}{3}\right) \cdot 3x = \left(\frac{1}{3}\right) \cdot 9$$ **Multiplication property**
$$x = 3 \quad \square$$

Remember: The solution set to an equation contains the number that makes the equation a true statement of equality. This fact provides us with a convenient way to check our work. Thus we should always know when we have correctly solved an equation.

■ **EXAMPLE 21**   Solve for x: $\dfrac{x}{5} - 3 = 7$.

$$\frac{x}{5} - 3 = 7$$
$$\frac{x}{5} - 3 + (3) = 7 + (3)$$
$$\frac{x}{5} = 10$$
$$5 \cdot \frac{x}{5} = 5 \cdot 10$$
$$x = 50$$

Check: If $x = 50$,

$$\frac{50}{5} - 3 = 7$$
$$10 - 3 = 7 \quad \square$$

As defined earlier, an equation is a statement of equality between two algebraic expressions. Frequently, both the left side and the right side of the equation contain expressions involving the unknown. When this occurs, we need to combine all terms involving the unknown on one side of the equation and all terms not involving the unknown on the opposite side in order to make it easier to solve. Terms are moved from one side to the other by using the addition property of equality. Before doing so, however, we may need to use any of the steps for simplifying algebraic expressions, discussed earlier in the chapter. (See Example 22c.)

■ **EXAMPLE 22**

Solve each of the following for x.

a)
$$3x - 4 = 2x + 5$$
$$3x - 4 + (4) = 2x + 5 + (4)$$
$$3x = 2x + 9$$
$$3x + (-2x) = 2x + (-2x) + 9$$
$$x = 9$$

Check:
$$3(9) - 4 = 2(9) + 5$$
$$27 - 4 = 18 + 5$$
$$23 = 23$$

b)
$$5(x - 2) = 2x + 8$$
$$5x - 10 = 2x + 8$$
$$5x - 10 + 10 = 2x + 8 + 10$$
$$5x = 2x + 18$$
$$5x - 2x = 2x - 2x + 18$$
$$3x = 18$$
$$\left(\frac{1}{3}\right) \cdot 3x = \left(\frac{1}{3}\right) \cdot 18$$
$$x = 6$$

Check:
$$5(6 - 2) = 2(6) + 8$$
$$5(4) = 12 + 8$$
$$20 = 20$$

c)
$$3(2x - 4) = 14 - (x - 2)$$
$$6x - 12 = 14 - x + 2$$
$$6x - 12 = 16 - x$$
$$6x = 16 + 12 - x$$
$$6x = 28 - x$$
$$6x + x = 28$$
$$7x = 28$$
$$\left(\frac{1}{7}\right) \cdot 7x = \left(\frac{1}{7}\right) \cdot 28$$
$$x = 4 \quad \square$$

Check:
$$3(2 \cdot 4 - 4) = 14 - (4 - 2)$$
$$3(8 - 4) = 14 - (2)$$
$$3(4) = 14 - 2$$
$$12 = 12$$

Inequalities

Although equations are extremely important in problem solving, many relationships are expressed as inequalities rather than equalities. Inequalities are written using the symbols $<$, $>$, $\leq$, and $\geq$.

■ **EXAMPLE 23**

a) $6 < 9$ is read "6 is less than 9."
b) $9 > 6$ is read "9 is greater than 6."
c) $x \leq 3$ is read "x is less than or equal to 3."
d) $x \geq 5$ is read "x is greater than or equal to 5."
e) $-1 < x < 2$ is read "-1 is less than x, which is less than 2."   □

The addition and multiplication properties of equality may also be applied to inequalities, with one exception.

PROPERTIES OF INEQUALITIES

If a, b, and c are real numbers, and:

1. If $a < b$, then $a + c < b + c$.
2. If $a < b$, then $a - c < b - c$.
3. If $a < b$ and c is positive, then $a \cdot c < b \cdot c$.
4. If $a < b$ and c is negative, then $a \cdot c > b \cdot c$.
5. If $a < b$ and c is positive, then $\dfrac{a}{c} < \dfrac{b}{c}$.
6. If $a < b$ and c is negative, then $\dfrac{a}{c} > \dfrac{b}{c}$.

Similar results occur for the inequalities $>$, $\leq$, and $\geq$. These properties show that multiplication and division by a negative number cause the sense of the inequality to be reversed. This is a step the student must remember to perform.

■ **EXAMPLE 24**

a) If $6 < 9$, then

$$6 + 3 < 9 + 3$$
$$9 < 12. \qquad \textbf{Property 1}$$

b) If $x + 3 > 7$, then

$$x + 3 + (-3) > 7 + (-3)$$
$$x > 4. \qquad \textbf{Property 2}$$

c) If $2 < 5$, then

$$3 \cdot 2 < 3 \cdot 5$$
$$6 < 15. \qquad \textbf{Property 3}$$

d) If $5 \geq 2$, then

$$3 \cdot 5 \geq 3 \cdot 2$$
$$15 \geq 6. \qquad \textbf{Property 3}$$

e) If $2 < 5$, then

$$(-3) \cdot 2 \not< (-3) \cdot 5 \qquad \textbf{$\not<$ means is not less than.}$$
since
$$-6 > -15. \qquad \textbf{Property 4}$$

f) If $5 > 2$, then

$$(-3) \cdot 5 \not> (-3) \cdot 2$$
since
$$-15 < -6. \qquad \textbf{Property 4} \quad \square$$

■ **EXAMPLE 25**   Find the solution set of each of the following.

a)
$$-3x < 9$$
$$\left(-\frac{1}{3}\right)(-3x) > \left(-\frac{1}{3}\right)(9)$$
$$x > -3 \qquad \textbf{Property 4}$$

b)
$$\left(-\frac{1}{2}\right)x \geq 3$$
$$(-2)\left(-\frac{1}{2}\right)x \leq (-2)(3)$$
$$x \leq -6 \qquad \textbf{Property 4} \quad \square$$

Since they are infinitely large, the solution sets to inequalities can be shown on a number line. We use an open dot to show that a particular number is the starting point for the solution set, but does not belong to the set. We use a solid dot to show that the number does belong. The arrow points in the direction of the solutions.

■ **EXAMPLE 26**   Graph the solution set of $x < 3$ on the number line.

Solution

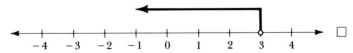

■ **EXAMPLE 27**   Graph the solution set of $x \geq -2$ on the number line.

Solution

■ **EXAMPLE 28**   Graph the solution set of $-1 \leq x < 2$ on the number line.

Solution

■ **EXAMPLE 29**   Graph the solution set of $x < -2$ or $x \geq 1$ on the number line.

Solution

Exercise Set 0.6

Solve each of the following and check your solution.

1. $x + 7 = 9$ **2.** $x - 8 = 11$

3. $2x = 10$ **4.** $\dfrac{x}{2} = 8$

5. $3x + 1 = 10$ **6.** $2x - 5 = -11$

7. $6x + 5 = -13$ **8.** $-8x - 3 = -19$

9. $-3x + 19 = 19$ **10.** $-8x + 3 = -29$

11. $8x + 5 = 4x + 13$ **12.** $7x + 4 = 6x + 7$

13. $x - 2 = 6 - 3x$ **14.** $2x - 10 = 7x$

15. $8x = 3x + 20$ **16.** $-10x + 4 = -x - 14$

17. $\dfrac{x}{3} = 8$ **18.** $\dfrac{x}{2} + 4 = 7$

19. $\dfrac{x}{4} - 5 = -3$ **20.** $5 - (9 - 6x) = 2x - 2$

21. $7 - (5 - 8x) = 4x + 3$ **22.** $2x - 5 = 3(4x + 5)$

23. $2(3x + 1) - 4 = 16$ **24.** $6x + 2(2x + 3) = 16$

25. $3x - 7 = 5(2x + 7)$ **26.** $5x + 2(3x + 1) = 3x + 5$

27. $3[2 - 4(y - 1)] = 3(2y + 8)$ **28.** $5[2 - (2x - 4)] = 2(5 - 3x)$

29. $5 + 3[1 + 2(2x - 3)] = 6(x + 5)$

Graph the solution set for each of the following.

30. $x < 5$ **31.** $x > -3$ **32.** $x \leq -2$

33. $x \geq 0$ **34.** $-x < -3$ **35.** $-x \geq 4$

36. $x > 3$ and $x < 7$ **37.** $x > -2$ and $x < 3$ **38.** $x < -2$ or $x \geq 3$

Solve each of the following and graph the solution set.

39. $x + 5 > 3$ **40.** $5x + 3 > 4x + 5$

41. $7x - 14 \leq 6x - 16$ **42.** $6x + 4 \geq 5x - 2$

43. $7x - 3 \geq 6x - 2$ **44.** $3x - 2 < 5x + 4$

45. $2x - 9 \geq 5x + 4$ **46.** $5(x - 2) > 9x - 3(2x - 4)$

47. $4(3x - 1) > 3(2 - 5x)$ **48.** $3(3 - 2x) \geq -5x - 2(3 - x)$

49. $5(2 - x) > 3(2x - 5)$ **50.** $3x - 2(3x - 5) > 4(2x - 1)$

Graph both inequalities on the same number line. Where do the two solution sets overlap?

51. $3x - 12 \leq 6$ and $2x - 5 \geq -2$

52. $5x + 3 < 4x + 5$ and $7x - 14 \geq 6x - 16$

53. $7x - 3 \geq 6x - 2$ and $6x - 2 \leq 5x + 4$

0.7
Factoring Polynomials

In the preceding section we were finding *products* by carrying out various types of multiplication. Sometimes a product is known and we want to find the *factors* that yielded that product. The process of finding those factors is called *factoring*. Factoring is the *inverse operation of multiplication*. That is, a product is found by multiplying factors. On the other hand, a number can be expressed as the product of its factors. We know that $2 \times 5 = 10$. Multiplying the factors 2 and 5 gives the product 10. As another example, $15 = 3 \times 5$ since 3 and 5 are factors of 15.

We are going to examine several types of factoring. Although a certain amount of trial and error may be necessary, some general forms and rules and familiarity with multiplication will help cut down on this. Keep in mind that if an algebraic expression is factorable, *there is only one correct factorization.* Also, remember that the factoring can be checked by carrying out the multiplication to see if the original expression results. Generally, we will factor only those expressions that leave integral exponents after the factoring has been completed.

■ TYPES OF FACTORING

1. THE COMMON FACTOR

When an algebraic expression has factors common to each term, the *common factors* can be removed using the distributive property.

■ EXAMPLE 30

Factor the common factor(s) in each of the following.

a) $6x - 12y = 6(x - 2y)$
b) $a^2x - ay = a(ax - y)$
c) $-10a^2b + 15ab^2 - 20ab = -5ab(2a - 3b + 4)$ □

■

2. THE DIFFERENCE OF PERFECT SQUARES (DOPS)

When the sum of two terms and the difference between the same two terms are multiplied, the result is the difference between two perfect squares.

■ EXAMPLE 31

Solution

Find the product of the sum and difference of x and y.

$$(x + y)(x - y) = x^2 + xy - xy - y^2$$
$$= x^2 - y^2 \quad □$$

Therefore, when the difference between two perfect squares is factored, the factors will be two binomials: one the sum and the other the difference between the square roots of the perfect squares.

■ EXAMPLE 32

Factor each of the following DOPS.

a) $x^2 - y^2 = (x + y)(x - y)$
b) $a^2 - 9 = (a + 3)(a - 3)$
c) $4x^2 - 25y^2 = (2x + 5y)(2x - 5y)$ □

Sometimes the difference between perfect squares results after a common factor has been removed.

■ **EXAMPLE 33**   Factor: $12a^2 - 27y^2$.

Solution $12a^2 - 27y^2 = 3(4a^2 - 9y^2) = 3(2a + 3y)(2a - 3y)$ □

■

> **3.** PERFECT-SQUARE TRINOMIAL (PST)
>
> When a binomial is squared, the result is a perfect-square trinomial. Its characteristics are that the first and last terms are perfect squares and the middle term is *twice* the product of the square roots of the first and last terms.

■ **EXAMPLE 34**   **a)** Expand $(x + y)^2$.

$(x + y)^2 = x^2 + 2xy + y^2$ **Note that x^2 and y^2 are perfect squares and that $2xy = 2\sqrt{x^2} \cdot \sqrt{y^2}$.**

b) Expand $(x - y)^2$.

$(x - y)^2 = x^2 - 2xy + y^2$ **Note the difference between the sign arrangement of this example and that of Example 34(a).**   □

The factors of a PST will be two identical binomials or, in other words, a binomial squared. The first term of the binomial will be the square root of the first term of the PST; the second term of the binomial will be the square of the last term of the PST; and the sign of the binomial will be the sign of the middle term of the PST.

■ **EXAMPLE 35** $a^2 + 2ab + b^2$ factors into $(a + b)(a + b) = (a + b)^2$.   □

■ **EXAMPLE 36**   Factor each of the following PSTs.

a) $a^2 - 6a + 9$ is a PST because a^2 and 9 are perfect squares and $6a = 2\sqrt{a^2} \cdot \sqrt{9}$. Therefore, $a^2 - 6a + 9 = (a - 3)^2$.
b) $x^2 + 10xy + 25y^2 = (x + 5y)^2$
c) $4x^2 - 12xy + 9y^2 = (2x - 3y)^2$ □

■

> **4.** QUADRATIC TRINOMIAL (QT)
>
> A quadratic trinomial is the result of multiplying two binomials in the same literal terms.

■ **EXAMPLE 37**   Multiply the binomials $(x + 3)(x + 2)$.

Solution
$$(x + 3)(x + 2) = x^2 + 2x + 3x + 6$$
$$= x^2 + 5x + 6 \quad \square$$

Note that a quadratic trinomial is different from a perfect-square trinomial in that any one of the three characteristics of the PST will not be satisfied. Also, note where the middle term of the QT comes from. After all four products have been found through multiplication, the middle term, $5x$, is the result of combining the like terms $2x$ and $3x$. This is important to the factoring process. Also note that the first term of the QT, x^2, is the result of multiplying the first terms of the two binomials, x and x. The last term of the QT results from multiplying the last terms of the two binomials, 3 and 2. In factoring a QT, some trial and error may be necessary. We will attempt to find factors of the last term of the QT that will combine to give us the middle term. The middle term is *never* factored. In Example 37, $(3)(2) = 6$ and $3 + 2 = 5$. Look closely at the following example.

■ **EXAMPLE 38**   Examine the following products.

a) $x^2 + 5x + 6 = (x + 3)(x + 2)$: $(3)(2) = 6; 3 + 2 = 5$
b) $x^2 - 5x + 6 = (x - 3)(x - 2)$: $(-3)(-2) = 6$;
                                          $(-3) + (-2) = -5$
c) $x^2 + 7x + 12 = (x + 3)(x + 4)$: $(3)(4) = 12; 3 + 4 = 7$
d) $x^2 - 7x + 12 = (x - 3)(x - 4)$: $(-3)(-4) = 12$;
                                          $(-3) + (-4) = -7$ $\quad \square$

We can see in Examples 38(c) and (d) that 2 and 6 will not work, even though $(2)(6) = 12$, because $2 + 6 \neq 7$. We must also pay close attention to the sign arrangement in the QT so that we can correctly place the signs in the binomial factors. In Example 38, if the last term of the QT is +, the signs of the factors will be the same, either both + or both −. Whether the factors are + or − depends on the sign of the middle term of the QT.

Now consider the following example.

■ **EXAMPLE 39**   Look at the following QTs and their factors.

a) $x^2 + 5x - 6 = (x + 6)(x - 1)$: $(6)(-1) = -6$;
                                          $(+6) + (-1) = +5$
b) $x^2 - 5x - 6 = (x - 6)(x + 1)$: $(-6)(+1) = -6$;
                                          $(-6) + (+1) = -5$
c) $x^2 + x - 12 = (x - 3)(x + 4)$: $(-3)(+4) = -12$;
                                          $(-3) + (+4) = +1$

d) $x^2 - x - 12 = (x + 3)(x - 4)$: $(+3)(-4) = -12$;
$(+3) + (-4) = -1$ □

In Example 39, you can see that if the sign of the last term in the QT is $-$, the signs of the binomial factors must be different. One will be $+$ and one $-$. A good point to remember is that, in a QT of this type, the largest factor of the last term of the QT will have the same sign as the middle term of the QT. In Example 39(c) and (d), the largest factor of 12 is 4. In the binomial factors, 4 is $+$ when the middle term is $+$ and $-$ when the middle term is $-$.

■ **EXAMPLE 40**

a) Factor $x^2 + 3x - 10$.

$x^2 + 3x - 10 = (x + 5)(x - 2)$: $(+5)(-2) = -10$;
$(+5) + (-2) = +3$

b) Factor $x^2 - 3x - 10$.

$x^2 - 3x - 10 = (x - 5)(x + 2)$: $(-5)(+2) = -10$;
$(-5) + (+2) = -3$ □

All of the QTs we have looked at thus far had 1 as the coefficient on the first term. If the coefficient is not 1, the factoring becomes a little more difficult. We now consider the factors of the first term multiplied by the factors of the last term and then combined to give the middle term.

■ **EXAMPLE 41**

a) Factor $2x^2 + 11x + 12$.

$2x^2 + 11x + 12 = (2x + 3)(x + 4)$: $(2x)(4) + (3)(x) = 11x$

b) Factor $3x^2 - 11x + 10$.

$3x^2 - 11x + 10 = (3x - 5)(x - 2)$: $(3x)(-2) + (-5)(x) = -11x$
□

Note in Example 41 that if the last term of the QT is positive, the signs of the binomials are determined as before. However, if the sign of the last term is negative, some trial and error may be necessary to correctly place the signs.

■ **EXAMPLE 42**

a) Factor $2x^2 + 5x - 12$.

$2x^2 + 5x - 12 = (2x - 3)(x + 4)$: $(2x)(4) + (-3)(x) = +5x$

b) Factor $3x^2 - x - 10$.

$3x^2 - x - 10 = (3x + 5)(x - 2)$: $(3x)(-2) + (5)(x) = -x$ □

As the number of factors of the first and last terms of the QT increases, the factoring becomes more difficult. Consider Example 43 and the number of trial-and-errors that may be attempted.

■ EXAMPLE 43

Factor $4x^2 + 13x - 12$.

Solution

$$4x^2 + 13x - 12 \neq (2x - 3)(2x + 4) \text{ since } (2x)(4) + (-3)(2x) \neq 13x$$
$$\neq (2x + 3)(2x - 4) \text{ since } (2x)(-4) + (3)(2x) \neq 13x$$
$$\neq (2x - 6)(2x + 2) \text{ since } (2x)(2) + (-6)(2x) \neq 13x$$
$$\neq (2x + 6)(2x - 2) \text{ since } (2x)(-2) + (6)(2x) \neq 13x$$
$$\neq (2x - 1)(2x + 12) \text{ since } (2x)(12) + (-1)(2x) \neq 13x$$
$$\neq (2x + 1)(2x - 12) \text{ since } (2x)(-12) + (1)(2x) \neq 13x$$
$$\neq (4x - 6)(x + 2) \text{ since } (4x)(2) + (-6)(x) \neq 13x$$
$$\neq (4x + 6)(x - 2) \text{ since } (4x)(-2) + (6)(x) \neq 13x$$
$$\neq (4x - 1)(x + 12) \text{ since } (4x)(12) + (-1)(x) \neq 13x$$
$$\neq (4x + 1)(x - 12) \text{ since } (4x)(-12) + (1)(x) \neq 13x$$
$$\neq (4x + 3)(x - 4) \text{ since } (4x)(-4) + (3)(x) \neq 13x$$

Finally, we have

$$4x^2 + 13x - 12 = (4x - 3)(x + 4) \text{ since } (4x)(4) + (-3)(x) = 13x. \quad \square$$

Sometimes more than one type of factoring may be involved in any one problem. A good rule to follow is to *always* look for the common factors first, and then attempt to factor whatever quantity remains. As we stated earlier, factor common factors only if the coefficients remain integers. Remember, too, that not every algebraic expression is factorable, just as not every real number is factorable. Any algebraic expression that is not factorable is called a *prime expression*.

■ EXAMPLE 44

$x^2 + x + 1$ is not factorable with integer coefficients. $\quad \square$

Exercise Set 0.7

Factor each of the following algebraic expressions.

1. $12x - 15y$
2. $16ac + 24bc$
3. $20a^2b - 45a^3bc$
4. $21w - 7w^2$
5. $a^2b^2 + a^2b + ab^2$
6. $12xy - 20x$
7. $24x^2 - 12xy + 18xy^2$
8. $(x - y)a - (x - y)b$
9. $x^{n+1} + x^n$
10. $x^{2a+1} - x^{2a}$
11. $a^2 - b^2$
12. $4x^2 - 9y^2$
13. $x^2 - 36$
14. $9a^2 - 25^2$
15. $x^2 - 1$
16. $3x^2 - 27$
17. $2a^2 - 32b^2$
18. $4a^2 - 12y^2$
19. $x^4 - y^4$
20. $x^8 - y^8$

Fill in the blank to make each of the following a perfect-square trinomial.

21. $x^2 + \underline{\hspace{1cm}} x + 16$ **22.** $x^2 - \underline{\hspace{1cm}} x + 36$

23. $9x^2 - \underline{\hspace{1cm}} xy + 25y$ **24.** $x^2 + 6x + \underline{\hspace{1cm}}$

25. $x^2 - 10x + \underline{\hspace{1cm}}$ **26.** $4x^2 - 12xy + \underline{\hspace{1cm}}$

Factor each of the following perfect-square trinomials.

27. $a^2 + 6a + 9$ **28.** $x^2 - 16x + 64$ **29.** $y^2 - 22yz + 121z^2$

30. $4x^2 + 40x + 100$ **31.** $9x^2 - 6xy + y^2$ **32.** $32x^2 + 144x + 162$

Factor each of the following quadratic trinomials.

33. $x^2 - 7x + 12$ **34.** $x^2 + 8x + 15$ **35.** $y^2 + y - 20$

36. $z^2 - 2x - 15$ **37.** $a^2 - 9ab - 36b$ **38.** $x^2 - 11xy - 42y^2$

39. $y^2 + 12y + 35$ **40.** $a^2 - 6a - 72$ **41.** $2x^2 + 3x + 1$

42. $3y^2 + 10y + 3$ **43.** $8b^2 + 18b + 9$ **44.** $7a^2 - 12a - 4$

45. $4x^3 - 8x^2 - 32x$ **46.** $6z^2 - 18z + 12$

0.8
Operations with Algebraic Fractions

When we are dealing with algebraic expressions, it is frequently necessary to work with fractions. Since an algebraic expression is just a representation of some number, the techniques used in the arithmetic of fractions will form the basis for our operations on algebraic fractions. An important application of factoring, covered in the preceding section, also occurs in the operations on the algebraic fractions.

One important concept to remember is that a fraction can be changed into an equivalent fraction by either multiplying both the numerator and denominator by the same nonzero number or dividing both the numerator and denominator by the same nonzero number.

■ **EXAMPLE 45**
$$\frac{6}{8} = \frac{6 \times 2}{8 \times 2} = \frac{12}{16} \quad \text{and} \quad \frac{6}{8} = \frac{6 \div 2}{8 \div 2} = \frac{3}{4} \quad \square$$

Reducing Fractions

We use the principle of dividing both the numerator and the denominator by the same nonzero number to reduce a fraction. We first factor both the numerator and denominator, and then divide out any common factors.

■ **EXAMPLE 46**   Reduce each of the following fractions.

a) $\dfrac{x^2 - 9}{x^2 - 6x + 9} = \dfrac{(x + 3)(x - 3)}{(x - 3)(x - 3)} = \dfrac{x + 3}{x - 3}$

b) $\dfrac{x^2 + 5x + 6}{x^2 + 7x + 12} = \dfrac{(x + 2)(x + 3)}{(x + 4)(x + 3)} = \dfrac{x + 2}{x + 4}$

c) $\dfrac{a - b}{b - a} = \dfrac{a - b}{-(-b + a)} = \dfrac{a - b}{-(a - b)} = \dfrac{1}{-1} = -1$ □

Multiplication and Division of Fractions

The key to multiplying algebraic fractions is your ability to factor. The first step is to factor all the polynomials, and then divide out any common factors. The remaining factors should then be multiplied. If the fractions are made up of only monomials, there is no need to factor.

■ **EXAMPLE 47**   Multiply each of the following pairs of fractions.

a) $\dfrac{10ab^2c}{14x^3y} \cdot \dfrac{7x^2yz}{2ab} = \dfrac{5bcz}{2x}$

b) $\dfrac{a^2 - 9}{a^2 - 7a + 12} \cdot \dfrac{a^2 - 6a + 8}{2a + 6} = \dfrac{(a + 3)(a - 3)}{(a - 3)(a - 4)} \cdot \dfrac{(a - 2)(a - 4)}{2(a + 3)}$

$$= \dfrac{a - 2}{2}$$

c) $\dfrac{3x - 12}{x^2 - 8x + 16} \cdot \dfrac{x^2 - 16}{x^2 - x - 20} = \dfrac{3(x - 4)}{(x - 4)(x - 4)} \cdot \dfrac{(x + 4)(x - 4)}{(x - 5)(x + 4)}$

$$= \dfrac{3}{x - 5}$$ □

The division of one algebraic fraction by another requires an initial step before we begin the factoring. Just as in arithmetic, we change the division into multiplication by multiplying the first fraction by the reciprocal of the second fraction and proceed as we did above.

■ **EXAMPLE 48**   Find the following quotient: $\dfrac{15ax}{x^2 - 9} \div \dfrac{25a^2bx}{x^2 + x - 12}$.

Solution $\dfrac{15ax}{x^2 - 9} \div \dfrac{25a^2bx}{x^2 + x - 12} = \dfrac{15ax}{x^2 - 9} \cdot \dfrac{x^2 + x - 12}{25abx}$

$$= \dfrac{15ax}{(x + 3)(x - 3)} \cdot \dfrac{(x + 4)(x - 3)}{25a^2bx}$$

$$= \dfrac{3(x + 4)}{5ab(x + 3)}$$ □

Addition and Subtraction of Fractions

The procedure for adding and subtracting algebraic fractions is very similar to the procedure for adding and subtracting in arithmetic. Only fractions with the same denominators can be added or subtracted. Those with different denominators can be converted into equivalent fractions by first finding the *least common denominator* (LCD) of the given denominators and making the corresponding changes in the numbers.

■ **EXAMPLE 49**   Find the sum of each of the following groups of fractions.

a) $\dfrac{3a}{8} + \dfrac{2b}{9} - \dfrac{a + b}{12}$

$= \dfrac{9 \cdot 3a}{72} + \dfrac{8 \cdot 2b}{72} - \dfrac{6(a + b)}{72}$

$= \dfrac{27a + 16b - 6a - 6b}{72}$

$= \dfrac{21a + 10b}{72}$
$\boxed{\begin{aligned} 8 &= 2 \cdot 2 \cdot 2 \\ 9 &= 3 \cdot 3 \qquad \text{LCD} = 2 \cdot 2 \cdot 2 \cdot 3 \cdot 3 \\ 12 &= 2 \cdot 2 \cdot 3 \qquad\qquad = 72 \end{aligned}}$

b) $\dfrac{3}{a + b} + \dfrac{2a}{a^2 - b^2}$

$= \dfrac{3}{a + b} + \dfrac{2a}{(a + b)(a - b)}$

$= \dfrac{3(a - b)}{(a + b)(a - b)} + \dfrac{2a}{(a + b)(a - b)}$

$= \dfrac{3a - 3b + 2a}{(a + b)(a - b)} = \dfrac{5a - 3b}{(a + b)(a - b)}$
$\qquad \boxed{\text{LCD} = (a + b)(a - b)}$

c) $\dfrac{x + 2}{x^2 - 9} + \dfrac{x - 5}{x^2 + 6x + 9}$

$= \dfrac{x + 2}{(x + 3)(x - 3)} + \dfrac{x - 5}{(x + 3)(x + 3)}$

$= \dfrac{(x + 2)(x + 3)}{(x + 3)(x + 3)(x - 3)} + \dfrac{(x - 5)(x - 3)}{(x + 3)(x + 3)(x - 3)}$

$= \dfrac{x^2 + 5x + 6 + x^2 - 8x + 15}{(x + 3)(x + 3)(x - 3)}$

$= \dfrac{2x^2 - 3x + 21}{(x + 3)^2(x - 3)}$
$\qquad \boxed{\text{LCD} = (x + 3)(x + 3)(x - 3)} \qquad \square$

Exercise Set 0.8

Reduce each of the following fractions.

1. $\dfrac{8a - 12b}{16a + 24b}$

2. $\dfrac{am + an}{ab - ac}$

3. $\dfrac{2x + 4y}{x^2 - 4y^2}$

4. $\dfrac{6x + 9y}{4x^2 - 9y^2}$

5. $\dfrac{3a - 9b}{6b - 2a}$

6. $\dfrac{9 - x^2}{x^2 - 3x}$

Perform the indicated operations.

7. $\dfrac{x^2 - y^2}{16} \cdot \dfrac{12}{3x - 3y}$

8. $\dfrac{2a - 4}{3a + 2} \cdot \dfrac{6a + 4}{12}$

9. $\dfrac{x^2 - 8x + 16}{x^2 - 16} \cdot \dfrac{x^2 + 7x + 12}{x^2 - 9}$

10. $\dfrac{x + 3y}{a^2 - 1} \cdot \dfrac{a^2 - 7a + 6}{x^2 - 9y}$

11. $\dfrac{ax + x^2}{b + cx} \cdot \dfrac{a + x}{x^3} \cdot \dfrac{bx + cx^2}{(a + x)^2}$

12. $\dfrac{x^2 + 13x + 42}{x^2 + 2x - 15} \cdot \dfrac{x^2 + 6x + 5}{x^2 + 4x - 12}$

13. $\dfrac{x^2 - y^2}{2x + y} \div \dfrac{3x^2y - 3y^3}{10x + 5y}$

14. $\dfrac{a^2 - 12a + 35}{a^2 + 3a + 2} \div \dfrac{a^2 - 25}{a^2 - 4}$

15. $\dfrac{x^2 - y^2}{x} \div \dfrac{x - y}{y}$

16. $\dfrac{14a}{a - 4} \div \dfrac{7a}{a^2 - 11a + 28}$

17. $\dfrac{a^2 - 6a + 8}{a^2 + 6a + 9} \div \dfrac{a - 2}{a + 3}$

18. $\dfrac{x^2 + 3x - 4}{x^2 - 5x + 6} \div \dfrac{x^2 + 6x + 8}{x^2 + x - 6}$

19. $\dfrac{x + 5}{5} - \dfrac{m + 6}{6}$

20. $\dfrac{y - 3}{6} + \dfrac{y - 2}{9}$

21. $\dfrac{2c}{9ab} + \dfrac{a}{12bc} - \dfrac{5b}{6ac}$

22. $\dfrac{a + 5}{6} + \dfrac{3a + 1}{9} - \dfrac{a + 9}{2}$

23. $\dfrac{3}{b + c} + \dfrac{c}{b(b - c)} - \dfrac{2b}{b^2 - c^2}$

24. $\dfrac{x + 6y}{x^2 - 4y^2} + \dfrac{3}{x + 2y} - \dfrac{2}{x - 2y}$

25. $\dfrac{3}{x^2 - 4y^2} + \dfrac{1}{x^2 + 3xy + 2y^2} - \dfrac{3}{x^2 - xy - 2y^2}$

26. $\dfrac{2}{a^2 + 3ab + 2b^2} + \dfrac{3}{a^2 + ab - 2b^2} - \dfrac{2}{a^2 - b^2}$

0.9
Solving Fractional Equations

In Section 0.8, we discussed operations with algebraic fractions. Determining the LCD can also be useful when we are faced with the problem of solving an equation that contains fractions. We can use the

LCD to eliminate the fractions from the equations by taking advantage of two facts.

■

1. By the multiplication property of equality, both sides of the equation can be multiplied by the LCD.
2. The LCD contains *all* factors of each individual denominator.

When each term on both sides of the equation is multiplied by the LCD, the individual denominators divide out with the factors in the LCD. This process is called *clearing of fractions.*

■ **EXAMPLE 50**   Solve $\dfrac{3x - 1}{4} + \dfrac{2}{3} = \dfrac{7}{6}$ for x.

Solution   By inspection, we see that the LCD is 12. Thus we have

$$12 \cdot \left(\frac{3x - 1}{4} + \frac{2}{3} \right) = 12 \cdot \frac{7}{6} \qquad \textbf{Multiply both sides by 12.}$$

$$12 \cdot \frac{3x - 1}{4} + 12 \cdot \frac{2}{3} = 12 \cdot \frac{7}{6}$$

$$3(3x - 1) + 4 \cdot 2 = 2 \cdot 7 \qquad \textit{Reminder:} \textbf{ This equation has the}$$
$$9x - 3 + 8 = 14 \qquad\qquad \textbf{same solution as the original. We have}$$
$$9x + 5 = 14 \qquad\qquad \textbf{changed the form but not the}$$
$$9x = 9 \qquad\qquad\qquad \textbf{solution.}$$
$$x = 1.$$

Check: If $x = 1$, then

$$\frac{3(1) - 1}{4} + \frac{2}{3} = \frac{7}{6}$$

$$\frac{3 - 1}{4} + \frac{2}{3} = \frac{7}{6}$$

$$\frac{2}{4} + \frac{2}{3} = \frac{7}{6}$$

$$\frac{6}{12} + \frac{8}{12} = \frac{14}{12}$$

$$\frac{14}{12} = \frac{14}{12}. \quad \square$$

■ **EXAMPLE 51**   Solve for x in $2 + \dfrac{5}{x} = 7$.

Solution   The LCD is x. Thus we have

$$x\left(2 + \frac{5}{x}\right) = 7 \cdot x \qquad\qquad \textbf{Multiply both sides by } x.$$
$$2x + 5 = 7x$$
$$5 = 7x - 2x$$
$$5 = 5x$$
$$1 = x.$$

Check: If $x = 1$, then

$$2 + \frac{5}{1} = 7$$
$$2 + 5 = 7$$
$$7 = 7. \quad \square$$

Note: If our steps had produced $x = 0$, the equation would have no solution, since that result would have caused division by zero in the original equation and we know division by zero is undefined. We could have decided, by inspection of the original equation, that if we arrived at $x = 0$, the equation would have no solution.

■ **EXAMPLE 52**   Solve for x in $\dfrac{5}{x + 3} = \dfrac{3}{x - 1}$.

Solution   The LCD is $(x + 3)(x - 1)$. Thus

$$(x + 3)(x - 1) \cdot \frac{5}{x + 3} = \frac{3}{x - 1} \cdot (x + 3)(x - 1)$$
$$(x - 1)5 = 3(x + 3)$$
$$5x - 5 = 3x + 9$$
$$5x = 3x + 14$$
$$2x = 14$$
$$x = 7.$$

Check: $$\frac{5}{7 + 3} = \frac{3}{7 - 1}$$
$$\frac{5}{10} = \frac{3}{6}$$
$$\frac{1}{2} = \frac{1}{2} \quad \square$$

Note: If our steps had produced $x = -3$ or $x = 1$, the equation would have no solution since both would cause division by zero. Once again, we could have seen this possibility before starting the problem.

■ **EXAMPLE 53**

Solve for x in $\dfrac{x}{x+4} = 3 - \dfrac{4}{x+4}$.

Solution

The LCD is $x + 4$. Therefore,

$$(x+4) \cdot \frac{x}{x+4} = 3(x+4) - \frac{4}{x+4} \cdot (x+4)$$
$$x = 3x + 12 - 4$$
$$x = 3x + 8$$
$$-2x = 8$$
$$x = -4.$$

Check: If $x = -4$, then

$$\frac{-4}{-4+4} = 3 - \frac{4}{-4+4}$$
$$\frac{-4}{0} = 3 - \frac{4}{0}.$$

Since division by 0 is not defined, the original equation has no solution. ☐

Exercise Set 0.9

Solve each of the following equations for x. First determine the LCD. If the answer does not have a solution, state "no solution" as your answer.

1. $\dfrac{x}{3} - \dfrac{1}{4} = \dfrac{1}{12}$

2. $\dfrac{2x}{3} - \dfrac{5}{2} = -\dfrac{1}{2}$

3. $\dfrac{x}{3} - \dfrac{1}{4} = \dfrac{x}{4} - \dfrac{1}{6}$

4. $\dfrac{2x}{9} - \dfrac{1}{6} = \dfrac{x}{9} + \dfrac{1}{6}$

5. $\dfrac{3}{4} + \dfrac{5}{8} = \dfrac{x}{12}$

6. $\dfrac{x}{8} - \dfrac{x}{12} = \dfrac{1}{8}$

7. $\dfrac{x+1}{3} - \dfrac{x-1}{2} = 1$

8. $\dfrac{x+1}{3} + \dfrac{x+2}{7} = 5$

9. $\dfrac{2}{3x} + \dfrac{1}{x} = 10$

10. $\dfrac{1}{2x} + \dfrac{1}{x} = -12$

11. $\dfrac{x-7}{x+2} = \dfrac{1}{4}$

12. $\dfrac{x-2}{x+3} = \dfrac{3}{8}$

13. $\dfrac{2}{x+3} = \dfrac{5}{x}$

14. $\dfrac{4}{5x-1} = \dfrac{2}{2x-1}$

15. $2 + \dfrac{3}{x-3} = \dfrac{x}{x-3}$

16. $\dfrac{3x}{x-4} = 5 + \dfrac{12}{x-4}$

17. $\dfrac{x-2}{x-3} = \dfrac{x-1}{x+1}$

18. $\dfrac{6x-2}{2x-1} = \dfrac{9x}{3x+1}$

19. $\dfrac{3}{x-5} + \dfrac{1}{x+5} = \dfrac{2}{x^2-25}$

20. $\dfrac{1}{x+3} + \dfrac{1}{x-3} = \dfrac{1}{x^2-9}$

Summary

The purpose of this chapter is to provide a review of elementary algebraic concepts for those students who feel there might be some weaknesses in their algebraic preparation.

We began by developing the *real-number system* beginning with the *natural numbers* and progressing through the *integers, rationals,* and *irrationals,* and finally arriving at the real-number system. We also examined the *real-number properties* in examples and in exercises.

We then reviewed division by zero and discussed why it cannot be defined. Next, we looked at the laws of integral exponents, including the definitions of zero and negative exponents.

We discussed operations with algebraic expressions, including simplification, addition, subtraction, multiplication, and division in preparation for finding the solutions to *linear equations* and *inequalities.* We discovered that some equations are always true (*identities*) and some are true only on the condition that the variable assumes a certain value (*conditional equations*). We also found out that some equations, such as $x + 1 = x$, have no solutions.

We then reviewed various forms of factoring, including common factors, the difference between perfect squares, perfect-square trinomials, and quadratic trinomials. The importance of factoring became clear as we looked at operations with algebraic fractions. We reviewed reducing fractions, multiplying and dividing fractions, and adding and subtracting fractions.

Finally, we learned to *clear fractions* as a step in solving linear equations that contain fractions.

Review Exercises

Perform the indicated operations.

1. $3 \cdot 4 \cdot 0 \cdot 2$

2. $15 \cdot 2 \cdot 0$

3. $5 \cdot x^3 \cdot x^{-3}$

4. $(-7x^2)^3$

5. $7(x^2y^3)^0$

6. $-3(x^{-2}y^4)^3$

Combine like terms and simplify.

7. $4x^2y - 3xy^2 - x^2y + 9xy^2$

8. $4x^2yz - (5xyz - 3x^2yz + 2xyz)$

9. $5 - [3x - (x - y) + 2]$

10. $7 - 3\{x - 2[4 - (x - 5)]\}$

11. $-3\{x - 2[5 - 2(x - 3)]\}$

12. $-\{8 - 2[3 - (x - 4)]\}$

Perform the indicated operations.

13. $(-4xy)(-3x^2y)$

14. $-(2ab)(-3a^2b)$

15. $2ay(3x - 4y)$

16. $3xy^2(2x^2y - 3x^2y^2)$

17. $(2x - 3y)(x - 2y)$

18. $(x - 4y)(2x + 3y)$

19. $24x^2y^2 \div 3xy$

20. $-18ab \div 3a$

21. $(15xy^2 - 5x^2y + 5xy) \div 5xy$

22. $(36abc - 12ab - 18ac) \div 6a$

Solve each of the following equations and inequalities. Check your solutions.

23. $5x - 4 = 2x + 5$

24. $2x + 5 = 3(x - 4)$

25. $3x - 2(x - 4) = 7$

26. $3x - 2(x - 4) = 2x - 3$

27. $2x - 3 > 5$

28. $4x - 3 \le 2x + 3$

29. $x - 3 < 2(x - 4)$

30. $2(x - 2) \ge 3(x + 4)$

31. $\dfrac{12}{3x - 2} = 3$

32. $\dfrac{5}{x + 3} = \dfrac{3}{x - 1}$

Factor each of the following algebraic expressions.

33. $24xy - 8x$

34. $4ab - 2abc$

35. $4x^2 - 9y^2$

36. $24a^2 - 6b^2$

37. $x^2 + 20x + 100$

38. $2x^2 - 28x + 98$

39. $x^2 - 2x - 15$

40. $x^2 + 8x - 7$

41. $y^2 - 3y - 28$

42. $a^2 - 3a + 2$

Perform the indicated operations.

43. $\dfrac{x^2 - 9}{24} \cdot \dfrac{18}{3x + 9}$

44. $\dfrac{x^2 - 6x + 9}{x^2 - 9} \cdot \dfrac{x^2 + 7x + 12}{x^2 - 16}$

45. $\dfrac{x - 3y}{a^2 - 1} \cdot \dfrac{a^2 + 7a + 6}{x^2 - 9y}$

46. $\dfrac{x^2 - 8x + 12}{x^2 - 4} \cdot \dfrac{x^2 + 4x + 4}{x^2 - 36}$

47. $\dfrac{x^2 - y^2}{x + y} \div \dfrac{x - y}{x}$

48. $\dfrac{x^2 - 3x - 4}{x^2 - 5x + 6} \div \dfrac{x^2 + 6x + 8}{x^2 - x - 6}$

49. $\dfrac{a + b}{3} + \dfrac{2a + 3b}{9}$

50. $\dfrac{2xy}{4} + \dfrac{3xy}{12} - \dfrac{xy}{8}$

51. $\dfrac{3}{x^2 - y^2} + \dfrac{2}{x^2 - 2xy + y^2}$

52. $\dfrac{2}{x^2 - 5x + 6} - \dfrac{4}{x^2 - 4}$

Mathematics: A Necessary Tool for Data Processing and Computer Programming

CHAPTER 1 OBJECTIVES

Before beginning a formal study of the mathematical concepts in this text, we present an informal overview of the book. By reading this introductory commentary, you should be able to:

1. Gain a perspective, or framework, within which subsequent concepts and/or developments can be placed.

By seeing the relevancy of some of the mathematical ideas to data processing and computer work as illustrated, you should be able to:

2. Answer the often-asked question: "Why are we studying this?"

Also, by reading this introduction you should be able to:

3. Acquire a thread of continuity for the emerging topics. Keep in mind, however, that not all the terminology will be immediately familiar to you upon the first reading. We hope that as your study of the material progresses, several rereadings of this chapter will help to unify the many mathematical topics into a more cohesive whole.

1

1.1
Introduction

As we near the beginning of the twenty-first century, it is clear that our lives have been influenced more by computers than by any other technology of civilization. Our need for knowledge about this continually evolving field has grown exponentially. Mastery of specific skills cannot be our only goal, for these skills will soon become obsolete. Instead, we must focus on understanding the principles beneath these skills. In any science, not only computer science, the fundamental principles and theorems can only be understood through the medium of mathematics. The mathematics taught to students of computer technology has changed radically since the early days of the discipline. As the field of computer technology has evolved, the uses of mathematics have become more sophisticated. The mathematics has developed a distinctive character, incorporating aspects of set theory, logic, and universal algebra. Current views by such organizations as DPMA (Data Processing Managers Association) and ACM (Association for Computing Machinery Education Board) stress the need for substantial training and education in mathematics in order to understand the tools and use them well. In summary, computer science draws—and will continue to draw—heavily on mathematical ideas and methods.

The field of data processing is extremely dynamic, ever growing, and becoming more and more diverse, as well as more encompassing. Absolute definitions of data processing, computer science, or information science are difficult to state. For our discussion here, we will informally define "data processing" as the discipline concerned with the representation and processing of information with the aid of computers. As such, this area lies somewhere between the field of mathematics and the field of computer technology itself—close enough to each area to be profoundly affected by the recently accelerated developments in each.

1.2
Questions

What are the particular mathematical tools relevant to data processing? Is the mathematics needed for data processing the same as that needed for computer science, for programming, for systems analysis, for data processing management?

1.3
An Outline for Some Answers

First we shall discuss several general ideas and types of mathematical knowledge directly beneficial to those studying data processing. Then the remainder of this chapter will point to particular content areas in mathematics—areas containing topics necessary for a successful venture into data processing and computer programming.

Dedicated to the development of information processing as a discipline, the Association for Computing Machinery is one of the most important organizations in the field of data processing. One of the group's explicit purposes is to advance both the sciences and the art of information processing. In ACM's *Recommended Curricula for Computer Science and Information Processing Programs in Colleges and Universities, 1968–1981*, a strong recommendation is made for the attainment of at least a minimal level of mathematical skill and vocabulary. An understanding of, and capability to use, mathematical concepts and techniques is vitally important! A solid mathematical background should be acquired by those students enrolled in data processing. Mathematics serves as a communication language. Also—and more important—logical thinking and reasoning skills can be improved and organizational skills can be strengthened by studying, understanding, and using the structure of mathematics.

1.4
Broad General Mathematical Ideas as Tools

As a minimal requirement, at least one course in mathematics that includes such topics as set theory, logic, functions, linear equations, and matrix algebra should be taken. A study of probability and statistics develops the required tools for the measurement and evaluation of programs and systems. Much of the mathematics in a discrete mathematics course is closely associated with the development of computer science concepts and techniques (with application in areas such as algorithm analysis and testing, as well as data structure design). Students in data processing should be able to use and apply to their computer work the appropriate vocabulary and techniques from such topics as Venn diagrams, summation notation, logical propositions, and frequency distributions, to name just a few.

ACM's *Recommended Curricula* (p. 170, cited above) states, "Perhaps more than any other area, the mathematics background and expertise of the graduate determine the number of future options available and the variety of directions open as career paths."

> Because of the rapid change in the computer technology, training with specific mathematical tools is not quite as important as understanding principles, models, and the problem-solving process.

Knowledge varies from simple classification to an understanding based on a system of principles. Useful knowledge results from the ability to recognize similarities in different events, to isolate important factors, and to generalize. Generalizations allow and provide for effective operation in new environments by using information from the past.

Mathematical models are models or "analogies" based on mathematically stated principles. Often these models are a mathematical characterization of a phenomenon or process. They can be used to present information, to aid in investigation and prediction, to offer simulations, and to provide a means for performing analysis and computation. Because of its rigor and lack of ambiguity, mathematics has provided—and continues to provide—a good language for the expression of principles and models.

Inherent in the study of mathematics should be the development of an awareness of these principles and models and their proper usage. Accompanying this development (and perhaps the most important tool to be obtained from a study of mathematics) is an understanding of, and the ability to use, the problem-solving process. For most computer problems there is a prescribed pattern for obtaining the solution. Defining the problem, developing the solution, constructing an appropriate algorithm, coding the problem, testing and debugging, and then documentation are the major components in this process. In the development and understanding of the problem-solving process, mathematical reasoning plays an essential role. As you develop mathematical reasoning and mathematical maturity, you will be able to apply more and more sophisticated mathematics to your data processing courses and professional environment.

1.5
Some Specific Mathematical Tools

With a developed awareness of the need for the more general mathematical skills, we can now focus our attention on a more specific listing of the mathematical tools that are provided in this text. The list does not try to be exhaustive, but rather to instill awareness of the many tools necessary for data processing and computer programming.

Computer Arithmetic Tools

In order to understand computer arithmetic, we must master such specific concepts as scientific notation, truncation, overflow, integers vs. floating point, binary numbers, and base systems. A prerequisite to understanding the inner workings of the computer (storage and processing) is a working knowledge of base systems—primarily bases 2, 8, and 16. This is essential for the study of assembly language. Did you know that the word *bit* is short for Binary digIT? Did you know that the computer will convert a decimal number (14) to a binary number (14 = 1110 in base 2) in order to perform computations? Did you know that subtraction is accomplished by using binary numbers and their complements? Were you aware that division is done by repeated subtraction? For locating errors and analyzing program flow, a study is made of storage locations and addresses, and the printout is in either octal (base 8) form or hexadecimal (base 16) form.

Algorithms and Decision-making Skills

For the process of programming, probably no tool is more important than the efficient and effective use of algorithms. Essentially the step-by-step outline for solving a problem, an algorithm can be expressed in many ways; two of the more common forms are a flowchart and pseudocode. Regardless of the format, an algorithm should be precise and unambiguous (just as the language of mathematics is). The development and use of algorithms corresponds to the analysis part of the problem-solving process in mathematics. Along with this process should be the development of reasoning skills and the use of inferences, and of the ability to think logically. All of these tools are a prerequisite to the many iterative computing and processing procedures present in data processing.

The Mathematical Language of Sets

Our interest in sets is twofold: (1) Sets pervade all of mathematics, and (2) sets are useful tools in modeling and investigating problems in data processing. The field of business continually analyzes and classifies elements by groups or sets. Study is often made of subsets (for example, the subset of all hourly employees who have been with a company for more than ten years). Files and data structures also require the use of sets. To understand the complex interrelationships among sets, models such as Venn diagrams can prove to be useful analytic tools. Sets are also indirectly important in the study of symbolic logic and even in the design of the digital computers.

Basic Tools Needed from Algebra

Specific tools from real-number algebra include a working knowledge of the hierarchy of operations, the manipulation and evaluation of expressions, the ability to analyze and then translate a problem into the symbolism of mathematics, and the use of equations in problem solving. The process of proceeding from understanding a problem or task to formulating the logic to complete the task in terms of an algorithm is virtually identical to that needed in data processing.

A mature understanding of algebra provides several beneficial tools:

1. The simplification and evaluation of mathematical expressions— for example, ((hours × rate + overtime × 1.5) × 0.876).
2. The solution of equations through the manipulation of algebraic symbols—for example, solving the interest formula

$$A = P(1 + r/k)^{kt}$$

for the interest rate r.
3. The analysis of profit and cost relationships to determine break-even values.
4. The translation of verbal statements to the corresponding algebraic statements (this eventually leads to programming statements in BASIC or COBOL or Pascal).

Many of the problems facing programmers and decision makers are concerned with computing an unknown, given several unknowns. With the use of algebra and equation solving, we can program formulas or approximations to formulas to compute optimal production levels or other unknowns. Although computer software does exist to solve various types and forms of equations, the process of reading, analyzing, and translating a problem to its algebraic counterpart still relies quite heavily on the human element! Development of this capability requires much work, but the dividends paid are directly proportional to the amount of work invested!

In addition, a good working knowledge of algebra is indirectly a prerequisite for learning other (and often higher) mathematical concepts useful in business data processing, such as linear programming, forecasting, and decision analysis.

The Idea of Functional Relationships

Intimately related to mathematics is the concept of a function. Data processing problems such as cost analysis, profit ratios, production of goods, or supply and demand investigate the relationships between quantities or sets of quantities. An understanding of functional relationships helps in the creation of mathematical models to represent given problems. A knowledge of types of functions (constant, linear, quadratic, exponential) and their characteristics is essential in the development or selection of an appropriate model.

Developing techniques to graphically represent functional relationships between variables of interest is a useful tool. Using graphical methods to illustrate proposed models or existing relationships does much to further the understanding and application of the concept of functions.

Relations are also useful in the characterization of structures. Composite data structures such as lists, arrays, or trees represent sets of objects with a relation that holds among members. Conceptually understanding relations and functions also plays an important role in the theory of computation, program structure, analysis of algorithms, numerical application, information retrieval, and network problems.

Logical Thinking as a Requisite Programming Tool

No study of mathematics for data processing would be complete without some study of logic. Symbolic logic is the basis of the problem-solving activities of computer work. An understanding of logic will usually produce both a more efficient and a more effective programmer. A programmer programs decision-making abilities: (IF SALARY > 820 AND (AGE < 35 OR NUM DEPEND < 2) THEN COMPUTE TAX AT 28.2%). The high-level computer languages such as COBOL, Pascal, and BASIC are capable of handling very complex logical statements. To make effective and correct decisions, the programmer needs an understanding of the basic concept of mathematical logic, the capability to correctly use the binary logical operators (AND and OR), and the ability to simplify and evaluate complex logical expressions.

The procedures linking the input to the computer to produce reliable output from the computer make up the logical sequence of instructions to the computer. The logical path of a program is developed by the programmer. Algorithms and flowcharts are tools to aid in the development of this logical step-by-step procedure toward the solution of the problem. Decision tables also help data processors to convey information quickly and clearly to a programmer or to a reader of system specifications. These programming tools all require logical thinking!

Applications of Logic: Circuits and Boolean Algebra

At times Boolean algebra is nearly synonymous with the algebra of propositional logic. The topic of Boolean algebra—often a course in and of itself—finds application not only in the switching circuits of digital computers, but also in computer design and computer programming. As a tool, Boolean algebra is used to construct circuits and logic patterns and to analyze and simplify the seemingly complex logic circuit networks and programs associated with computers. Techniques such as fundamental products (to create logical circuits) and sum-of-products form and Karnaugh maps (to facilitate the analysis and simplification process) are tools a practitioner in data processing should not overlook.

An important outcome of the study of Boolean algebra is an awareness of the close similarities between sets, logic, circuitry, and Boolean algebra. The concept of properties, common to each of these four seemingly different areas, allows a person to begin to understand the principles behind mathematics. And as we've commented several times before, the importance of understanding principles as a tool, rather than relying merely on specific skills, cannot be underestimated.

Solving Systems of Equations

From algebra, we learn processes to solve an equation for a particular unknown. Quite often we seek a common solution to several equations at the same time (simultaneously). For example: (1) to find the break-even point in business, we must determine a point of production at which total cost will equal total revenue, and (2) to achieve market equilibrium, we know that supply should equal demand. To solve these problems and many others like them, a useful tool is the ability to solve a set of equations (system) simultaneously.

Of the different techniques for solving systems of simultaneous equations, the methods of forming linear combinations and Gaussian elimination have the greatest usefulness. First, these methods serve as the basis for learning and effectively using the simplex method in linear programming—a procedure that is used extensively in business data processing. Second, because of the repetitive nature of solving systems of equations, computers can quite easily be programmed to solve systems by the simplex method. To program the computer, or to even use such programs effectively, the programmer must have a sound working knowledge of these techniques.

The Ability to Use Arrays, Matrices, and Determinants

Much of business data processing and management is concerned with the manipulation of quantities of data. Models of the financial well-being of a company, cost accounting systems, inventory parts listing updates, and compilation of nationwide sales figures to illustrate trends are but a few situations in which huge amounts of data must be processed, and often processed as a single entity. Frequently the operations performed on the data are repetitive (for example, providing a daily update of sales figures). With the use of computers and the mathematical tools of arrays, matrices, and determinants, enormous quantities of data can be processed very efficiently. We can regard an array, or matrix, as a "big table" of data, such as that shown in Table 1.1. The same table in matrix form would appear as shown in Table 1.2; these figures might represent quarterly sales totals for 1985. Total sales for each person, for each quarter, for the years 1981–1985 could be found easily by the following computer statement:

MAT TOTSALES = SALES81 + SALES82 + SALES83
+ SALES84 + SALES85

assuming matrices were available for 1981–1985. Commissions for 1983 could be computed simplistically by

MAT COM = SALES83 * 0.24

In these examples, MAT, as you may have guessed, stands for matrix.

Table 1.1
Quarterly sales totals for 1985

	Q1	Q2	Q3	Q4
Smith	40	505	650	346
Jones	70	950	750	874
Rich	80	820	740	954

Table 1.2
Quarterly sales:
Matrix form

$$\begin{pmatrix} 40 & 505 & 650 & 346 \\ 70 & 950 & 750 & 874 \\ 80 & 820 & 740 & 954 \end{pmatrix}$$

Matrices and arrays are tools that allow the manipulation of entire quantities of data as a single entity. As such, arrays and matrices are also extremely useful tools in the manipulation and solution of systems of equations.

A Working Knowledge of Linear Programming

Arrays and matrices are useful tools in another area of mathematics called *linear programming*. Through linear programming, the best or optimal solution can be found to a problem with many restrictions on the variables involved. For example, often in business the objective is to achieve a maximum profit or a minimum cost as a large number (often several hundred) of variables are manipulated. Money, time, plant space, machine hours, availability of raw materials, etc., all impose restrictions on what can and cannot be done. This in turn affects the quantity manufactured, which in turn affects the profit or cost. In such situations, the application of linear programming is useful. While developed rather recently (the 1940s), linear programming is perhaps the single most widely applied topic in all of mathematics, especially in the area of business.

To develop the ability to use linear programming effectively, the graphical method and the algebraic method should first be studied and understood. Then a knowledge is gained for the principles behind the simplex method, the method used most frequently by high-speed digital computers for an optimal solution to a business problem.

The Language and Concepts of Statistics and Probability

In business data processing, masses of information, or data, are often gathered, analyzed, and processed. To summarize these data and make inferences (predict future trends, for example), we need a working knowledge of statistical vocabulary and concepts. Graphical methods and descriptive statistics (measures of central tendency, measures of dispersion) are indispensable tools in the succinct presentation of gathered data. Also quite often in business applications, there is no way of knowing an exact answer (for example, determining the amount of a commodity needed for sale next month). At this point, mathematical models are constructed using the statistical tools of probability and expected values. Probability is also useful in auditing, error checking, and quality control as the idea of sampling is used to make inferences. Forecasting and linear regression are yet two more statistical skills used to manipulate past records and predict (or make inferences) for future periods of time. Statistical tools are used increasingly in business data processing. The language and concepts of statistics and probability are yet other examples of the many indispensable mathematical tools needed in a computer-related world.

Computer Arithmetic

CHAPTER 2 OBJECTIVES

After completing this chapter, you should be able to:

1. Determine which digits are significant in a number.
2. Compare two numbers and determine which one is the more precise.
3. Apply the rules for rounding numbers.
4. Convert numbers expressed in standard form into scientific notation, and vice versa.
5. Perform arithmetic operations in integer arithmetic.
6. Express numbers in the normalized and normalized E forms.
7. Perform arithmetic operations with numbers in their normalized form.

2

2.1

Significant Digits and Approximation

The study of mathematics is very similar to the study of any language. Just as there are words, phrases, concepts, and syntax to worry about in a language, there are words, phrases, concepts, symbols, and syntax to learn in mathematics. Communication between people requires a working knowledge of a language. Since mathematics is a universal language, it is also necessary to have a working knowledge of mathematics if you are ever going to be able to supply answers to the "hows": how often, how far, how many, how much, how close, how thin, etc.

Mathematics plays a major role in our communication with the computer, since input to, and output from, the computer is frequently in the form of numbers. In fact, one of the most valuable aspects of the modern computer is its ability to perform many mathematical computations on numerical input with tremendous speed and accuracy. The only drawback is that a certain amount of error is unavoidable, because of limitations on the size of the numbers that the computer can handle at any given time. A computer can process only a finite number of digits during any given moment. Therefore, the programmer must be aware of the computer's limitations and take the appropriate steps to work around those limitations.

Computers use numbers in two different forms: *fixed point* and *floating point*. Fixed-point numbers are *whole* numbers in which the decimal point is *fixed* after the rightmost digit. In mathematics, these numbers are called *integers*. Floating-point numbers are numbers that contain a decimal point, sometimes referred to as *real* numbers. Integers (whole numbers) are also real numbers; thus fixed-point numbers can also be used in a floating-point mode. However, not all real numbers are integers. Integers are numbers encountered from data gathered through counting, whereas real numbers are a result of some form of measurement. Therefore, floating-point numbers (real numbers) *cannot* be used in a fixed-point mode (as integers). In a computer program, a programmer determines whether the numbers are fixed-point or floating-point and how many digits a number can have. In other words, a programmer is forced to use approximations for some numbers by limiting the total number of digits to a predetermined size.

Significant Digits

By limiting the size of a number to a specified number of decimal positions, we must be concerned with the *accuracy,* as well as the precision, of the number. These concepts involve *significant digits.* The rules used to determine significant digits are listed on the following page.

■ **DEFINITION**

ACCURACY

The number of significant digits is a measure of the *accuracy* of a number.

■ **EXAMPLE 1**

a) $\sqrt{3} = 1.732050808$ is accurate to ten significant digits.

b) $\frac{1}{4} = 0.25$ is accurate to two significant digits.   □

■ **DEFINITION**

PRECISION

The position of the last significant digit to the right indicates the *precision* of a number.

RULES FOR DETERMINING SIGNIFICANT DIGITS

1. All nonzero digits are significant.

    For example, 325, 42.5, and 0.256 each contain three significant digits.

2. All zero digits occurring between nonzero digits are significant.

    For example, 3204, 0.4005, and 7.055 each contain four significant digits.

3. The ending zeros following the decimal point are significant.

    For example, 24.00 and 4.300 each contain four significant digits.

4. The zeros between the decimal point and the first nonzero digit in a number less than one are *not* significant digits.

    For example, 0.00352, 0.0352, and 0.352 all have three significant digits.

5. The trailing zeros in an integer are not generally considered significant digits.

    For example, 4730, 47,300, and 4,730,000 all have three significant digits. The zeros are not significant digits. If the trailing zero(s) is meant to be a significant digit, it should be explicitly stated, hence, "4730 has four significant digits."

■ **EXAMPLE 2**

a) 1.73205 has greater precision than 0.25.

b) 0.25 has greater precision than 0.2.   □

Truncating versus Rounding

The programmer determines the size of the numbers in a program. Normally, the computer will drop any digits that *overflow* the reserved spaces, a process known as *truncating*. To illustrate, consider 0.473 × 1.55 = 0.73315. Suppose this arithmetic operation were performed on a computer that was programmed to print answers to an accuracy of four significant digits. The printed answer would be 0.7331, with the least significant digit 5 truncated.

Since truncating by the computer can result in considerable error, it does not always provide a desirable result. This is especially true when dealing with monetary figures. For example, if $73.99 were truncated to the nearest dollar amount, the 99¢ would be lost. If this were to take place in a computer program using a loop, the error would be compounded each time the program ran through the loop.

A process that can be used to minimize the error caused by truncating and still reduce the number of significant digits in an answer is called *rounding*. Under the process of rounding, undesired digits are dropped, but the last desired significant digit is adjusted accordingly.

RULES FOR ROUNDING

1. ROUNDING DOWN (similar to truncating). If the digit immediately to the right of the last desired significant digit is less than 5, the last desired digit is unchanged and the remaining digits are dropped (truncated).

   For example, if 34.52 is rounded to three significant digits, the result is 34.5, with the 2 being dropped.

2. ROUNDING UP. If the digit to the right of the last desired digit is 5 or larger, the last desired digit is incremented by 1, and the remaining digits are dropped.

   For example, if 34.57 is rounded to three significant digits, the result is 34.6. The number 34.75 is rounded to 34.8. The number 34.95 is rounded to 35.0.

In the previous illustration in which $73.99 was truncated to two significant digits and the 99¢ was lost, the programmer could have made adjustments for rounding rather than truncating. Then the error would have been 1¢ rather than 99¢, since $73.99 rounds to $74.00. Rounding is more advantageous than truncating because roundoff errors tend to balance themselves out in any group of numbers, since some numbers are being rounded up as others are rounded down. Truncating does not provide this balancing effect. Suppose $73.99 and $74.20 are the results of computations, but we are looking for the whole-dollar amount. Truncating would give us $73 and $74, respectively, whereas rounding would give us $74 and $74. The latter result, $74, is a closer approximation of both $73.99 and $74.20.

Exercise Set 2.1

What is the accuracy (number of significant digits) of each of the following?

1. 342

2. 402

3. 40.2

4. 4.003

5. 27.00

6. 276.0

7. 25,000

8. 240

9. 10.000

10. 1.00

11. 0.0402

12. 0.00430

13. 100.2

14. 0.000537

15. 75,000,000

16. 0.07002

Which number in each of the following pairs of numbers is the more precise?

17. 3.14159; 2.7182

18. 0.3333; 0.66667

19. 4.5402; 4.540

20. 7.747; 8.32

Truncate each of the following to three significant digits.

21. 3.14159

22. 4.754

23. 0.73678

24. 0.234790

25. 24.3724

26. 7.218

Round each of the following off to three decimal places.

27. 3.14159

28. 17.1938

29. 0.00452

30. 247.3423

31. 24.3724

32. 0.0129

33. 14.295

34. 1.473499

2.2
Scientific Notation

The place-value characteristic of decimal numbers allows us to write a number in several different forms. One such form was initially developed as a means of dealing with very large and very small numbers. This shorthand notation was used when all computations were performed by pencil and paper. Even now that computations can be done electronically with a handheld calculator, scientific notation is still a necessary and convenient shorthand notation. Floating-point numbers are quite often written in scientific notation.

> ■ **DEFINITION**
>
> SCIENTIFIC NOTATION
> A number written in the form $N \times 10^{P}$, where $1 \leq N < 10$ and P is an integer, is written in *scientific notation*.

Scientific notation is important when we are dealing with computers because of the limitations on exactly how many digits of a number can be displayed and/or stored. Most computers will accept numbers in integer form, in decimal form, or in scientific notation. Scientific notation will be in a slightly different form than that illustrated above. Typically, E will take the place of "× 10."

STEPS FOR WRITING A NUMBER IN SCIENTIFIC NOTATION

1. The decimal point is moved left or right until only one nonzero digit remains to the left of the decimal point.
2. The resulting number is then multiplied by a power of 10 corresponding to the position value of the leftmost significant digit. The exponent is equal to that position value if the decimal point is moved left. The exponent is negative if the decimal point is moved right and again is equal to the position value of the leftmost significant digit. For example:

   a) $93,000,000 = 9.3 \times 10^7$;
   b) $0.00125 = 1.25 \times 10^{-3}$;
   c) $4.27 = 4.27 \times 10^0$.

■ **EXAMPLE 3**   The examples in the box above would appear as follows:

a) $93,000,000 = 9.3E7$ or $9.3E + 7$
b) $0.00125 = 1.25E - 3$
c) 4.27 would appear as is.   □

Although we can normally enter a number in any form, the computer will print it in scientific notation if the value of the number exceeds a certain range. The exact limits of the range vary from computer to computer. In the computer language BASIC, any number containing six or fewer digits will be displayed in standard form, whereas those with more than six digits will be displayed in scientific notation. The range allowable for the exponent in scientific notation is usually -99 to $+99$, inclusive, with some variation from computer to computer.

Exercise Set 2.2

Express each of the following numbers in scientific notation.

1. 7406   **2.** 30,452

3. 9274.3   **4.** 1257.04

5. 93.704   **6.** 2.7432

Express each of the following numbers in the form of scientific notation displayed on the computer.

7. 6.34×10^4

8. 2.70×10^5

9. 8.321×10^{-2}

10. 5.54×10^{-8}

11. 76,000,000

12. 43,200,000,000

13. 0.00000742

14. 0.0000000021

15. 332,000,000

16. 0.0000405

Express each of the following numbers in standard notation.

17. 3.42×10^4

18. 5.07×10^{-4}

19. 4.735×10^{-7}

20. 3.2×10^0

21. $7.26E + 3$

22. $3.2E + 7$

23. $2.3754E - 4$

24. $7.98E - 10$

25. $4.327E + 5$

26. $1.25E - 6$

2.3
Fixed-point and Floating-point Numbers

Fixed-point Numbers

As you recall, fixed-point numbers are integers. The decimal point is fixed after the rightmost digit. Arithmetic performed on the computer in a fixed-point mode presents problems of which the programmer must be aware.

Before we consider these problems, let us review the hierarchy of operations. The order in which arithmetic operations are performed in most computer languages is as follows, with the highest level listed first and the lowest level last.

■

HIERARCHY OF OPERATIONS
1. Exponentiation
2. Multiplication and division
3. Addition and subtraction

When two or more operations on the same level are encountered, the operations are performed from left to right. Grouping symbols, such as parentheses, are used to change the natural order and must be dealt with first.

■ **EXAMPLE 4**   Find the value of the arithmetic expression $4^2 \times 3 + 5 \times 2 \div 2$.

Solution

Find 4^2:	$16 \times 3 + 5 \times 2 \div 2$
Find 16×3:	$48\ \ \ + 5 \times 2 \div 2$
Find 5×2:	$48\ \ \ +\ \ \ 10\ \ \ \div 2$
Find $10 \div 2$:	$48\ \ \ +\ \ \ \ 5$ **Note that addition is the last**
Find $48 + 5$:	53 **operation performed.** □

■ **EXAMPLE 5**   Find the value of the arithmetic expression

$$2 \times 3 \times 4 \div (3 + 5) \times 3^2.$$

Solution

Find $(3 + 5)$:	$2 \times 3 \times 4 \div 8 \times 3^2$
Find 3^2:	$2 \times 3 \times 4 \div 8 \times 9$
Find 2×3:	$6\ \ \times 4 \div 8 \times 9$
Find 6×4:	$24\ \ \ \ \ \div 8 \times 9$
Find $24 \div 8$:	$3\ \ \ \ \ \ \times 9$
Find 3×9:	27 □

If the computer is operating in a fixed-point arithmetic mode, the programmer must realize that only integer results are possible. This fact actually presents a problem only when division is used. If the quotient is not an integer, the computer keeps only the whole-number quotient and truncates or loses any remainder. This is known as *integer arithmetic*.

■ **EXAMPLE 6**   In integer arithmetic:

a) $9/4 = 2$, with the remainder 1 being lost;
b) $12/7 = 1$, with the remainder 5 lost.   □

Although division is the one operation that is affected by integer arithmetic, another rule involving addition of fractions generally does not hold in integer arithmetic either. That is,

$$\frac{a}{c} + \frac{b}{c} \neq \frac{a + b}{c}.$$

Division has a definite priority over addition and must be performed first.

■ **EXAMPLE 7**   **a)** In ordinary arithmetic:

$$\frac{4}{3} + \frac{5}{3} = \frac{4 + 5}{3} = \frac{9}{3} = 3; \qquad \frac{2}{3} + \frac{4}{3} = \frac{6}{3} = 2.$$

b) In integer arithmetic:

$$\frac{4}{3} + \frac{5}{3} = 1 + 1 = 2; \qquad \frac{2}{3} + \frac{4}{3} = 0 + 1 = 1.$$ □

Floating-point Numbers

Floating-point numbers contain a decimal point and are also called real numbers. (A more extensive discussion of the real-number system occurs in Chapter 0. In short, real numbers are all whole numbers and all decimal fractions.) They are usually written in scientific notation or in a *normalized* notation in which the significant digits are expressed as a fraction between 0.1 and 1.0 (including 0.1 but not 1.0) times a power of 10. That is, the significant digits are expressed with the decimal point written immediately to the left of the first significant digit on the extreme left. The normalized notation is also displayed in the E form mentioned earlier.

■ **EXAMPLE 8**

Standard notation	*Normalized notation*	*Scientific notation*
a) 43,000	0.43×10^5	4.3×10^4
	or 0.43E + 5	or 4.3E + 4
b) 0.0043	0.43×10^{-2}	4.3×10^{-3}
	or 0.43E − 2	or 4.3E − 3
c) 43	0.43×10^2	4.3×10^1
	or 0.43E + 2	or 4.3E + 1 □

The computer generally handles floating-point numbers in the normalized form. The part of the number containing the significant digits is called the *mantissa* and the power of 10 is the *exponent*. In Example 8, when we write 43,000 as 0.43×10^5, 0.43 is the mantissa and 5 is the exponent.

Arithmetic is performed in the normalized form by operating on the mantissas and the exponents.

Adding Normalized Numbers

1. If the exponents are the same, we add the mantissas and use the common exponent.

$$\text{The E form:}$$

$$(0.573 \times 10^4) + (0.348 \times 10^4) \qquad (0.573 \times 10^4) + (0.348 \times 10^4)$$
$$(0.573 + 0.348) \times 10^4 \qquad (0.573 + 0.348)E + 4$$
$$0.921 \times 10^4 \qquad 0.921E + 4$$

2. If the exponents are different, we adjust one of the numbers to make the exponents equal by moving the decimal point to the left as required and increasing the exponent; then we proceed as above.

$$(0.573 \times 10^4) + (0.348 \times 10^3)$$
$$(0.573 \times 10^4) + (0.0348 \times 10^4)$$
$$(0.573 + 0.0348) \times 10^4$$
$$0.6078 \times 10^4$$

If the sum is not in the normalized notation, it must be normalized. If the sum of the mantissas contains more significant digits than spaces allowed, the mantissa must be rounded or truncated.

$$(0.873 \times 10^4) + (0.964 \times 10^4)$$
$$(0.973 + 0.964) \times 10^4$$
$$1.837 \times 10^4$$
$$0.1837 \times 10^1 \times 10^4$$
$$0.1837 \times 10^5$$
$$0.184 \times 10^5$$

Subtracting Normalized Numbers
We follow the procedure we used for addition, except that we subtract the mantissas. The exponents must be the same and the answer must be normalized.

a) $(0.573 \times 10^4) - (0.348 \times 10^4)$
$$(0.573 - 0.348) \times 10^4$$
$$0.225 \times 10^4$$

b) $(0.573 \times 10^4) - (0.485 \times 10^4)$
$$(0.573 - 0.485) \times 10^4$$
$$0.088 \times 10^4$$
$$0.88 \times 10^3$$

Multiplying Normalized Numbers
We multiply the mantissas and *add* the exponents. The exponents do *not* have to be the same.

$$(0.573 \times 10^4) \times (0.348 \times 10^3)$$
$$(0.573 \times 0.348) \times 10^4 \times 10^3$$
$$0.199404 \times 10^7$$

Notice that the answer is in the normalized form, but the mantissa contains more significant digits than the original mantissas. Thus the mantissa of the product would be rounded or truncated depending on what the computer had been programmed to do.

$$(0.573 \times 10^4) \times (0.21 \times 10^3)$$
$$(0.573 \times 0.21) \times 10^4 \times 10^3$$
$$0.12033 \times 10^7$$
$$0.12 \times 10^7$$

Dividing Normalized Numbers

We divide the mantissas and *subtract* the exponents, subtracting the divisor exponent from the dividend exponent. The quotient may have to be normalized and possibly rounded or truncated.

a) $(0.3 \times 10^4) \div (0.15 \times 10^4)$ **(0.15×10^4) is the divisor.**
$$(0.3 \div 0.15) \times 10^{4-4}$$
$$2 \times 10^0$$
$$0.2 \times 10^1$$

b) $(0.15 \times 10^4) \div (0.3 \times 10^3)$
$$(0.15 \div 0.3) \times 10^{4-3}$$
$$0.5 \times 10^1$$

c) $(0.15 \times 10^3) \div (0.3 \times 10^4)$
$$(0.15 \div 0.3) \times 10^{3-4}$$
$$0.5 \times 10^{-1}$$

Exercise Set 2.3

Find the value of each arithmetic expression following the hierarchy of operations.

1. $5 \times 3 + 2 \times 4$

2. $8 \div 4 - 3 \times 2$

3. $2^3 + 4 \times 2 - 5$

4. $3^2 + 5 \times 2 - 4$

5. $4^3 + 2 \times 8 \div 4 \times 2$

6. $2^4 \times 3 + 4 \times 2 \div 2$

7. $2 \times 4 \times 5 \div (3 + 5) \times 2^3$

8. $3 \times 2 \times 4 \div (1 \times 2)^2 + 4$

9. $2 + 4 \times 5 \div 10 \times 2$

10. $3 \times 2 + 4 \div (1 \times 2)^3 + 4$

Perform the following operations using integer arithmetic.

11. $13/4 - 5/4$

12. $3/4 + 1/4 + 7/4$

13. $3 \times 2 \times 4/5$

14. $2 \times 3/4 \times 2$

15. $3 \times 4/5 \times 2$

16. $2 \times 2 \times 3/4$

17. $3 \times (4/5) \times 2$

18. $2 \times (3/4) \times 2$

19. $4 \times 3 \times 2/5/2$

20. $4 \times 5 \times 2/6/3$

21. $2 \times 3/4 \times 5/4$

22. $3 \times 5/6 \times 2/3$

Express each of the following numbers in the normalized notation with powers of 10.

23. 32.7

24. 173.0

25. 8.45

26. 0.743

27. 0.000745

28. 0.0063

Express each of the following numbers in the normalized E notation.

29. 72.4   **30.** 285.0

31. 743,000   **32.** 0.647

33. 0.643×10^2 **34.** 0.143×10^{-4}

35. 0.0004172   **36.** 0.00214

Perform the indicated operations. Make sure your answers are normalized.

37. $(0.3748 \times 10^2) + (0.4235 \times 10^2)$ **38.** $(0.4837 \times 10^2) + (0.9427 \times 10^2)$

39. $(0.842 \times 10^3) + (0.473 \times 10^2)$ **40.** $(0.738 \times 10^4) + (0.532 \times 10^3)$

41. $(0.6324 \times 10^3) - (0.3247 \times 10^3)$ **42.** $(0.4378 \times 10^3) - (0.8342 \times 10^2)$

43. $(0.3748 \times 10^2) \times (0.4235 \times 10^2)$ **44.** $(0.4837 \times 10^2) \times (0.9427 \times 10^2)$

45. $(0.842 \times 10^3) \times (0.473 \times 10^2)$ **46.** $(0.274 \times 10^{-4}) \times (0.562 \times 10^{-3})$

47. $(0.8 \times 10^6) \div (0.2 \times 10^2)$ **48.** $(0.12 \times 10^{-4}) \div (0.4 \times 10^2)$

49. $(0.125 \times 10^3) \div (0.25 \times 10^{-2})$ **50.** $(0.375 \times 10^4) \div (0.125 \times 10^{-2})$

Summary

In this chapter, we learned that despite the computer's ability to perform a multitude of arithmetic operations at a high rate of speed, it does have limitations in the number of digits that it can process at any given time. For this reason, the programmer must be aware of and know how to work with the *significant digits* of a number, as well as understand the concepts of *precision* and *accuracy* when dealing with numbers.

We classified numbers as either *fixed point* (integers) or *floating point* (real numbers). We discussed how the programmer must be aware of and/or specify which type of mode is being used because of the different results that are possible in arithmetic. Fixed-point numbers can be used in a floating-point mode, but floating-point numbers cannot be used in the fixed-point mode. Such is the case because all integers are real numbers, but not all real numbers are integers.

Since the computer is limited to dealing with numbers of a prespecified finite size, we discussed the differences between *truncating* and *rounding*. Again, these are concepts the programmer must deal with in order to limit possible errors evolving from computations.

Two new methods of writing numbers were introduced because of the nature in which numbers are stored within the computer. These were *scientific notation* and the *normalized notation,* including the associated E form of each. The chapter concluded with a discussion of addition, subtraction, multiplication, and division performed with numbers in the normalized form.

Review Exercises

State the number of significant digits of each of the following.

1. 4730   **2.** 40,730   **3.** 473.0

4. 0.00702   **5.** 0.03040   **6.** 27.04

7. 93,000,000   **8.** 7000.00   **9.** 0.276

Determine which number has the greater precision.

10. 27.425; 27.4253   **11.** 0.4325; 0.432

12. 3.14159; 2.7182818   **13.** 0.125; 0.4752

Truncate to four significant digits.

14. 3.14159   **15.** 2.7182818

16. 0.476305   **17.** 0.204763

Round off to four decimal places.

18. 3.14159   **19.** 2.7182818

20. 0.0005724   **21.** 24.71523

Express each of the following numbers in standard notation.

22. 3.476E + 4   **23.** 6.745E − 3

24. 4.53×10^4 **25.** 3.7524E + 6

26. 1.09×10^{-5} **27.** 2.71×10^9

Express each of the following numbers in scientific notation.

28. 0.000000708   **29.** 93,000,000

30. 0.00214

Perform the indicated operations in integer arithmetic.

31. $2 \times (5/3) \times 3$ **32.** $2 \times 5/3 \times 3$

33. $3 \times 2 \times 15/7/5$ **34.** $4 \times 12/5/3$

35. $15/4 - 11/4$ **36.** $4 \times (3/4) \times 2$

Perform the indicated operations. Make sure your answers are in the normalized notation.

37. $(0.3715 \times 10^3) + (0.4625 \times 10^3)$ **38.** $(0.832 \times 10^2) + (0.735 \times 10^4)$

39. $(0.7043 \times 10^3) - (0.6421 \times 10^3)$ **40.** $(0.475 \times 10^4) - (0.905 \times 10^3)$

41. $(0.305 \times 10^2) \times (0.12 \times 10^3)$ **42.** $(0.731 \times 10^3) \times (0.265 \times 10^{-1})$

43. $(0.1075 \times 10^2) \div (0.43 \times 10^{-1})$ **44.** $(0.882 \times 10^3) \div (0.245 \times 10^2)$

Algorithms, Flowcharts, Pseudocode, and Decision Tables

CHAPTER 3 OBJECTIVES

After completing this chapter, you should be able to:

1. List the steps required to solve a problem or complete a task.
2. Convert the list of steps required to solve a problem into a diagram called a flowchart.
3. Express the list of steps to solve a problem in pseudocode language.
4. Identify the conditions and actions in a problem and prepare a decision table showing all combinations of actions possible from the various combinations of conditions.

3

■ CHAPTER OUTLINE

3.1 Algorithms and Flowcharts
3.2 Pseudocode
3.3 Decision Tables
Summary
Review Exercises

3.1
Algorithms and Flowcharts

Algorithms

If you have ever written a set of directions to follow in order to get from one place to another, or if you have ever followed the step-by-step directions to assemble a toy model, then you have used an *algorithm*.

■ DEFINITION

ALGORITHM
A step-by-step list of instructions for accomplishing a specified task is an *algorithm*.

Mathematicians have realized the value of a logical step-by-step approach to problem solving for centuries. This same approach, so necessary and useful in mathematics, is also an integral part of computer programming. Generally speaking, the detailed instructions given to the computer in the form of a computer program are really the basic steps of an algorithm. The lines of the program are the steps of the algorithm that the computer must execute in order to perform the desired task.

■

CHARACTERISTICS OF AN ALGORITHM
1. Each instruction must be clear and exact.
2. There is a finite number of instructions.
3. One or more of the instructions may be repeated by looping within the algorithm.
4. The algorithm must be effective, that is, it must solve the problem.

■ **EXAMPLE 1**

The following is an algorithm to find the area of a rectangle by the formula $A = l \times w$.

    step 1   Input a value for l (length)
    step 2   Input a value for w (width)
    step 3   Multiply l and w
    step 4   Write the result as the area   □

■ **EXAMPLE 2**

The following algorithm is used to compute simple interest in an account, using the formula $i = prt$.

1. Input a value for p
2. Input a value for r
3. Input a value for t
4. Compute $i = p \times r \times t$
5. Write the value for i □

■ **EXAMPLE 3**

The following is an algorithm to pay a bonus to employees whose hourly wage is below $6.00. Note that the sequence of steps is repeated for several different input values.

1. Read in hourly wage
2. Compare hourly wage to $6.00: if > $6.00 goto step 5
3. Read gross pay
4. Add bonus to gross pay
5. More data? If yes goto step 1
6. End   □

Flowcharts

Algorithms can be expressed in a method other than a numbered list of steps, that is, by a **flowchart** in which the instructions of the algorithm are presented in a diagram format. The flowchart provides a visual description of the steps necessary to perform a certain task or to solve a problem.

If we examine the algorithms above, we see that each involves *input, processing,* and *output* instructions. The flowchart for an algorithm uses various geometric shapes in which instructions can be written, and indicates by arrows the *flow* from symbol to symbol. Some algorithms may involve decisions, as in Example 3 above. Each algorithm also must have a beginning as well as an end.

Below are the accepted flowchart symbols with a brief description of each.

Flow direction: The arrows indicate the direction of flow of the steps. Usually, an attempt is made to keep the main logic flow moving downward and the secondary logic flow from left to right, shown here.

Terminal: This symbol is used to indicate the beginning and end of the algorithm.

Input/output: This symbol is used at the point at which information is input prior to processing and again after processing at which time the results are recorded.

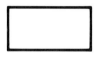

Processing: This symbol is used to indicate the calculating step(s) of the algorithm. Generally, some arithmetic action occurs at this point.

Decision: This symbol is used whenever a decision must be made before continuing further. At this point *branching* to other instructions, or returning to an earlier instruction, is made possible by using the various exit points of the decision symbol.

Connector: This symbol occurs in complex flowcharts and is used to indicate exit from one point of the flowchart and entry to another part.

As an illustration, let us consider how we might diagram flowcharts for the algorithms in Examples 1–3.

■ **EXAMPLE 4**   Draw a flowchart to find the area of a rectangle by the formula $A = l \times w$.

Solution   The flowchart is shown in Fig. 3.1. □

■ **EXAMPLE 5**   Draw a flowchart showing how to compute the interest on an account, using the formula $i = prt$.

Solution   The flowchart is shown in Fig. 3.2. □

■ **EXAMPLE 6**   Draw a flowchart to show how to pay a bonus if the wage is below $6.00 per hour.

Solution   The flowchart is shown in Fig. 3.3. □

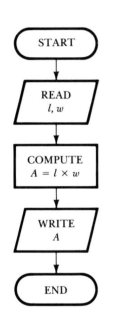

Figure 3.1

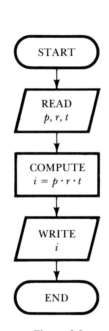

Figure 3.2

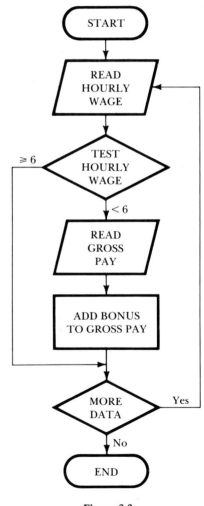

Figure 3.3

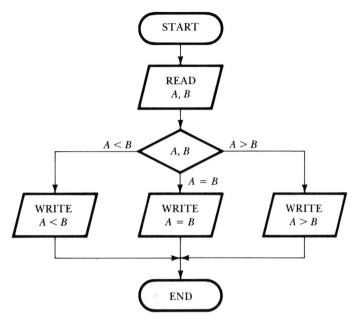

Figure 3.4

■ **EXAMPLE 7**   For another example using the decision symbol, consider the flow-chart shown in Fig. 3.4, in which two numbers A and B are compared. □

  Although computer programs can be extremely complex and their related flowcharts also quite complex, our interest here is to give an introduction to the symbols used in flowcharting and to illustrate the logic flow with several simple examples. Additional work with flow-charting will be encountered in Chapters 5 and 8.

Exercise Set 3.1

1. Draw a flowchart to read in a set of whole numbers and print only those divisible by 2.

2. Each incoming new student must take a placement test in algebra. Draw a flowchart to read in the student scores and print one list of those students who scored > 25 and another list of those students who scored ≤ 25.

3. The Federated Department Stores bill customers according to the following plan.

  If a customer's balance is > \$15, the customer is billed for \$15 plus 1/2 the difference between the current balance and \$15. If the balance is ≤ \$15, the customer is billed for the entire amount. If the balance is zero, no bill is sent.

Draw a flowchart for this plan.

4. Draw a flowchart to read in values for A and B and print the result y if $y = \dfrac{A}{A - B}$. Watch out if $A = B$!

5. Draw a flowchart to print the solutions to the quadratic equation $Ax^2 + Bx + C = 0$, where $A \neq 0$, using the quadratic formula

$$x = \frac{-B \pm \sqrt{D}}{2A}, \qquad \text{where } D = B^2 - 4AC.$$

Read in A, B, and C. Then:

   Print "Two solutions": x_1 and x_2 if $D > 0$.
   Print "Unique solution": x if $D = 0$.
   Print "No real solutions": if $D < 0$.

6. The following chart lists the dues that each boy or girl must pay to a local fellowship club for children. The dues are dependent on the age of the boy or girl.

Age	Annual dues
< 12	$5.00
between 12 and 18	$7.00
> 18	$10.00

Draw the flowchart showing the dues each age group must pay.

3.2

Pseudocode

In the preceding section we discussed that an algorithm can be expressed by a flowchart rather than simply a numbered list of instructions. An algorithm may also be expressed by still another method, called **pseudocode** or *pseudocode language*. Pseudocode language is nothing more than short, simple statements or commands written in English. These statements very closely resemble the higher-level computer languages called PASCAL and BASIC. With pseudocode, we may see some of the same instructions as those used in flowcharts, such as READ and WRITE statements.

Pseudocode language is generally presented in three different forms of logic organization.

1. *Sequence logic:* Instructions are executed in order from top to bottom.
2. *Selection logic:* This form sets up conditions using the IF form, which can be single, double, or even multiple choice, as below.

   a) Single choice: IF–THEN
   b) Double choice: IF–THEN–ELSE
   c) Multiple choice: ELSE–IF

3. *Iteration logic:* This involves setting up loops of one of three types, as follows.

   a) DO
   b) DOWHILE
   c) DOUNTIL

Sequence Logic

With sequence logic, the instructions are executed in order from top to bottom. Some people prefer to list a START statement at the beginning and an END statement as the last statement, but since the instructions are executed from top to bottom anyway, these statements are not necessary.

■ **EXAMPLE 8**   The pseudocode form of the algorithm shown earlier in Example 1 is as follows:

$$\text{READ length, width}$$
$$\text{area} = \text{length} \times \text{width}$$
$$\text{WRITE area}$$
$$\text{END} \quad \square$$

■ **EXAMPLE 9**   Example 2 would be written as follows:

$$\text{READ principal, rate, time}$$
$$\text{interest} = \text{principal} \times \text{rate} \times \text{time}$$
$$\text{WRITE interest}$$
$$\text{END} \quad \square$$

Selection Logic

Single choice
The single-choice form of selection logic uses the IF–THEN statement. As in English, THEN may be implied and need not be written. ENDIF is used to signal the end of the condition. In this form, when the IF condition holds, the THEN instruction is executed. When the IF condition does not hold, the first instruction following the ENDIF is the next instruction executed.

■ **EXAMPLE 10**   This example illustrates the single-choice selection logic.

READ name, hourly wage, gross pay, bonus
IF hourly wage < $6.00,
        THEN gross pay + bonus
ENDIF
WRITE name, gross pay
END □

Double choice

The double-choice form uses the IF–THEN–ELSE statement. The ELSE instruction separates the two executable instructions. If the condition holds, the command following the THEN is executed. Otherwise, the command following the ELSE is executed.

■ **EXAMPLE 11**   The following illustrates the double-choice selection logic.

READ A, B
IF A = B,
        THEN A = A + 2
                B = B + 1
ELSE
        A = A + 1
        B = B + 2
ENDIF
WRITE A, B
END □

Multiple choice

This form of the selection logic involves more than two choices. One IF form is contained in the command portion of another IF statement. Thus we say that the IF forms are nested. Such a form may appear as an ELSEIF form, in which the ELSEIF is located between the IF and ELSE commands. With this, as with the other forms, only one command will be executed, the one following the first condition that holds. If no condition holds, the command following the ELSE statement is executed.

■ **EXAMPLE 12**   This example illustrates the multiple-choice selection logic.

READ A, B
IF A = B,
        THEN A = A + 2
                B = B + 1
        WRITE A, B
ELSEIF A < B,
        THEN A = A + 1

$$B = B + 2$$
$$\text{WRITE A, B}$$
$$\text{ELSE}$$
$$A = A - 1$$
$$B = B - 2$$
$$\text{WRITE A, B}$$
$$\text{ENDIF}$$
$$\text{END} \quad \square$$

Iteration Logic

DO

This logic form sets up a loop and executes a command a specified number of times. The statement following the DO sets the specifications for the loop, beginning with the first value given and continuing until the last value given. The steps or increments are automatically one unless otherwise specified. Examples 13 and 14 illustrate the DO loop.

■ **EXAMPLE 13**

$$\text{DO } x = 1 \text{ to } 5$$
$$y = x^2$$
$$\text{WRITE } y$$
$$\text{ENDDO}$$
$$\text{END} \quad \square$$

The numbers 1, 2, 3, 4, 5 will be squared.

■ **EXAMPLE 14**

$$\text{READ N}$$
$$\text{DO } x = 1 \text{ to N step 2}$$
$$y = x^2$$
$$\text{WRITE } y$$
$$\text{ENDDO}$$
$$\text{END}$$

The value of N is unspecified in the program, which gives the program greater flexibility. The step has been changed to skip a number. The numbers 1, 3, 5, 7, ..., N will be squared. $\square$

DOWHILE

In the loop set up by this logic form, a statement must be inserted before the DOWHILE statement to initialize the condition that controls the loop. Another command must be included within the loop to cause the looping process to terminate at some point. Looping in the DOWHILE structure continues until the condition is *not* met. The condition of this logic form is always tested at the beginning of the loop. Consequently, the DOWHILE loop may never actually be executed. This happens when the condition is not satisfied. Control passes on to the first statement following the DOWHILE loop.

■ **EXAMPLE 15**

The following example of a DOWHILE loop will compute the value of y in $y = 2x - 3$ for values of x from 1 to 4.

$x = 1$ **Initializes the condition**
DOWHILE $x < 4$ **Tests the condition**
    $y = 2x - 3$ **Finds the value of y for the x-values**
    WRITE x, y **1, 2, 3, 4**
    $x = x + 1$
ENDDO
END □

■ **EXAMPLE 16**

This example computes the value of y in $y = x^2 - 2x + 1$ for values of x from 1 to 4 in steps of 0.5.

$x = 1$
DOWHILE $x < 4$
    $y = x^2 - 2x + 1$
    WRITE x, y
    $x = x + 0.5$ **Increments in steps of 0.5**
ENDDO
END □

DOUNTIL

An initializing statement is required for the DOUNTIL form, since this logic structure is also a loop. In the DOUNTIL structure, looping continues until the condition is met. Another difference between this loop and the DOWHILE is that the condition is tested at the end of the DOUNTIL loop rather than at the beginning. This means that the DOUNTIL loop must be executed at least one time.

■ **EXAMPLE 17**

We can rewrite Example 15 using a DOUNTIL loop, as follows.

$x = 1$ **Initializes the condition**
DOUNTIL $x > 4$
    $y = 2x - 3$ **Finds the value of y for the x-values**
    WRITE x, y **1, 2, 3, 4. The loop ends when $x > 4$.**
    $x = x + 1$
ENDDO
END □

■ **EXAMPLE 18**

Rewriting Example 16 gives us the following.

$x = 1$
DOUNTIL $x > 4$
    $y = x^2 - 2x + 1$
    WRITE x, y
    $x = x + 0.5$
ENDDO
END □

Exercise Set 3.2

1. Determine the number of loops and the value of N for each DO loop.

   a) DO N = 1 to 6
   b) DO N = 1 to 12 step 2
   c) DO N = 2 to 5 step 0.5
   d) DO N = −3 to 5 step 2

2. Write a pseudocode program using the IF single-choice selection logic. Calculate the commission if a salesperson receives 10% of sales over $5000.

3. Write a pseudocode program using the IF–ELSE double-choice selection logic. Calculate the salary of a salesperson if a bonus of $100 plus 3% of sales is paid when sales are less than or equal to $5000, or a straight 5% commission is paid when sales exceed $5000.

4. Write a pseudocode program using the ELSE–IF multiple-choice selection logic. Calculate the salary of a salesperson if 3% commission is paid for sales between $5000 and $7000, a 5% commission is paid for sales exceeding $7000, and a flat $250 salary is paid for sales less than $5000.

5. Write a pseudocode program using the DO iteration logic to find the value of y in $y = x^2 - 3x$ for $x = 1$ to 5.

6. Write a pseudocode program using the DOWHILE iteration logic to find the value of y if $y = x^2 - 3x + 2$ for $x = 1$ to 5 step 0.25.

7. Write a pseudocode program using the DOUNTIL iteration logic for the information in Exercise 6.

3.3
Decision Tables

A **decision table** is a chart that shows what actions are to be taken for various combinations of conditions. When we use a decision table, we record the conditions applying to a specific situation along with the actions that come from the different combinations of conditions.

A *condition* is a fact that has at least two values, such as true or false, and in some manner influences the actions to be taken. The condition corresponds to the statement contained in a decision symbol when flowcharting. It also corresponds to the statement following the IF statement in selection logic of pseudocode programming.

An *action* is the operation to be executed either alone or as part of a sequence. It corresponds to the processing step in a flowchart. In pseudocode selection logic, the action is what follows the THEN statement.

Each decision table consists of a list of conditions, a list of actions, and the rules that show the relationship between the combinations of conditions and the appropriate actions to be performed. IF certain conditions are true, THEN a certain action or actions are necessary.

A decision table is divided into four parts: (1) the condition stub; (2) the condition entry; (3) the action stub; and (4) the action entry, as shown in Fig. 3.5. The stub portions contain the conditions and actions and the entry portion contains the rules. The condition stub lists the conditions that exist and the action stub lists the actions to be taken for various combinations of condition entries.

	Condition Stub	Condition Entry
IF	Action Stub	Action Entry
THEN		

Figure 3.5

■ EXAMPLE 19

Construct a decision table for the following situation: If I do not get at least one letter grade higher on this test, I will drop the course.

Solution

Condition: I get at least one letter grade higher.
               yes or no
Actions: I drop the course.
               I do not drop the course.

The decision table is shown in Fig. 3.6. The symbols N and Y in the condition entry stand for No and Yes, respectively, while the **x**'s in the action entry represent action completed. The symbol — represents no action or irrelevant action.   □

Condition entry

One grade higher	N	Y
I drop	x	—
I do not drop	—	x

Action entry   Figure 3.6

■ EXAMPLE 20

The following is an example of the use of a decision table to compute the value of y in an algebraic function, using different rules for various values of x (see Fig. 3.7):

$x < 0$	Y	N	N
$x = 0$	N	Y	N
$x > 0$	N	N	Y
$y = x^2$	x	—	—
$y = x$	—	x	—
$y = -x^2$	—	—	x

$$y = \begin{cases} x^2 & \text{if } x < 0, \\ x & \text{if } x = 0, \\ -x^2 & \text{if } x > 0. \end{cases}$$

Conditions	Actions
$x < 0$	$y = x^2$
$x = 0$	$y = x$
$x > 0$	$y = -x^2$

Figure 3.7

■ **EXAMPLE 21** Construct a decision table for the following situation: If pay is greater than $250, multiply pay by 0.04, giving tax; else multiply pay by 0.03, giving tax. See Fig. 3.8.

Solution

Pay > 250	N	Y
Pay ≤ 250	Y	N
(Pay)(0.04)	—	x
(Pay)(0.03)	x	—

Figure 3.8

Conditions	Actions
pay > $250	(pay)(0.04)
pay ≤ $250	(pay)(0.03)

■ **EXAMPLE 22** A day of snow skiing is planned, but the weather forecast is indefinite at this time. It is necessary to determine what combinations of clothes and accessories are required for whatever weather conditions exist. The decision table is shown in Fig. 3.9.

Sunny: temp > 35°	Y	N	N	N	N
Sunny: temp ≤ 35°	N	Y	N	N	N
Overcast: temp > 35°	N	N	Y	N	N
Overcast: temp ≤ 35°	N	N	N	Y	N
Snowing	N	N	N	N	Y
Wear powder jacket	x	—	x	—	—
Wear down jacket	—	x	—	x	x
Wear sunglasses	x	x	—	—	—
Wear goggles	—	—	x	x	x
Wear extra sweater	—	x	—	x	x
Wax skis: soft snow	x	—	x	—	x
Wax skis: hard pack	—	x	—	x	—

Figure 3.9

■ **EXAMPLE 23** Figure 3.10 illustrates the use of a decision table to determine the weekly payroll for a small company that has three types of employees: (1) parttime, < 40 hours per week; (2) fulltime, 40 hours per week; (3) overtime, > 40 hours per week. Overtime is paid at 1.5 times the normal hourly rate. Employees fall into the 16% wage bracket for federal income tax (FIT) withholding if they are married and the 19% bracket if they are single. Social security tax (FICA) is withheld at the rate of 6.7%. All income earned will be taxed.

$$\text{Gross pay (GP)} = \text{Standard pay (ST)} + \text{Overtime pay (OT)}$$

$$\text{Net pay} = \text{Gross pay} - (\text{FIT} + \text{FICA})$$

Parttime (< 40)	Y	Y	N	N	N	N
Fulltime (40)	N	N	Y	Y	N	N
Overtime (> 40)	N	N	N	N	Y	Y
Single	Y	N	Y	N	Y	N
Married	N	Y	N	Y	N	Y
ST = hrs (≤ 40) $\times$ rate	x	x	x	x	x	x
OT = (hrs $-$ 40) $\times$ 1.5	—	—	—	—	x	x
GP = ST + OT	x	x	x	x	x	x
FIT = 0.16 $\times$ GP	—	x	—	x	—	x
FIT = 0.19 $\times$ GP	x	—	x	—	x	—
FICA = 0.067 $\times$ GP	x	x	x	x	x	x
Net = GP $-$ (FIT + FICA)	x	x	x	x	x	x

Figure 3.10

Exercise Set 3.3

Prepare a decision table for each of the following algebraic functions. Identify the conditions and actions.

1. $y = \begin{cases} (x + 1) & \text{if } x < 0, \\ -(x + 1) & \text{if } x \geq 0 \end{cases}$

2. $y = \begin{cases} x^2 - 3x & \text{if } x < 0, \\ x^2 + 3 & \text{if } x = 0, \\ x^2 + 3x & \text{if } x > 0 \end{cases}$

3. Prepare a decision table to compute the sales tax on an item if the tax is 5% on items costing $1.00 to $10.00, 5.5% on items costing $10.01 to $50.00, and 6% on items costing more than $50.00.

4. Prepare a decision table to compute total pay if total pay = $250 + commission and commission is 3% for sales < $3000, 4% for sales between $3000 and $5000, and 5% for sales > $5000.

5. Prepare a decision table to compute total pay for the commission percentages in Exercise 4 if pay = $250 + commission for sales ≤ $5000 and pay = $175 + commission for sales > $5000.

6. Change the decision table in Example 23 if the hourly rate of pay for parttime employees is $10.00 and for fulltime employees $15.00 and the only classifications are parttime (< 40) and fulltime (≥ 40) and single or married.

Summary

In this chapter we discussed several topics that help show a relationship between mathematics and computer programming. *Algorithms* have been used in problem solving tasks in mathematics for over a thousand years. We discovered that an algorithm is merely a list of steps leading to the completion of a task or the solving of a problem. A computer program is such a list since it provides steps for the computer to perform in order to accomplish a particular task. We defined flowcharts as diagrams or pictures of algorithms. Sometimes a flowchart is easier to follow than a list of written steps. This is especially true if looping is involved, that is, where some part of the algorithm is repeated more than once.

We examined *pseudocode programming* as still another way to present an algorithm. In this method special words or combinations of words provide simple looping and branching structures, and we began to see for the first time different logic forms that are used in programming.

Finally, we were introduced briefly to *decision tables,* a means of examining a complex problem to break it down into a list of *conditions* and *actions*. We found that any actions required are actually determined by various combinations of conditions. The table provides a clearer picture of the different combinations. Decision tables are tools that are used by both mathematicians and computer programmers in an effort to simplify problems.

Review Exercises

Draw a flowchart for each of the following exercises in Exercise Set 3.3.

1. Exercise 1
2. Exercise 2
3. Exercise 3
4. Exercise 4

5. Write a pseudocode program for Exercise 1 in Exercise Set 3.3, using IF–ELSE double-choice selection logic.

6. Write a pseudocode program for Exercise 2 in Exercise Set 3.3, using ELSE–IF multiple-choice selection logic.

7. Write a pseudocode program using the DO iteration logic to find the value of y in $y = 2x - 3$ for $x = 1$ to 10 step 0.5. How many loops are there? What is the value of y after each loop?

8. Write a pseudocode program using the DOWHILE iteration logic for Review Exercise 7.

9. Write a pseudocode program using the DOUNTIL iteration logic for Review Exercise 7.

10. Prepare a decision table for Exercise 4 in Exercise Set 3.2 to determine salary.

11. Prepare a decision table for the following algebraic function:

$$y = \begin{cases} -x & \text{if } x < 0, \\ x^2 & \text{if } x = 0, \\ -x^2 & \text{if } x > 0. \end{cases}$$

12. Change the decision table for Example 23 if FICA is withheld for only the first $37,800 a person earns in a year. (All other information for the problem is the same.)

Decimal
and Nondecimal
Numeration Systems

After completing this chapter, you should be able to:

1. Apply the characteristics of the decimal numeration system to other numeration systems.
2. Convert between binary and decimal numbers.
3. Convert between octal and decimal numbers.
4. Convert between hexadecimal and decimal numbers.
5. Perform binary, octal, and hexadecimal arithmetic.
6. Convert between binary and octal numbers.
7. Convert between binary and hexadecimal numbers.
8. Convert between octal and hexadecimal numbers.
9. Perform complementary arithmetic.

4

4.1
Modular Arithmetic

We are accustomed to operating with real numbers and using the real-number properties, reviewed in Chapter 0, without needing to stop and think about what we are doing. Now, however, we are going to look at a numeration system using addition and multiplication that yields quite different results from what we usually see. We will have to stop and think as we do the arithmetic. We will also compare this new system to the real-number properties.

A finite mathematical system can be created using the face of a clock. We call this type of system **clock arithmetic**, or **modular arithmetic**. The **modulus** is the number of symbols in the system and is abbreviated as mod.

81

■ **EXAMPLE 1**

a) Mod 4 consists of only the symbols (0, 1, 2, 3).

b) Mod 5 consists of only the symbols (0, 1, 2, 3, 4). □

To illustrate why we can think of modular arithmetic as clock arithmetic, consider the standard nonmilitary 12-hour clock. If it is now 10:00, what time will it be in 4 hours? Since $10 + 4 = 2$ (on a clock), it will be 2:00. (See Fig. 4.1).

We perform addition and multiplication in modular arithmetic in the following manner. We position the numerals as though they were on the face of a clock, placing the 0 where the 12 would normally be. We can now add two numbers by starting at 0 and counting in a clockwise direction to the first numeral, and then counting from that point an amount equal to the second number. The sum is the number at which we finish.

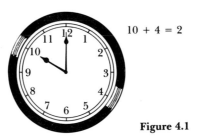

$10 + 4 = 2$

Figure 4.1

■ **EXAMPLE 2**

a) The Mod 4 face

b) The Mod 5 face

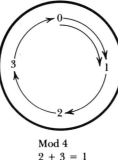

Mod 4
$2 + 3 = 1$

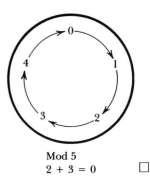

Mod 5
$2 + 3 = 0$ □

In Example 2, we can see that $2 + 3$ is not the same in these two systems and certainly not equal to $2 + 3$ in the real-number system. To show that we are dealing with different numeration systems, we write the addition symbol as $\oplus$ and identify the modular system as shown in Example 3, using the equivalence symbol $\equiv$, rather than the equal symbol $=$.

■ **EXAMPLE 3**   **a)** $2 \oplus 3 \equiv 1$, Mod 4   **b)** $2 \oplus 3 \equiv 0$, Mod 5   □

Multiplication is performed using the fact that multiplication is a short form of addition, that is, $2 \times 3 = 3 + 3$ and $3 \times 2 = 2 + 2 + 2$. In order to find $2 \otimes 3$ on the clock, we begin at 0 and count the number 3 twice. Example 4 illustrates the process.

■ **EXAMPLE 4**   **a)**   **b)**

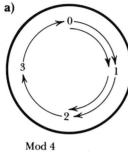   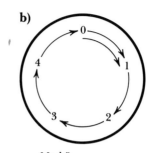

Mod 4   Mod 5
$2 \otimes 3 \equiv 2$, Mod 4 $2 \otimes 3 \equiv 1$, Mod 5   □

Because it is not really convenient to draw a clock whenever we want to add or multiply in modular arithmetic, we use the following method for finding sums and products in modular arithmetic.

■

TO FIND A SUM:	TO FIND A PRODUCT:
1. Add the numbers in base 10.	1. Multiply the numbers in base 10.
2. Divide the sum by the modulus.	2. Divide the product by the modulus.
3. Discard the quotient.	3. Discard the quotient.
4. The remainder is the answer.	4. The remainder is the answer.

■ **EXAMPLE 5**   In Mod 4 arithmetic, $2 \otimes 3 \equiv ?$

Solution $2 \otimes 3 = 6$

$$
\begin{array}{r}
1 \\
4 \overline{)6} \\
-4 \\
\hline
2
\end{array}
$$

Thus $2 \otimes 3 \equiv 2$, Mod 4   □

■ **EXAMPLE 6**   In Mod 5 arithmetic, $2 \otimes 3 \equiv$?

Solution $2 \otimes 3 = 6$

$$\begin{array}{r} 1 \\ 5\overline{)6} \\ -5 \\ \hline 1 \end{array}$$

Thus $2 \otimes 3 \equiv 1$, Mod 5   □

Because these modular systems are finite, every possible sum and product can be found as shown in Fig. 4.2. We can test these modular systems to see whether they satisfy any of the eleven real-number properties. Some interesting conclusions can be drawn.

Mod 4

$\oplus$	0	1	2	3
0	0	1	2	3
1	1	2	3	0
2	2	3	0	1
3	3	0	1	2

(a)

$\otimes$	0	1	2	3
0	0	0	0	0
1	0	1	2	3
2	0	2	0	2
3	0	3	2	1

(b)

Mod 5

$\oplus$	0	1	2	3	4
0	0	1	2	3	4
1	1	2	3	4	0
2	2	3	4	0	1
3	3	4	0	1	2
4	4	0	1	2	3

(c)

$\otimes$	0	1	2	3	4
0	0	0	0	0	0
1	0	1	2	3	4
2	0	2	4	1	3
3	0	3	1	4	2
4	0	4	3	2	1

(d)

Figure 4.2

■ **EXAMPLE 7**   Are these systems closed with respect to addition and multiplication? Since these are finite systems it is easy to examine all sums and products, as in Fig. 4.2. Since no other numerals exist, no other sums or products are possible. These systems present a perfect illustration of the closure property.   □

■ **EXAMPLE 8**   Does the order in which we add or multiply matter? In other words, do we have commutativity? Does $2 \oplus 3 \equiv 3 \oplus 2$? Does $2 \otimes 3 \equiv 3 \otimes 2$? It is easy to see that regardless of the modular system, commutativity does exist.   □

■ **EXAMPLE 9**   Zero is the identity element for addition for both Mod 4 and Mod 5. Does each number in Mod 4 and Mod 5 have an additive inverse? If

$a + b = 0$, then a and b are additive inverses of each other. Looking at each row of the Mod 4 and Mod 5 addition tables (Fig. 4.2a and 4.2c), note the existence of a zero in each row. For example, $3 \oplus 2 \equiv 0$, Mod 5. This means that each number *does* have an additive inverse. □

We leave the testing of the remaining properties to the exercises.

Exercise Set 4.1

Find the value for each of the following. Pay attention to the modulus. Put your answer in the form $a \equiv b$, Mod n.

1. $4 \oplus 7$, Mod 8   2. $6 \oplus 3$, Mod 7

3. $14 \oplus 12$, Mod 16   4. $6 \oplus 3$, Mod 9

5. $4 \otimes 7$, Mod 8   6. $5 \otimes 4$, Mod 16

7. $5 \otimes 4$, Mod 6   8. $7 \otimes 5$, Mod 9

Construct the addition and multiplication tables for each of the following.

9. Mod 3

10. Mod 6

11. Mod 7

12. Mod 8

13. Mod 9

14. Do any of the systems in Exercises 9–13 satisfy the eleven properties of real numbers? Can you make a general statement about all modular numeration systems and the properties of real numbers? (*Hint*: Concentrate on the multiplicative inverse property.)

15. Consider the mathematical system represented by the following tables.

$\oplus$	a	b	c	d	e
a	a	b	c	d	e
b	b	c	d	e	a
c	c	d	e	a	b
d	d	e	a	b	c
e	e	a	b	c	d

$\otimes$	a	b	c	d	e
a	a	a	a	a	a
b	a	b	c	d	e
c	a	c	e	b	d
d	a	d	b	e	c
e	a	e	d	c	b

a) Is this system closed with respect to addition? to multiplication?
b) What is the identity element for addition? for multiplication?
c) Identify the additive inverse of each element that has an additive inverse.
d) Identify the multiplicative inverse of each element that has a multiplicative inverse.

16. Consider a Mod 7 system in which the days of the week correspond to numbers: 0–Sunday; 1–Monday; 2–Tuesday; 3–Wednesday; 4–Thursday; 5–Friday; 6–Saturday. If Memorial Day, May 30, is the 150th day of the year and falls on Monday in a particular year, on what day of the week does July 4th, the 185th day of the year, fall in the same year? On what day of the week does Halloween, the 304th day of the year, fall in the same year? On what day of the week does Christmas, the 359th day of the year, fall in that year?

4.2
The Decimal Numeration System: Base Ten

The numeration system we encounter most frequently is the *decimal numeration system*. The use of a base-ten numeration system can be explained in part by the fact that we have a total of ten thumbs and fingers on our hands. In early times, the fingers provided a convenient source of digits for counting. As a result, people became adept at using the grouping by ten in their counting process. Thus it was only natural that 10 symbols would be created to symbolize from 0 to 9 objects. (If people had only six fingers, it is quite possible that we would be using a base-six system today!)

The *base*, or *radix* of a numeration system is the number of symbols used in that system. Any positive integer greater than 1 can be used as the base of a numeration system. The ten symbols of the base-ten system are 0, 1, 2, 3, 4, 5, 6, 7, 8, 9.

Probably the two most important characteristics of the decimal system are (1) place value or positional value, and (2) the principle of cipherization. The word *cipher*, as a noun, stands for the symbol 0. As a verb, to cipher means to compute with numbers. The creation of the symbol 0 allowed people to write any number, large or small, by combining the ten basic symbols. When we reach the tenth item while counting, we group the items, which gives us one group of ten with no units left over. Thus the combination of symbols appears as 10, 11, 12, 13, etc. The concept of place value means that each position in a multidigit number is given a value corresponding to a positive or negative integral power of base ten, as shown below:

$$\underline{10^5} \ \underline{10^4} \ \underline{10^3} \ \underline{10^2} \ \underline{10^1} \ \underline{10^0} . \underline{10^{-1}} \ \underline{10^{-2}} \ \underline{10^{-3}} \ \underline{10^{-4}}$$

■ **EXAMPLE 10**   A multidigit number such as 5732 has:

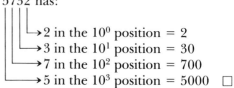

$\rightarrow$ 2 in the 10^0 position = 2

$\rightarrow$ 3 in the 10^1 position = 30

$\rightarrow$ 7 in the 10^2 position = 700

$\rightarrow$ 5 in the 10^3 position = 5000   □

■ **CONCLUSION**

Regardless of the base of the system, place value applies, and each position is a positive or negative integral power of the base.

We can also express the positional notation of 5732 in the cipherization form.

■ **EXAMPLE 11**

a) $5732 = 5000 + 700 + 30 + 2$
$= (5 \times 1000) + (7 \times 100) + (3 \times 10) + (2 \times 1)$
$= (5 \times 10^3) + (7 \times 10^2) + (3 \times 10^1) + (2 \times 10^0)$

b) $7630.4 = (7 \times 1000) + (6 \times 100) + (3 \times 10) + (0 \times 1)$
$+ (4 \times 0.1)$
$= (7 \times 10^3) + (6 \times 10^2) + (3 \times 10^1) + (0 \times 10^0)$
$+ (4 \times 10^{-1})$

c) $12{,}045.32 = (1 \times 10^4) + (2 \times 10^3) + (0 \times 10^2) + (4 \times 10^1)$
$+ (5 \times 10^0) + (3 \times 10^{-1}) + (2 \times 10^{-2})$ □

Positional notation is not unique to the decimal numeration system. We will see it applied to other numeration systems elsewhere in this chapter. Each position will represent a power of whatever base with which we are dealing.

Exercise Set 4.2

Express each of the following numbers in the cipherization form.

1. 7406
2. 30,452
3. 9274.3
4. 1257.04
5. 93.704
6. 2.7432
7. 324,507
8. 3,803,360
9. 10.0378
10. 402.60403

Express each of the following in standard notation.

11. $(2 \times 10^4) + (0 \times 10^3) + (6 \times 10^2) + (0 \times 10^1) + (7 \times 10^0)$

12. $(8 \times 10^3) + (6 \times 10^2) + (2 \times 10^1) + (0 \times 10^0) + (1 \times 10^{-1})$

13. $(3 \times 10^2) + (4 \times 10^1) + (5 \times 10^0) + (0 \times 10^{-1}) + (7 \times 10^{-2})$

14. $(2 \times 10^1) + (7 \times 10^0) + (2 \times 10^{-1}) + (0 \times 10^{-2}) + (4 \times 10^{-3})$

15. $(4 \times 10^0) + (3 \times 10^{-1}) + (0 \times 10^{-2}) + (3 \times 10^{-3}) + (9 \times 10^{-4})$

16. $(1 \times 10^{-1}) + (5 \times 10^{-2}) + (3 \times 10^{-3}) + (0 \times 10^{-4}) + (2 \times 10^{-5})$

17. We know that $347 = (3 \times 10^2) + (4 \times 10^1) + (7 \times 10^0)$ in base ten. How would we write 347 in cipherization form in base eight?

18. How would we write 347 in cipherization form in base b, if $b \geq 8$?

4.3
The Binary Numeration System: Base Two

To communicate with someone who speaks only a foreign language would require one of us to learn the other's language. So it is with the computer. Digital computers consist of electronic circuits through which electricity is flowing or not flowing. This means the circuitry can indicate only two states, either ON or OFF. This two-state condition can be best represented by a numeration system that consists of just two symbols.

The computer uses the **binary numeration system,** which consists of only the symbols 0 and 1. These two symbols are used to represent the presence of an electrical impulse 1 (ON), or the lack of the impulse 0 (OFF). In dealing with other numeration systems, we will continue to use the Arabic symbols used in base ten.

There is evidence that the binary numeration system was used by the Chinese 5000 years ago and the Egyptians 3000 years ago. However, the best documented historical work with binary numbers is credited to Baron Gottfried von Leibniz approximately 300 years ago.

Borrowing from what we know about the decimal numeration system and its characteristics of positional notation and place value, we can see that any numeration system exhibits the same characteristics:

$$\ldots, b^4, b^3, b^2, b^1, b^0 . b^{-1}, b^{-2}, b^{-3}, b^{-4}, \ldots$$

The binary numeration system is such a system. Each binary position moving to the left of the *binary point* represents a nonnegative integral power of base two and each position to the right is a negative integral power of base two:

$$\ldots, 2^4, 2^3, 2^2, 2^1, 2^0 . 2^{-1}, 2^{-2}, 2^{-3}, 2^{-4}, \ldots$$

The binary number 110101_2 has a one in the 2^5, 2^4, 2^2, and 2^0 positions, while there is a zero in the 2^3 and 2^1 positions.

A subscript is used to distinguish a number in one base from a number in another base. If no subscript is used, the number is understood to be a base-ten number; that is, $22 = 22_{10}$ but $10 \neq 10_2$.

Before learning to convert between base ten and base two, we should look at some binary–decimal relationships in a counting mode, as shown here:

Base ten	Base two			
	2^3	2^2	2^1	2^0
0	0	0	0	0
1	0	0	0	1
2	0	0	1	0
3	0	0	1	1
4	0	1	0	0
5	0	1	0	1
6	0	1	1	0
7	0	1	1	1
8	1	0	0	0
9	1	0	0	1
10	1	0	1	0
11	1	0	1	1
12	1	1	0	0
13	1	1	0	1
14	1	1	1	0
15	1	1	1	1

These relationships will be important as we progress.

Binary-to-Decimal Conversion

The place-value characteristic allows for an easy conversion from binary to decimal using the principle of cipherization. See the front endpapers for positive and negative powers of 2.

■ **EXAMPLE 12** Convert 110101_2 to base ten. (The table shown in the front endpapers can be used to derive the positive and negative powers of 2.)

Solution
$$110101_2 = (1 \times 2^5) + (1 \times 2^4) + (0 \times 2^3) + (1 \times 2^2) + (0 \times 2^1)$$
$$+ (1 \times 2^0)$$
$$= (1 \times 32) + (1 \times 16) + (0 \times 8) + (1 \times 4) + (0 \times 2)$$
$$+ (1 \times 1)$$
$$= 32 + 16 + 0 + 4 + 0 + 1$$
$$= 53 \quad \square$$

■ **EXAMPLE 13** Convert 10110_2 to base ten.

Solution
$$10110_2 = (1 \times 2^4) + 0 + (1 \times 2^2) + (1 \times 2^1) + 0$$
$$= 16 + 0 + 4 + 2 + 0$$
$$= 22 \quad \square$$

■ **EXAMPLE 14**   Convert 11011.01_2 to base ten.

Solution $$11011.01_2 = (1 \times 2^4) + (1 \times 2^3) + 0 + (1 \times 2^1) + (1 \times 2^0) + 0$$
$$+ (1 \times 2^{-2})$$
$$= 16 + 8 + 0 + 2 + 1 + 0 + 0.25$$
$$= 27.25 \quad \square$$

Another method of converting from binary to decimal is called the "multiply by two and add" method:

Multiply the leftmost binary digit by two and add the product to the next digit to the right. Multiply this result by two and add to the next digit to the right. Keep progressing to the right until the last digit has been added.

With a little practice, you'll be able to calculate mentally using this procedure.

■ **EXAMPLE 15**   Convert 10110_2 to base ten.

Solution

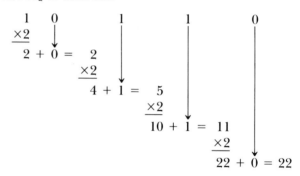

Thus $10110_2 = 22$. $\square$

■ **EXAMPLE 16**   Convert 110101_2 to base ten.

Solution

Thus $110101_2 = 53$. $\square$

Decimal-to-Binary Conversion

Although several techniques exist for converting decimal numbers into binary numbers, we will present just one here. It is a method that works just as conveniently for converting any decimal integer or decimal fraction into any other number base.

When converting an integer, begin by dividing the decimal number by the largest power of two that is smaller than the given decimal number. Record a 1 if divisible and a 0 if not divisible. Continue dividing by decreasing powers of two until the final division by 2^0.

■ EXAMPLE 17   Convert 22 into a binary number.

Solution   The largest power of 2 smaller than 22 is 16.

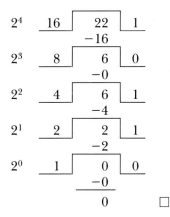

Since 8 does not divide into 6, we record a 0.

Reading from top to bottom, we see that we have $22 = 10110_2$. □

■ EXAMPLE 18   Convert 53 into a binary number.

Solution

$$
\begin{array}{rcl}
2^5 & 32\,)\,53 & 1 \\
    & -32 & \\
2^4 & 16\,)\,21 & 1 \\
    & -16 & \\
2^3 & 8\,)\,5 & 0 \\
    & -0 & \\
2^2 & 4\,)\,5 & 1 \\
    & -4 & \\
2^1 & 2\,)\,1 & 0 \\
    & -0 & \\
2^0 & 1\,)\,1 & 1 \\
    & -1 & \\
    & 0 &
\end{array}
$$

Thus, $53 = 110101_2$. □

When converting a decimal fraction, continue dividing by negative powers of two until a remainder of zero is reached or until a sequence of binary digits repeats. Not every terminating decimal fraction will terminate as a binary fraction. However, if it does not terminate, a sequence of digits will repeat. Example 23 illustrates this.

■ **EXAMPLE 19**   Convert 6.625 into a binary number.

Solution

$$
\begin{array}{lllll}
2^2 & 4 & 6.625 & 1 \\
 & & -4 \\
2^1 & 2 & 2.625 & 1 \\
 & & -2 \\
2^0 & 1 & 0.625 & 0 \\
 & & -0 \\
2^{-1} & 0.5 & 0.625 & .1 \\
 & & -0.5 \\
2^{-2} & 0.25 & 0.125 & 0 \\
 & & -0 \\
2^{-3} & 0.125 & 0.125 & 1 \\
 & & -0.125 \\
 & & 0
\end{array}
$$

Therefore, $6.625 = 110.101_2$. □

■ **EXAMPLE 20**   Convert 6.4375 into a binary number.

Solution

$$
\begin{array}{lllll}
2^2 & 4 & 6.4375 & 1 \\
 & & -4 \\
2^1 & 2 & 2.4375 & 1 \\
 & & -2 \\
2^0 & 1 & 0.4375 & 0 \\
 & & -0 \\
2^{-1} & 0.5 & 0.4375 & .0 \\
 & & -0 \\
2^{-2} & 0.25 & 0.4375 & 1 \\
 & & -0.25 \\
2^{-3} & 0.125 & 0.1875 & 1 \\
 & & -0.125 \\
2^{-4} & 0.0625 & 0.0625 & 1 \\
 & & -0.0625 \\
 & & 0
\end{array}
$$

Thus we have $6.4375 = 110.0111_2$.   □

To illustrate the fact that not all terminating decimal fractions will convert into terminating binary fractions, let's consider a different method for converting decimal fractions into binary fractions, as follows.

> Multiply the decimal fraction by two. If the product yields the integer 1, consider this to be the first digit to the right of the binary point. If the integer 1 did not appear, consider this the binary digit 0. Now multiply the remaining decimal fraction by two again. If the integer 1 appears, this is the next binary digit moving to the right. If no 1 occurred, this is the binary zero. Again multiply the fractional part by two. Keep repeating the process until a fractional remainder of zero occurs or the desired accuracy is obtained.

■ **EXAMPLE 21**

Solution

Convert 6.625 into a binary number using the above method.

$$6.625 = 6 + 0.625$$
$$6 = 110_2$$
$$0.625 = ?$$

$$0.625 \times 2 = 1.25 \longrightarrow 1$$
$$0.25 \ \times 2 = 0.50 \longrightarrow 0$$
$$0.50 \ \times 2 = 1.00 \longrightarrow 1$$

Thus $6.625 = 6 + 0.625 = 110_2 + 0.101_2 = 110.101_2$ □

■ **EXAMPLE 22**

Solution

Convert 6.4375 into a binary number.

$$6.4375 = 6 + 0.4375$$
$$= 110_2 + ?$$

$$0.4375 \times 2 = 0.875 \longrightarrow 0$$
$$0.875 \ \times 2 = 1.75 \ \longrightarrow 1$$
$$0.75 \ \ \times 2 = 1.50 \ \longrightarrow 1$$
$$0.50 \ \ \times 2 = 1.00 \ \longrightarrow 1$$

Thus $6.4375 = 6 + 0.4375 = 110_2 + 0.0111_2 = 110.0111_2$ □

■ **EXAMPLE 23**

Solution

Convert 6.1 into a binary number.

$$6.1 = 6 + 0.1 =$$
$$= 110_2 + ?$$

$$0.1 \times 2 = 0.2 \longrightarrow 0$$
$$0.2 \times 2 = 0.4 \longrightarrow 0$$
$$0.4 \times 2 = 0.8 \longrightarrow 0$$ **Repeating**
$$0.8 \times 2 = 1.6 \longrightarrow 1$$ **sequence**
$$0.6 \times 2 = 1.2 \longrightarrow 1$$

$$
\left.
\begin{aligned}
0.2 \times 2 &= 0.4 \longrightarrow 0 \\
0.4 \times 2 &= 0.8 \longrightarrow 0 \\
0.8 \times 2 &= 1.6 \longrightarrow 1 \\
0.6 \times 2 &= 1.2 \longrightarrow 1
\end{aligned}
\right\} \quad
\begin{aligned}
&\textbf{Repeating} \\
&\textbf{sequence}
\end{aligned}
$$

Thus $6.1 = 6 + 0.1 \approx 110_2 + 0.0001100111_2 \approx 110.0001100111_2$ $\quad\square$

This method will work for converting a decimal fraction into any other base b. In each case, the multiplier is b (that is, in base eight, multiply by eight, etc.).

Exercise Set 4.3

Convert each of the following binary numbers into a decimal number.

1. 11101_2
2. 101011_2
3. 110011.01_2
4. 11001.101_2
5. 1110.011_2
6. 110101.10111_2
7. 1011.1011_2
8. 10001.0101_2
9. 11011.1011_2
10. 110101.11011_2

Convert each of the following decimal numbers into a binary number. If the decimal fraction does not terminate as a binary fraction, carry out the conversion until the repeating digits can be established on the right of the binary point.

11. 27
12. 91
13. 135
14. 320
15. 12.75
16. 24.375
17. 40.3125
18. 7.625
19. 5.3
20. 1.9
21. 0.6
22. 0.138

4.4

The Octal Numeration System: Base Eight

Through our discussion of the binary numeration system, we have seen that long strings of binary zeros and ones are needed to express even small decimal numbers (for example, $27.25 = 11011.01_2$). Such long strings of binary digits are difficult to read and handle. Also, although the conversion between binary and decimal is not necessarily difficult, it is certainly not a simple substitution either.

What is needed is a system that is compatible with binary, that allows for a quick and easy conversion, and that does not require a long string of digits to express a number. Any system that has a base whose powers match the powers of two and also has an equivalent numeral for all possible combinations of a group of binary digits would satisfy these requirements.

The **octal numeration system** is such a system. Consider the following:

$$
\begin{aligned}
101110001_2 &= (1 \times 2^8 + 0 \times 2^7 + 1 \times 2^6) \\
&\quad + (1 \times 2^5 + 1 \times 2^4 + 0 \times 2^3) \\
&\quad + (0 \times 2^2 + 0 \times 2^1 + 1 \times 2^0) \\
&= (1 \times 2^2 + 0 \times 2^1 + 1 \times 2^0)2^6 \\
&\quad + (1 \times 2^2 + 1 \times 2^1 + 0 \times 2^0)2^2 \\
&\quad + (0 \times 2^2 + 0 \times 2^1 + 0 \times 2^0)2^0 \\
&= (4 + 0 + 1)2^6 + (4 + 2 + 0)2^3 + (0 + 0 + 1)2^0 \\
&= (5)2^6 + (6)2^3 + (1)2^0
\end{aligned}
$$

Since

$$
\begin{aligned}
2^6 &= 8^2, \\
2^3 &= 8^1, \\
2^0 &= 8^0,
\end{aligned}
$$

then

$$
\begin{aligned}
(5)2^6 + (6)2^3 + (1)2^0 &= (5)8^2 + (6)8^1 + (1)8^0 \\
&= 561_8.
\end{aligned}
$$

Therefore, $101110001_2 = 561_8$.

The octal numeration system contains the numerals 0, 1, 2, 3, 4, 5, 6, 7. (Remember, we are using some of the base-ten symbols.) Since $2^3 = 8$, each octal digit corresponds to a group of three binary digits according to the following format:

Base eight	Base two
0	000
1	001
2	010
3	011
4	100
5	101
6	110
7	111

Given a number such as 743_8, we can quickly convert it into binary by replacing each octal digit by the proper binary sequence. This substitution is possible because base eight and base two "align" themselves at every third position.

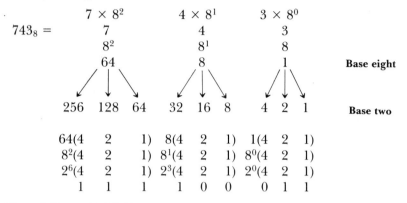

$$743_8 = $$

7×8^2	4×8^1	3×8^0	
7	4	3	
8^2	8^1	8	
64	8	1	**Base eight**

| 256 128 64 | 32 16 8 | 4 2 1 | **Base two** |

$$
\begin{array}{ccccccccc}
64(4 & 2 & 1) & 8(4 & 2 & 1) & 1(4 & 2 & 1) \\
8^2(4 & 2 & 1) & 8^1(4 & 2 & 1) & 8^0(4 & 2 & 1) \\
2^6(4 & 2 & 1) & 2^3(4 & 2 & 1) & 2^0(4 & 2 & 1) \\
1 & 1 & 1 & 1 & 0 & 0 & 0 & 1 & 1
\end{array}
$$

Thus, $743_8 = 111100011_2$.

■ **EXAMPLE 24**   Convert each of the following to base two.

a) $743_8 = \overset{7}{111}\ \overset{4}{100}\ \overset{3}{011} = 111100011_2$

b) $620.5_8 = \overset{6}{110}\ \overset{2}{010}\ \overset{0}{000}.\overset{5}{101} = 110010000.101_2$ □

To convert from binary to octal, simply reverse the process, starting with the binary point and moving left and right, grouping by threes.

■ **EXAMPLE 25**   Convert each of the following to base eight.

a) $1011101_2 = \overset{1}{001}\ \overset{3}{011}\ \overset{5}{101} = 135_8$

b) $10111101.001_2 = \overset{2}{010}\ \overset{7}{111}\ \overset{5}{101}.\overset{1}{001} = 275.1_8$

c) $1011.01101_2 = \overset{1}{001}\ \overset{3}{011}.\overset{3}{011}\ \overset{2}{010} = 13.32_8$ □

Octal-to-Decimal Conversion

Because the programmer must live and work in a "base-ten world," it is also necessary to be able to convert between octal and decimal number systems. We can use the process of cipherization from Section 4.2 to do so. Each position of the octal number is a power of eight. See the front endpapers for positive and negative powers of 8.

■ EXAMPLE 26   Convert each of the following to base ten.

a) $135_8 = (1 \times 8^2) + (3 \times 8^1) + (5 \times 8^0)$
$= (1 \times 64) + (3 \times 8) + (5 \times 1)$
$= 64 + 24 + 5$
$= 93$

b) $275.1_8 = (2 \times 8^2) + (7 \times 8^1) + (5 \times 8^0) + (1 \times 8^{-1})$
$= (2 \times 64) + (7 \times 8) + (5 \times 1) + (1 \times 0.125)$
$= 128 + 56 + 5 + 0.125$
$= 189.125$ □

The multiply-and-add method can also be used to convert from octal into decimal. The multiplier is eight. The process is left to the student to try. (See Appendix B for a conversion table.)

Decimal-to-Octal Conversion

The procedure for converting from base ten to base eight is the same as converting from base ten to base two. We divide the decimal number by the largest power of eight not larger than the given base-ten number. Then we continue dividing by decreasing powers of eight until the final division by 8^0.

■ EXAMPLE 27

Solution

$$
\begin{array}{r|r|r}
8^2 \quad 64 & 93 & 1 \\
 & -64 & \\
\hline
8^1 \quad 8 & 29 & 3 \\
 & -24 & \\
\hline
8^0 \quad 1 & 5 & 5 \\
 & -5 & \\
\hline
 & 0 &
\end{array}
$$

Thus, $93 = 135_8$.   □

■ EXAMPLE 28   Convert 375 into base eight.

Solution

$$
\begin{array}{r|r|r}
8^2 \quad 64 & 375 & 5 \\
 & -320 & \\
\hline
8^1 \quad 8 & 55 & 6 \\
 & -48 & \\
\hline
8^0 \quad 1 & 7 & 7 \\
 & -7 & \\
\hline
 & 0 &
\end{array}
$$

Thus, $375 = 567_8$.   □

When converting a decimal fraction, we continue dividing by negative powers of eight until either a remainder of zero or the desired accuracy is reached. A terminating decimal fraction may not convert to a terminating octal fraction. We use the method of multiplying the fractional part by the given base (eight, in this case). (See Example 31.)

■ **EXAMPLE 29**   Convert 189.125 into base eight.

Solution

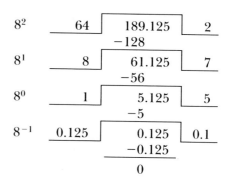

Thus, $189.125 = 275.1_8$.   □

■ **EXAMPLE 30**   Convert 612.328125 into base eight.

Solution

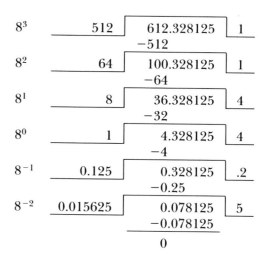

Therefore, $612.328125 = 1144.25_8$.   □

■ **EXAMPLE 31**   Convert 0.3 into base eight.

Solution

$$0.3 \times 8 = 2.4 \longrightarrow 2$$
$$\left.\begin{array}{l} 0.4 \times 8 = 3.2 \longrightarrow 3 \\ 0.2 \times 8 = 1.6 \longrightarrow 1 \\ 0.6 \times 8 = 4.8 \longrightarrow 4 \\ 0.8 \times 8 = 6.4 \longrightarrow 6 \end{array}\right\} \quad \text{Repeating sequence}$$
$$\left.\begin{array}{l} 0.4 \times 8 = 3.2 \longrightarrow 3 \\ 0.2 \times 8 = 1.6 \longrightarrow 1 \\ 0.6 \times 8 = 4.8 \longrightarrow 4 \\ 0.8 \times 8 = 6.4 \longrightarrow 6 \end{array}\right\} \quad \text{Repeating sequence}$$

Therefore, $0.3 \approx 0.23\overline{1463146}_8$. □

Exercise Set 4.4

Convert each of the following octal numbers into a decimal number.

1. 162_8

2. 7172_8

3. 5116_8

4. 16.64_8

5. 123.16_8

6. 321.124_8

Convert each of the following decimal numbers into an octal number. If the conversion does not terminate, carry out the conversion until the repeating digits are determined.

7. 120

8. 449

9. 1603

10. 12.75

11. 123.6875

12. 138.0625

13. 0.6

14. 0.1

15. 0.3

16. 0.9

Convert between binary and octal.

17. $1101110_2 =$ _____$_8$

18. $1111100110.01_2 =$ _____$_8$

19. $1010110.001_2 =$ _____$_8$

20. $101011000.1001_2 =$ _____$_8$

21. $457_8 =$ _____$_2$

22. $7362.35_8 =$ _____$_2$

23. $7602_8 =$ _____$_2$

24. $1347.02_8 =$ _____$_2$

25. $5431.32_8 =$ _____$_2$

26. $1101101.011_2 =$ _____$_8$

4.5
The Hexadecimal Numeration System: Base Sixteen

If the octal numeration system represents a convenient device for representing binary quantities, the **hexadecimal numeration system** provides an even better representation. Since its base is sixteen, it requires sixteen distinct symbols. This presents a minor problem since only ten symbols are available from the decimal number system, 0 through 9. Although any group of symbols could be used, the most common representation for the hexadecimal numbers that follow 9 are A, B, C, D, E, F. The complete group of hexadecimal symbols is

0, 1, 2, 3, 4, 5, 6, 7, 8, 9, A, B, C, D, E, F.

Note that 10_{16} does not follow 9 since 10_{16} in base sixteen stands for one group of sixteen with zero units. The number 10_{16} in base sixteen will follow F, since F stands for the number fifteen.

If we count through the first sixteen numbers in binary we see that the binary digits are grouped by fours. We should probably have expected this, since $2^4 = 16$. Below are the relationships between base ten, base two, and base sixteen.

Base ten	Base two	Base sixteen
0	0000	0
1	0001	1
2	0010	2
3	0011	3
4	0100	4
5	0101	5
6	0110	6
7	0111	7
8	1000	8
9	1001	9
10	1010	A
11	1011	B
12	1100	C
13	1101	D
14	1110	E
15	1111	F

The counting process now follows the place-value concept, as shown below:

0, 1, 2, ..., 8, 9, A, B, C, D, E, F, 10, 11, 12, ...,

19, 1A, 1B, 1C, ..., 1F, 20, ..., 2F, 30.

Grouping by fours provides the method for converting between base two and base sixteen. To convert a base-sixteen number into a binary number, we replace each hexadecimal digit by the appropriate sequence of four binary digits:

A = 1010,
B = 1011,
C = 1100,
D = 1101,
E = 1110,
F = 1111.

■ **EXAMPLE 32**   Convert each of the following to base two.

$$\overset{7E3}{\textbf{a)}\ 7E3_{16}} = 0111\ 1110\ 0011 = 111111100011_2$$

$$\overset{2BAD4}{\textbf{b)}\ 2BAD.4_{16}} = 0010\ 1011\ 1010\ 1101.0100 = 10101110101101.01_2\ \square$$

We merely reverse the process to convert from binary to hexadecimal.

■ **EXAMPLE 33**   Convert each of the following to base sixteen.

$$\overset{5D}{\textbf{a)}\ 1011101_2} = 0101\ 1101 = 5D_{16}$$

$$\overset{6FA}{\textbf{b)}\ 1101111.101_2} = 0110\ 1111.1010 = 6F.A_{16}\ \ \square$$

Occasionally it may be necessary to convert from octal to hexadecimal and vice versa. The binary system provides us with a very convenient process. When converting from base eight to base sixteen, we replace each octal digit by the equivalent group of three binary digits and then regroup these in groups of four.

■ **EXAMPLE 34**   Convert the octal number 637_8 to base sixteen.

Solution

$$\overset{63719F}{637_8} = 110\ 011\ 111 = 110011111_2 = 0001\ 1001\ 1111$$
$$637_8 = 19F_{16}\ \ \square$$

To convert from hexadecimal into octal, we replace each base-sixteen digit by the equivalent four binary digits and then regroup these in groups of three.

■ **EXAMPLE 35**   Convert the hexadecimal number $19F_{16}$ to base eight.

Solution

$$\begin{array}{ccc} 1 & 9 & F \end{array} \qquad\qquad \begin{array}{ccc} 6 & 3 & 7 \end{array}$$
$$19F_{16} = 0001\ 1001\ 1111 = 110011111_2 = 110\ 011\ 111$$
$$19F_{16} = 637_{16} \quad \square$$

Hexadecimal-to-Decimal Conversion

We can use the process of cipherization to convert any hexadecimal number into a decimal number. Each position to the left of the hexadecimal point is a positive power of 16 and each position to the right is a negative power of 16 (see the front endpapers for positive and negative powers of 16):

$$\ldots, 16^3, 16^2, 16^1, 16^0 . 16^{-1}, 16^{-2}, 16^{-3}, \ldots$$

■ **EXAMPLE 36**   Convert each of the following hexadecimal numbers to base ten.

a) $3AE_{16} = (3 \times 16^3) + (A \times 16^2) + (E \times 16^1)$
$\qquad\quad = (3 \times 256) + (10 \times 16) + (14 \times 1)$
$\qquad\quad = 768 + 160 + 14$
$\quad 3AE_{16} = 942$
b) $1D3B.C_{16} = (1 \times 16^3) + (D \times 16^2) + (3 \times 16^1) + (B \times 16^0)$
$\qquad\qquad\quad + (C \times 16^{-1})$
$\qquad\qquad = (1 \times 4096) + (13 \times 256) + (3 \times 16) + (11 \times 1) +$
$\qquad\qquad\quad (12 \times 0.0625)$
$\qquad\qquad = 4096 + 3328 + 48 + 11 + 0.75$
$\quad 1D3B.C_{16} = 7483.75 \quad \square$

The multiply-and-add method can also be used to convert from hexadecimal into decimal. The multiplier is sixteen. The process is left to the student to try. See Appendix A for a conversion table.

Decimal-to-Hexadecimal Conversion

In order to convert from decimal into hexadecimal, we divide the decimal number by the largest power of sixteen, not larger than the given base-ten number. We then continue dividing by decreasing powers of sixteen until the final division by 16^0. Whenever the quotient is 10, 11, 12, 13, 14, or 15, we replace it by the appropriate base-sixteen equivalent.

■ **EXAMPLE 37**   Convert 942 into base sixteen.

Solution

$$
\begin{array}{rrrll}
16^2 & \underline{256} & \underline{\big|\ 942\ } & \underline{\big|\ 3\ } \\
& & -768 \\
16^1 & \underline{16} & \underline{\big|\ 174\ } & \underline{\big|\ 10\ } & = A \\
& & -160 \\
16^0 & \underline{1} & \underline{\big|\ 14\ } & \underline{\big|\ 14\ } & = E \\
& & -14 \\
& & 0
\end{array}
$$

Thus, $942 = 3AE_{16}$.   □

■ **EXAMPLE 38**   Convert 7483.75 into base sixteen.

Solution

$$
\begin{array}{rrrll}
16^3 & \underline{4096} & \underline{\big|\ 7483.75\ } & \underline{\big|\ 1\ } \\
& & -4096 \\
16^2 & \underline{256} & \underline{\big|\ 3387.75\ } & \underline{\big|\ 13\ } & = D \\
& & -3328 \\
16^1 & \underline{16} & \underline{\big|\ 59.75\ } & \underline{\big|\ 3\ } \\
& & -40 \\
16^0 & \underline{1} & \underline{\big|\ 11.75\ } & \underline{\big|\ 11\ } & = B \\
& & -11 \\
16^{-1} & \underline{0.0625} & \underline{\big|\ 0.75\ } & \underline{\big|\ 0.12\ } & = .C \\
& & -0.75 \\
& & 0
\end{array}
$$

Thus, $7483.75 = 1D3B.C_{16}$.   □

Once again we can convert the fractional part by multiplying by the base, sixteen, and using the integer part of the products.

■ **EXAMPLE 39**   Convert 0.78125 into base sixteen.

Solution
$$0.78125 \times 16 = 12.5 \longrightarrow 12 \longrightarrow C$$
$$0.5 \times 16 = \ \ 8.0 \longrightarrow \ \ 8$$

Therefore, $0.78125 = C8_{16}$.   □

■ **EXAMPLE 40**   Convert 0.3 into base sixteen.

Solution
$$0.3 \times 16 = \ \ 4.8 \longrightarrow \ \ 4$$
$$0.8 \times 16 = 12.8 \longrightarrow 12 \longrightarrow C$$
$$0.8 \times 16 = 12.8 \longrightarrow 12 \longrightarrow C$$

Therefore, $0.3 \approx 0.4\overline{CC}_{16}$.   □

Exercise Set 4.5

Convert each of the following hexadecimal numbers into a decimal number.

1. $1B9_{16}$ **2.** $2FF_{16}$

3. $21D0_{16}$ **4.** $1F.C_{16}$

5. $3DAD.2_{16}$ **6.** $1BAD.E_{16}$

Convert each of the following base-ten numbers into a hexadecimal number.

7. 120   **8.** 273

9. 4296   **10.** 8764

11. 3249   **12.** 446.1875

13. 0.328125   **14.** 0.6328125

15. 0.1   **16.** 0.7

Make the following conversions between octal, hexadecimal, and binary.

17. $1101110_2 = $ _____ $_{16}$ **18.** $110011.0011_2 = $ _____ $_{16}$

19. $101011.101_2 = $ _____ $_{16}$ **20.** $100110101.011_2 = $ _____ $_{16}$

21. $3DEC_{16} = $ _____ $_2$ **22.** $9ABE.C_{16} = $ _____ $_2$

23. $4AE.C_{16} = $ _____ $_2$ **24.** $2FAB_{16} = $ _____ $_2$

25. $5047_8 = $ _____ $_{16}$ **26.** $507B_{16} = $ _____ $_8$

27. $3A7C_{16} = $ _____ $_8$ **28.** $243.2_8 = $ _____ $_{16}$

4.6

Arithmetic Operations in Different Bases

Binary Arithmetic

All possible sums of two single-digit binary numbers are:

$$
\begin{array}{cccc}
0 & 0 & 1 & 1 \\
\underline{+0} & \underline{+1} & \underline{+0} & \underline{+1} \\
0 & 1 & 1 & 10
\end{array}
$$

The process of addition in binary follows closely the process of addition in base ten. The rightmost digit (the units digit) is recorded in each sum, and the higher-order digits are carried.

■ **EXAMPLE 41**   Add the binary numbers 1011 and 1110.

Solution

```
  1   1
  1  0  1  1
+ 1  1  1  0
1  1  0  0  1₂  □
```

■ **EXAMPLE 42**   Add the binary numbers 11101, 10111, and 111.

Solution

```
  1   10   1   1
  1   1    1   0   1
  1   0    1   1   1
+           1   1   1
1  1   1    0   1   1₂  □
```

The process of subtraction also follows from the base-ten system, including the need to borrow. The possible binary differences are:

$$
\begin{array}{cccc}
0 & 1 & 1 & 10 \\
-0 & -0 & -1 & -1 \\
\hline
0 & 1 & 0 & 1
\end{array}
$$

Remember, when you borrow in base two you are borrowing a group of two, which is written as 10.

■ **EXAMPLE 43**   Subtract the binary number 10 from the binary number 100.

Solution

```
 0   10
 ̶1   ̶1̶0  0
−      1  0
    1  0₂  □
```

■ **EXAMPLE 44**   Subtract the binary number 1111 from the binary number 111110.

Solution

```
         10  10  10
     0   0   0   0   10
  1  ̶1   ̶1   ̶1   ̶1   ̶1̶0
−          1   1   1   1
  1  0   1   1   1   1₂  □
```

Note that in base ten, as it is necessary to borrow across a string of zeros, the zeros become nines. This happens because a group of ten is being brought back to the next lowest position and then one is borrowed from that ten. The process is repeated until ten is brought back to the desired position.

■ EXAMPLE 45

Solution

Subtract 38 from 2000 in base ten.

$$
\begin{array}{r}
\overset{\scriptstyle 1}{\cancel{2}}\ \overset{\scriptstyle 9}{\cancel{\cancel{0}}}\ \overset{\scriptstyle 9}{\cancel{\cancel{0}}}\ 10 \\
-\qquad\ \ 3\ \ 8 \\
\hline
1\ \ 9\ \ 6\ \ 2\ \ \square
\end{array}
$$

The same thing happens in the binary system except that the string of zeros becomes ones. This happens because a group of two (written 10) is borrowed and then one must be taken from that group, that is, $10 - 1 = 1$.

■ EXAMPLE 46

Solution

Subtract 11_2 from 10000_2.

$$
\begin{array}{r}
0\ \ \overset{\scriptstyle 1}{\cancel{\cancel{1}}}\ \overset{\scriptstyle 1}{\cancel{\cancel{0}}}\ \overset{\scriptstyle 1}{\cancel{\cancel{0}}}\ 10 \\
\cancel{1}\ \ \cancel{0}\ \ \cancel{0}\ \ \cancel{0}\ \ \cancel{0} \\
-\qquad\qquad\ \ 1\ \ 1 \\
\hline
1\ \ 1\ \ 0\ \ 1_2\ \ \square
\end{array}
$$

Although multiplication in base two is performed as in base ten, binary multiplication is easier since all products are either 0 or 1. The partial products are combined as in addition.

■ EXAMPLE 47

Solution

Multiply the binary numbers 1101 and 11.

$$
\begin{array}{r}
1\ \ 1\ \ 0\ \ 1 \\
\times\qquad\ \ 1\ \ 1 \\
\hline
1\ \ 1\ \ 0\ \ 1 \\
1\ \ 1\ \ 0\ \ 1\qquad \\
\hline
1\ \ 0\ \ 0\ \ 1\ \ 1\ \ 1_2\ \ \square
\end{array}
$$

■ EXAMPLE 48

Solution

Multiply the binary number 1101 by the binary number 110.

$$
\begin{array}{r}
1\ \ 1\ \ 0\ \ 1 \\
\times\qquad\ \ 1\ \ 1\ \ 0 \\
\hline
1\ \ 1\ \ 0\ \ 1\ \ 0 \\
1\ \ 1\ \ 0\ \ 1\qquad \\
\hline
1\ \ 0\ \ 0\ \ 1\ \ 1\ \ 1\ \ 0_2\ \ \square
\end{array}
$$

Octal Arithmetic

Since the numerals used in the octal number system are the same as those used in base ten (0, 1, 2, 3, 4, 5, 6, 7), octal addition is performed by finding the base-ten sum of pairs of octal digits and immediately converting that sum into base eight. Remember, we record the right-most digit and carry the higher-order digit(s).

■ **EXAMPLE 49**   Add the octal numbers 743 and 567.

Solution

$$
\begin{array}{r}
\overset{1}{}\;\overset{1}{} \\
7\;\;4\;\;3 \\
+\;5\;\;6\;\;7 \\
\hline
1\;\;5\;\;3\;\;2_8
\end{array}
\qquad
\begin{array}{r}
3 \\
+7 \\
\hline
10 = 12_8
\end{array}
\qquad
\begin{array}{r}
\overset{1}{4} \\
+6 \\
\hline
11 = 13_8
\end{array}
\qquad
\begin{array}{r}
\overset{1}{7} \\
+5 \\
\hline
13 = 15_8
\end{array}\;\;\square
$$

■ **EXAMPLE 50**   Add the octal numbers 17652 and 76046.

Solution

$$
\begin{array}{r}
\overset{1}{}\;\overset{1}{}\;\overset{1}{} \\
1\;\;7\;\;6\;\;5\;\;2 \\
+\;7\;\;6\;\;0\;\;4\;\;6 \\
\hline
1\;\;1\;\;5\;\;7\;\;2\;\;0_8 \;\;\square
\end{array}
$$

Base-eight subtraction is performed by borrowing a group of eight from the next higher order to the left. Any string of zeros will become sevens through the process. In general, subtraction in any base P will result in strings of zeros becoming the digits $P - 1$.

■ **EXAMPLE 51**   Subtract the octal number 567 from the octal number 743.

Solution

$$
\begin{array}{r}
6\;\;\overset{11}{\cancel{}}\;\;11 \\
\cancel{7}\;\;\cancel{4}\;\;\cancel{3} \\
-5\;\;6\;\;7 \\
\hline
1\;\;5\;\;4_8 \;\;\square
\end{array}
$$

11 is the base-ten value, not an octal number.

■ **EXAMPLE 52**   Subtract the octal number 17652 from the octal number 76046.7.

Solution

$$
\begin{array}{r}
6\;\;\overset{13}{\cancel{5}}\;\;\overset{7}{\cancel{8}}\;\;12 \\
\cancel{7}\;\;\cancel{6}\;\;\cancel{0}\;\;\cancel{4}\;\;6 \\
-1\;\;7\;\;6\;\;5\;\;2 \\
\hline
5\;\;6\;\;1\;\;7\;\;4_8 \;\;\square
\end{array}
$$

■ **EXAMPLE 53**   Subtract the octal number 4736 from the octal number 50002.

Solution

$$
\begin{array}{r}
4\;\;7\;\;7\;\;7\;\;10 \\
\cancel{5}\;\;\cancel{0}\;\;\cancel{0}\;\;\cancel{0}\;\;\cancel{2} \\
-\;\;\;\;4\;\;7\;\;3\;\;6 \\
\hline
4\;\;3\;\;0\;\;4\;\;4_8 \;\;\square
\end{array}
$$

Octal multiplication is performed by finding the base-ten product of pairs of octal digits with the product converted right back into base eight, much as we did with octal addition. The units digit is recorded and the higher-order digits are carried just as in base-ten multiplication.

■ **EXAMPLE 54**   Find the product of the octal numbers 43 and 56.

Solution

$$
\begin{array}{rr}
 & 4\ 3 \\
\times & 5\ 6 \\
\hline
 & 3\ 2\ 2 \\
 & 2\ 6\ 4 \\
\hline
3\ & 1\ 6\ 2_8
\end{array}
\qquad
\begin{array}{l}
3 \\
\times 6 \\
\hline
18 = 22_8
\end{array}
\qquad
\begin{array}{l}
4 \\
\times 6 \\
\hline
24 + 2 = 26 = 32_8
\end{array}
$$

□

■ **EXAMPLE 55**   Find the product of the octal numbers 273 and 52.

Solution

$$
\begin{array}{rr}
 & 2\ 7\ 3 \\
\times & 5\ 2 \\
\hline
 & 5\ 6\ 6 \\
1\ & 4\ 4\ 7 \\
\hline
1\ 5\ & 2\ 5\ 6_8
\end{array}
$$
□

Hexadecimal Arithmetic

Hexadecimal arithmetic is made more difficult by the six additional symbols A, B, C, D, E, and F. Addition is performed as in octal addition. That is, we find the sum of the two digits and convert it to base sixteen. Note the difference between a sum of 11, which is B_{16} and the conversion of 17, which is 11_{16}.

■ **EXAMPLE 56**   Add the hexadecimal numbers 478 and 239.

Solution

$$
\begin{array}{rr}
 & 1 \\
 & 4\ 7\ 8 \\
+ & 2\ 3\ 9 \\
\hline
 & 6\ B\ 1_{16}
\end{array}
\qquad
\begin{array}{l}
8 \\
+9 \\
\hline
17 = 11_{16}
\end{array}
$$

Note that there was no carry since $1 + 7 + 3 = B$.   □

■ **EXAMPLE 57**   Add the hexadecimal numbers A76 and 2BC.

Solution

$$
\begin{array}{rr}
 & 1\quad 1 \\
 & A\ 7\ 6 \\
+ & 2\ B\ C \\
\hline
 & D\ 3\ 2_{16}
\end{array}
$$
□

In hexadecimal subtraction, we can borrow a group of sixteen, and strings of zeros become F's, which can be written as 15's.

■ **EXAMPLE 58** Subtract the hexadecimal 2BF from the hexadecimal 6AC.

Solution

$$
\begin{array}{ccc}
 & 25 & \\
5 & 9 & 28 \\
\cancel{6} & \cancel{A} & \cancel{C} \\
-2 & B & F \\
\hline
3 & E & D
\end{array}
\qquad
\begin{array}{c}
28 \\
-\ F \\
\hline
13 = D \quad \square
\end{array}
$$

■ **EXAMPLE 59** Subtract the hexadecimal 2A7B from the hexadecimal A004B.

Solution

$$
\begin{array}{ccccc}
9 & 15 & 15 & 20 & \\
\cancel{A} & \cancel{0} & \cancel{0} & \cancel{4} & B \\
- & 2 & A & 7 & B \\
\hline
9 & D & 5 & D & 0_{16} \quad \square
\end{array}
$$

Multiplication is performed as in base eight. That is, we multiply the two digits in base ten, and then convert the answer to base sixteen.

■ **EXAMPLE 60** Multiply the hexadecimal 47B by the hexadecimal 2A.

Solution

$$
\begin{array}{ccc}
 & 4 & 7 & B \\
\times & & 2 & A \\
\hline
2 & C & C & E \\
8 & F & 6 & \\
\hline
B & C & 2 & E_{16}
\end{array}
$$

$$
\begin{array}{l}
B \\
\times A \\
\hline
110 = 6E_{16}
\end{array}
\qquad
\begin{array}{l}
7 \\
\times A \\
\hline
70 + 6 = 76 \\
\qquad = 4C_{16}
\end{array}
\qquad
\begin{array}{l}
4 \\
\times 8 \\
\hline
40 + 4 = 44 \\
\qquad = 2C_{16}
\end{array}
\qquad \square
$$

■ **EXAMPLE 61** Multiply the hexadecimal 2C31 by the hexadecimal 9B.

Solution

$$
\begin{array}{cccccc}
 & 2 & C & 3 & 1 \\
\times & & & 9 & B \\
\hline
 & 1 & E & 6 & 1 & B \\
1 & 8 & D & B & 9 & \\
\hline
1 & A & C & 1 & A & B \quad \square
\end{array}
$$

Exercise Set 4.6

Perform each of the following additions.

1. $\begin{array}{r} 11011_2 \\ +\ 1110_2 \\ \hline \end{array}$ **2.** $\begin{array}{r} 101110_2 \\ +110111_2 \\ \hline \end{array}$ **3.** $\begin{array}{r} 11110111_2 \\ +10110111_2 \\ \hline \end{array}$

4. 10110111_2
$+11001111_2$

5. 11101_2
11110_2
$+ 1111_2$

6. 11011_2
11101_2
$+10111_2$

7. 763_8
$+526_8$

8. 1476_8
$+3764_8$

9. 5764_8
3205_8
$+2154_8$

10. 46013_8
3276_8
$+ 4753_8$

11. $2BAD_{16}$
$+ BC04_{16}$

12. $3E76_{16}$
$+C0AB_{16}$

13. FAD_{16}
$+ 274_{16}$

14. $DA91_{16}$
$+AC9A_{16}$

Perform each of the following subtractions.

15. 1010_2
$- 111_2$

16. 101101_2
$- 11111_2$

17. 110010_2
-11111_2

18. 11000_2
$- 1111_2$

19. 674_8
-456_8

20. 723_8
-567_8

21. 6004_8
-3765_8

22. 30200_8
$- 10745_8$

23. $A69_{16}$
$-39B_{16}$

24. $19CB_{16}$
$- 43D_{16}$

25. $A50B0_{16}$
$- 3C43F_{16}$

26. $C8000_{16}$
$-5D47A_{16}$

Perform each of the following multiplications.

27. 1110_2
$\times 11_2$

28. 1110_2
$\times 10_2$

29. 111011_2
$\times 101_2$

30. 101101_2
$\times 111_2$

31. 473_8
$\times 25_8$

32. 1753_8
$\times 47_8$

33. 2056_8
$\times 247_8$

34. 3267_8
$\times 752_8$

35. 763_{16}
$\times 2A_{16}$

36. $4AB_{16}$
$\times 27_{16}$

37. $BAD3_{16}$
$\times 20B_{16}$

38. $32BC1_{16}$
$\times 4A3_{16}$

Something extra: Try these binary divisions.

39. $10\overline{)10100}$

40. $10\overline{)10110}$

41. $10\overline{)110011}$

42. $101\overline{)1000001}$

43. $111\overline{)1001111}$

44. $111\overline{)1001101}$

4.7
Complements

A temperature of three degrees below zero is written as $-3°$. A temperature of 97 degrees above zero is $+97°$. A deposit to a checking account is recorded as $+\$200$, whereas a withdrawal is $-\$40.00$. Not a day goes by that we are not faced with dealing with numbers and their opposites. We distinguish between these signed quantities by the use of the negative sign $(-)$ and the positive sign $(+)$. We generally have no problem using this notation. However, the computer cannot handle negative numbers the same way we do. Most machines—whether they are adding machines, calculators, or computers—store negative numbers in their complement form. Subtraction is also performed using complements.

A number, in any base, has a complement. The **complement** of a number is the difference between that number and the next higher power of the given base. For example, the complement of 763 in base ten is $1000 - 763 = 237$. The complement of 1476 is 8524 ($10,000 - 1476 = 8524$). This process of determining the complement of a number is difficult for the internal mechanics of the computer, since to subtract 1476 from 10,000 requires borrowing from each position. Another method for finding complements that is more mechanically feasible is illustrated in the following example.

■ **EXAMPLE 62**

Find the complement of the decimal number 763.

Solution

$(1000 - 1) - 763 + 1$
$\quad 999 \quad\ - 763 + 1$
$\quad 236 \quad\quad + \quad\ 1$
$\quad\quad\ \ 237$

Thus, 237 is the complement of 763. We reduced the power of the base by 1. This turned the zeros into nines, which eliminated the need to borrow. We then subtracted, and added 1 back in as a correction. In effect, we added 0 to $1000 - 763$, but in the form $(-1) + (+1)$. That simple step removes the problem of borrowing from each position. □

■ **EXAMPLE 63**

Find the complement of 1476.

Solution

$(10,000 - 1) - 1476 + 1$
$9999 - 1476 + 1$
$8523 + 1$
8524 □

■ **EXAMPLE 64**   Find the complement of 223.

Solution $(1000 - 1) - 223 + 1$
$999 - 223 + 1$
$776 + 1$
777 □

Since the computer operates in the binary mode, we need to look at finding the complement of a binary number. We use the same process as that shown above.

■ **EXAMPLE 65**   Find the complement of 1101.

Solution $(10,000 - 1) - 1101 + 1$
$1111 - 1101 + 1$
$10 + 1$
11 □

■ **EXAMPLE 66**   Find the complement of 1010_2.

Solution $(10,000 - 1) - 1010 + 1$
$1111 - 1010 + 1$
$101 + 1$
110_2 □

If you examine Examples 65 and 66 closely, you should note a remarkable fact. The complement of 1101 is simply $0010 + 1$. Each 1 in the given number is converted to 0 and each 0 is converted to 1. Then 1 is added to the result. Example 66 could be found as follows:

The complement of 1010_2 is $0101 + 1 = 110_2$.

In other words, the complement is the reverse of the original number, plus 1.

■ **EXAMPLE 67**   a) The complement of 11001_2 is $00110 + 1 = 111_2$.
b) The complement of 1010111_2 is $0101000 + 1 = 101001_2$.
c) The complement of 1100101_2 is $0011010 + 1 = 11011_2$.   □

Using complements, we can perform binary subtraction without the confusing borrowing that we encountered in Section 4.6. Example 68 illustrates how the subtraction works. On the left, subtraction is performed using borrowing. On the right, the same subtraction is performed using the complement of the subtrahend (the number being subtracted). Note that the problem then becomes an addition problem.

EXAMPLE 68   Subtract 101_2 from 1000_2 using the complement of 101_2.

Solution
$$
\begin{array}{cccc}
1 & 0 & 0 & 0 \\
-0 & 1 & 0 & 1 \\
\hline
0 & 0 & 1 & 1_2
\end{array}
$$
The complement of 0101 is 1011.

$$
\begin{array}{cccc}
1 & 0 & 0 & 0 \\
+1 & 0 & 1 & 1 \\
\hline
\end{array}
$$
$1\ 0\ 0\ 1\ 1_2$ □

True, $0011 \neq 10011$. But in the computer, the digits are stored in a register that has size limitations. Thus, if the above example were being performed in a register that could hold only four digits, the higher-order 1 on the left would be lost. This would be called an overflow condition in a 4-bit register.

Consider a computer with an 8-bit register. The first position or first bit on the left is reserved for the sign of the number. A 0 bit indicates a positive number and a 1 indicates a negative number. Negative numbers are stored in their complement form.

EXAMPLE 69
a) The positive decimal number 59 written in binary in an 8-bit register is 00111011. The number −59 written in its binary complement form is 11000101 (that is, 11000100 + 1).
b) The decimal number −112 written in the 8-bit register binary complement form is 10010000. The number 112 is 01110000. □

Let us again consider subtraction.

EXAMPLE 70   Subtraction performed using the complement of the subtrahend (the number being subtracted) becomes an addition problem, as shown below.

a)
$$
\begin{array}{l}
\ \ 0\,1\,0\,1\,0\,0\,1\,0 \\
-\,0\,1\,1\,1\,0\,1\,1\,1 \\
\end{array}
\quad \text{becomes} \quad
\begin{array}{l}
\ \ 0\,1\,0\,1\,0\,0\,1\,0 \\
+\,1\,0\,0\,0\,1\,0\,0\,1 \\
\hline
\ \ 1\,1\,0\,1\,1\,0\,1\,1
\end{array}
$$

b)
$$
\begin{array}{l}
\ \ 1\,0\,0\,0\,1\,0\,0\,1 \\
-\,1\,0\,0\,1\,1\,1\,1\,1 \\
\end{array}
\quad \text{becomes} \quad
\begin{array}{l}
\ \ 1\,0\,0\,0\,1\,0\,0\,1 \\
+\,0\,1\,1\,0\,0\,0\,0\,1 \\
\hline
\ \ 1\,1\,1\,0\,1\,0\,1\,0
\end{array}
$$

c)
$$
\begin{array}{l}
\ \ 1\,0\,1\,1\,0\,0\,0\,0 \\
-\,0\,1\,1\,1\,0\,1\,1\,1 \\
\end{array}
\quad \text{becomes} \quad
\begin{array}{l}
\ \ 1\,0\,1\,1\,0\,0\,0\,0 \\
+\,1\,0\,0\,0\,1\,0\,0\,1 \\
\hline
1\,0\,0\,1\,1\,1\,0\,0\,1
\end{array}
$$

The leftmost digit is lost. □

■ **EXAMPLE 71**

Solution

a) Find $43 - 27$ in binary complement form using an 8-bit register.

$$43 = 00101011_2$$
$$27 = 00011011_2 \qquad \text{Remember, } 43 - 27 = 43 + (-27).$$

$$
\begin{array}{ll}
00101011 & (43) \\
- \ 00011011 & (-27) \\
\hline
\end{array}
\qquad \text{becomes} \qquad
\begin{array}{ll}
00101011 & (43) \\
+ \ 11100101 & (-27) \\
\hline
100010000_2 &
\end{array}
$$

The leftmost digit is lost. The answer is $00010000_2 = 16$ ($43 - 27 = 16$). □

■ **EXAMPLE 72**

Solution

Find $59 - 112$ in binary complement form using an 8-bit register.

$$59 = 00111011_2$$
$$112 = 01110000_2 \qquad \text{Remember, } 59 - 112 = 59 + (-112).$$

$$
\begin{array}{l}
00111011 \\
- \ 01110000 \\
\hline
\end{array}
\qquad \text{becomes} \qquad
\begin{array}{l}
00111011 \\
+ \ 10010000 \\
\hline
11001011_2
\end{array}
$$

The leftmost digit indicates a negative number. The answer is the complement of 11001011. The complement is $00110101_2 = 53$ ($59 - 112 = -53$). □

Exercise Set 4.7

Perform each of the following subtractions using complementary arithmetic. Check your results by subtracting using the method given in Section 4.6.

1. $\begin{array}{r} 10110 \\ - \ 1101 \\ \hline \end{array}$

2. $\begin{array}{r} 10011 \\ - \ 1110 \\ \hline \end{array}$

3. $\begin{array}{r} 1101100 \\ - \ 110110 \\ \hline \end{array}$

4. $\begin{array}{r} 1011101 \\ -0110011 \\ \hline \end{array}$

5. $\begin{array}{r} 10100000 \\ -01100001 \\ \hline \end{array}$

6. $\begin{array}{r} 10000000 \\ -00111001 \\ \hline \end{array}$

Write each of the following decimal numbers in 8-bit binary form. Use the leftmost digit to indicate the $+$ or $-$ sign. Express any negative number in complement form.

7. 52   **8.** 125   **9.** 64   **10.** 132

11. -29 **12.** -52 **13.** -64 **14.** -128

Perform the indicated operations in binary form using 8-bit registers. Make note of the overflow condition when it occurs.

15. $(+48) + (+24)$ **16.** $(-32) + (-12)$

17. $(+52) - (+20)$ **18.** $(+27) - (+64)$

19. $(+29) - (-52)$ **20.** $(-38) - (+42)$

21. $(-40) - (-32)$ **22.** $(-40) - (-54)$

23. What are the largest and smallest positive and negative numbers that can be stored in an 8-bit register? (Remember, the first bit is for the $+$ or $-$ sign.) What about in a 16-bit register?

Express each of the following in 16-bit binary form. Use the leftmost digit to indicate the $+$ or $-$ sign. Express any negative number in complement form.

24. 52   **25.** 125   **26.** -64 **27.** -128

Summary

We began this chapter by looking at finite numeration systems in *modular arithmetic*. We found that the arithmetic, in the form of addition and subtraction, performed in these numeration systems yielded results quite different from the results we are accustomed to seeing when we perform arithmetic calculations with real numbers. We also discovered that some of these modular systems obeyed the eleven real-number properties. When the *modulus* is a prime number, all eleven real-number properties are obeyed. When the modulus is *not* a prime number, the multiplicative inverse property does not hold for that system.

Next we reviewed the decimal numeration system with its characteristics of *place value* and *cipherization*. We discovered that *positional notation* is not unique to the decimal numbers, and in fact exists regardless of the *base* of the system. Because of their use in the programming and the operation of computers, we then learned to use positional notation in the *binary, octal,* and *hexadecimal numeration systems*. We learned to convert between *base ten, base two, base eight,* and *base sixteen* systems, and to perform addition, subtraction, and multiplication within them.

We concluded the chapter with a discussion of *complements*. We learned to perform subtraction in the binary complement form. In this process, the subtraction becomes an addition when we use the complement form of the subtrahend. This was necessary because of the storage method of negative numbers in the computer and the operation of the computer in the binary mode.

Review Exercises

Express each of the following numbers in the cipherization form. Pay close attention to the number base.

1. 3240_{10} **2.** 47213_8

3. 101101_2 **4.** $4BAC_{16}$

5. 210.24_{10} **6.** 34.71_8

7. 101.101_2 **8.** $2A.CB_{16}$

Convert each of the following numbers to base ten.

9. 10101_2 **10.** 1101.101_2

11. 7430_8 **12.** 321.76_8

13. $C3B_{16}$ **14.** $A03.C_{16}$

Convert each of the following decimal numbers into the base indicated.

15. $43 = $ _____ $_2$ **16.** $403.3125 = $ _____ $_2$

17. $116 = $ _____ $_8$ **18.** $45.75 = $ _____ $_8$

19. $112 = $ _____ $_{16}$ **20.** $4096.1875 = $ _____ $_{16}$

21. $3.1 = $ _____ $_2$ **22.** $6.3 = $ _____ $_2$

23. $0.4 = $ _____ $_8$ **24.** $0.975 = $ _____ $_8$

25. $0.4 = $ _____ $_{16}$ **26.** $0.975 = $ _____ $_{16}$

Perform each of the following conversions.

27. $101110_2 = $ _____ $_8$ **28.** $1101110_2 = $ _____ $_{16}$

29. $7453_8 = $ _____ $_2$ **30.** $3B2C.D_{16} = $ _____ $_2$

31. $7036_8 = $ _____ $_{16}$ **32.** $F0B2.3_{16} = $ _____ $_8$

Perform the indicated operations.

33. $\begin{array}{r} 1011_2 \\ +\ 101_2 \end{array}$ **34.** $\begin{array}{r} 11011_2 \\ +\ 1111_2 \end{array}$ **35.** $\begin{array}{r} 7632_8 \\ +\ 477_8 \end{array}$

36. $\begin{array}{r} 75.32_8 \\ +21.76_8 \end{array}$ **37.** $\begin{array}{r} 4EB2_{16} \\ +79AC_{16} \end{array}$ **38.** $\begin{array}{r} 2B04_{16} \\ +7ABC_{16} \end{array}$

39. $\begin{array}{r} 10011_2 \\ -\ 1011_2 \end{array}$ **40.** $\begin{array}{r} 10001_2 \\ -\ 1111_2 \end{array}$ **41.** $\begin{array}{r} 4732_8 \\ -3777_8 \end{array}$

42. $\begin{array}{r} 70063_8 \\ -\ 2276_8 \end{array}$ **43.** $\begin{array}{r} 2EC3_{16} \\ -1BD5_{16} \end{array}$ **44.** $\begin{array}{r} 4372_{16} \\ -1ADC_{16} \end{array}$

45. $\begin{array}{r} 10111_2 \\ \times\ 1011_2 \end{array}$ **46.** $\begin{array}{r} 11101_2 \\ \times\ 111_2 \end{array}$ **47.** $\begin{array}{r} 632_8 \\ \times\ 47_8 \end{array}$

48. $\begin{array}{r} 6347_8 \\ \times\ 243_8 \end{array}$ **49.** $\begin{array}{r} 23B_{16} \\ \times\ 2A_{16} \end{array}$ **50.** $\begin{array}{r} 47BD_{16} \\ \times\ AB_{16} \end{array}$

Write each of the following decimal numbers in both 8-bit binary form and 16-bit binary form. Remember, the leftmost digit should indicate the + or − sign. Express all negative numbers in complement form.

51. 75   **52.** 112   **53.** −32   **54.** −125

Perform the indicated operations in binary form using an 8-bit register. Indicate the overflow condition when it exists.

55. (+24) + (+32)   **56.** (−24) + (−32)

57. (+52) − (−20)   **58.** (+29) − (+64)

Sets

After completing this chapter, you should be able to:

1. Distinguish between finite and infinite sets.
2. Identify the elements of a well-defined set.
3. Form the union of two sets.
4. Form the intersection of two sets.
5. Form the complement of a set relative to a universal set.
6. Given a set expression, draw the Venn diagram and write the set expression given the Venn diagram.
7. Draw a flowchart from either a Venn diagram or a set expression.
8. Verify Boolean properties for any given sets.

5

5.1
Basic Concepts of Sets: The Language

Classifying objects in various ways is an important component of practically all logical thought whether its application be scientific, business, or mathematical. This classification process consists of placing the items into sets, a concept that appears in all branches of mathematics. Set theory has many interesting applications, one of which is that it is fundamental in the design of digital computers. It is also used in symbolic logic, which we will discuss in Chapter 8, and in Boolean algebra, which we will find in Chapter 9. Consider the following problem as an illustration.

> The personnel director of a data processing firm is responsible for hiring a new computer programmer. The departmental manager has instructed the personnel director to consider only those applicants who are two-year college graduates or at least 25 years old and to eliminate from consideration anyone 25 or older who is not a two-year college graduate.

Georg Cantor

Although this problem sounds confusing, it actually has a very simple solution. We will reduce the problem to a statement using set relationships later in the chapter.

The theory of sets was developed by the German mathematician Georg Cantor (1845–1918) around 1875. Very simply put, a **set** is any collection of similar objects. These objects may be numbers, people, tangible items, or abstract ideas. The individual objects that belong to a set are called its *elements*, or *members*. The basic property of a set of objects is the identification of its elements. A set is *well defined* if it is possible to determine whether or not an object belongs to the set. Being able to determine whether an element belongs to a set may not be as simple as it may seem, as illustrated by the following version of "Russell's Paradox" attributed to the British mathematician Bertrand Russell in 1901:

> In the town of Seville lives Joe the Barber who boasts, "Although I don't shave those people who shave themselves, I do shave all those people of Seville who do not shave themselves."

Does Joe shave himself?

Let A be the set of all people of Seville who shave themselves. Then the question is "Is Joe the Barber an element of A?" If we say that Joe is an element of A, then Joe is a person who shaves himself. However, Joe claims that he does not shave the people who shave themselves. Hence, Joe the Barber does not shave himself and therefore cannot be an element of A.

If we say that Joe is not an element of A, then Joe is not a person who shaves himself. Since Joe the Barber shaves all persons who do not shave themselves, he must shave himself and therefore must be an element of A.

The paradox is that if we say Joe the Barber is an element of A, then he is not. If we say that he is not an element of A, then he is. Either beginning leads to a contradiction and it is impossible to determine whether Joe the Barber of Seville is an element of set A. As you can see, it is important to make sure that all sets in a discussion are well defined.

If it is possible to count all elements of a set, we say that the set is *finite*. If the elements of the set cannot be counted, the set is *infinite*. The number of elements in the set is its **cardinal number**. Finite sets will always have a counting number as a cardinal number.

The elements of a set may be specifically listed, as in a roster, or the set may be described. Braces are used to enclose the members of a set and capital letters are used as names for sets. The symbol $\in$ is called the *set inclusion symbol*. If the number 3 is an element of set A, we can write this as $3 \in A$. If 4 is not an element of set A, it is written $4 \notin A$.

■ **EXAMPLE 1**   The set of the three greatest novels of the twentieth century is an example of a set that is *not well defined* since we probably would not have agreement on either the three novels or on the criteria for choosing them.   □

■ **EXAMPLE 2**   The set of letters of the English alphabet is a well-defined set.   □

■ **EXAMPLE 3**   {1, 2, 3, 4, 5, . . .} is the set of natural numbers. This is an infinite set since the elements cannot all be counted.   □

■ **EXAMPLE 4**   {January, February, . . . , November, December} is the set of months of the year and is finite. The cardinal number of the set is 12.   □

■ **EXAMPLE 5**   The list of elements showing the possible outcomes of one roll of a die is finite. The cardinal number of the set is 6. The set = {1, 2, 3, 4, 5, 6}.   □

■ **EXAMPLE 6**   The four possible outcomes for two flips of a coin is the set {HH, HT, TH, TT}. The cardinal number is 4.   □

■ **EXAMPLE 7**   If A is the set of integers, then $5 \in A$ but $0.5 \notin A$.   □

Any set that does not have any elements is called the *empty set*, or **null set,** and is denoted by either $\emptyset$ or { }. However, {$\emptyset$} is not an empty set.

When all elements of one set are contained in another set, the first set is a *subset* of the second set.

■ **EXAMPLE 8**   a) If A = {1, 2, 3} and B = {1, 2, 3, 4}, then $A \subset B$. Set A is a subset of set B.
    b) If C = {1, 3, 5}, then $C \not\subset B$. Set C is not a subset of set B.   □

If there is at least one element of the second set that is not in the first set, the first set is a *proper subset* of the second.

■ **EXAMPLE 9**   If A = {1, 2, 3} and B = {1, 2, 3, 4}, then set A is a proper subset of set B.   □

In any discussion of sets, there must be a set containing all elements possible in the discussion. This set is called the **universal set** and is denoted by U. All of the employees at the home office of Nationwide Insurance Company constitute a universal set. All female employees and all employees in the data processing department are subsets of the universal set. By agreement, the null set is a subset of every set, that is, $\emptyset$ is a subset of any set A. Since $\emptyset$ is empty, it contains no elements that are not in A. Thus there is no contradiction.

■ **EXAMPLE 10**　　List all subsets of $\{1, 2, 3\}$.

Solution

$\{1\}$　$\{1, 2\}$　$\{1, 2, 3\}$
$\{2\}$　$\{1, 3\}$　　$\emptyset$
$\{3\}$　$\{2, 3\}$

Remember, the null set is a subset *and* the original set is a subset of itself. Note that there are eight subsets.　□

● Question　　Can we predetermine how many subsets a given set will have? How many subsets are there for $\{a, b, c, d\}$? List them. How does the number of subsets compare to 2^4?

Answer　　The number of subsets for a given set can be determined from 2^n, where n is the cardinal number for the set.

Exercise Set 5.1

1. Give two examples of finite sets and two examples of infinite sets.

2. Use the roster method to describe each of the following sets. If the set is finite, give the cardinal number of the set.

   a) The set of letters in the word Tennessee

   b) The set of odd natural numbers

   c) The set of vowels in the English alphabet

   d) The set of numbers satisfying $2x + 4 = -6$

   e) The set of real numbers satisfying $x + 1 = x$

   f) The set of even numbers between 2 and 24

   g) The set of powers of 3

   h) The set of positive integers divisible by 5

   i) The set of days of the week

3. Given the set $\{a, b, c\}$, determine how many subsets it has. List them.

4. Given the set $\{1, 2, 3, 4, 5\}$, determine how many subsets it has. List them.

5. Let $U = \{4, \sqrt{3}, 2/3, -1/2, -5, 17, \sqrt{9}, \pi\}$. List the sets for each of the following.

   a) Integers in U

   b) Rational numbers in U

   c) Irrational numbers in U

   d) Natural numbers in U

6. Determine whether each of the following is true or false. (In order to be true, the statement must *always* be true.) Let U = {1, 2, 3, 4, 5, 6, 7, 8, 9, 10}, A = {2, 4, 6, 8, 10}, and B = {1, 3, 6, 7, 8}.

a) $2 \in A$ b) $11 \in B$

c) $4 \notin B$ d) $A \subset U$

e) $B \subset A$ f) $A \in U$

5.2

The Pictorial Representation of Sets: Venn Diagrams

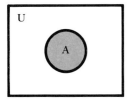

Figure 5.1

One convenient method of picturing sets and relationships between sets was first invented by Leonard Euler, a Swiss mathematician of the eighteenth century. In 1876 the British logician John Venn further refined and extended Euler's work. We call the pictures *Venn diagrams*.

In a Venn diagram, the universal set is shown as the set of points inside a rectangle. Subsets of the universal set are shown as points inside circles or disks. In Fig. 5.1, set A is a subset of U, whereas in Fig. 5.2, set B is a subset of set A. Sets that are not subsets of each other are shown in Fig. 5.3.

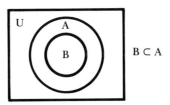

Figure 5.2

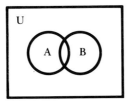 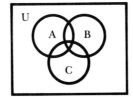 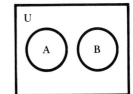

Figure 5.3

Crosshatching or shading are used to show various combinations of sets. For example, the shaded region in Fig. 5.4 represents the elements that are in both sets A *and* B, whereas the shaded region in Fig. 5.5 illustrates the points or elements that are in set A but not in set B.

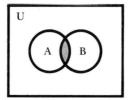

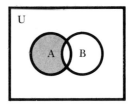

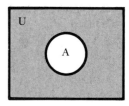

Figure 5.4   Figure 5.5   Figure 5.6

Let's consider as an example letting the universal set be the set of all employees at the home office of Nationwide Insurance Company. Let's also say that set A is all employees of the data processing department and set B is all female employees at Nationwide's home office. Figure 5.4 would then show all employees of the data processing department who are female. The shaded region in Fig. 5.5 then represents all employees of the data processing department who are not female; that is, the male employees who work in the data processing department. We also see that in Fig. 5.6 the shaded region represents all employees at the home office who do not work in the data processing department.

5.3
Operations on Sets: Union, Intersection, Complements

Our discussion of sets thus far has dealt primarily with set descriptions, set inclusion, and set relationships, in other words, subsets. We now turn our attention to operations that can be performed on sets. Although we cannot add, subtract, multiply, or divide sets, operations unique to sets are possible, with each operation forming a new set.

■ DEFINITION

UNION
The *union* of two sets is the set formed by joining the elements of the two sets. If an element belongs to both sets, it is listed only once in the union.

The symbol $\cup$ is used to indicate the union operation.

■ **EXAMPLE 11**   If A = {1, 3, 5} and B = {2, 4, 5, 6}, then A $\cup$ B = {1, 2, 3, 4, 5, 6}.   □

■ **EXAMPLE 12**   If A = {all female employees of Nationwide} and B = {all employees at Nationwide between the ages of 21 and 30}, then a list for A ∪ B would include the names of all female employees together with all male employees between the ages of 21 and 30.   □

■ **EXAMPLE 13**   A ∪ B is represented by the shaded region below.

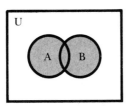

□

_____

■ **STATEMENT**

In order for an element to belong to A ∪ B, it must belong to either A *or* B, *or* both.

_____

■ **DEFINITION**

INTERSECTION The *intersection* of two sets is the set formed by joining *only* the common elements of the two sets. In order to belong to the intersection, the element(s) must be in both sets.

The symbol ∩ is used to represent intersection.

■ **EXAMPLE 14**   If A = {1, 3, 5} and B = {2, 4, 5, 6}, then A ∩ B = {5}.   □

■ **EXAMPLE 15**   If A = {all female employees of Nationwide} and B = {all employees of Nationwide between the ages of 21 and 30}, then a printed list for A ∩ B would contain the names of only female employees between the ages of 21 and 30.   □

■ **EXAMPLE 16**   A ∩ B is the shaded region shown below.

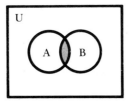

□

■ **STATEMENT**

In order for an element to belong to A ∩ B, it must belong to both A *and* B.

If A ∩ B = ∅, that is, if A and B have no common elements, then sets A and B are said to be **disjoint**.

Union and intersection are operations performed on two sets. The following operation is performed on one set, but in reference to the universal set.

■ **DEFINITION**

COMPLEMENT

The *complement* of a given set is the set containing all elements from the universal set that are not contained in the given set.

The prime, ′, is used as the complement sign. Therefore, A′ contains all elements not in A, with regard to U.

■ **EXAMPLE 17**   If U = {1, 2, 3, 4, 5} and A = {1, 3, 5}, then A′ = {2, 4}.   □

■ **EXAMPLE 18**   If A = {all female employees of Nationwide}, then A′ = {all male employees of Nationwide}.   □

■ **EXAMPLE 19**   A′ is the shaded region outside of A in the figure below.

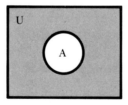

□

By definition of the term complement, A and A′ are disjoint, that is, A ∩ A′ = ∅.

The union and intersection operations are binary operations. In other words, they can be performed on two sets at a time. When more than two sets are involved, the hierarchy of operations is (1) quantities, (2) complement, (3) intersection, and (4) union.

We are aided in our work with sets by a group of properties that provide insight into the algebra of sets. There are some peculiar differences between these properties and the real-number properties discussed in Chapter 0; note, for example, the existence of two distributive properties in the algebra of sets.

PROPERTIES OF THE ALGEBRA OF SETS

1. IDEMPOTENT PROPERTIES
   a) $A \cup A = A$ b) $A \cap A = A$

2. COMMUTATIVE PROPERTIES
   a) $A \cup B = B \cup A$ b) $A \cap B = B \cap A$

3. ASSOCIATIVE PROPERTIES
   a) $(A \cup B) \cup C = A \cup (B \cup C)$
   b) $(A \cap B) \cap C = A \cap (B \cap C)$

4. DISTRIBUTIVE PROPERTIES
   a) $A \cup (B \cap C) = (A \cup B) \cap (A \cup C)$
   b) $A \cap (B \cup C) = (A \cap B) \cup (A \cap C)$

5. IDENTITY PROPERTIES
   a) $A \cup \emptyset = A$ b) $A \cap \emptyset = \emptyset$
   c) $A \cup U = U$ d) $A \cap U = A$

6. INVOLUTION PROPERTY
   $(A')' = A$

7. COMPLEMENT PROPERTIES
   a) $A \cup A' = U$ b) $A \cap A' = \emptyset$
   c) $U' = \emptyset$ d) $\emptyset' = U$

8. DeMORGAN'S PROPERTIES
   a) $(A \cup B)' = A' \cap B'$ b) $(A \cap B)' = A' \cup B'$

We leave it to the student to verify some of the above properties in Exercise Set 5.3.

Let us restate, using set theory, the problem introduced earlier in the chapter in which the personnel director of the data processing firm was to consider applicants for a programming position. He was told to consider only those applicants who are two-year college graduates or at least 25 years old, *and* should not consider anyone 25 or over who is not a two-year college graduate.

If A = the set of applicants who are two-year college graduates and B = the set of applicants who are 25 or over, then:

A ∪ B = the set of applicants who are two-year graduates *or* greater than or equal to 25 years of age;

A' ∩ B = the set of applicants who are not two-year college graduates but are greater than or equal to 25 years of age.

The personnel director should consider the set $(A' \cap B)'$, that is, those *not* in $A' \cap B$.

$$
\begin{aligned}
(A \cup B) \cap (A' \cap B)' &= (A \cup B) \cap (A \cup B') && \textbf{DeMorgan's law} \\
&= A \cup (B \cap B') && \textbf{Distributive law} \\
&= A \cup \emptyset && \textbf{Complement law} \\
&= A && \textbf{Identity law}
\end{aligned}
$$

Conclusion: The personnel director should consider only two-year college graduates.

Exercise Set 5.3

1. If $U = \{1, 2, 3, 4, 5, 6\}$, $A = \{1, 2, 4, 6\}$, $B = \{2, 3, 5\}$, and $C = \{4, 6\}$, list the members of each of the following.

   a) $A \cup B$
   b) $A \cap B$
   c) A'
   d) $A' \cap C$
   e) $A' \cap B'$
   f) $B' \cup C'$
   g) $(A \cup B)'$
   h) $(B \cap C)'$
   i) $(A \cup B) \cap C$
   j) $(A \cap C) \cup B'$

2. If $U = \{a, b, c, d, e\}$, $A = \{a, b\}$, $B = \{d, e\}$, $C = \{a, c, e\}$, and $D = \{b, c, d\}$, list the members of each of the following.

   a) $A \cap B$
   b) $A \cup C$
   c) B'
   d) $B' \cup C$
   e) $A' \cap D'$
   f) $(A \cup D)'$
   g) $(A' \cap B) \cup D$
   h) $(A \cup B') \cap D$

3. Given that $U = \{$all employees in Nationwide's home office$\}$, $A = \{$all female employees in the home office$\}$, $B = \{$all employees of Nationwide between 21 and 30 years old$\}$, $C = \{$all employees of the data processing department$\}$, $D = \{$all employees of Nationwide 21 years of age and over$\}$, and $E = \{$all employees of Nationwide 50 years of age and under$\}$. Describe each of the following. Think of a list being printed in each case.

   a) A'
   b) $A \cup C$
   c) $A' \cap B$
   d) $D \cap E$
   e) E'
   f) $(D \cap E)'$
   g) $C \cup E$
   h) $A \cap C \cap E$
   i) $(A' \cap B) \cup C$
   j) $(D \cap E)' \cap C$

4. Determine whether each of the following is true or false. (To be true, the statement must always be true.)

   a) If A = {1, 3, 4} and A ∩ B = {3}, then 3 ∈ B.

   b) If 2 ∈ (A ∩ B), then 2 ∈ A.

   c) A ∩ B is a subset of A.

   d) A ∪ B is a subset of B.

   e) If A = {1, 3, 4} and A ∪ B = {1, 3, 4, 5}, then 3 ∈ B.

   f) If A = {1, 3, 4} and A ∪ B = {1, 3, 4, 5}, then 5 ∈ B.

   g) If A = {1, 3, 4} and A ∪ B = {1, 3, 4, 5}, then 3 ∈ (A ∩ B).

   h) If A ⊂ B and 2 ∈ A, then 2 ∈ B.

   i) If A ∩ B = ∅ and 2 ∈ A, then 2 ∉ B'.

Figure 5.7

5. Using a Venn diagram such as that shown in Fig. 5.7, shade the set represented by each of the following set expressions. Use a different Venn diagram for each expression.

   a) A ∪ B   b) A' ∪ B

   c) A ∩ B'   d) (A ∩ B)'

   e) A' ∪ B'   f) (A ∪ B)'

   g) A' ∩ B'   h) (A ∪ B')'

6. Use a Venn diagram to verify each of the following distributive laws.

   a) A ∩ (B ∪ C)   b) A ∪ (B ∩ C)

   c) (A ∩ B) ∪ (A ∩ B)   d) (A ∪ B) ∩ (A ∪ C)

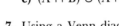

Figure 5.8

7. Using a Venn diagram such as the one in Fig. 5.8, shade the set represented by each of the following set expressions. Use a different Venn diagram for each expression.

   a) A ∩ B ∩ C   b) A ∩ (B' ∪ C)

   c) A ∩ B' ∩ C'   d) B' ∩ (A ∪ C)

   e) C' ∪ (A ∩ B)   f) B' ∪ A ∪ C

8. Write the set expression for each of the given Venn diagrams.

   a) 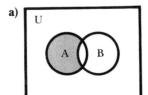   b)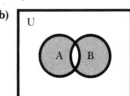

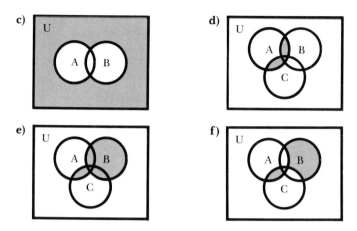

5.4

Sets and Flowcharts

As we learned in Chapter 3, an algorithm is an orderly list of the step-by-step process required to solve a problem. These steps outline the solution process from the statement of the problem through the analysis of the problem to the final output. We also saw that a flowchart is an alternative method to just listing the steps in the algorithm. The flowchart can actually be a picture of the algorithm itself.

We will now see how we can use flowcharts to show set relationships that lead to some type of result or conclusion. In other words, the intersection of two sets may form a list of those elements that actually do belong to both sets.

We make no attempt here to teach formal flowcharting. Rather, we use flowcharts as another means of visualizing set relationships just as we used Venn diagrams in Section 5.3. This time, however, we will emphasize more the final outcome or output and what it really means. In doing so we will assume that each flowchart must have a "start" and an "end." Therefore, we will not bother to include the symbols for these steps. We will use instead only the input/output and decision symbols, as shown in Fig. 5.9. Although the flowcharts will not be technically complete, the critical portions showing the set relationships will be accurate.

We are interested in whether or not an element belongs to a set (yes or no), so we will use only two of the three outputs of the decision symbol. We will also follow the convention that main flow through the flowchart is from top to bottom, with secondary flow from left to right and tertiary flow returning to the original input. The input will be to read data from a record or file. The decision will be whether the input

(a)

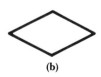

(b)

Figure 5.9
(a) Input/output;
(b) decision.

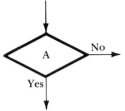

Figure 5.10

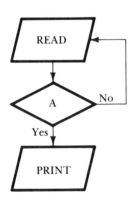

Figure 5.11

should be kept or discarded, and the output will be a printed list of those elements satisfying the conditions of the set relationships.

Since the main flow is downward and we are interested in whether an element belongs to a set, the downward flow out of the decision symbol is YES and the secondary flow to the right NO. Thus the question of whether the element belongs to A will appear as shown in Fig. 5.10.

If we place the input/output symbols together with the decision symbol, read a file, and print only those elements that belong to A, we will have the flowchart shown in Fig. 5.11.

Since we must frequently deal with complements and are interested in the element that does not belong to A, we will flowchart complements as shown in Fig. 5.12, which still allows the main flow to be downward.

The flowchart in Fig. 5.13 would provide a list of those elements that do not belong to A, that is, A'.

We are now ready to consider the flowcharts of two or more sets combined by union and/or intersection. Each set requires a decision symbol. Both outputs from all decision symbols must lead somewhere. No "dangling" or unconnected outputs are acceptable.

Recall from Section 5.2 that in order for an element to get into the set formed by the *union*, it must belong to one set *or* the other. Therefore, if we check set A first and we find that the element belongs (that is, the answer is yes), there is no need to check to see whether the element is in set B. The element is simply printed. Set B need be checked only if the answer to A is no. Figure 5.14 shows the flowchart of the union operation. Note that an element can still be on the printed list even if it does not belong to set A, but does belong to set B.

Also recall from Section 5.2 that in order for an element to belong to the set formed by the *intersection*, it must belong to *both* sets A *and* B. Therefore, even if we know that the element belongs to set A (the

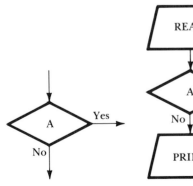

Figure 5.12

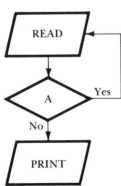

Figure 5.13

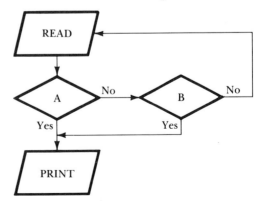

Figure 5.14

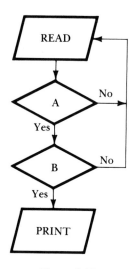

Figure 5.15

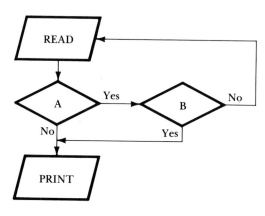

Figure 5.16

answer is yes), we must still check set B; if the answer to B is yes, then the element is printed. If the answer is no to either, the element is not printed. If the answer is no to the first set checked, there is no need to check the other set and the element is rejected. This is illustrated in Fig. 5.15.

Note that in both A ∪ B and A ∩ B, the commutative property holds and we get exactly the same list using B ∪ A as we do when we use A ∪ B, and the same list for B ∩ A as we do for A ∩ B.

■ **EXAMPLE 20**   Figure 5.16 shows the flowchart for A′ ∪ B. Using this, we get a list of elements that are not in A together with those that are in B.   □

■ **EXAMPLE 21**   The following figure is a flowchart of A ∩ B′. Here we get only those elements that belong to A and do not belong to B.

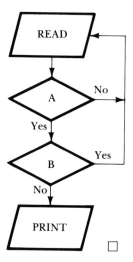

□

Now let us consider some expressions in which three sets are involved. If both unions and intersections are involved, one method is to draw a flowchart according to the hierarchy of set operations. Another method is to attempt to draw a flowchart that will provide us with the list as quickly as possible.

■ **EXAMPLE 22**   Draw a flowchart of A ∪ B ∪ C.

Solution

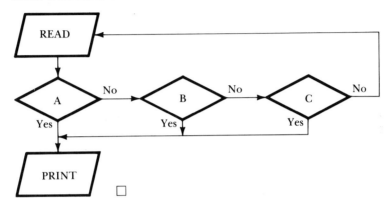

■ **EXAMPLE 23**   Draw a flowchart of A ∩ B ∩ C.

Solution

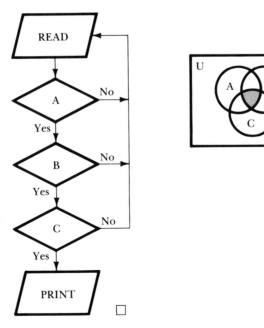

■ **EXAMPLE 24**   Draw a flowchart of A ∪ (B ∩ C).

Solution

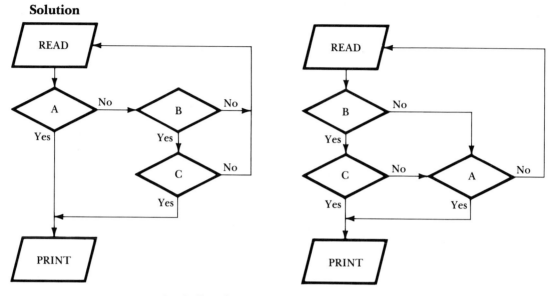

Both flowcharts above will provide the same printed list. However, it is reasonable to expect that the flowchart on the left might provide the list quicker since some elements (those belonging to A) will be printed after only one check. The flowchart on the right requires at least two checks before the first element is printed.   □

■ **EXAMPLE 25**   Draw a flowchart of A ∩ (B ∪ C).

Solution

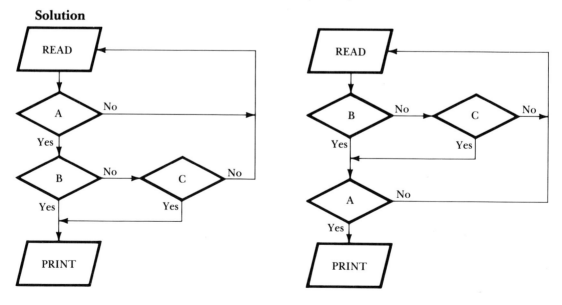

Again in this example, the two flowcharts above would give us the same list. However, without knowing any additional information, it would be difficult to decide whether one is better than the other in this case. A minimum of two checks is required in each before the first element can be printed. □

When the right kind of additional information is known, it is possible to determine when one flowchart is better than another even though both produce the same desired list. In fact, generalizations can be made to cover any situation in which the proper information is available. Consider the following problem.

A list is desired of those employees in the Nationwide Insurance Company home office who are female and who are two-year college graduates. If there are 2000 employees of Nationwide in Columbus, 1200 of them female, and 300 employees who are two-year college graduates, 175 of whom are female, we have the universal set U with 2000 members and two intersecting subsets, as shown in Fig. 5.17. That is,

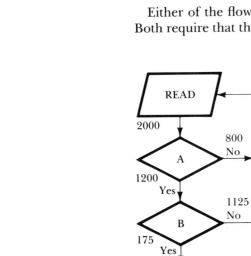

Figure 5.17

> U: 2000 employees,
> A: 1200 female employees,
> B: 300 two-year graduates,
> A ∩ B: 175 female two-year graduates.

Either of the flowcharts in Fig. 5.18 would lead to the correct list. Both require that the entire universal set be checked. There is no way

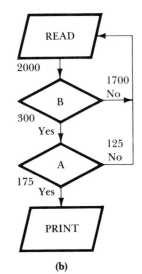

(a)   (b)

Figure 5.18

around this. Every name in the master file must be read. However, the flowchart in Fig. 5.18(a) requires a total of 3200 (2000 + 1200) checks to produce the list of 175 names, whereas the flowchart in Fig. 5.18(b) requires only 2300 (2000 + 300) checks in order to produce the same list.

If we follow the flow in Fig. 5.18(a), we see that after a name from the master file is read, it is checked to see if it belongs to set A. If the name is in set A, it is checked again to see if it belongs to set B. A total of 1200 (the female employees at Nationwide's home office) belong to set A and are therefore checked to see if they belong to set B.

When we follow the flow in Fig. 5.18(b), we see that the first question asked is, "Is the name in set B?" Only 300 fall into this category and must be checked again. If the smaller set is checked first, the greater number of rejections occur immediately, which results in fewer names requiring further checks. Thus the flowchart in Fig. 5.18(b) requires fewer total checks and, therefore, less time to produce the required list. Saving time on the computer saves money.

● Question

After having examined this problem, can you conclude generally which of two sets should be examined first when flowcharting their intersection, if you know the cardinal number of each?

Before we answer this question, let us look at the next problem.

Consider again the same group of employees. Let's say that this time we want a list of those employees who are female or two-year college graduates. Recall that

$$
\begin{array}{rl}
\text{U:} & 2000 \text{ elements,} \\
\text{A:} & 1200 \text{ elements,} \\
\text{B:} & 300 \text{ elements,} \\
\text{A} \cap \text{B:} & 175 \text{ elements.}
\end{array}
$$

This time we are interested in A ∪ B. The Venn diagram is shown in Fig. 5.19, and the flowcharts in Fig. 5.20. Again, we see that both flowcharts will produce the same list. How many names does the list contain?

In this case, if a name belongs to the first set checked, that name is

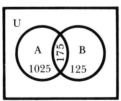

Figure 5.19

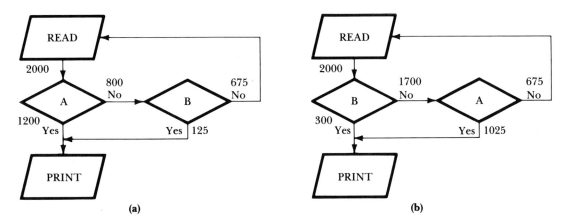

Figure 5.20

printed without any further checks being required. If the answer is no, the name is checked to see if it belongs to the second set. The flowchart in Fig. 5.20(a) requires 2800 (2000 + 800) checks, whereas the flowchart in Fig. 5.20(b) requires 3700 (2000 + 1700) checks. Which flowchart saves computer time and money?

The answer to the question raised above is given in the following conclusion.

■ CONCLUSION

1. When constructing a flowchart for intersection, begin with the smallest set and progress to the largest.
2. When constructing a flowchart for union, begin with the largest set and progress to the smallest.

Keep in mind that in order for these rules to apply, the cardinal number of the sets must be known. Often this information cannot be easily determined. If this is the case, the programmer must use his or her best judgment.

Exercise Set 5.4

1. Draw a flowchart for each of the following set expressions. The Venn diagrams were drawn in Exercise Set 5.3. In each case, consider a list to be printed.

   a) $A' \cup B$ **b)** $A \cap B'$

   c) $A' \cup B'$ **d)** $A' \cap B'$

e) A ∩ (B′ ∪ C)   **f)** C′ ∪ (A ∩ B)

g) A ∩ B′ ∩ C′   **h)** B′ ∩ (A ∪ C)

2. Draw a flowchart for each of the given Venn diagrams. You may want to write the set expression first. The printed list is to contain only those elements in the shaded region.

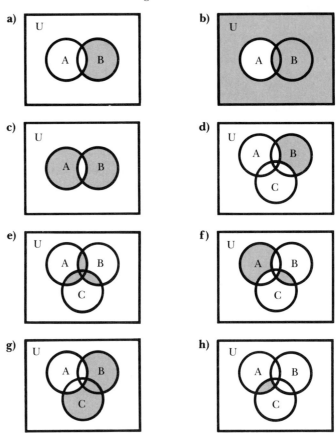

3. Using Fig. 5.21, draw the appropriate flowcharts and answer the following questions.

a) How many checks would be made in A ∪ B? in B ∪ A?

b) How large is the printed list in each case?

c) Which flowchart should be used?

d) How many checks would be made in A ∩ B? in B ∩ A?

e) How large is the printed list in each case?

f) Which flowchart should be used?

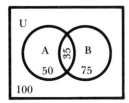

Figure 5.21

4. Repeat Exercise 3 for each of the following.

    a) A′ ∪ B and B ∪ A′

    b) A ∩ B′ and B′ ∩ A

5. Using Fig. 5.22, draw the appropriate flowcharts. Determine the number of checks required and the size of the printed list.

    a) (A ∩ B) ∩ C   **b)** (A ∪ B) ∪ C

    c) (A ∩ B) ∪ C   **d)** (A ∪ B) ∩ C

    e) A′ ∪ (B ∩ C)   **f)** A′ ∩ (B ∪ C)

    g) B′ ∪ (A ∪ C)   **h)** B′ ∩ (A ∩ C)

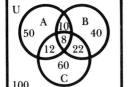

Figure 5.22

5.5
Sets and Boolean Properties

Although we will be studying Boolean algebra, an area of mathematics very important to the study of computers, in detail in Chapter 9, it is appropriate to point out here that a very close relationship exists between many properties from real-number algebra, the algebra of sets, and Boolen algebra. We will encounter many of these same properties when we study symbolic logic in Chapter 8.

To illustrate, consider a property that you first encountered in elementary school, when you added two numbers and then checked the sum by performing the addition in reverse order.

■ **EXAMPLE 26**   The order in which two numbers are added does not matter:

$$6 + 7 = 7 + 6. \quad \square$$

This property is called the *commutative property of addition* and is stated for the general case in algebra by writing $a + b = b + a$. The commutative property exists in the algebra of sets also, as we found when we examined the flowcharts of A ∪ B and B ∪ A. In other words, we saw that for any sets A and B, A ∪ B = B ∪ A and A ∩ B = B ∩ A.

The distributive properties and DeMorgan's rules (Section 5.3) are examples of rules that are not consistent with real-number algebra. The real numbers do obey the property $a * (b + c) = a * b + a * c$, called the *distributive law*, in which multiplication is distributed across addition.

■ **EXAMPLE 27** $2(3 + 5) = 2(3) + 2(5)$
$2(8) = 6 + 10$
$16 = 16$ $\square$

If we attempt to write a property in which addition is distributed across multiplication, we get a result that is not consistent with real-number algebra.

■ **EXAMPLE 28** $2 + (3 * 5) \neq (2 + 3) * (2 + 5)$
$2 + 15 \neq 5 * 7$
$17 \neq 35$ $\square$

In the algebra of sets, we do have both of the distributive properties.

■

> DISTRIBUTIVE PROPERTIES
> 1. $A \cup (B \cap C) = (A \cup B) \cap (A \cup C)$
> 2. $A \cap (B \cup C) = (A \cap B) \cup (A \cap C)$

To illustrate the first property, let $A = \{1, 2, 5\}$, $B = \{3, 4, 5\}$, and $C = \{1, 4, 7, 8\}$ in the following example.

■ **EXAMPLE 29** $A \cup (B \cap C) = (A \cup B) \cap (A \cup C)$
$A \cup \{4\} = \{1, 2, 3, 4, 5\} \cap \{1, 2, 4, 5, 7, 8\}$
$\{1, 2, 4, 5\} = \{1, 2, 4, 5\}$ $\square$

Two other properties inconsistent with real-number algebra are DeMorgan's rules.

■

> DEMORGAN'S PROPERTIES
> 1. $(A \cup B)' = A' \cap B'$
> 2. $(A \cap B)' = A' \cup B'$

Exercise Set 5.5

1. Given that $A = \{a, b, d\}$, $B = \{c, d, f, g\}$, $C = \{b, c\}$, and $U = \{a, b, c, d, e, f, g\}$, verify each of the following properties.

a) $A \cup (B \cap C) = (A \cup B) \cap (A \cup C)$ **Distributive property**

b) $A \cap (B \cup C) = (A \cap B) \cup (A \cap C)$ **Distributive property**

c) $(A \cup B) \cup C = A \cup (B \cup C)$ **Associative property**

d) $(A \cap B) \cap C = A \cap (B \cap C)$ **Associative property**

e) $(A \cup B)' = A' \cap B'$ **DeMorgan's rule**

f) $(A \cap B)' = A' \cup B'$ **DeMorgan's rule**

2. Given that $A = \{1, 2, 3, 4\}$, $B = \{1, 3, 4\}$, $C = \{2, 4\}$, and $U = \{1, 2, 3, 4, 5\}$, verify each of the properties in Exercise 1.

3. Given that $A = \{a, c, d, e\}$, $B = \{a, b\}$, $C = \{b, c, f\}$, and $U = \{a, b, c, d, e, f\}$, verify each of the properties in Exercise 1.

4. Given that $A = \{1, 3, 4, 5\}$, $B = \{3, 4\}$, $C = \{2, 5\}$, and $U = \{1, 2, 3, 4, 5, 6\}$, verify each of the properties in Exercise 1.

Summary

In this chapter we have learned various concepts about sets, including the definition of a *well-defined* set. We also learned to distinguish between *finite* and *infinite* sets and were able to determine set membership. We found that the *cardinal number* of a set is the number of elements in the set. We discussed subsets, especially in terms of a *universal* set. We also found out that some sets are empty, and these are called *null* sets.

After learning the terminology associated with sets, we were introduced to operations that can be performed on sets. We discovered that the *union* of two sets is also a set that contains all of the elements from the two sets. If an element is in both sets initially, it is listed in the union only once. We talked about the *intersection* of two sets. This intersection is also a set, but it contains only those elements common to both sets. Whereas the union and intersection operations are performed on two sets at a time, which makes them binary operations, the *complement* is formed by looking at one set, but in relation to the known *universal* set. The complement of a set is a set containing all elements from the universal set that are not members of the original set.

After learning how to form the union, intersection, and complement of sets, we were introduced to a method of picturing sets and set operations known as *Venn diagrams*. Another method for picturing set relationships, especially when the end result is a printed list of all of those elements that satisfy the stated conditions, is a *flowchart*. A flowchart allows us to visualize the "flow" in deriving a desired list of elements.

Finally, we were introduced to several *Boolean properties*. These are basically properties of the algebra of sets and many of them are not consistent with real-number algebra. We will encounter Boolean properties again in Chapters 8 and 9.

Review Exercises

1. List the elements for each of the following sets.

   a) The set of integers between 4 and 15

   b) The set of integers greater than 10

   c) The set of natural numbers less than 10

   d) The set of integers less than 0

2. Which of the sets described in Exercise 1 are finite? Which are infinite? Give the cardinal number for those that are finite.

3. Given that $U = \{1, 2, 3, 4, 5, 6, 7, 8, 9, 10\}$, $A = \{1, 2, 4, 6, 8\}$, $B = \{3, 6, 7, 8, 9, 10\}$, $C = \{1, 3, 5, 7, 9\}$, $D = \{2, 4, 6, 8, 10\}$, and $E = \{3, 7\}$, find each of the following.

   a) $A \cap B$ **b)** $A \cup C$ **c)** $A' \cap B$

   d) $A \cup D'$ **e)** $A \cap B' \cup C$ **f)** $B' \cap (C \cup B)$

   g) $(A \cup E)' \cap D$ **h)** $(D \cap C) \cap E$

4. For each of the following, draw a Venn diagram and shade the given set.

   a) $(A \cup B)'$ **b)** $(A \cap B)'$ **c)** $(A \cup B)' \cap C$

   d) $(A' \cap C) \cup B'$ **e)** $A' \cap B \cap C$ **f)** $A \cup B' \cup C$

5. At a Fourth of July picnic attended by 100 children, the following observations were made: 50 children participated in the pizza-eating contest, 65 participated in the three-legged race, and 70 participated in the egg toss. Of these, 25 participated in both the egg toss and pizza-eating contest, 20 participated in both the pizza-eating contest and the three-legged race, and 40 participated in both the egg toss and the three-legged race. Five children did not participate in any event and ten participated in all three events.

   a) How many children participated in at least one event?

   b) How many children participated in only one event?

   c) How many children participated in exactly two events?

   (*Hint:* Draw a Venn diagram.)

6. Given the following information:

   $A = \{50$ students enrolled in English$\}$,
   $B = \{60$ students enrolled in mathematics$\}$,
   $C = \{40$ students enrolled in principles of data processing$\}$,
   5 students are not enrolled in any of the three courses,
   $A \cap B = \{15$ students enrolled in English and mathematics$\}$,
   $B \cap C = \{20$ enrolled in mathematics and data processing$\}$,
   $A \cap C = \{10$ enrolled in English and data processing$\}$,
   $A \cap B \cap C = \{5$ enrolled in all three courses$\}$,

draw the flowchart for each of the following.

a) $A \cup B$ **b)** $A \cap C$ **c)** $A' \cap B$

d) $B \cup C'$ **e)** $(A \cap B)' \cap C$ **f)** $(A \cup B)' \cap C$

g) $A \cup B \cup C$ **h)** $A \cap B \cap C$

7. Determine the number of checks required to come up with the desired list for each of the flowcharts in Exercise 6. How large is the printed list in each flowchart?

Intermediate Algebra

After completing this chapter, you should be able to:

1. Convert between fractional-exponent form and radical form.
2. Perform simplifications on radical expressions.
3. Solve a quadratic equation either by factoring, by completing the square, or by using the quadratic formula.
4. Solve an equation containing radical terms.
5. Solve quadratic inequalities that are factorable.

6

■ **CHAPTER OUTLINE**

6.1
Fractional Exponents and Radicals

In Chapter 0, we reviewed expressions of the form $2^3 = 8$. At that time, we called 8 the third power of 2. By definition, 2 is said to be the third *root* of 8. When talking about roots instead of powers, we use the notation $\sqrt[n]{b} = a$, where $\sqrt{}$ is called the radical symbol, n is the index number, and b is the radicand.

■ **DEFINITION**

If $a^n = b$, then a is called the *n*th *root* of b.

From $2^3 = 8$, we see the relationship $\sqrt[3]{8} = 2$. The expression $2^3 = 8$ means that 2 is one of the three equal factors of 8. The radical symbol and index number can be thought of as asking the question, "What is one of the three equal factors of 8?" In general, if $\sqrt[n]{b} = a$, then a is one of the n equal factors of b.

■ **EXAMPLE 1**

a) $\sqrt{49} = 7$ since $7^2 = 49$. If the index number is not written, it is understood to be 2. Also, $(-7)^2 = 49$. Seven is called the *principal square root* of 49. The principal square root is implied anytime we write $\sqrt{a}$. If we are looking for the negative square root, then it must be written $-\sqrt{a}$ (for example, $-\sqrt{49} = -7$).

b) $\sqrt[3]{125} = 5$ since $5^3 = 125$

c) $\sqrt[4]{16} = 2$ since $2^4 = 16$

d) $\sqrt[3]{-8} = -2$ since $(-2)^3 = -8$

e) $\sqrt{-9}$ is not real since the square of any real number is positive, that is, $\sqrt{-9} \neq -3$ since $(-3)^2 = 9$.   □

■ **DEFINITION**

1. For b positive and n even or odd, $\sqrt[n]{b}$ is positive.
2. For b negative and n odd, $\sqrt[n]{b}$ is negative.
3. For b negative and n even, $\sqrt[n]{b}$ does not exist in the real-number system.
4. $\sqrt[n]{0} = 0$

Recall the laws and definitions of integer exponents that were reviewed in Chapter 0.

■

LAWS	DEFINITIONS
1. $a^m \cdot a^n = a^{m+n}$	1. $a^0 = 1, \quad a \neq 0$
2. $\dfrac{a^m}{a^n} = a^{m-n}, \quad a \neq 0$	2. $a^{-n} = \dfrac{1}{a^n}, \quad a \neq 0$
3. $(a^m)^n = a^{mn}$	
4. $(ab)^n = a^n b^n$	
5. $\left(\dfrac{a}{b}\right)^n = \dfrac{a^n}{b^n}, \quad b \neq 0$	

The definition of a^n as the multiplication of n factors of a (that is, $a \cdot a \cdot a \cdot \cdots \cdot a$ for n factors of a) implies only positive-integer exponents. The laws listed above are actually valid for any real-number exponents, and in practice, we frequently encounter fractional exponents. The only problem is to establish a meaning for a fractional exponent. Consider the following argument.

In $9^2 = 81$, we know that 9 is one of the two equal factors of 81. In $2^3 = 8$, 2 is one of three equal factors of 8. Applying the same meaning to $5^{1/2}$, however, does not make any sense. Although 9^2 means 9 taken as a factor two times, $5^{1/2}$ does not mean 5 taken as a factor one-half time.

On the other hand, if we use the first law listed above, we have that $5^{1/2} \cdot 5^{1/2} = 5^{1/2+1/2} = 5^1 = 5$. Since $5^{1/2} \cdot 5^{1/2} = 5$, we can say that $5^{1/2}$ is one of the two equal factors of 5.

We will also see that $\sqrt{5} \cdot \sqrt{5} = 5$, which makes $\sqrt{5}$ one of the two equal factors of 5. Thus we can say that since $5^{1/2} \cdot 5^{1/2} = 5$ and $\sqrt{5} \cdot \sqrt{5} = 5$, $5^{1/2}$ can be defined as $\sqrt{5}$.

■

PROPERTIES OF RADICALS

In general, if a is a real number, then:

1. $a^{1/n} = \sqrt[n]{a}, \quad n \neq 0$;
2. $a^{m/n} = \sqrt[n]{a^m}, \quad n \neq 0$.

If $\sqrt[n]{a^m}$ is a real number, then $\sqrt[n]{a^m} = (\sqrt[n]{a})^m$.

■ EXAMPLE 2

a) $8^{1/3} = \sqrt[3]{8} = 2$
b) $9^{1/2} = \sqrt{9} = 3$
c) $8^{2/3} = \sqrt[3]{8^2} = (\sqrt[3]{8})^2 = (2)^2 = 4$ □

■ EXAMPLE 3

a) $x^{1/3} \cdot x^{2/3} = x^{1/3+2/3} = x^{3/3} = x$
b) $x^{1/2} \cdot x^{1/3} = x^{1/2+1/3} = x^{3/6+2/6} = x^{5/6} = \sqrt[6]{x^5}$
c) $(x^{2/3})^3 = x^2$
d) $\dfrac{(2x)^{1/2}}{(2x)^{1/3}} = (2x)^{1/2-1/3} = (2x)^{3/6-2/6} = (2x)^{1/6} = \sqrt[6]{2x}$ □

Exercise Set 6.1

Find the value of each of the following.

1. $\sqrt{25}$
2. $\sqrt{36}$
3. $\sqrt[3]{27}$
4. $\sqrt[3]{64}$
5. $\sqrt[3]{-27}$
6. $\sqrt[5]{-32}$
7. $\sqrt[4]{81}$
8. $\sqrt[4]{625}$
9. $49^{1/2}$
10. $81^{1/2}$
11. $25^{-1/2}$
12. $8^{-1/3}$
13. $(-8)^{5/3}$
14. $-\left(\dfrac{1}{16}\right)^{-1/2}$
15. $\left(\dfrac{1}{8}\right)^{1/3}$
16. $\left(\dfrac{1}{4}\right)^{5/2}$
17. $(-64)^{-2/3}$
18. $\left(\dfrac{1}{16}\right)^{-3/4}$
19. $(-125)^{-2/3}$
20. $\left(\dfrac{9}{16}\right)^{-1/2}$

Apply the laws of exponents to simplify each of the following.

21. $x^{1/4} \cdot x^{1/3}$
22. $2x^{2/3} \cdot 2x^{1/2}$

23. $(x^{4/3})^{3/2}$

24. $(3x^{3/5})^{2/3}$

25. $\dfrac{x^{3/2}}{x^{1/3}}$

26. $\dfrac{(3x)^{3/4}}{(3x)^{1/3}}$

27. $(x^{-3/5})^{-1/3}$

28. $[(2x)^{-2/3}]^{3/2}$

29. $(x^{4/5}y^{1/3})^{3/4}$

30. $(2x^{2/3}y^{1/4})^{3/2}$

6.2

Operations on Radical Expressions

In algebra, we are frequently required to simplify expressions that contain radicals. Several properties exist that can help us with these simplifications. The properties are derived from the fractional-exponent forms studied in Section 6.1.

■

PROPERTIES OF RADICALS

If a is a positive real number, then:

1. $\sqrt[n]{a^n} = a$, since $\sqrt[n]{a^n} = a^{n/n} = a^1 = a$;
2. $\sqrt[n]{a} \cdot \sqrt[n]{b} = \sqrt[n]{ab}$, since $\sqrt[n]{a} \cdot \sqrt[n]{b} = a^{1/n} \cdot b^{1/n} = (ab)^{1/n}$;
3. $\dfrac{\sqrt[n]{a}}{\sqrt[n]{b}} = \sqrt[n]{\dfrac{a}{b}}$, since $\dfrac{\sqrt[n]{a}}{\sqrt[n]{b}} = \dfrac{a^{1/n}}{b^{1/n}} = \left(\dfrac{a}{b}\right)^{1/n}$, $b \neq 0$.

■ **EXAMPLE 4**

a) $\sqrt[3]{8} = \sqrt[3]{2^3} = 2$

b) $\sqrt[3]{3} \cdot \sqrt[3]{2} = \sqrt[3]{3 \cdot 2} = \sqrt[3]{6}$

c) $\dfrac{\sqrt[3]{24}}{\sqrt[3]{3}} = \sqrt[3]{\dfrac{24}{3}} = \sqrt[3]{8} = 2$ □

As is true with any equality (if $a = b$, then $b = a$), the above properties can be applied from left to right or from right to left.

■ **EXAMPLE 5**

We can use the second property listed above to simplify radicals that are not perfect powers, but contain factors that are perfect powers. For example, let us simplify $\sqrt{32}$.

Solution

Although 32 is not a perfect square, $32 = 16 \times 2$ and 16 is a perfect square. Therefore,

$$\sqrt{32} = \sqrt{16 \cdot 2} = \sqrt{16} \cdot \sqrt{2}$$
$$= 4\sqrt{2}.$$

Notice too that $32 = 4 \times 8$. Therefore,

$$\sqrt{32} = \sqrt{4 \cdot 8}$$
$$= \sqrt{4} \cdot \sqrt{4 \cdot 2}$$
$$= \sqrt{4} \cdot \sqrt{4} \cdot \sqrt{2}$$
$$= 2 \cdot 2 \cdot \sqrt{2}$$
$$= 4\sqrt{2}. \quad \square$$

■ **EXAMPLE 6**   Simplify each of the following.

a) $\sqrt[3]{54} = \sqrt[3]{27 \cdot 2} = \sqrt[3]{27} \cdot \sqrt[3]{2} = 3\sqrt[3]{2}$

b) $\sqrt[4]{48} = \sqrt[4]{16 \cdot 3} = \sqrt[4]{16} \cdot \sqrt[4]{3} = 2\sqrt[4]{3}$

c) $\sqrt{xy^3z^2} = \sqrt{x \cdot y^2 \cdot y \cdot z^2} = yz\sqrt{xy}$

d) $\sqrt[3]{24x^5y^3z^4} = \sqrt[3]{8 \cdot 3 \cdot x^3 \cdot x^2 \cdot y^3 \cdot z^3 \cdot z}$
$$= 2xyz\sqrt[3]{3x^2z} \quad \square$$

■ **EXAMPLE 7**

a) $\sqrt{\dfrac{10}{9}} = \dfrac{\sqrt{10}}{\sqrt{9}} = \dfrac{\sqrt{10}}{3}$

b) $\sqrt[3]{\dfrac{8}{27}} = \dfrac{\sqrt[3]{8}}{\sqrt[3]{27}} = \dfrac{2}{3}$

c) $\sqrt[3]{\dfrac{7}{8}} = \dfrac{\sqrt[3]{7}}{\sqrt[3]{8}} = \dfrac{\sqrt[3]{7}}{2} \quad \square$

Other uses of the properties of radicals are illustrated below.

■ **EXAMPLE 8**

a) $\sqrt{3x} \cdot \sqrt{2xy} = \sqrt{6x^2y} = x\sqrt{6y}$

b) $\sqrt[3]{12} \cdot \sqrt[3]{4} = \sqrt[3]{48} = \sqrt[3]{8 \cdot 6} = 2\sqrt[3]{6}$

c) $\dfrac{\sqrt{14}}{\sqrt{2}} = \sqrt{\dfrac{14}{2}} = \sqrt{7}$

d) $\dfrac{\sqrt[3]{2w^5}}{\sqrt[3]{2w^2}} = \sqrt[3]{\dfrac{2w^5}{2w^2}} = \sqrt[3]{w^3} = w \quad \square$

Exercise Set 6.2

Simplify each of the following radicals.

1. $\sqrt{9x^2}$

2. $\sqrt{25x^2y^2}$

3. $-\sqrt{(3x)^2}$

4. $-\sqrt{(-5x)^2}$

5. $\sqrt[3]{40}$

6. $\sqrt[3]{250}$

7. $\sqrt[3]{24x^4y^5}$

8. $\sqrt[3]{108x^5y^3}$

9. $\sqrt{\dfrac{48x}{3x}}$

10. $\sqrt[3]{\dfrac{3}{x^2}}$ **11.** $\sqrt[3]{\dfrac{5x^2}{8x^3}}$ **12.** $\sqrt[3]{\dfrac{4}{27}}$

Perform the indicated operations and simplify if possible.

13. $\sqrt{5}\cdot\sqrt{5}$ **14.** $\sqrt{3}\cdot\sqrt{6}$ **15.** $\sqrt{5}\cdot\sqrt{10}$

16. $\sqrt[3]{4}\cdot\sqrt[3]{12}$ **17.** $\sqrt[3]{5x}\cdot\sqrt[3]{25x^2y}$ **18.** $\sqrt[3]{2x^2y^2}\cdot\sqrt[3]{4xy^4}$

19. $\dfrac{\sqrt{xy^3}}{\sqrt{xy}}$ **20.** $\dfrac{\sqrt[3]{54x^5y}}{\sqrt[3]{3xy^2}}$ **21.** $\sqrt{5xy}\cdot\sqrt{20x^3y}$

22. $\sqrt[3]{3xy^2}\cdot\sqrt[3]{9x^3y}$ **23.** $(3\sqrt{2})^2$ **24.** $(2\sqrt{3x})^2$

6.3
Quadratic Equations

In Chapter 0, solving linear equations was reviewed. As you recall, a linear equation is any equation that can be put in the form $ax + b = 0$, where $a \neq 0$.

In this section we will discuss methods of solving another form of equation. Any equation that can be put in the form $ax^2 + bx + c = 0$, where $a \neq 0$, is called a **quadratic equation**. In Chapter 0, we discussed factoring quadratic trinomials. If a quadratic trinomial is set equal to zero, we have a quadratic equation.

Consider the quadratic trinomial $x^2 + 5x + 6$. Its factors are $(x + 2)$ and $(x + 3)$. Setting the quadratic equal to zero and then its factorization equal to zero, we have

$$x^2 + 5x + 6 = 0$$
$$(x + 2)(x + 3) = 0.$$

Recall from Chapter 0 that if the product of two real numbers is zero, then one or the other or both of the numbers is zero. That is,

$$\text{if}\quad a\cdot b = 0,$$
$$\text{then}\quad a = 0,$$
$$\text{or}\quad b = 0,$$
$$\text{or}\quad a = 0 \text{ and } b = 0.$$

Applying this principle to our problem will provide a method of solution for a quadratic equation. We set each factor equal to zero and solve the resulting two linear equations:

$$x^2 + 5x + 6 = 0$$
$$(x + 2)(x + 3) = 0$$
$$x + 2 = 0 \quad\text{or}\quad x + 3 = 0$$
$$x = -2 \quad\text{or}\quad x = -3.$$

Remember, if -2 and -3 are the solutions, then each must make the original equation true.

Check: If $x = -2$,

$$x^2 + 5x + 6 = 0$$
$$(-2)^2 + 5(-2) + 6 = 0$$
$$4 - 10 + 6 = 0$$
$$0 = 0.$$

If $x = -3$,

$$x^2 + 5x + 6 = 0$$
$$(-3)^2 + 5(-3) + 6 = 0$$
$$9 - 15 + 6 = 0$$
$$0 = 0.$$

■ **EXAMPLE 9**   Solve $x^2 - x - 12 = 0$ by factoring.

Solution
$$x^2 - x - 12 = 0$$
$$(x + 3)(x - 4) = 0$$
$$x + 3 = 0 \qquad x - 4 = 0$$
$$x = -3 \qquad x = 4 \quad \square$$

In order to solve a quadratic equation by factoring, we must be sure that the equation is written in the form $ax^2 + bx + c = 0$.

■ **EXAMPLE 10**   Solve $2x^2 + 12 = x^2 + 7x$.

Solution   After transposing all terms to the left side, we can solve as follows:

$$2x^2 - x^2 - 7x + 12 = 0$$
$$x^2 - 7x + 12 = 0$$
$$(x - 3)(x - 4) = 0$$
$$x - 3 = 0 \quad x - 4 = 0$$
$$x = 3 \qquad x = 4. \quad \square$$

Completing the Square

If all quadratics were able to be factored, quadratic equations would be easy to study. Unfortunately, however, that is not the case. What we need then is a method for solving any quadratic equation whether it can be factored or not. This method is called *completing the square*. It utilizes the fact that *the square of a binomial is equal to a perfect-square trinomial.* Recall that

$$(x + 3)^2 = x^2 + 6x + 9,$$
$$(x - 5)^2 = x^2 - 10x + 25.$$

We can see that half of the coefficient of the middle term, squared, equals the constant term of the perfect-square trinomial. Note that

$$\left(\frac{1}{2}\right)(6) = 3, \quad \text{and } 3^2 = 9;$$

$$\left(\frac{1}{2}\right)(-10) = -5, \quad \text{and } (-5)^2 = 25;$$

$$\left(\frac{1}{2} \cdot \text{the coefficient of } x\right)^2 = \text{the constant.}$$

The following example illustrates how the method of completing the square is used.

■ **EXAMPLE 11**

Given the quadratic $x^2 - 6x - 5 = 0$, solve by completing the square. (Note that it is not factorable.)

Solution

Step 1: Divide the equation by the coefficient of x, in this case 1.

Step 2: Transpose the constant:

$$x^2 - 6x = 5.$$

Step 3: Find half the coefficient of x, square it, and add to both sides:

$$\frac{1}{2} \cdot (-6) = -3, \quad (-3)^2 = 9$$
$$x^2 - 6x + 9 = 14.$$

Step 4: We see that the left side is a PST, and we can factor it:

$$(x - 3)^2 = 14.$$

Step 5: Take $\pm$ the square root of both sides:

$$x - 3 = \pm\sqrt{14}.$$

Step 6: When we solve for x, we find that the solutions are $3 + \sqrt{14}$ and $3 - \sqrt{14}$.

Check: If $x = 3 + \sqrt{14}$,
$$x^2 - 6x - 5 = 0$$
$$(3 + \sqrt{14})^2 - 6(3 + \sqrt{14}) - 5 = 0$$
$$9 + 6\sqrt{14} + 14 - 18 - 6\sqrt{14} - 5 = 0$$
$$0 = 0.$$

$$\text{If } x = 3 - \sqrt{14},$$
$$x^2 - 6x - 5 = 0$$
$$(3 - \sqrt{14})^2 - 6(3 - \sqrt{14}) - 5 = 0$$
$$9 - 6\sqrt{14} + 14 - 18 + 6\sqrt{14} - 5 = 0$$
$$0 = 0. \quad \square$$

Any quadratic equation can be solved by completing the square. However, since completing the square is not as easy as factoring, you should use it only when you cannot factor.

The Quadratic Formula

If the process of completing the square is performed on the general form of the quadratic equation, $ax^2 + bx + c = 0$, an interesting result occurs.

Step 1: Since the coefficient of x is not 1, divide through by a:

$$x^2 + \frac{b}{a}x + \frac{c}{a} = \frac{0}{a}.$$

Step 2: Transpose the constant:

$$x^2 + \frac{b}{a}x = -\frac{c}{a}.$$

Step 3: Find

$$\left(\frac{1}{2} \cdot \frac{b}{a}\right)^2$$

and add to both sides. Then simplify the right side:

$$x^2 + \frac{b}{a}x + \frac{b^2}{4a^2} = \frac{b^2}{4a^2} - \frac{c}{a}.$$

Step 4: Factor the left side:

$$x^2 + \frac{b}{a}x + \frac{b^2}{4a^2} = \frac{b^2 - 4ac}{4a^2}$$

$$\left(x + \frac{b}{2a}\right)^2 = \frac{b^2 - 4ac}{4a^2}.$$

Step 5: Take $\pm$ the square root of both sides:

$$x + \frac{b}{2a} = \pm\sqrt{\frac{b^2 - 4ac}{4a^2}}.$$

Simplify the right side:

$$x + \frac{b}{2a} = \frac{\pm \sqrt{b^2 - 4ac}}{2a}.$$

Step 6: Solve for x:

$$x = -\frac{b}{2a} \pm \frac{\sqrt{b^2 - 4ac}}{2a}$$

$$x = \frac{-b \pm \sqrt{b^2 - 4ac}}{2a}.$$

The result in step 6 is called the **quadratic formula**. Although this formula can be used to solve *any* quadratic equation, it is particularly useful for solving those equations that cannot be factored.

■

THE QUADRATIC FORMULA

$$x = \frac{-b \pm \sqrt{b^2 - 4ac}}{2a}$$

■ **EXAMPLE 12**

Solution

Solve $x^2 - 6x - 5 = 0$ by the quadratic formula.

In our example, $a = 1$, $b = -6$, $c = -5$. We solve as follows:

$$x^2 - 6x - 5 = 0$$

$$x = \frac{-(-6) \pm \sqrt{(-6)^2 - 4(1)(-5)}}{2(1)}$$

$$x = \frac{6 \pm \sqrt{36 + 20}}{2}$$

$$x = \frac{6 \pm \sqrt{56}}{2}$$

$$= \frac{6 \pm \sqrt{4 \cdot 14}}{2} = \frac{6 \pm 2\sqrt{14}}{2}$$

$$x = 3 \pm \sqrt{14}.$$

We have already checked this result and know that $3 + \sqrt{14}$ and $3 - \sqrt{14}$ are the solutions to $x^2 - 6x - 5 = 0$. □

■ **EXAMPLE 13**

Solve $21x^2 - 47x + 20 = 0$ by the quadratic formula.

Solution

In $21x^2 - 47x + 20 = 0$, $a = 21$, $b = -47$, and $c = 20$. Thus we have

$$21x^2 - 47x + 20 = 0$$

$$x = \frac{-(-47) \pm \sqrt{(-47)^2 - 4(21)(20)}}{2(21)}$$

$$x = \frac{47 \pm \sqrt{2209 - 1680}}{42}$$

$$x = \frac{47 \pm \sqrt{529}}{42} = \frac{47 \pm 23}{42}$$

$$x = \frac{47 + 23}{42} \quad \text{and} \quad x = \frac{47 - 23}{42}$$

$$x = \frac{70}{42} \qquad\qquad x = \frac{24}{42}$$

$$x = \frac{5}{3} \qquad\qquad x = \frac{4}{7}.$$

Check:

If $x = \dfrac{5}{3}$,

$$21x^2 - 47x + 20 = 0$$

$$21\left(\frac{5}{3}\right)^2 - 47\left(\frac{5}{3}\right) + 20 = 0$$

$$21\left(\frac{25}{9}\right) - \frac{235}{3} + 20 = 0$$

$$\frac{175}{3} - \frac{235}{3} + \frac{60}{3} = 0$$

$$0 = 0.$$

If $x = \dfrac{4}{7}$,

$$21x^2 - 47x + 20 = 0$$

$$21\left(\frac{4}{7}\right)^2 - 47\left(\frac{4}{7}\right) + 20 = 0$$

$$21\left(\frac{16}{49}\right) - \frac{188}{7} + 20 = 0$$

$$\frac{48}{7} - \frac{188}{7} + \frac{140}{7} = 0$$

$$0 = 0.$$

Thus, $\frac{5}{3}$ and $\frac{4}{7}$ are the solutions to $21x^2 - 47x + 20 = 0$.   □

Some conclusions can be drawn about the solutions to a quadratic equation by evaluating the radicand of the radical portion of the quadratic formula. This radicand, $b^2 - 4ac$, is called the **discriminant** and its value provides a clue to the type of solutions the quadratic equation will have.

■

If $b^2 - 4ac > 0$, the two roots are real numbers and unequal.
If $b^2 - 4ac = 0$, the two roots are real numbers and equal.
If $b^2 - 4ac < 0$, the two roots are complex numbers and unequal.

Complex numbers are such numbers as $3 + \sqrt{-16}$, where the radicand is negative.

■ **EXAMPLE 14**   Describe the roots of $x^2 - 3x - 5 = 0$ without finding the roots.

Solution   In $x^2 - 3x - 5 = 0$, $a = 1$, $b = -3$, and $c = -5$. Evaluating the discriminant, we have:

$$b^2 - 4ac = (-3)^2 - 4(1)(-5)$$
$$= 9 + 20 = 29.$$

Since $29 > 0$, there are two unequal real-number roots.   □

Exercise Set 6.3

Solve each of the following by factoring.

1. $x^2 - 2x - 15 = 0$ **2.** $x^2 - 8x + 16 = 0$

3. $x^2 - 4x = 0$ **4.** $2x^2 + 6x = 0$

5. $x^2 - 49 = 0$ **6.** $2x^2 - 72 = 0$

7. $x^2 - 12 = -4x$ **8.** $x^2 + 9 = 8x - 11$

Solve each of the following by completing the square.

9. $x^2 + 2x - 3 = 0$ **10.** $x^2 + 7x - 8 = 0$

11. $x^2 + 4x + 5 = 0$ **12.** $x^2 + 10x = -7$

13. $x^2 - 5 = 3x$ **14.** $x^2 - x - 2 = 0$

15. $2x^2 - 2x + 6 = 3$ **16.** $2x^2 - 2x - 9 = -8$

Describe the nature of the roots of each of the following quadratic equations, using the discriminant.

17. $9x^2 - 12x + 4 = 0$ **18.** $x^2 + 3x + 4 = 0$

19. $x^2 - 2x + 4 = 0$ **20.** $3x^2 + 5x + 1 = 0$

21. $x^2 + 3x + 5 = 0$ **22.** $2x^2 - 4x + 1 = 0$

Solve each of the following using the quadratic formula.

23. $x^2 - x - 2 = 0$ **24.** $x^2 - 2x + 4 = 0$

25. $x^2 - 3x + 8 = 0$ **26.** $5x^2 - 8x - 3 = 0$

27. $3x^2 + 2x - 7 = 0$ **28.** $x^2 - 6x + 13 = 0$

29. $2x^2 + 3x - 6 = -4x - 2$ **30.** $x^2 - 9x - 1 = -6x + 7$

6.4
Radical Equations

In Chapter 0 we reviewed methods of solving linear equations and in Section 6.4, we saw how to solve quadratic equations. We will now take a look at still another type of equation.

Any equation that contains a variable in the radicand of a radical is called a **radical equation**. In order to solve this type of equation, we need the following property.

■ PROPERTY

If two numbers are equal, the squares of those two numbers are equal.

Finding the solution to a radical equation involves isolating the radical term, squaring both sides of the equation, and solving for the unknown, as shown in the following example.

■ **EXAMPLE 15**   Solve each of the following for x.

a) $\sqrt{x-2} = 3$ *Check:* $\sqrt{11-2} = 3$
   $(\sqrt{x-2})^2 = 3^2$ $\sqrt{9} = 3$
   $x - 2 = 9$ $3 = 3$
   $x = 11$

b) $\sqrt{x} - 3 = 5$ *Check:* $\sqrt{64} - 3 = 5$
   $\sqrt{x} = 5 + 3$ $8 - 3 = 5$
   $(\sqrt{x})^2 = (8)^2$ $5 = 5$
   $x = 64$

c) $\sqrt{x} + 5 = 0$ *Check:* $\sqrt{25} + 5 \neq 0$
   $\sqrt{x} = -5$ $5 + 5 \neq 0$
   $(\sqrt{x})^2 = (-5)^2$ $10 \neq 0$
   $x = 25$

Since $10 \neq 0$, 25 is not a solution to $\sqrt{x} + 5 = 0$. In fact, this equation does not have a solution.   □

As we can see from Example 15, it is very important to check your answer to a radical equation. The number 25, which appeared to be a solution to Example 15(c), is called an **extraneous root**, or false root, to the equation.

If the equation contains more than one radical term, the process of isolating the radical may have to be repeated.

■ **EXAMPLE 16**   Solve $\sqrt{2x - 3} = \sqrt{x + 4}$ for x.

Solution
$$\sqrt{2x - 3} = \sqrt{x + 4}$$
$$(\sqrt{2x - 3})^2 = (\sqrt{x + 4})^2$$
$$2x - 3 = x + 4$$
$$x = 7$$

Check:
$$\sqrt{2(7) - 3} = \sqrt{7 + 4}$$
$$\sqrt{14 - 3} = \sqrt{11}$$
$$\sqrt{11} = \sqrt{11}$$

In this example, both radicals were eliminated at one time.   □

■ **EXAMPLE 17**   Solve $\sqrt{x - 3} = 4 - \sqrt{x + 5}$ for x.

Solution
$$\sqrt{x - 3} = 4 - \sqrt{x + 5}$$
$$(\sqrt{x - 3})^2 = (4 - \sqrt{x + 5})^2$$ **Note that the expression on the right of the equals sign is a binomial.**
$$x - 3 = 16 - 8\sqrt{x + 5} + x + 5$$
$$x - 3 = 21 + x - 8\sqrt{x + 5}$$ **Isolate the remaining radical.**
$$-24 = -8\sqrt{x + 5}$$ **Divide both sides by -8.**
$$3 = \sqrt{x + 5}$$
$$3^2 = (\sqrt{x + 5})^2$$ *Check:* $\sqrt{4 - 3} = 4 - \sqrt{4 + 5}$
$$9 = x + 5$$ $1 = 4 - 3$
$$4 = x$$ $1 = 1$ □

■ **EXAMPLE 18**   Solve $\sqrt{4x + 1} = 3 + \sqrt{x - 2}$ for x.

Solution
$$\sqrt{4x + 1} = 3 + \sqrt{x - 2}$$
$$(\sqrt{4x + 1})^2 = (3 + \sqrt{x - 2})^2$$
$$4x + 1 = 9 + 6\sqrt{x - 2} + x - 2$$
$$4x + 1 = 7 + x + 6\sqrt{x - 2}$$
$$3x - 6 = 6\sqrt{x - 2}$$ **Divide both sides by 3.**
$$x - 2 = 2\sqrt{x - 2}$$
$$(x - 2)^2 = (2\sqrt{x - 2})^2$$
$$x^2 - 4x + 4 = 4(x - 2)$$
$$x^2 - 4x + 4 = 4x - 8$$
$$x^2 - 8x + 12 = 0$$ **Notice the quadratic equation.**
$$(x - 2)(x - 6) = 0$$
$$x = 2 \quad \text{or} \quad x = 6 \quad □$$

In this case, both 2 and 6 are solutions. It is necessary to check both numbers since one or possibly both could be extraneous.

Exercise Set 6.4

Solve each of the following. Be sure to check your answer.

1. $\sqrt{2x} = 8$

2. $\sqrt{5x} = 10$

3. $\sqrt{2x + 1} = 9$

4. $\sqrt{2x} - 4 = -2$

5. $\sqrt{5x} + 7 = 22$

6. $\sqrt{x + 1} - 5 = 8$

7. $\sqrt{x + 2} = -3$

8. $\sqrt{4x} + 3 = 1$

9. $\sqrt{2x - 1} = \sqrt{x + 4}$

10. $\sqrt{3x + 1} = \sqrt{2x + 6}$

11. $\sqrt{3x + 4} = \sqrt{5x - 6}$

12. $\sqrt{5x - 3} = \sqrt{2x + 3}$

13. $\sqrt{x - 3} + \sqrt{x} = 3$

14. $\sqrt{x - 7} + \sqrt{x} = 1$

15. $\sqrt{4x + 1} - \sqrt{x - 2} = 3$

16. $4 + \sqrt{12 - x} = 6 + \sqrt{4 - x}$

6.5

Quadratic Inequalities

In Section 6.3, to help solve a quadratic equation that is factorable, we used the property of real numbers that states, "If the product of two numbers is zero, then at least one of the numbers must be zero." At that time we found that if the product is zero, then one or the other of the numbers must be zero, or both numbers might be zero.

If the quadratic is expressed as an inequality, other properties must be used.

■

PROPERTIES OF INEQUALITIES

1. If $a \cdot b > 0$, then either $a > 0$ *and* $b > 0$ *or* $a < 0$ *and* $b < 0$.
2. If $a \cdot b < 0$, then either $a < 0$ *and* $b > 0$ *or* $a > 0$ *and* $b < 0$.

In general terms, we say that if the product of two numbers is positive, both numbers must be positive or both numbers must be negative. If the product of two numbers is negative, one number must be negative and one must be positive.

To solve a quadratic inequality, we factor the quadratic and make use of the appropriate property.

■ **EXAMPLE 19**   Solve $x^2 - 6x + 8 > 0$.

Solution
$$x^2 - 6x + 8 > 0 \qquad \textbf{Factor}$$
$$(x - 2)(x - 4) > 0 \qquad \textbf{Since the product is positive, both quantities must be positive or both must be negative.}$$

Thus either

$$x - 2 > 0 \quad \text{and} \quad x - 4 > 0 \qquad or \qquad x - 2 < 0 \quad \text{and} \quad x - 4 < 0$$
$$x > 2 \quad \text{and} \qquad x > 4 \qquad or \qquad x < 2 \quad \text{and} \qquad x < 4.$$

Figure 6.1

Consider $x > 2$ and $x > 4$. Any number > 4 is also > 2, but a number > 2 is not necessarily > 4. Therefore, $x > 2$ and $x > 4$ can be shortened to just $x > 4$.

Consider $x < 2$ and $x < 4$. Any number < 2 is also < 4, but a number < 4 is not necessarily < 2. The combined statement $x < 2$ and $x < 4$ can be shortened to $x < 2$.

The solution to $x^2 - 6x + 8 > 0$ is $x < 2$ or $x > 4$. We can show the solution on a number line (see Fig. 6.1). Any number less than 2 or greater than 4 satisfies the original inequality. A number between 2 and 4 does not satisfy the inequality.

Check: Choose a number less than 2, such as 0. If $x = 0$, then

$$0^2 - 6(0) + 8 > 0$$
$$8 > 0$$

is true.

Choose a number greater than 4, such as 5. If $x = 5$, then

$$5^2 - 6(5) + 8 > 0$$
$$25 - 30 + 8 > 0$$
$$3 > 0$$

is true.

Choose a number between 2 and 4, such as 3. If $x = 3$, then

$$3^2 - 6(3) + 8 > 0$$
$$9 - 18 + 8 > 0$$
$$-1 > 0$$

is false.   □

We could have saved some of the work in Example 19 if we had gone directly to the number line once we determined that 2 and 4 were the *critical* values of the inequality. The numbers 2 and 4 divide the number line into three "zones," that is, (1) numbers < 2, (2) numbers between 2 and 4, and (3) numbers > 4. After arbitrarily selecting a number from one of these zones, we test it in the original inequality to see whether the result is true or false. If the result is true, the zone from which the number came belongs to the solution set. If the result is false, that zone does not belong to the solution set. Whatever the first result, the zones must alternate between true and false results so that no other number need be tried. In Example 19, when $x = 0$ (numbers < 2), the result is true. When $x = 3$ (numbers between 2 and 4), the result is false. When $x = 5$ (numbers > 4), the result is true. (See Fig. 6.2.)

Figure 6.2

We will solve Example 20 using this approach.

■ EXAMPLE 20   Solve $x^2 - 6x + 8 < 0$.

Solution
$$x^2 - 6x + 8 < 0 \qquad \textbf{Factor}$$
$$(x - 2)(x - 4) < 0$$
$$x - 2 = 0 \qquad x - 4 = 0$$
$$x = 2 \qquad x = 4$$

Thus, 2 and 4 are the critical values. The three zones are:

1. $x < 2$;
2. $2 < x < 4$;
3. $x > 4$.

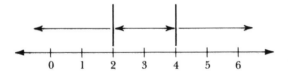

Test: Choose a number < 2, such as 0. If $x = 0$, then

$$0^2 - 6(0) + 8 < 0$$
$$8 < 0$$

is false.

Automatically, we know that the numbers between 2 and 4 belong to the solution set and the numbers < 2 and > 4 do not.

Test: Choose a number between 2 and 4, such as 3. If $x = 3$, then

$$3^2 - 6(3) + 8 < 0$$
$$9 - 18 + 8 < 0$$
$$-1 < 0$$

is true.

Test: Choose a number > 4, such as 5. If $x = 5$, then

$$5^2 - 6(5) + 8 < 0$$
$$25 - 30 + 8 < 0$$
$$3 < 0$$

is false. ☐

■ **EXAMPLE 21**   Solve $x^2 - 3x - 10 \geq 0$.

Solution

$$x^2 - 3x - 10 \geq 0 \qquad \textbf{Factor}$$
$$(x - 5)(x + 2) \geq 0$$
$$x - 5 = 0 \qquad x + 2 = 0$$
$$x = 5 \qquad\quad x = -2$$

Thus, 5 and -2 are the critical values. The three zones are:

1. < -2;
2. $-2 < x < 5$;
3. > 5.

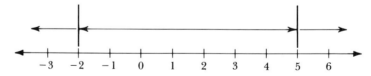

Test: Choose a number < -2, such as -3. If $x = -3$ then,

$$(-3)^2 - (3)(-3) - 10 \geq 0$$
$$9 + 9 - 10 \geq 0$$
$$8 \geq 0$$

is true. Note what happens when a critical value is chosen. If $x = -2$,

$$(-2)^2 - (3)(-2) - 10 = 0.$$

Therefore, the solution set consists of the numbers ≤ -2 and ≥ 5. The numbers between -2 and 5 are not in the solution set. $\square$

Exercise Set 6.5

Solve each of the following quadratic inequalities. Show the solution on a number line. Check your solution set.

1. $x^2 - 8x + 15 > 0$
2. $x^2 - 3x - 10 > 0$
3. $x^2 + 4x - 21 \geq 0$
4. $x^2 + 5x - 14 \geq 0$
5. $x^2 + 5x + 4 < 0$
6. $x^2 - 5x + 6 < 0$
7. $x^2 - 7x - 8 \leq 0$
8. $x^2 - 5x - 24 \leq 0$
9. $x^2 - 3x \leq 0$
10. $2x^2 - 8x < 0$
11. $x^2 + 2x \geq 0$
12. $x^2 + 10x > 0$
13. $x^2 - 16 > 0$
14. $x^2 - 25 \geq 0$
15. $x^2 - 9 < 0$
16. $2x^2 - 8 < 0$
17. $x^2 + 3x - 28 > -18$
18. $x^2 - 5x - 6 \geq -10$
19. $x^2 + 13x + 30 \leq -10$
20. $x^2 + 3x - 4 < 14$
21. $2x^2 - 7x + 3 > 0$
22. $3x^2 - 5x - 2 \leq 0$

Summary

We began the chapter by learning of the relationship between fractional exponents and radicals and learned how to change from one form to the other. We saw that the laws of exponents reviewed in Chapter 0 for integer exponents are also valid for fractional exponents.

DEFINITIONS

1. $a^{1/n} = \sqrt[n]{a}$ 2. $a^{m/n} = (\sqrt[n]{a})^m$

In the next section we learned how to perform operations on radical expressions, including multiplying and dividing radicals with equivalent index numbers. We also learned to simplify radicals by removing roots.

■

> **LAWS OF RADICALS**
>
> 1. $\sqrt[n]{a^n} = a$
> 2. $\sqrt[n]{a} \cdot \sqrt[n]{b} = \sqrt[n]{ab}$
> 3. $\dfrac{\sqrt[n]{a}}{\sqrt[n]{b}} = \sqrt[n]{\dfrac{a}{b}}$

We encountered two new types of equations in this chapter, *quadratic* and *radical*. We found that if a quadratic equation could not be solved by factoring, it could be solved by completing the square or by using the *quadratic formula*. We also discovered that the radicand portion of the quadratic formula, called the *discriminant*, can be used to determine the nature of the roots of the quadratic equation. Solutions to radical equations and methods for finding them were explored. We found that the existence of *extraneous roots* emphasizes the importance of checking our answers in the radical equations to make sure they are valid solutions.

The last topic discussed was *quadratic inequalities*. Although the discussion was limited to those quadratics that are factorable, it is a very simple matter to generalize the method of expressing the solution to any quadratic inequality. We also found that expressing the solutions on a number line helped us to understand the specific solution set.

Review Exercises

Simplify each of the following.

1. $\sqrt{121}$
2. $\sqrt{225}$
3. $\sqrt[3]{4x} \cdot \sqrt[3]{2x^2}$

4. $\sqrt[3]{25x} \cdot \sqrt[3]{5x^2}$
5. $\sqrt[5]{(2x)^5}$
6. $\sqrt[3]{(2x)^4}$

7. $\dfrac{\sqrt[3]{16}}{\sqrt[3]{2}}$
8. $\dfrac{\sqrt[3]{32x}}{\sqrt[3]{2x}}$
9. $\dfrac{\sqrt[3]{48x^4}}{\sqrt[3]{6x}}$

10. $\dfrac{\sqrt{12xy}}{\sqrt{3xy}}$
11. $\sqrt[4]{32x^5}$
12. $\sqrt[3]{24x^3}$

13. $\sqrt[3]{4x^2y^2} \cdot \sqrt[3]{6x^4y^2}$
14. $\sqrt[4]{4x^3y^2} \cdot \sqrt[4]{4xy^2}$
15. $-(4)^{-1/2}$

16. $\left(\dfrac{1}{8}\right)^{-1/3}$
17. $x^{2/5} \cdot x^{2/3}$
18. $x^{3/4} \cdot x^{-1/3}$

Solve each of the following quadratic equations. If the equation is not factorable, use the quadratic formula.

19. $x^2 + 4x = 0$
20. $3x^2 - 9x = 0$

21. $2x^2 - 18 = 0$
22. $4x^2 - 100 = 0$

23. $5x^2 = 25$ **24.** $2x^2 = 98$

25. $x^2 + 2x - 15 = 0$ **26.** $x^2 + 3x + 2 = 0$

27. $x^2 - 10x + 18 = 0$ **28.** $x^2 - 8x - 2 = 0$

29. $x^2 + 7x = -x - 15$ **30.** $x^2 - x = -4x + 4$

31. $x^2 - 2x + 1 = 4x + 12$ **32.** $x^2 - 2x + 1 = 2x - 4$

33. $2x^2 - 3x + 6 = 0$ **34.** $2x^2 - 3x + 7 = 0$

Solve each of the following radical equations. Check your answer.

35. $\sqrt{x + 1} = 5$ **36.** $\sqrt{2x - 1} = 5$

37. $\sqrt{x + 2} = -4$ **38.** $\sqrt{2x + 1} + 6 = 0$

39. $\sqrt{x + 2} = \sqrt{2x - 1}$ **40.** $\sqrt{3x + 7} = \sqrt{x + 1}$

41. $\sqrt{2x + 1} = \sqrt{2x - 4}$ **42.** $\sqrt{x - 3} = -\sqrt{2x + 6}$

43. $\sqrt{x - 3} + \sqrt{x} = 6$ **44.** $\sqrt{x - 5} + \sqrt{x} = 1$

45. $\sqrt{4x + 1} - \sqrt{x - 2} = 6$ **46.** $\sqrt{5 - x} = 6 + \sqrt{1 - x}$

Solve each of the following quadratic inequalities. Graph your answer on a number line. Check your solution set.

47. $x^2 + 3x > 0$ **48.** $3x^2 - 12x \le 0$

49. $x^2 - 49 \ge 0$ **50.** $2x^2 - 18 < 0$

51. $x^2 + 8x + 12 \le 0$ **52.** $x^2 + x - 6 > 0$

53. $2x^2 + x - 15 \ge 0$ **54.** $2x^2 - 9x + 9 < 0$

Functions

After completing this chapter, you should be able to:

1. Graph points on a Cartesian coordinate system if given a table of values, an equation, a formula, or a mathematical model of a function.
2. Define what is meant by a function, as well as define related terms such as domain, range, polynomial function, linear function, and quadratic function.
3. Identify and graph linear functions by using the ideas of slopes, x-intercepts, and y-intercepts.
4. Graph linear inequalities.
5. Identify and graph quadratic functions by locating the x- and y-intercepts and the vertex.

7

7.1
The Cartesian Coordinate System and Graphing (A Review)

The Cartesian Coordinate System

Ever notice how street plans are laid out in many major cities? The streets in the city map shown in Fig. 7.1 are in the form of a rectangular grid. If we wanted to give someone directions from the center of town (Upper James St. and Stone Church Rd.) to Macassa Park, we could give two numbers: Travel 3 blocks east and then 2 blocks north. The point on the map could be represented by the pair of numbers (3, 2). Each intersection is represented by a pair of numbers, and for each pair of numbers there is exactly one intersection. If we agree to always travel first east, and then north, the point (3, 2) is not the same as the point (2, 3). The pairs are *ordered* in that the first number always indicates movement east (or west), while the second number indicates

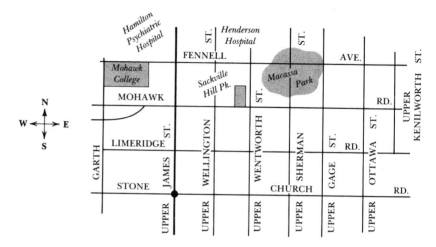

Figure 7.1 Macassa Park: at the intersection of Sherman St. and Mohawk Rd.

movement north (or south). This city map is patterned after a *rectangular*, or *Cartesian coordinate system*.

   To construct a Cartesian coordinate system, we form perpendicular lines from two number lines (see Fig. 7.2a, b, and c). The horizontal number line is called the *x*-axis, the vertical number line the *y*-axis. These axes divide the plane into four *quadrants*, as shown in Fig. 7.2(d). The point of intersection of the two axes is labeled (0, 0) and referred to as the *origin* (or starting point). Any point in the plane, *P*, is located, or represented, by an ordered pair of numbers (*x*, *y*); these values together are called the *coordinates* of point *P*. The first value of the ordered pair, the *x*-value or the *abscissa*, is the horizontal distance of *P* from the *y*-axis (see Fig. 7.3a). The second value of the ordered pair, the *y*-value or the *ordinate*, is the vertical distance of *P* from the *x*-axis (see Fig. 7.3b).

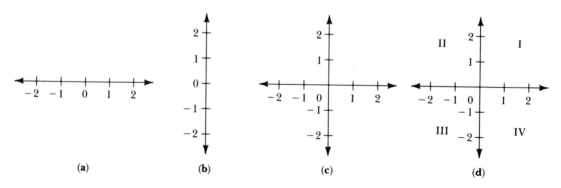

Figure 7.2

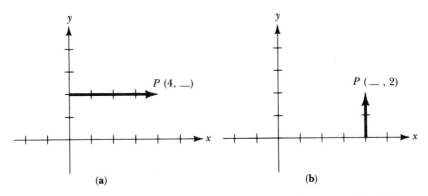

Figure 7.3   (a) The abscissa, or x-value, is 4: (b) the ordinate, or y-value, is 2.

Each ordered pair of real numbers is geometrically associated with a point in the plane.

To locate the point $A(3, 2)$, we start at the origin and count 3 units to the right (the perpendicular distance from the y-axis), and then count 2 units up (the perpendicular distance from the x-axis). (This is shown in Fig. 7.4.) To locate $B(-3, 2)$, we go to the left 3 units and up 2 units. To locate $C(-3, -2)$, we go to the left 3 units and down 2 units. To locate $D(3, -2)$, we go to the right 3 units and down 2 units.

All these points are graphed, with their coordinates indicated, in Fig. 7.4.

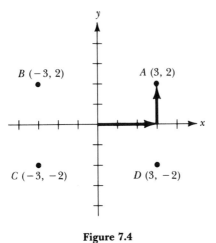

Figure 7.4

Remember: For x positive, move right; for x negative, move left. For y positive, move up; for y negative, move down.

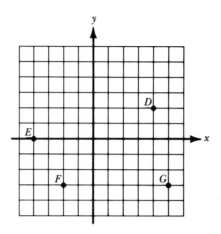

Figure 7.5

■ **EXAMPLE 1**   What are the coordinates of the points D, E, F, and G in Fig. 7.5?

Solution   The coordinates are (4, 2), (−4, 0), (−2, −3), and (5, −3), respectively. □

■ **EXAMPLE 2**   Graph (4, 0), (0, −2), and (0, 0).

Solution   See Fig. 7.6. For (4, 0), we move 4 units to the right and 0 units vertically. For (0, −2), we move 0 units horizontally, and then 2 units down. To graph (0, 0), we move 0 units horizontally and 0 units vertically! □

In summary, for each ordered pair of numbers there is associated a point in the plane; conversely, associated with each point in the plane is a pair of real numbers. A one-to-one matching such as this is referred to as a **one-to-one correspondence**.

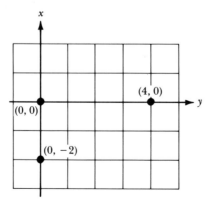

Figure 7.6

Graphing Equations

Using the Cartesian coordinate system, we can draw graphs to represent equations. The graph of an equation is actually the graph of the solution set for the equation.

■

> To graph an equation in two variables:
>
> *Step 1:* Find an ordered pair of numbers that satisfies the equation by substituting a real number for the first variable, and then solving the resulting equation for the second number.
> *Step 2:* Repeat step 1 using a different first number.
> *Step 3:* Repeat step 1 using a different first number.
>
> $\vdots$
>
> Continue until enough points are plotted so that the graph can be accurately drawn.

■ **EXAMPLE 3**

Graph $y = x^2$.

Solution

One at a time, we use each value of x, substituting into the equation to find the corresponding y-coordinate.

$$y = x^2 \qquad\qquad y = x^2 \qquad\qquad y = x^2$$
$$y = (-3)^2 \qquad\quad y = (-2)^2 \qquad\quad y = (3)^2 \quad \text{etc.}$$
$$y = 9; \qquad\qquad y = 4; \qquad\qquad y = 9;$$

x	-3	-2	-1	0	1	2	3
y							

x	-3	-2	-1	0	1	2	3
y	9	4	1	0	1	4	9

Plotting the points to correspond to each ordered pair,

$(-3, 9), (-2, 4), (-1, 1), (0, 0),$
$(1, 1), (2, 4), (3, 9),$

we obtain the graph shown in Fig. 7.7. □

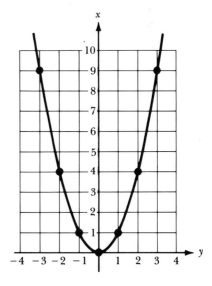

Figure 7.7

■

Two observations should be noted concerning the relationship between the *graph* of an equation and the *equation* itself.
1. If a point satisfies the equation, the point lies on the graph of the
    equation.
2. Conversely, if a point lies on the graph of an equation, then the
    coordinates of the point satisfy the equation.

Exercise Set 7.1

1. Graph each of the following points on the rectangular (Cartesian) coordinate system in the figure.

$A(4, 3)$ $B(3, 4)$

$C(-2, 4)$ $D(-5, 0)$

$E(2, 0)$ $F(4, -3)$

$G(2.5, -6)$ $H(-3, -3)$

$I(0, -4)$ $J(-6, 1)$

$K(-6, -1)$ $L(-1, -6)$

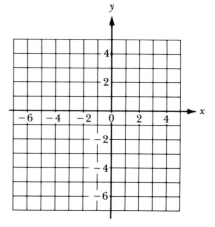

2. For each of the points graphed in the figure, write the coordinates.

$A(\ ,\ )$ $B(\ ,\ )$

$C(\ ,\ )$ $D(\ ,\ )$

$E(\ ,\ )$ $F(\ ,\ )$

$G(\ ,\ )$ $H(\ ,\ )$

$I(\ ,\ )$

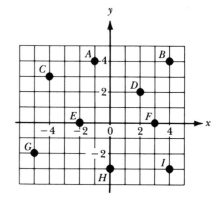

3. Indicate all the ordered pairs that satisfy each equation.

a) $x + 2y = 8$ (1, 3), (2, 3), (0, 4), (4, 0), (3, 2), (8, 0),

b) $2x + 3y = 12$ (0, 6), (6, 0), (3, 2), (2, 3), (−3, 6), (9, −2),

c) $1.2x - 3.6y = 7.2$ (6, 0), (0, −2), (2, 0), (3, −2), (1, −3),

d) $y = 3x - 2$ (1, 1), (4, 10), (2, 3), (0, 2), (0, −2), (−1, 1)

4. Graph each of the following equations by first completing the accompanying table of values.

a) $y = 2x - 3$

x	−3	−2	−1	0	1	2	3
y							

b) $y = x^2 - 4$

x	−3	−2	−1	0	1	2	3
y							

c) $y = 2^x$

x	−3	−2	−1	0	1	2	3
y							

For parts (d) through (g), supply the x-coordinates, and then determine the y-coordinates.

d) $y = x^2 + 1$ **e)** $y = 4 - x^2$

f) $y = (1/2)(4 - x^2)$ **g)** $y = 1/x$

5. For each equation in Exercise 4:

a) At what point(s) does the graph intersect the x-axis?

b) Where does the graph intersect the y-axis?

6. Describe, or graph, the location of all points:

a) whose abscissas are 2;   **b)** whose ordinates are 0;

c) whose abscissas equal the   **d)** for which $x > 0$;
ordinates;

e) for which $y > 0$;   **f)** for which $x \geq 0$ *and* $y \geq 0$.

7.2
Functions Defined and the Algebra of Functions

Relations and Functions

One of the most important basic concepts in mathematics is the **function**. Often in business, in computer programming, and in mathematics, an intensive study is made between two variables. As the number of units of production increases, how is the unit cost affected? How are units of production related to unit cost? Or, is the amount of money spent in advertising reflected in, or *related* to, a change in sales?

In most situations it is easier and preferable to express a *relationship* between variables by an equation or formula:

$$NP = 200(P - 34.5).$$

In the equation above, net profit (NP) is *related* to price per unit (*P*). In the following equation, a mathematical model is given *relating* the price of a new car *P* to the demand *x*, where *x* represents the number of cars:

$$P = -0.15x + 16{,}200.$$

■ **DEFINITION**

RELATION

A set of ordered pairs (x, y) is called a *relation*; (3, 4), (2, 4), (2, 5), and (8, 7) represent a relation between *x* and *y*.

In the development of mathematical models, the most important type of relationships are those that are also *functions*.

■ **DEFINITION**

FUNCTION

Whenever a relationship exists between two variables *x* and *y* such that for each value of *x* there is only one corresponding value of *y*, then *y* is said to be a *function* of *x*.

Figure 7.8 shows a function. The *function* is composed of a set of values (called the **domain**) assigned to the independent variable *x*, a set of values (called the **range**) assigned to the dependent variable *y*, and a correspondence that produces a *unique* value of *y* for each value of *x*. (Occasionally, a function is also referred to as a *mapping*, where the members of the domain are matched, or mapped, onto the members of the range.)

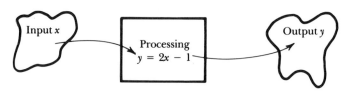

Figure 7.8

In the net profit equation, NP $= 200(P - 34.5)$, NP depends on, or *is a function of*, the variable P. Operations are performed on P to arrive at a value for NP. This is stated, "NP is a function of P," and is written in mathematical notation as NP $= f(P)$. Likewise, from the other example, we see that the price of a new car P is a function of demand x; that is, $P = f(x)$. The symbol $f(x)$ is read "f of x" and refers to the value of P obtained after substituting a value for x into the correspondence of the function. The notation $f(x)$ does *not* mean f times x!

■ **EXAMPLE 4**

In the mathematical model relating the price of a car (P) to the demand for cars (where x represents number of cars),

$$P = f(x) = -0.15x + 16{,}200,$$

if $x = 10{,}000$, then

$$f(10{,}000) = -0.15(10{,}000) + 16{,}200 \quad \text{or} \quad 14{,}700.$$

Hence, 10,000 is a member or an element of the domain and 14,700 is an element of the range. Likewise, if $x = 20{,}000$, then

$$f(20{,}000) = -0.15(20{,}000) + 16{,}200 = 13{,}200.$$

This mapping is shown in Fig. 7.9, where we see that for each value of x that is selected, we will find *exactly one* value for P. □

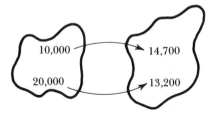

Figure 7.9

It seems in Example 4 that as demand increases, the price decreases. Would it be possible that if enough cars were demanded, the price per car would be less than $100.00?

Examples 5 and 6 list other examples of functional relationships.

■ **EXAMPLE 5**

In the equation $y = f(x) = x^2 - 5$, y is a function of x. Specifically, to obtain a specific value for y, we square the value for the independent variable and then subtract five.

$$f(3) = 3^2 - 5 = 4$$
$$f(2) = 2^2 - 5 = -1$$
$$f(10) = 10^2 - 5 = 95$$
$$f(-2) = (-2)^2 - 5 = -1 \quad \square$$

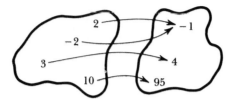

Figure 7.10

Note that two or more x-values may "share" some y-value in a function (as is the case in Example 5 with both 2 and -2 yielding -1). However, any given x-value may have no more than one y-"partner."

Some members of the domain and the range of Example 5 are shown in Fig. 7.10. If 3, 2, 10, and -2 are some elements of the domain, what is the domain? In general, if an equation or formula is given without specifying the domain, the domain is understood to be the set of all real numbers for which the function is *defined*.

For a function such as $y = f(x) = \sqrt{x - 4}$, if $x < 4$, then $x - 4$ will be less than 0, and $\sqrt{x - 4}$ is an imaginary number. Since $x - 4$ cannot be a negative number, $x - 4$ must be $\geq$ to 0, or $x \geq 4$. The *domain* then is the set of all values of x that are greater than or equal to 4!

For a function such as $y = f(x) = 1/(x - 2)$, if $x = 2$, then the denominator is 0 and y is undefined. Here, then, the domain is the set of all real numbers other than 2.

Although the range of a function *may be determined* in advance, often it is easier to determine a number of corresponding values for the dependent variable *and then* describe the range.

● **Question**

For the function $y = f(x) = x^2$, what are the domain and the range?

Answer

The domain is the set of all real numbers. (For this function, any real number can be entered for x; thus the domain is the set of all real numbers.) The range is the set of all real numbers greater than or equal to 0. (After a value of x is entered and then squared, the result (or y-value) will always be nonnegative.)

■ **EXAMPLE 6**

The formula for converting from Fahrenheit to Celsius temperature is $C = f(F) = (\frac{5}{9})(F - 32)$. Find the Celsius values for Fahrenheit readings of 212, 122, and 32.

Solution

With this formula, we see that Celsius temperature is *functionally* related to Fahrenheit by the following correspondence:

1. We begin with the Fahrenheit temperature (element of the domain),
2. we subtract 32, and then
3. we multiply the result by $\frac{5}{9}$.

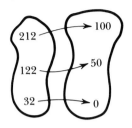

Several values are computed below and then diagrammed in Fig. 7.11:

$$f(212) = (\tfrac{5}{9})(212 - 32) = 100,$$
$$f(122) = (\tfrac{5}{9})(122 - 32) = 50,$$
$$f(32) \ = (\tfrac{5}{9})(32 - 32) \ = 0.$$

What are the domain and the range for $C = (\tfrac{5}{9})(F - 32)$? $\square$

Figure 7.11

The function $C = (\tfrac{5}{9})(F - 32)$ tells us to subtract 32 from the value of F and then multiply by $\tfrac{5}{9}$. An analogy is a computer program that would take a number when entered (input), would subtract 32 and multiply by $\tfrac{5}{9}$ (processing), and then yield an answer (output), as shown in Fig. 7.12.

In computer programming, the concept of the function is of such major importance that most high-level computer languages have special reserved words for special predefined functions, as well as words to define the programmer's own function. For example, the above function could be defined as follows:

In BASIC, $\quad$ $FN(x) = (\tfrac{5}{9}) * (x - 32)$
In FORTRAN, $\quad$ $FNC(x) = (\tfrac{5}{9}) * (x - 32)$
In Pascal, $\quad$ $F := (\tfrac{5}{9}) * (x - 32)$

In a statement such as $FN(x) = (\tfrac{5}{9}) * (F - 32)$, we refer to x as the *argument* of the function.

Although functional relationships are often expressed as equations or formulas, this is not always the case. Sometimes an expression or formula is difficult, if not impossible, to construct. For example, the amount of food you consume determines your weight. While weight is

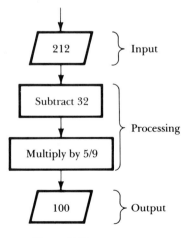

Figure 7.12

Table 7.1
Tax chart relating
amount of state
income tax to income

$ 0 – 4,999	0.625% of taxable income
$ 5,000 – 9,999	$ 31.25 plus 1.25% of excess over $5,000
$10,000 –14,999	$ 93.75 plus 2.50% of excess over $10,000
$15,000 –19,999	$218.75 plus 3.125% of excess over $15,000

a function of food intake, W = f(FI), an exact expression for f(FI) would be nearly impossible to derive.

Table 7.1 is a tax chart showing the amount of state income tax to be paid for various amounts of adjusted gross income, which illustrates a *functional relationship*. Note that for each value of adjusted gross income, there will be *only one* corresponding tax value. For example, for an adjusted gross income of $10,000, the tax is $93.75, for $12,000, the tax is $143.75 [$93.75 + 0.025(2000)], and so on. For this function, what would be the domain? What would be the range?

● Question   Are all relations functions?

Answer   We can answer the question best by examining the graph of a relation that is also a function and the graph of a relation that is *not* a function (see Fig. 7.13).

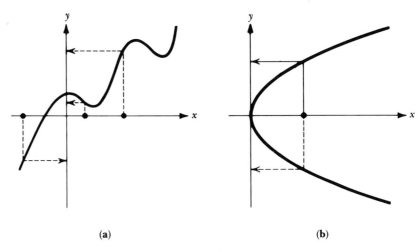

(a)   (b)

Figure 7.13

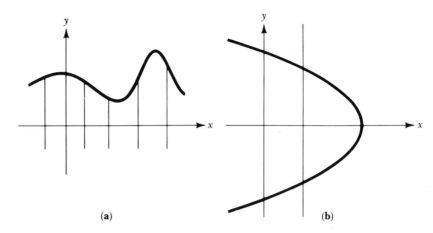

Figure 7.14   (a) A function; (b) not a function.

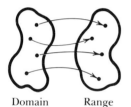

Domain   Range

Figure 7.15

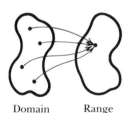

Domain   Range

Figure 7.16

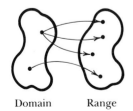

Domain   Range

Figure 7.17

In Fig. 7.13(a), for each value of x, there is only one corresponding value of y; this graph *does* represent a function.

In Fig. 7.13(b), for the value of x, there is more than one value of y. Thus the graph does *not* represent a function!

Graphically, Fig. 7.13 leads us to an easy test, called the **vertical-line test**, to determine whether a given relation is a function. If each vertical line drawn intersects the graph in at most one place, then the graph *does* represent a function. (In other words, for each x-coordinate, there is only one y-coordinate.) If a vertical line intersects the graph of a relation at more than one point, then the relation is *not* a function. (See Fig. 7.14.)

In Fig. 7.15, each x-coordinate has only one y-coordinate. This relation is not only a function, but also a one-to-one function. Each x has only one y-value, and each y has only one x-value.

Figure 7.16 *does* show a function. Any particular x has only one corresponding y-value. This function illustrates a many-to-one relationship. A mathematical example of this is $y = 7$, or $y = x^2$.

The relation shown in Fig. 7.17 does *not* represent a function because the x-value at the top is associated with *more than one* corresponding y-value.

The Algebra of Functions

Now that we have defined functions, if we can learn to combine several types of functions, our ability to use them is greatly increased. While "f" is the more common way to represent a function, other lower-case letters, or Greek letters, may also be used.

Let's take the following functions, for example:

$$f(x) = x^2 - 2,$$
$$g(x) = 3x - 1,$$
$$\theta(x) = x^3 - x^2 + 19.86,$$
$$h(x) = \sqrt{x^2 - 4}.$$

Because $f(x)$, $g(x)$, $\theta(x)$, and $h(x)$ represent the y-values, we can perform algebraic operations on these values. To add the functions $f(x) + g(x)$, where $f(x) = x^2 - 2$ and $g(x) = 3x - 1$, we have

$$f(x) + g(x) = (x^2 - 2) + (3x - 1) \quad \text{or} \quad x^2 + 3x - 3.$$

■ **EXAMPLE 7**

Perform these algebraic operations using the functions given above.

a) Find $f(x) - g(x)$.
b) Find $f(x) \cdot g(x)$.
c) If $x = 1$, find $f(x)/g(x)$.

Solution

a) $f(x) - g(x) = (x^2 - 2) - (3x - 1)$
$$= x^2 - 2 - 3x + 1$$
$$= x^2 - 3x - 1$$

b) $f(x) \cdot g(x) = (x^2 - 2)(3x + 1)$
$$= 3x^3 + x^2 - 6x - 2$$

c) $\dfrac{f(1)}{g(1)} = \dfrac{(1^2 - 2)}{(3 - 1)} = -0.5$ □

In addition to the operations of addition, subtraction, multiplication, and division on functions, functions may be combined by "composition." If $f(x)$ and $g(x)$ are the values of two functions, $f(g(x))$ is a *composite function*, which is formed by substituting $g(x)$ for x in the formula for $f(x)$.

■ **EXAMPLE 8**

Given $f(x) = x^2$ and $g(x) = x + 2$, find $f(g(x))$.

Solution

The f-function squares given values (or quantities), while the g-function adds 2 to given quantities. The composite function $f(g(x))$ is evaluated working "from the inside out." Thus if $x = 3$, for instance, we first compute $g(3)$:

$$g(3) = (3) + 2 = 5.$$

The g-function adds 2 to given quantities. Thus $f(g(3))$ becomes $f(5)$ and $f(5) = 5^2$ or 25, because the f-function squares given quantities. In summary, we have $f(g(3)) = 25$. Thus for any value of x,

$$f(g(x)) = f(x + 2)$$

and by "squaring" we get

$$f(g(x)) = (x + 2)^2$$
$$= x^2 + 4x + 4. \quad \square$$

● Question

Does $f(g(x))$ mean $f(x)$ times $g(x)$?

Answer

No! To see that this is not so, multiply $f(x)$ times $g(x)$ in Example 8 and compare your answer to the expression for $f(g(x)) = x^2 + 4x + 4$.

● Question

Does $f(g(x))$ equal $g(f(x))$?

Answer

If we use $f(x) = x^2$ and $g(x) = x + 2$, for example, we see that $f(g(x))$ yields the expression $x^2 + 4x + 4$. To obtain an expression for the composite function $g(f(x))$, we substitute x^2 for $f(x)$. Since the f-function squares quantities,

$$g(f(x)) = g(x^2).$$

Next the g-function will operate on x^2. The g-function adds 2 to given values or quantities, so

$$g(f(x)) = (x^2) + 2.$$

In summary, the composite function would be

$$g(f(x)) = x^2 + 2.$$

For these functions, $f(g(x)) \neq g(f(x))$.

For functions in general, $f(g(x))$ will not equal $g(f(x))$; however, under special circumstances some functions will exhibit the property $f(g(x)) = g(f(x))$.

Classification of Functions

The functions $y = f(x) = x^2 - 5$, $C = f(F) = (5/9)(F - 32)$, $y = f(x) = x^2 - 2$, and $y = g(x) = 3x - 1$ are all examples of the simplest type of function: the **polynomial function**.

■ DEFINITION

POLYNOMIAL FUNCTION

A function is a *polynomial function* if it is defined by an equation of the form

$$y = f(x) = a_n x^n + a_{n-1} x^{n-1} + \cdots + a_1 x + a_0,$$

where n is a positive integer or zero, and the coefficients of each term are represented by real numbers $a_n, a_{n-1}, \ldots$, and a_0.

■ **EXAMPLE 9**   For $y = f(x) = x^2 - 5$,

$$n = 2, \qquad a_2 = 1, \qquad a_1 = 0, \quad \text{and} \quad a_0 = -5. \quad \square$$

The highest power n of the independent variable x is called the *degree* of the polynomial. The polynomial in Example 9 has degree 2. Most of our work will focus on first-degree and second-degree polynomial functions. First-degree equations such as $y = f(x) = 3x - 5$ are usually called **linear functions** because the graph of a linear function will always be a *straight line*. Second-degree polynomial functions such as $y = f(x) = x^2 - 5$ are called *quadratic functions*. A *cubic* polynomial is a third-degree polynomial, etc. Finally, a polynomial such as $y = f(x) = 2$ (degree 0) is called a *constant function*. (Why? Try graphing $y = 2$.)

Exercise Set 7.2

For each relationship, list the ordered pairs, assuming the domain for each to be $\{-3, -2, -1, 0, 1, 2, 3\}$. Graph each relation and give its range. Indicate which of the relations are functions.

1. $y = 2x - 3$ **2.** $y = x^2 - 1$

3. $y = 4 - x^2$ **4.** $2y - x = 6$

5. $y = 2$ **6.** $y = 2/(x + 4)$

7. $xy = 1$ **8.** $y = 2^x$

9. $y = \pm \sqrt{x}$ **10.** $y = \begin{cases} x^2 & \text{if } x \leq 0, \\ x & \text{if } x > 0 \end{cases}$

State, in words, the instructions for each of the following functions.

11. $f(x) = x^2 + 3$ **12.** $f(x) = 2 - x^3$

13. $f(x) = \dfrac{\sqrt{x}}{4}$ **14.** $g(x) = \sqrt{x^2 - 2}$

Identify which of the following graphs *do* represent functions.

15.

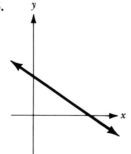

16.

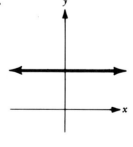

17.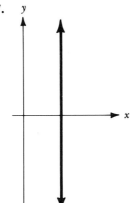

18.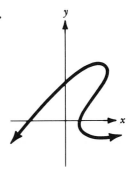

19.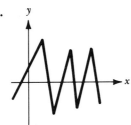

20.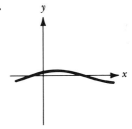

21.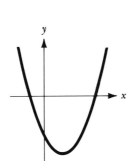

22.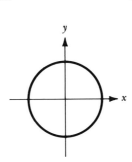

23.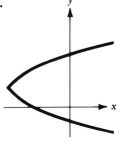

For each of the following functions, find $f(2), f(-2), f(0)$, and $f(w)$.

24. $f(x) = 3x - 1$

25. $f(x) = 4 - 2x$

26. $f(x) = x^2$

27. $f(x) = x^2 - 4$

28. $f(x) = 4 - x^2$

29. $f(x) = x^2 - 4x$

30. $f(x) = \sqrt{x^2 + 5}$

31. $f(x) = |x|$

For each of the following functions, identify the domain and the range.

32. $y = x + 4$ **33.** $y = x^2 - 9$

34. $y = |x|$ **35.** $y = -x^2$

36. $y = x^2 - 9$ **37.** $y = \dfrac{x}{(x + 1)}$

38. $y = \sqrt{9 - x^2}$ **39.** $y = 7.1$

40. The demand for a product is a function of the price. The demand function is given by the formula:

$$D(p) = p^2 - 25p + 200 \quad \text{for } p \geq 5.$$

   a) Find $D(5)$, $D(10)$, $D(15)$, and $D(20)$.

   b) Graph the relation.

   c) For what value of p will the demand be at a minimum?

Given that $f(x) = 2x^2 - 8$, $g(x) = 2x + 4$, and $h(x) = |x|$, find each of the following.

41. $f(x) + g(x)$ **42.** $f(x) \cdot g(x)$

43. $f(x) - g(x)$ **44.** $\dfrac{f(x)}{g(x)}$

45. $h(2)$ **46.** $h(-2)$

47. $f(g(3))$ **48.** $g(f(3))$

49. $f(g(h(1)))$ **50.** $f(g(h(-1)))$

51. (*Optional*). In this section, we studied the functional relationship of two variables. Often problems arise in which more than one independent variable must be considered. For example, in $C = P(1 + r)^t$, the compound amount of money is a function of the principal invested, the rate, and the time, that is, $C = f(P, r, t)$. Find each of the following.

   a) $f(1000, 0.08, 2)$ **b)** $f(2000, 0.04, 2)$

   c) $f(8000, 0.01, 2)$ **d)** $f(3452.85, 0.0825, 14)$

7.3

Formulas, Models, and Graphs

Functions express and succinctly describe relationships between quantities. Often, by using formulas and equations, we can create mathematical models to describe, or reflect, real-world situations. To convey

Table 7.2
Water usage in thousands of gallons for a 24-hour period

Time	12 P.M.	1:00	2:00	3:00	4:00	5:00	6:00	7:00	8:00	9:00	10:00	11:00
Usage	1.4	0.6	0.25	0.31	0.48	1.7	2.8	3.4	4.5	5.3	4.8	4.6

Time	Noon	1:00	2:00	3:00	4:00	5:00	6:00	7:00	8:00	9:00	10:00	11:00
Usage	4.2	4.0	3.4	3.2	3.0	3.1	2.6	2.9	3.8	4.3	4.2	2.2

these "functional relationships" at a glance, graphs are frequently used in the processing of data. In this section, some examples of formulas, mathematical models, and graphs are used to provide us with a deeper insight into functions *and* a better understanding of their applications.

Note: In most practical examples, letters other than f and x are used to represent the function and its independent variable. In Example 10, for instance, water usage (W) is a function of the time of day (H), that is, $W = f(H)$.

■ **EXAMPLE 10** In order to determine the peak hours of water usage—and consequently to better regulate usage—the chart shown in Table 7.2 was constructed for a suburb in Arizona. However, the graph of the functional relationship (Fig. 7.18) provides a much more visual impact of the information than does a mere list of numbers. ☐

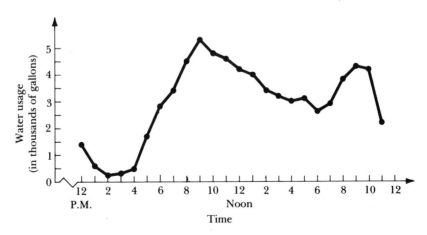

Figure 7.18 Water usage as a function of time.

Table 7.3
Compounded amount
of money for $1000
at 8%

t	0	1	2	3	4	5
A	1000.00	1080.00	1166.40	?	1360.49	1469.33

t	6	7	8	9	10
A	1586.87	?	1850.93	1999.00	2158.93

■ **EXAMPLE 11**

The formula $A = P(1 + r)^t$ can be used to determine the amount of money A after t years ($t \geq 0$) for a given principal P and a given interest rate r. The formula takes into consideration the fact that the interest is compounded yearly. Given that $1000 is placed into the account at 8%, show A plotted as a function of t for 10 years, that is, $A = f(t)$.

Solution

We evaluate $A = 1000(1.08)^t$ for selected values of t in Table 7.3. Although most values are computed, several have been left as an exercise. (Check *your* computed values for these against those already plotted.) The graph is shown in Fig. 7.19. □

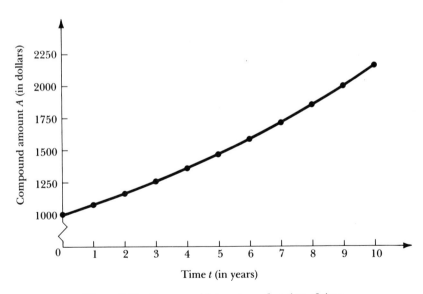

Figure 7.19 Compound interest as a function of time.

■ **EXAMPLE 12**

The total manufacturing costs of the BYTE Computer Company consist of a fixed overhead cost of $1200 plus a variable cost of $60 per unit constructed. Plot total cost as a function of the number of units made.

Solution

From a careful reading of the problem, we determine the formula for total cost:

Total cost (TC) = Fixed cost (F) plus Variable cost (V).

Total cost is a function of production units and is expressed then as TC = $f(u)$ = 1200 + 60u. By selecting several values for u and computing $f(u)$, or TC, we can complete the table shown in Fig. 7.20(a) and its corresponding graph in Fig. 7.20(b).

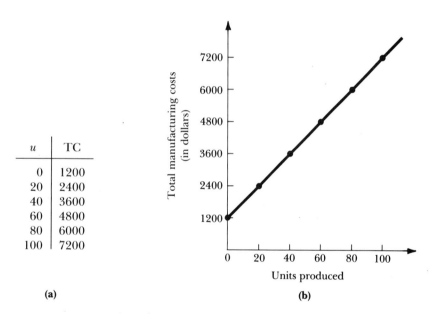

u	TC
0	1200
20	2400
40	3600
60	4800
80	6000
100	7200

(a)

(b)

Figure 7.20 Total cost per units produced.

■ **EXAMPLE 13**

A profit function for an industry is approximated by the the formula $P = f(i) = -i^2 + 10i - 9$. Find $f(0), f(1), f(2), f(3), ..., f(10)$; and plot profit as a function of i, the number of items sold. At what values of i will the profit be the greatest (maximum)? Are there any values for which profit is $0.00?

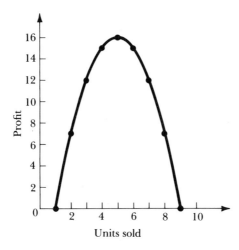

Figure 7.21   Profit as a function of units sold.

Solution   The graph is plotted in Fig. 7.21. By studying the figure, can you answer the questions above?   □

■ **EXAMPLE 14**   Advances in electronics technology have caused prices on personal microcomputers to drop drastically in the past months. In less than one year, the sale price of one particular brand was reduced by more than 70%! A market analyst derived a mathematical model to estimate the price P of a certain type of computer after m months:

$$P = f(m)$$
$$= \frac{8500}{6.5m + 2} + 250.$$

Plot the price over the next 12 months and then answer the following questions.

a) What is the current price of this micro?
b) What will be its price in one year?
c) What reduction in price can we expect for the next year?

Solution   By substituting values (0 to 12) in the function, we can complete the table in Fig. 7.22(a) and its accompanying graph in Fig. 7.22(b). From these, we can then answer the questions above as follows:

a) In the formula, m represents the number of months from now; the

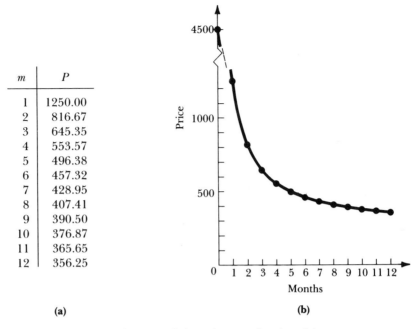

m	P
1	1250.00
2	816.67
3	645.35
4	553.57
5	496.38
6	457.32
7	428.95
8	407.41
9	390.50
10	376.87
11	365.65
12	356.25

(a)

(b)

Figure 7.22 Declining prices as a function of time.

time *now* is $m = 0$. Evaluating the formula with $m = 0$, we get $P = \$4500$.

b) By substituting $m = 12$ into the formula, we find that in one year, or in 12 months, $P = \$356.25$.

c) After the second year, when $m = 24$ months, the price will be $303.80; in other words, there will be a reduction of $52.45 from the end of the first year, and a reduction of $4196.20 from the original price. □

Exercise Set 7.3

1. Graph the sales figures (given in hundreds of dollars) from the following table by plotting years on the *x*-axis and sales on the *y*-axis.

Yearly sales figures

1970	1971	1972	1973	1974	1975	1976	1977	1978	1979
50	59	67	75	80	90	99	108	116	124

2. The hourly temperatures (in degrees Fahrenheit) have been recorded by a local weather station as shown in the following table. Represent the data graphically.

Hourly temperatures for a 24-hour period

Hour	6 A.M.	7	8	9	10	11	12	1 P.M.	2	3	4	5	6
Temperature	44	48	53	64	71	79	82	87	93	97	95	88	81

3. a) Plot sales revenue R as a function of the items produced i, where $R = 200i - 4i^2$.

   b) For what value of i will the revenue be the greatest?

   c) What will this maximum revenue be?

4. Display the data listed in the following table graphically. Determine which functional relationship given by either (a) $f(p) = 410 + 26x$, or (b) $f(p) = 400 + 25x$ best approximates these data by calculating $f(p)$ for each p.

P	1	3	6	8	9
$f(p)$	420	475	553	601	630

5. When q units of a commodity are produced, the cost is represented as a function of q by the formula $C = f(q) = 3q^2 + 5q + 75$. By computing $f(0)$ through $f(9)$ and plotting the points (q, C), determine at what level of production the cost will be the smallest.

6. The formula $A = P(1 + rt)$ can be used to determine the amount of money A for a given principal P invested at $r\%$ for a given period of time t. This formula reflects *simple interest*. Plot A as a function of t for $1000 invested at 8%. Compare your graph to the graph given in Example 11.

7. By plotting $R = f(t) = 49/t^2$, $(t \geq 0)$, determine at approximately what value of t that R will be less than 3.0.

8. Draw a graph showing total cost as a function of units produced if:

   a) the fixed overhead is $10,000 and production costs are $2400 per unit;

   b) the fixed overhead is $15,000 and production costs are $2400 per unit;

   c) the fixed overhead is $10,000 and production costs are $3600 per unit;

   d) the fixed overhead is $15,000 and production costs are $3600 per unit.

   Analyze and compare the four graphs above.

9. To better understand the relationships among types of functions, graph each of the following pairs of functions on the same axes, and then explain how the graphs are related.

   a) $y = f(x) = x^2$ and $y = g(x) = x^2 + 3$

   b) $y = f(x) = x^2$ and $y = g(x) = 2x^2$

   c) $y = f(x) = x^2$ and $y = g(x) = -x^2$

   d) $y = f(x) = x^2$ and $y = g(x) = (x + 2)^2$

10. By plotting both functions on the same axes, approximate the value of x for which $C(x)$ will equal $R(x)$ if:

$$C(x) = 60x + 200 \quad \text{and} \quad R(x) = 30x + 500.$$

7.4
Graphing Linear Functions

In this section our attention will be restricted to a subset of polynomial functions, those polynomials whose terms are either first-degree terms or constant terms. Examples of first-degree polynomials are $x + y = 5$, $56 = 4x + 8y$, $y = 3x + 7$, and $NP = -0.15p + 16{,}200$. Because the graph of each first-degree polynomial function is a straight line when graphed, first-degree functions are referred to as *linear functions*.

We begin our study of different kinds of functions by concentrating on linear functions for two reasons: (1) mathematically, linear functions are the easiest to handle, and (2) although a rather simple kind of function, linear functions can be used in an amazingly wide assortment of significant applications. In business (and in mathematics) linear functions serve as exact models for many relationships, for example, supply and demand functions. They also serve to approximate more complicated relationships with a high degree of accuracy, as you will see later in the text. Linear functions can also be the basis for describing many statistical patterns (such as linear regression).

Any relation of the form $ax + by = c$, where a, b, and c are real numbers ($b \neq 0$), is called a *linear function*. Written in this manner, the function is said to be in *standard form* (or, sometimes, *general form*).

■

To graph linear functions, there are several methods:

a) solving explicitly for y and substituting values for the independent variable;

b) using the x- and y-intercepts;

c) solving explicitly for y and using the slope, y-intercept form.

Method (a): Graphing Linear Functions by Solving Explicitly for y and Substituting Values

Given a linear function in standard form $ax + by = c$, it is often easier to graph the function by first putting the equation into explicit form (that is, solved for y). Then we can substitute values for the independent variable x and compute the corresponding values for y. Because we know that the graph of a linear function will be a straight line, we want to find three ordered pairs of numbers to graph the function (only two are needed). Two points will determine the line, and a third point will serve as a check for the other two points.

■ **EXAMPLE 15**   Graph $2x - y = 4$.

Solution   Solving for y, we have

$$-y = -2x + 4 \quad \text{or}$$
$$y = 2x - 4.$$

We then make a table by choosing convenient values for the independent variable x and then computing y.

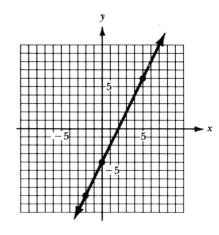

x	-2	0	5
y	?	?	?

x	-2	0	5
y	-8	-4	6

Figure 7.23

When we plot these ordered pairs, we have the graph of $2x - y = 4$, shown in Fig. 7.23. □

Note: Because any vertical line will intersect the graph in Fig. 7.23 in at most one point, $2x - y = 4$ is a *linear function.*

■ **EXAMPLE 16**   Graph $3x - 2y = 6$.

Solution

Solving explicitly for y, we have

$$-2y = -3x + 6 \quad \text{or}$$
$$y = (\tfrac{3}{2})x - 3.$$

Completing a table with *arbitrarily* chosen values for x and computed values for y, we have

x	0	4	6
y	-3	3	6

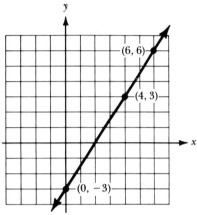

Figure 7.24

Plotting the ordered pairs gives us the linear graph shown in Fig. 7.24. ☐

Method (b): Graphing Linear Functions by the Intercept Method

Each linear function of the form $ax + by = c$, where $a \neq 0$ and $b \neq 0$, will intersect both the x- and y-axes. Some examples are given in Fig. 7.25.

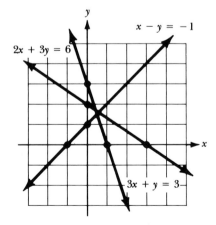

Figure 7.25

■ **DEFINITION**

INTERCEPTS

The point at which a line crosses the y-axis is called the y-*intercept*. The x-coordinate of that point is 0. The point at which a line crosses the x-axis is called the x-*intercept*. Its y-coordinate is 0.

■ **EXAMPLE 17**

Solution

Use intercepts to draw the graph of the linear equation $2x + 3y = 12$.

We first find the y-intercept by letting $x = 0$:

$$2(0) + 3y = 12$$
$$3y = 12$$
$$y = 4.$$

The y-intercept is then $(0, 4)$. For the x-intercept, we let $y = 0$:

$$2x + 3(0) = 12$$
$$2x = 12$$
$$x = 6.$$

The x-intercept is then $(6, 0)$.

While we could sketch the graph using only these two points, we would be better off obtaining a third point to serve as a "check" on the two intercepts. We choose any value for x, or for y, and then calculate the other corresponding coordinate. Let's suppose x were chosen to be 3; then

$$2(3) + 3y = 12$$
$$6 + 3y = 12$$
$$3y = 6$$
$$y = 2.$$

Our three points for graphing are

$(0, 4)$, $(6, 0)$,   and $(3, 2)$.

The three points are *collinear* (lie on the same straight line) so the probability is pretty high that all three points are correct. The line is sketched in Fig. 7.26.   □

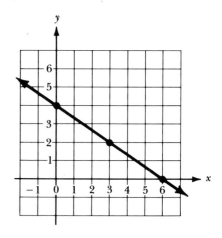

Figure 7.26

Method (c): Graphing Linear Functions by the Slope, y-intercept Form

Another method for graphing linear functions uses the *slope* or "steepness" of a line. For example, both ski slopes in Fig. 7.27 have a vertical height of 1000 feet. Are the slopes the same? Which slope would you be more willing to ski down as a beginner?

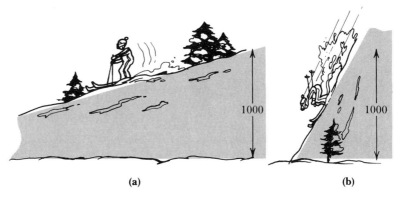

(a)   (b)

Figure 7.27

We can better determine the "steepness" or slope of a line if we know the horizontal distance as well as the vertical height. The **slope** is measured by the *ratio of the vertical distance to the horizontal distance*. It is sometimes referred to as the ratio of the *rise* to the *run*. The slope in Fig. 7.28(a) is 1000 to 5000, or 1 to 5. A slope of 1 to 5, or $\frac{1}{5}$, means that for every 5 units measured horizontally, there is an increase of 1 unit vertically. The slope in Fig. 7.28(b) is 1000 to 100, or 10 to 1, or simply $\frac{10}{1}$.

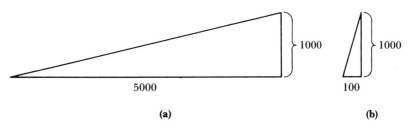

5000   100

(a)   (b)

Figure 7.28

● Question

Given the two points (2, 1) and (4, 5), how could we find the slope of the line passing through them?

Answer

Given the points and the associated line (shown in Fig. 7.29), we need a measure of the rise and run. A measure of the rise, or the change in the y-values, can be obtained by subtracting the y-coordinates, or 5 −

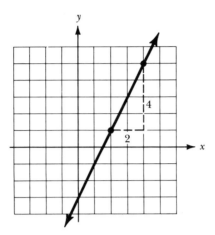

Figure 7.29

$1 = 4$. A measure of the run, or the change in the x-values, can be found by subtracting the x-coordinates, or $4 - 2 = 2$. The slope is the ratio of the rise to the run. Here, the slope $= \frac{4}{2}$ or $\frac{2}{1}$. For each 1 unit traveled to the right ($+1$), the vertical rise is 2 (upward because 2 is positive).

Now for a more mathematically precise definition of slope.

■ DEFINITION

SLOPE

The *slope* of a line through the two given points

$$P_1(x_1, y_1) \quad \text{and} \quad P_2(x_2, y_2)$$

is the change in y divided by the change in x; that is,

$$\text{Slope} = \frac{\text{change in } y}{\text{change in } x} = \frac{\Delta y}{\Delta x} = \frac{y_2 - y_1}{x_2 - x_1},$$

where Δ is a symbol (read "delta") to represent "change in," or "increment."

A *horizontal* line has a slope of *zero*; a *vertical* line has a slope that is *undefined*.

■ **EXAMPLE 18**   Find the slope of the line passing through the two points $(x_1, y_1) = (1, 3)$ and $(x_2, y_2) = (5, 4)$.

Solution   We have that

$$\text{Slope} = \frac{\Delta y}{\Delta x} = \frac{4 - 3}{5 - 1} = \frac{1}{4}.$$

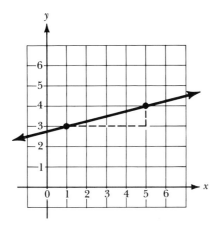

Figure 7.30

Figure 7.30 illustrates this ratio of 1 to 4 for the rise to the run. □

Three Applications of Slope

Applying Slope to Help Graph Linear Equations

Our original intent in learning about slope was to graph linear functions; this will be the first application. If we begin with a linear equation $2x - 3y = 6$, our only method of plotting this equation thus far has been to explicitly solve for y:

$$2x - 3y = 6$$
$$-3y = -2x + 6$$
$$y = (\tfrac{2}{3})x - 2.$$

First, let's plot some points and graph this line—then we'll discover something of surprising interest!

x	2	3	6
y	-2	0	2

Study the graph in Fig. 7.31 and then study the equation after it has been solved for y: $(y = (\tfrac{2}{3})x - 2)$.

a) What is the slope of the linear function?

$$\frac{0 - (-2)}{3 - 0} \quad \text{or} \quad \frac{2}{3}$$

b) What is the coefficient of x? $\tfrac{2}{3}$

c) Is this a coincidence? *No!*

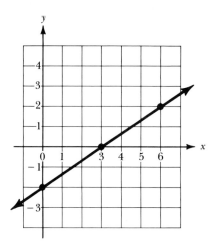

Figure 7.31

d) What is the significance of the -2 in $y = (\frac{2}{3})x - 2$?
e) Where does the line cross the y-axis? -2
f) Is this a coincidence?   *No!*
    (If $x = 0$, then $y = (\frac{2}{3})(0) - 2$ and $y = -2$!)

■ DEFINITION

SLOPE, y-INTERCEPT FORM

An equation solved explicitly for y,

$$y = mx + b,$$

is said to be in *slope, y-intercept form,* where m represents the slope and b is the y-intercept.

■ **EXAMPLE 19**

Using the definition above, graph $x + 2y = 6$ by means of the line's slope and y-intercept.

Solution

Solving for y, we have $y = -(\frac{1}{2})x + 3$. The slope is $-(\frac{1}{2})$ and the y-intercept is 3, or, written as an ordered pair, (0, 3). First, we plot the point (0, 3), and then we draw a line with slope $-(\frac{1}{2})$ through this point. To get a slope of $-(\frac{1}{2})$, we can use either $-1/+2$ or the other form, $+1/-2$. A slope of $-1/+2$ indicates a horizontal move of 2 units to the right, and then 1 unit down. A slope of $+1/-2$ indicates a horizontal move of 2 units to the left, and then 1 unit up. (See Fig. 7.32.) □

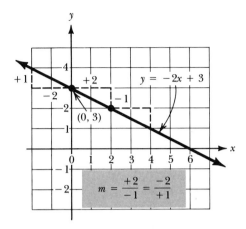

Figure 7.32

Using Slope to Determine Parallelism

A second application of slope is to determine whether two lines are *parallel*. For example, given the two equations $3x - 4y = 12$ and $6x - 8y = 16$, find the slope of each line:

$$3x - 4y = 12 \qquad\qquad 6x - 8y = 16$$
$$-4y = -3x + 12 \qquad\qquad -8y = -6x + 16$$
$$y = \tfrac{3}{4}x - 3; \qquad\qquad y = \tfrac{6}{8}x - 2.$$

The lines have the same slope ($\tfrac{3}{4} = \tfrac{6}{8}$) and the intercepts are different, so the lines are parallel, as we can see in Fig. 7.33.

Note: If the slope m is *positive*, the line "rises" as the line moves to the right. If the slope is *negative*, the line "falls" as the line moves to the right.

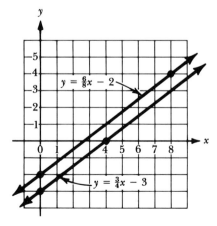

Figure 7.33

● Question

If two linear equations have equal slopes *and* identical *y*-intercepts, what can be said of the two equations?

Using Slope to Find Rate of Change

A third application of slope is of great importance in business because slope is a measure of the *rate of change* of one variable in relationship to a second variable. As an example, let's compare the change in stock prices over the ten-year period from 1970 to 1980, as shown in Fig. 7.34.

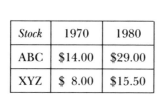

Stock	1970	1980
ABC	$14.00	$29.00
XYZ	$ 8.00	$15.50

(a)

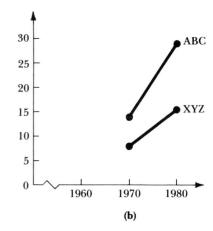

(b)

Figure 7.34

Over the ten-year period, ABC stock rose an average of

$$\frac{29 - 14}{1980 - 1970} = \frac{15}{10} \quad \text{or} \quad \$1.50 \text{ per year.}$$

(This reflects the change in price divided by the change in time.)
XYZ stock rose an average of

$$\frac{15.50 - 8}{1980 - 1970} = \frac{7.5}{10} \quad \text{or only} \quad \$0.75 \text{ per year.}$$

Through the concept of slope, we can quickly summarize many business situations and easily compare relative sizes.

A further illustration of the use of slope appears in *cost analysis*. Total cost is generally the sum of the variable cost and the fixed cost. A typical mathematical model used in cost analysis is $C = 2.5x + 1800$, where the variable cost is $2.50 per item and the fixed cost is $1800. Does $C = 2.5x + 1800$ look at all similar in form to the equation $y = mx + b$?

Exercise Set 7.4

For each of the following pairs of points, find the slope of the line passing through the points.

1. $(1, 2), (4, 5)$ **2.** $(1, 2), (4, 7)$

3. $(1, 2), (2, 7)$ **4.** $(0, 0), (4, 3)$

5. $(0, -2), (3, 0)$ **6.** $(-2, 4), (1, 5)$

7. $(-3, -1), (1, 4)$ **8.** $(-4, 2), (3, 2)$

9. $(1, 1), (3, 3)$ **10.** $(-2, 3), (100, -84)$

11. $(-3, -2), (-6, -5)$ **12.** $(3, 2), (-6, -5)$

For each of the following linear functions, (a) solve explicitly for y, (b) find the slope of the line, (c) find the y-intercept, and (d) graph the line.

13. $2x - 3y = 6$ **14.** $2x - 3y = 9$

15. $2x + 3y = 6$ **16.** $3x - 2y = 6$

17. $3x - 2y = 4$ **18.** $y - 2 = 0$

19. $x - 3y = 0$ **20.** $9x - 12 = 4y$

21. $5x - 2y = 0$ **22.** $2.6x - 1.3y = 5.2$

Graph each of the following lines with the given conditions.

23. Passing through the points $(-1, 3)$ and $(2, -1)$

24. Having a slope of $\frac{2}{3}$ and a y-intercept of $(0, 1)$

25. Having a slope of $-\frac{2}{3}$ and a y-intercept of $(0, 1)$

26. Having a slope of $\frac{3}{2}$ and passing through $(4, 1)$

27. Having an x-intercept of 3 and a y-intercept of 2

28. a) For $3x - 4y = 24$, the slope is _____, the x-intercept is (_____, _____), and the y-intercept is (_____, _____).

    b) For $3x - 4y = 12$, the slope is _____, the x-intercept is (_____, _____), and the y-intercept is (_____, _____).

    c) Plot both linear equations on the same set of axes. Any conclusions?

If a line has slope m and passes through a specific point (x_1, y_1), then the equation of the line is given by the formula

$$y - y_1 = m(x - x_1).$$

Find the linear equation, in standard form, satisfying each of the following conditions.

29. $m = 3, (x_1, y_1) = (4, 0)$

30. $m = \frac{1}{3}$, $(x_1, y_1) = (2, -2)$

31. $m = \frac{3}{4}$, and the y-intercept is 1

32. $m = -\frac{3}{4}$, and the x-intercept is 0

33. Parallel to the line $3x - 2y = 6$ and having a y-intercept of -2

34. Passing through the two points $(-3, 1)$ and $(5, 7)$

35. a) The line represented by $y = 2x + b$ passes through the point $(1, 7)$. Without graphing, find the value of the y-intercept.

    b) The line represented by $y = mx - 2$ passes through the point $(3, -4)$. Without graphing, find the slope m.

    c) Without graphing, find where the line $y = 3x - 2$ crosses the line $x = 2$.

36. The relationship between Fahrenheit and Celsius temperature is given by the formula

$$F = \tfrac{9}{5}C + 32.$$

Show this relationship graphically. What is the slope of the line you drew? Where does the line cross the y-axis (in this case, the F-axis)?

37. The total cost function of a certain product is given by

$$TC = f(u) = 0.3u + 1200,$$

where TC represents the total cost (in dollars) and u is the number of units produced.

    a) Find the total cost if 24,000 units are produced.

    b) What is the fixed cost per unit?

    c) What is the variable cost?

    d) What is the total cost if no units are produced?

    e) Sketch the function, identifying the slope and the u and the TC intercepts.

7.5

Graphing Linear Inequalities

In Section 7.4, we studied equations of the form $ax + by = c$ and found them to represent straight lines when graphed. Although linear equations have many direct applications, the most important indirect application is in determining the solution set for linear inequalities, a topic very useful in linear programming.

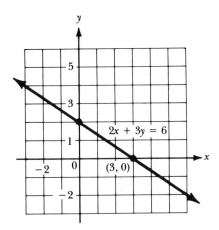

Figure 7.35

The graph of $2x + 3y = 6$, sketched in Fig. 7.35, represents *all* the points that *satisfy* the equation. For example, (3, 0) satisfies the equation and thus lies on the line. Conversely, any point on the line will in turn satisfy the equality $2x + 3y = 6$.

■

> In general, a linear equation, when graphed, divides the xy-plane into *three* areas:
>
> 1. all the points on the line,
> 2. all the points in the half-plane below the line,
> 3. all the points in the half-plane above the line.

Examining Fig. 7.35, let's plot the points (1, 3), (2, 2), (2, 3), (3, 1), (3, 2), and (3, 3). Substituting each ordered pair into $2x + 3y = 6$, we have

x	y	$2x + 3y$? 6	
1	3	11	> 6
2	2	10	> 6
2	3	13	> 6
3	1	9	> 6
3	2	12	> 6

We see that each of these points *satisfies the linear inequality* $2x + 3y > 6$.

A *linear inequality* in two variables, x and y, describes a region of the plane called a *half-plane*.

● Question

Where are all the points that satisfy the linear inequality $2x + 3y < 6$?

Answer

The linear equation $2x + 3y = 6$ separates the xy-plane into three distinct regions. All points *on* the line are represented by $2x + 3y$ *equals* 6. By our prior computation we see that all points chosen above the line are represented by the linear inequality $2x + 3y > 6$. Thus the only remaining region, the points below the line, must be represented by the linear inequality $2x + 3y < 6$. We can verify this by randomly checking some points in the shaded region shown in Fig. 7.36. The points $(0, 0)$, $(-3, 1)$, and $(2, -4)$, when substituted into $2x + 3y$, yield, respectively, 0, -3, and -8, all numbers *less* than 6.

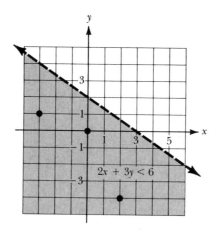

Figure 7.36

■ **EXAMPLE 20**

Graph and label the linear inequalities for

a) $x + y > 5$, and
b) $x + y < 5$.

Solution

a) We begin with the linear equation $x + y = 5$, which is sketched in Fig. 7.37. (*Hint:* The intercepts are $(5, 0)$ and $(0, 5)$.) All points *on* the line have coordinates each of *whose sum equals* 5. To graph $x + y < 5$, we choose *any* point on either side of the line. The origin $(0, 0)$ is a point that is easily seen to satisfy $x + y < 5$. The half-plane containing $(0, 0)$ then represents the linear inequality $x + y < 5$.
b) For the linear inequality $x + y > 5$, the region on the other side of the line is shaded. This and a few representative points are shown in Fig. 7.38.

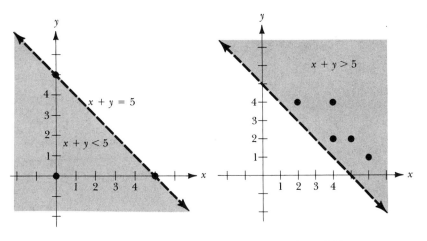

Figure 7.37   **Figure 7.38**

● Question

Where are the points located that satisfy $x - 2y \le 8$?

Answer

In this case, $x - 2y$ can either *equal* 8 *or* be *less than* 8. The graph is shown for $x - 2y = 8$ in Fig. 7.39.

  We obtain the graph for $x - 2y < 8$ by testing a point and shading the corresponding region. Let's arbitrarily choose the point $(0, 0)$, which *does* satisfy the inequality $x - 2y < 8$, so the half-plane is shaded as in Fig. 7.40. The union of these two sets,

$$x - 2y = 8 \quad \text{and} \quad x - 2y \le 8,$$

is formed for the answer to $x - 2y \le 8$. Points in the solution set can be either *on the line* or *in the shaded region*.

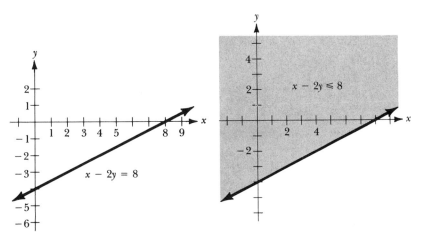

Figure 7.39   **Figure 7.40**

Generally, if the line is to be included as part of the solution set, it will be graphed as a solid line. For a strict linear inequality such as $x - 2y < 5$, the line itself is *not* part of the solution set, and is represented by a dotted or broken line.

Exercise Set 7.5

Graph each of the following linear inequalities.

1. $y > x + 2$
2. $y \leq x + 2$
3. $y \geq 7$
4. $y \geq 2x + 1$
5. $y \leq 2x + 1$
6. $3x - 2y < 6$
7. $3x + 2y < 6$
8. $y \leq 0$
9. $y \leq \frac{2}{3}x - 1$
10. $3x + 3y \leq 0$
11. $x \leq -2$
12. $x \geq 0$
13. $y > x$
14. $4x - 3y \leq -12$
15. $4.3x - 2.58y \leq 4.85$
16. $x \geq 1.4y - 5.6$

The solution set that contains ordered pairs *common* to two inequalities is located in the region where the respective two shaded areas overlap. Graph each of the following *groups* of linear inequalities on the same set of axes, and then shade the common intersection.

17. $y > 3x - 2$ and $x + y < 5$
18. $x + 2y \leq 12$ and $2x + y \leq 12$
19. $x + y < 4$, $y > x - 2$, and $y \geq 0$
20. $0.4x - 0.3y > 12$, $y < 4.2x - 5$, $y \leq 8$, and $x > 0$
21. Explain why most linear inequalities, such as $y \geq 3x - 2$ for example, are *not* functions.

7.6

Nonlinear Functions: Quadratic

Many situations in data processing are modeled by a linear relationship. However, there are as many—if not more—situations requiring formulas and equations that are expressed as nonlinear functions, that is, whose graphs are *not* straight lines. The graphs of polynomial functions of second degree or higher are nonlinear curves. Functional relationships between supply, demand, unit price, profit, production of goods, and compound interest are but a few examples of those requiring nonlinear models. Often the choice of a mathematical model (equation or formula) is predetermined or dictated by the situation. Most of the time, though, various models are tried until one is found that fits best, or most accurately describes, the functional relationship. With the high speeds of today's modern computers, the particular model can be determined in a very efficient and rather systematic manner.

■ **DEFINITION**

QUADRATIC FUNCTION

A model used quite frequently is called the *quadratic function* and is represented by an equation of the form

$$y = f(x) = ax^2 + bx + c,$$

where a, b, and c are constants and $a \neq 0$.

The righthand side of the equation above was studied in Chapter 6, where solutions were sought to such quadratic equations as $x^2 - 5x + 6 = 0$, $2x^2 + 5x - 3 = 0$, and $x^2 - 4 = 0$.

Two rather typical problems that are of interest to most businesses and that also use the quadratic function are (1) finding the minimum cost, and (2) finding the maximum profit.

Minimum Cost

An example from economics is the functional relationship between the cost per unit of production and the number of units produced. To a point, the more items produced, the lower the cost per unit to produce each item—hence the idea of mass production. The quadratic function $C = f(u) = 10u^2 - 60u + 100$ expresses cost as a function of the number of units produced. By substituting values for u, we can prepare a table such as the one shown in Fig. 7.41(a). We

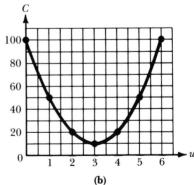

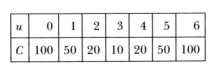

u	0	1	2	3	4	5	6
C	100	50	20	10	20	50	100

(a)

(b)

Figure 7.41

then use the points to plot the quadratic function. The graph produced by joining the tabulated points from a quadratic function is called a **parabola**. The parabola for this cost function is shown in Fig. 7.41(b).

As you can see from Fig. 7.41, the graph is a much more descriptive device than either the equation or the table. At a glance, how many units of production will yield the *minimum cost*? (At three units of production, the cost per unit will be $10.00.) Why would the cost per unit—after decreasing from zero to three units—again increase?

Maximum Profit

Who in business does not want the greatest possible profit? Many profit functions in business are quadratic.

■ **EXAMPLE 21**

Suppose the relationship between production units and total profit is given by the quadratic function

$$P = f(u) = -100u^2 + 600u - 500.$$

Graph the quadratic function, and from the graph, determine the number of units that will yield the greatest profit.

Solution

We begin by determining a table of values (Fig. 7.42a). Plotting these ordered pairs on a graph, we find the profit function as shown in Fig. 7.42(b). At a glance, we see that maximum profit occurs if three units are produced.

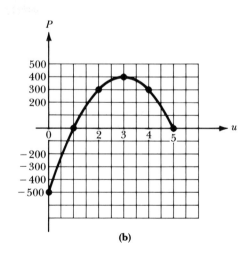

u	P
0	−500
1	0
2	300
3	400
4	300
5	0

(a)

(b)

Figure 7.42

The maximum profit appears to be \$400.00. This could be verified by substituting 3 into the profit function:

$$P = -100(3)^2 + 600(3) - 500, \quad \text{or} \quad P = \$400.00. \quad \square$$

From the illustrations on minimum cost and maximum profit, much can be learned about quadratic functions. As we have already stated, the graph of a quadratic function will be a U-shaped curve called a *parabola*. If the coefficient of the second-degree term in $y = f(x) = ax^2 + bx + c$ is positive, the parabola opens upward (the curve is said to be concave upward). In the example on minimum cost, the coefficient of u^2 is 10 and the curve opens upward.

If the coefficient of the second-degree term is negative, the parabola opens downward (concave downward). In Example 21 the coefficient of u^2 is -100 and the parabola opened downward. The lowest point (highest point) on a parabola opening upward (downward) is called the **vertex**. In our two examples, we see that the vertex of the quadratic function for minimum cost is (3, 10), and the vertex representing the maximum profit is (3, 400). (In the next few pages, we will derive a method for finding the vertex.)

A Method for Graphing Quadratic Functions

The methods used to graph the two preceding examples worked primarily because we could easily graph the quadratic functions by completing a table of values. For more complicated quadratic functions, we use a more general method to graph the corresponding parabola. Because we know that a quadratic function will be graphed as a parabola, we can often make a very efficient sketch by locating three points: the x-intercepts and the vertex.

We build on our earlier work on quadratic equations (you may wish to review quadratic equations in Chapter 6). At the x-intercepts (the points at which the graph intersects the x-axis), the y-coordinates must be zero. Thus $y = ax^2 + bx + c$ becomes $0 = ax^2 + bx + c$. The x-intercepts are called the *zeros* (solutions or roots) of the function. (The work done earlier in Chapter 6 on solving a quadratic equation for x was really a special application of finding the x-intercepts for the quadratic function, $y = ax^2 + bx + c$.)

■ **EXAMPLE 22**

Find the x-intercepts (zeros) of the quadratic function $y = x^2 - 6x + 5$.

Solution

Setting $y = 0$, we have

$$0 = x^2 - 6x + 5,$$

and by factoring, we get

$$0 = (x - 5)(x - 1).$$

Thus the zeros are

$$x = 5 \quad \text{and} \quad x = 1. \quad \square$$

What if the quadratic equation does not factor easily?

■ EXAMPLE 23   Find the x-intercepts (zeros) of the quadratic function $y = 4x^2 - 24x + 11$.

Solution   First, we let $y = 0$. We could try factoring, but if $y = 0$, then the quadratic equation $4x^2 - 24x + 11 = 0$ could be solved for x by use of the quadratic formula:

$$x = \frac{-b \pm \sqrt{b^2 - 4ac}}{2a}.$$

The two intercepts are then

$$\frac{-b + \sqrt{b^2 - 4ac}}{2a} = \frac{-(-24) + \sqrt{576 - 4(4)(11)}}{(2)(4)} = 5.5$$

and

$$\frac{-b - \sqrt{b^2 - 4ac}}{2a} = \frac{-(-24) - \sqrt{576 - 4(4)(11)}}{(2)(4)} = 0.5.$$

The parabola crosses the x-axis at the points (5.5, 0) and (0.5, 0). $\square$

If the roots (zeros) are *real and unequal* (as in the example above), the parabola crosses the x-axis at two distinct points. If both roots (zeros) are *real and equal,* the parabola will be tangent to the x-axis; that is, the vertex will be the only point touching the x-axis. (For an example of this, graph $y = x^2 - 6x + 9$.)

If the roots are *imaginary* (the value of the discriminant $b^2 - 4ac$ is less than zero), the graph of the parabola will not intersect the axis. (For an illustration of a quadratic function with imaginary solutions, try finding the zeros of the quadratic function given earlier to minimize cost, $C = f(u) = 10u^2 - 60u + 100$.)

The different types of possible roots to a quadratic equation are summarized in Table 7.4.

How is the *vertex* found? Let's begin by studying the graph of a "typical" quadratic function. For $y = ax^2 + bx + c$, the two zeros are given by

$$\frac{-b - \sqrt{b^2 - 4ac}}{2a} \quad \text{and} \quad \frac{-b + \sqrt{b^2 - 4ac}}{2a}.$$

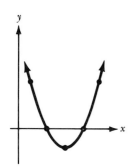

Figure 7.43
A quadratic function.

Locate the *vertex* in Fig. 7.43 in relationship to the *zeros.*

Table 7.4
Quadratic function
with real coefficients

DISCRIMINANT	Positive and perfect square	Positive and not perfect square	Zero	Negative
TYPE OF ROOTS	Real, un-equal, and rational	Real, unequal, irrational	Equal	Imaginary

The graph of a parabola should not appear as shown in Fig. 7.44(a), but should be *symmetrical* (Fig. 7.44b). By symmetrical we mean that the lefthand portion of the curve is a "mirror image" of the righthand portion.

To better understand symmetry, try this. On a light sheet of paper draw a parabola with heavy ink. Now fold this sheet of paper on a dotted line through the vertex (see Fig. 7.44b). If the sheet of paper is now held up to the light, the right side should (if the graph were drawn perfectly) coincide with the left side.

Symmetry is used in the following manner. If the two intercepts of a parabola are known to be

$$\frac{-b - \sqrt{b^2 - 4ac}}{2a} \quad \text{and} \quad \frac{-b + \sqrt{b^2 - 4ac}}{2a},$$

where is the vertex located? Figure 7.45 shows the intercepts. The

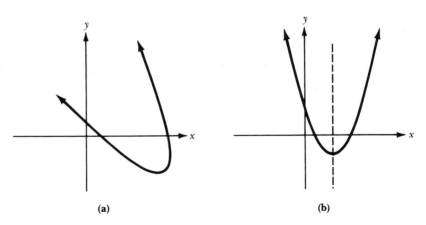

(a)   (b)

Figure 7.44 (a) Not a typical quadratic; (b) a typical quadratic (symmetrical).

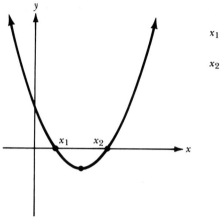

$$x_1 = \frac{-b - \sqrt{b^2 - 4ac}}{2a}$$

$$x_2 = \frac{-b + \sqrt{b^2 - 4ac}}{2a}$$

Figure 7.45

x-coordinate of the vertex is halfway between x_1 and x_2. To find the *x*-value midway between x_1 and x_2, we take the "average" of x_1 and x_2:

$$\frac{x_1 + x_2}{2} = \frac{(-b - \sqrt{b^2 - 4ac})/2a + (-b + \sqrt{b^2 - 4ac})/2a}{2}$$

$$= \frac{-2b/2a}{2} \quad \text{or} \quad \frac{-b}{2a} \, !$$

How do we find the *y-coordinate* of the vertex? By substituting $-b/2a$ into the quadratic equation $y = ax^2 + bx + c$, we get

$$y = a(-b/2a)^2 + b(-b/2a) + c$$
$$y = (4ac - b^2)/4a.$$

■

In summary, for the quadratic function

$$y = f(x) = ax^2 + bx + c,$$

the *x-intercepts* are given by

$$x_1 = \frac{-b + \sqrt{b^2 - 4ac}}{2a} \quad \text{and} \quad x_2 = \frac{-b - \sqrt{b^2 - 4ac}}{2a},$$

and the *vertex* is given by

$$\left(\frac{-b}{2a}, f\left(\frac{-b}{2a}\right)\right) \quad \text{or} \quad \left(\frac{-b}{2a}, \frac{(4ac - b^2)}{4a}\right).$$

■ EXAMPLE 24

Find the number of units of a product that must be sold in order to give the maximum profit if the profit function is given by

$$y = P(x) = -2x^2 + 40x - 150.$$

Solution

Before sketching the quadratic function, we note that the coefficient of x^2 is negative, so the parabola opens downward (concave down).

To find the x-intercepts, we use the quadratic formula with $y = 0$ to get

$$\frac{-40 \pm \sqrt{1600 - 1200}}{-4} \quad \text{or} \quad \frac{-40 \pm 20}{-4}$$

or

$$x_1 = \frac{-40 + 20}{4} = 5 \quad \text{and} \quad x_2 = \frac{-40 - 20}{4} = 15.$$

The x-intercepts (zeros) are $(5, 0)$ and $(15, 0)$.

To find the vertex, we note that halfway between $x = 5$ and $x = 15$ is $x = 10$. We could also have used

$$\frac{-b}{2a} = \frac{-40}{-4} = 10.$$

If $x = 10$, we substitute 10 into $y = -2x^2 + 40x - 150$ to get $y = 50$, which we could also have found by using

$$\frac{(4ac - b^2)}{4a} = \frac{-400}{-8} = 50.$$

The parabola $y = -2x^2 + 40x - 150$ with x-intercepts $(5, 0)$ and $(15, 0)$ and vertex $(10, 50)$ is graphed in Fig. 7.46.

The answer to the original question is that by producing 10 units, a maximum profit of $50.00 will be obtained!   □

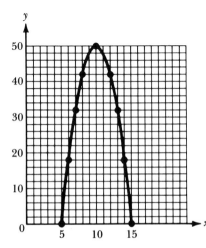

Figure 7.46

It is of interest to note that even if the parabola does not have x-intercepts, the vertex can still be found by using the same technique, that is, using the formula for x_1 and x_2 as above.

Exercise Set 7.6

1. Identify all ordered pairs from the list below that are solutions to the quadratic function $y = x^2 - 2x - 8$.

$$(3, 7), (-3, 7), (2, 0), (-2, 0), (0, -2), (1, -9),$$
$$(-1, 9), (0, -8), (4, 0), (5, 7), (1, 5), (-1, -5)$$

Then, with the *correct* ordered pairs, graph the function.

2. Plot the ordered pairs given in the table below. Determine which quadratic function "best" fits the graph of the ordered pairs.

x	0	1	2	3	4	5	6
y	12	5	0	-3	-4	-3	0

a) $y = 2x^2 - 16x + 24$

b) $y = x^2 - 8x + 12$

c) $y = 0.5x^2 - 4x + 12$

3. a) Graph the function defined by $y = x^2 - 4x - 5$. What are its zeros?

   b) Find the roots to the quadratic equation $0 = x^2 - 4x - 5$.

   c) Explain the relation between the answers to parts (a) and (b) and the graph in part (a).

4. Find the roots to each quadratic equation graphically: (1) Set each algebraic expression on the lefthand side equal to y. (2) Graph the function. (3) Estimate the roots by examining the graph.

   a) $x^2 + 2x - 3 = 0$ b) $x^2 - 4 = 0$

   c) $1 - x^2 = 0$ d) $6 + 5x - x^2 = 0$

Graph each of the following quadratic functions, and identify its x-intercepts and vertex.

5. $y = x^2 - 4$ 6. $y = -(x^2 - 4)$

7. $y = x^2 - 4x - 12$ 8. $y = -x^2 + 4x + 12$

9. $y = x^2 + 5x - 6$ 10. $y = x^2$

11. $y = 3x^2 - 12x$ 12. $y = f(x) = 7 + 4x - 2x^2$

13. $y = x^2 + 3x + 2$ 14. $y = x^2 - 3x + 2$

15. $y = 120x - 4x^2$ 16. $y = x - x^2$

17. $y = 5 - 8x - x^2$ 18. $y = 2(x^2 - 2)$

19. Explain how the value of the discriminant $b^2 - 4ac$ in

$$x = \frac{-b \pm \sqrt{b^2 - 4ac}}{2a}$$

characterizes the type of intercepts for the quadratic function

$$ax^2 + bx + c = y.$$

For each of the following quadratic functions, determine the values of k for which the graph of the function will (1) be tangent to the x-axis, (2) cross the x-axis at two distinct points, or (3) not intersect the x-axis at all.

20. $y = f(x) = 9x^2 + kx + 4$

21. $y = f(x) = 2x^2 - 4x + k$

22. $y = f(x) = 4x^2 + kx + 9$

23. a) Graph the equality $y = x^2 - 9$.

  b) Graph the inequality $y < x^2 - 9$.

  c) Graph the inequality $y > x^2 - 9$.

24. Sketch the graph of the quadratic function $y = f(x) = -(x^2 - 20x + 60)$. When will the y-value reach a maximum, and what will that maximum value be?

25. The total cost of manufacturing x units of a particular commodity is given by

$$C(x) = x^2 - 6x + 10.$$

Sketch the total cost function, and determine how many units will yield the *least* total cost.

26. Average cost is the total cost divided by the number of units. For Exercise 25, find the average cost function, and then sketch the average cost function on the same set of axes as the total cost.

27. Find the point of intersection (if any) of the linear function and the quadratic function by drawing the graph of each.

  a) $y = f(x) = x + 2$ and $y = g(x) = x^2 - 4$

  b) $2x + y = 15$ and $y = x^2$

28. The supply function for a product is given by

$$S = f(P) = P^2 + 2P,$$

where supply is a function of the price per unit. The demand function for the product is represented by

$$D = f(P) = 40 - P^2,$$

where demand is a function of the price per unit. Plot both quadratic functions on the same set of axes to determine the point(s) of equilibrium (where supply equals demand).

Summary

The concept of a function is one of the most basic, as well as widely applied topics in all of mathematics. In this chapter, we looked at relationships between two or more variables and systematically studied them with the use of functions. We learned that mathematical models (functions) are developed to approximate complex relationships and that graphs of functions simplify and provide an intuitive understanding of these relationships.

Like a computer, in which the three main components are input, processing, and output, a function also has three components: the *domain* (input), a *relationship* between the domain and the range, and the *range* (output). A *function* is a special type of relationship in that no two elements of the range may share the same element in the domain.

Special classifications of functions include polynomial functions, exponential functions, and logarithmic functions. *Polynomial functions* are those given by an equation of the form

$$y = a_n x^n + a_{n-1}x^{n-1} + a_{n-2}x^{n-2} + \cdots + ax + a_0,$$

where n is a nonnegative integer, and $a_n, a_{n-1}, \ldots, a_0$ are real-number constants. Usually business applications are limited to polynomials of degree two or less. First-degree polynomials, or *linear functions*, are widely used in mathematics and business; simple interest, simple discounts, linear depreciation, and linear regression are a few applications. Second-degree polynomials, or *quadratic functions*, are even more widely used. Determining maximum profit and/or minimum cost are but two of the many uses of quadratic functions.

Functions, in general, have a wide area of application. In computer programming, the computer is often programmed to evaluate expressions. Equating these expressions to a symbol uses the notion of functions, while interpretation of these expressions relies heavily on graphing functions.

Review Exercises

Plot each of the following ordered pairs and identify in which quadrant the pair belongs.

1. $(-3, 2)$ 2. $(5, 1)$

3. $(-4, -5)$ 4. $(6, -2)$

5. $(-3, -0.5)$ 6. $(-4, 1.8)$

Identify which ordered pairs satisfy the functional relationship between x and y.

7. $y = 3x - 2$ $\quad$ (1, 1), (0, 2), (2, 0), (0, −2), (2, 4)

8. $3x - 2y = 6$ $\quad$ (0, 3), (2, 0), (0, −3), (0, −2), (−2, 0)

9. $y = x^2 + 1$ $\quad$ (1, 1), (−1, 2), (1, 2), (0, 1), (1, 0)

10. $y = 2^x$ $\quad$ (1, 2), (2, 1), (2, 4), (3, 6), (3, 8)

For each of the following equations, complete a table of ordered pairs and then graph. For each function, describe the domain and the range.

11. $y = 3x - 2$ $\qquad\qquad$ **12.** $y = |x|$

13. $y = x^2 - 9$ $\qquad\qquad$ **14.** $y = 3^{-x}$

15. $y = \sqrt{4 - x^2}$ $\qquad\qquad$ **16.** $y = \sqrt{x^2 - 4}$

Identify which of the following relations represent functions.

17.

x	y
1	3
2	4
3	5
4	6

18.

x	y
1	3
2	4
1	5

19.

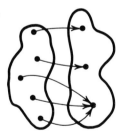

20.

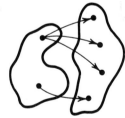

21.

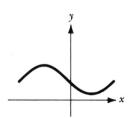

22.

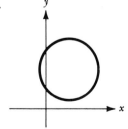

23.

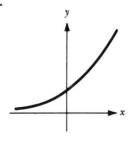

24.

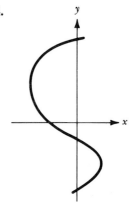

25. Using your best mathematical language, define the term *function*.

For the functions $f(x) = x^2 - 2$, $g(x) = 3x - 2$, and $h(x) = |x|$, find each of the following.

26. $f(3); f(-3); f(0)$ **27.** $g(3); g(-3); g(0)$

28. $h(3); h(-3); h(0)$ **29.** $f(3) + g(3); f(3) - g(3)$

30. $f(x) \cdot g(x); f(1) \cdot g(1)$ **31.** $f(g(1)); g(f(1))$

32. $f(g(h(2))); h(g(f(2)))$

Graph each of the following linear functions, labeling the slope, the x-intercept, and the y-intercept.

33. $y = 2x + 1$ **34.** $y = -\frac{2}{3}x + 2$

35. $2x - 3y = 12$ **36.** $1.3x - 3.9y = -5.2$

37. $x = \frac{2}{3}y + 1$ **38.** $y = x$

Find the common intersection of each of the following sets of linear inequalities.

39. $y \geq 2x - 1$ and $x + y \leq 5$

40. $y + 3 \geq \frac{2}{3}x$ and $y \leq 4 - x$

41. $x \geq 0, y \geq 0, 7x + 6y \leq 42,$ and $15x - 2y \leq 30$

Graph each of the following quadratic functions, labeling the intercepts (if any) and the vertex. Identify the function as being concave upward or concave downward.

42. $y = x^2 - 9$ **43.** $y = 9 - x^2$

44. $y = 2x^2 - 7x + 6$ **45.** $y = 2x^2 - 7x + 1$

46. $y = -x^2 + 6x$ **47.** $y = 1.6x^2 - 20.4x - 4.85$

48. Using the information in the following table, plot the yearly sales represented by s (in thousands) as a function of the rate of inflation (i).

i	3%	3.5%	5%	7%	9.5%	11%	10%	12%	9.4%
s	8.8	6.2	6.7	6.2	7.6	7.0	5.3	4.2	4.8

49. By sketching each linear function where $y = -3x + 10$ represents the cost
and $y = 2x - 5$ represents the revenue, find the point at which total cost
equals total revenue. (This is called the *break-even point.*)

50. If a cost function is given by the formula

$$y = C(x) = x^2 - 10x + 40,$$

determine the number of units that need to be produced so the total cost
will be a minimum.

51. For what values of x would the profit function

$$P(x) = -x^2 + 90x - 1000$$

yield a zero profit? For what values of x would the profit be the greatest?
Is it possible to obtain a profit of $20, and if so, for what value(s) of x?

Logic and Computer Programming

CHAPTER 8 OBJECTIVES

After completing this chapter, you should be able to:

1. Form compound statements using the logical connectives AND, OR, and NOT.
2. Construct truth tables for conjunction, disjunction, negation, conditional, and biconditional statements.
3. Draw a flowchart for a given complex logic statement.
4. Determine whether a complex logic statement is a tautology, or a contradiction, or possibly neither.

8

■ CHAPTER OUTLINE

8.1
Logical Operators: AND, OR, and NOT

Among the powerful capabilities of the computer is the ability to test conditions and implement steps on the basis of the conclusions drawn from those conditions. The steps are determined by the computer programmer. The fact that the computer can perform logical operations makes it far more powerful than any other type of calculating machine. The computer can compare two items to see whether they are equal, or whether one is greater than the other. It can also make decisions on the basis of whether a given statement is true or false. The programmer takes advantage of the computer's ability to compare and determine truth values by programming it to make decisions.

To be effective in utilizing the strengths of the computer, the programmer must understand some basic concepts of mathematical logic. He or she must comprehend the problem to be solved, see how the various components fit together, and then provide the computer with the proper sequence of steps that will lead to the solution of the problem. In mathematics, we say that this approach requires *deductive reasoning*. It is exactly the type of reasoning the programmer must use in instructing the computer so that a logical conclusion or solution can be made. The programmer must *input* valid information and provide the correct sequence of steps to the computer in order to have confidence in the *output,* or solution, provided by the computer.

The computer has been designed to use three types of *operators.* An operator is a symbol or code word that instructs the computer to perform a specific operation. The three types of operators are as follows.

■

ARITHMETIC OPERATORS	SYMBOL
1. Addition	+
2. Subtraction	−
3. Multiplication	*
4. Division	/
5. Exponentiation	**, ↑ , or ^

RELATIONAL OPERATORS	SYMBOL
1. Equality	=
2. Greater than	>
3. Less than	<
4. Greater than or equal to	>=
5. Less than or equal to	<=
6. Does not equal	<>

LOGICAL OPERATORS	SYMBOL
1. Conjunction	∧, AND
2. Disjunction	∨, OR
3. Negation	~, NOT

Logical operators allow us to construct complex test conditions for the computer to examine. In this section we will learn to use the logical operators by looking at their counterparts in mathematical logic. In mathematical logic, *conjunction, disjunction,* and *negation* are called *logical connectives.* We will also examine the conditional (IF–THEN) and biconditional (IF AND ONLY IF) connectives.

Logic is often referred to as the theory of reasoning. Logical reasoning involves combining two statements to form a new one using a connective or modifying a statement using the negation.

The most basic concept in logic is the **statement**. In the English language, sentences are classified as (1) interrogative, (2) exclamatory,

(3) imperative, and (4) declarative. Declarative sentences are unlike the other three forms in that they are the only type that can be determined as being true or false.

■ DEFINITION

STATEMENT

A declarative sentence, or mathematical equation, that has a truth value of either true or false, but not both, is called a *statement*.

■ **EXAMPLE 1**

a) "The number six is an even integer" is a true statement.
b) "The number -3 is a natural number" is a false statement.
c) "Today is really a nice day" expresses an opinion and does not, in general, have a truth value. □

The illustrations in Example 1(a) and 1(b) are called *simple* statements, since only one point is being made. Logical connectives can be used to combine simple statements into *compound* statements. The simple statements used to construct a compound statement are called the *components* of the statement. The truth value of a compound statement depends on the truth values of the component statements.

When the component statements of a compound statement are variables, the compound statement is called a *proposition*.

Negation

In Chapter 5, we studied the complement of a set. We found that the complement of a given set is a set containing all elements not in the given set, relative to the universal set. In logic, we have a similar situation, called **negation**. The negation of a statement has the opposite truth value of the given statement. In other words, the negation modifies the truth value of the original statement. This relationship can be shown by using a *truth table,* in which variables are used to represent statements and all possible cases are considered. If P is a statement, it may be either true or false depending on the situation. If we use $\sim$ to symbolize the negation, the table appears as shown in Table 8.1. We can also write the negation as NOT P. Note that NOT P is false when P is true and NOT P is true when P is false.

Table 8.1

P	$\sim P$
T	F
F	T

■ **EXAMPLE 2**

a) Given the statement $4 + 7 = 11$, the negation is $4 + 7 \neq 11$. The given statement is true. The negation is false.
b) Given the statement, "The moon is made of green cheese," the negation is, "The moon is NOT made of green cheese." The given statement is false. The negation is true. □

Conjunction

Two simple statements can be combined into one compound statement by the connective AND. Such a compound statement is called a **conjunction**. The conjunction closely resembles the intersection in set theory. As you recall, the intersection of two sets contained only those elements *common to both sets*. Similarly, the conjunction is true if and only if *both component statements are true*. Otherwise, the conjunction is false. Using the symbol $\wedge$ for conjunction, we can construct a truth table showing the possibilities that exist. When two statements, each with two possible truth values, are joined, we end up with four combinations of truth values. The conjunction truth table is shown in Table 8.2, and we have the following.

Case 1: P is true when Q is true.
Case 2: P is true when Q is false.
Case 3: P is false when Q is true.
Case 4: P is false when Q is false.

Table 8.2

	P	Q	$P \wedge Q$	
Case 1	T	T	T	T and T is T
Case 2	T	F	F	T and F is F
Case 3	F	T	F	F and T is F
Case 4	F	F	F	F and F is F

The conjunction is used to construct a complex problem in which several tests must be performed on a universal set in order to determine what particular action will be taken. Let's consider the following example.

■ EXAMPLE 3

Among other items, the records in a certain personnel department contain the following information on each employee.

1. Employee identification number
2. Employee name
3. Monthly salary
4. Age

A list of names and identification numbers of those employees who satisfy the following requirements is to be printed: Monthly salary greater than or equal to $2100.00 AND age greater than 35.

Solution

Only those individuals for whom the answer is true to both conditions will appear on the list. We can visualize the problem through a flow-chart similar to what we used in Chapter 5 when flowcharting A ∩ B.

(See Fig. 8.1.) We let

$$P = \text{Monthly} \geq 2100.00,$$
$$Q = \text{Age} > 35. \quad \square$$

Generally, the conjunction is expressed in English as AND. However, it may sometimes appear as "but," "however," or "nevertheless."

■ **EXAMPLE 4**   The statement $7 > 2$ AND $7 < 10$ has the same meaning as $7 > 2$ BUT $7 < 10$. $\square$

We have just seen how the flowchart of $P \wedge Q$ resembles the flowchart of $A \cap B$. We can also draw the flowchart of $\sim P$ by reversing the outputs of the decision symbol just as they are reversed for the flowchart of A'. (See Fig. 8.2.)

Disjunction

Another method of combining simple statements into a compound statement is called **disjunction.** The disjunction combines two statements with the connective OR, which is symbolized by $\vee$.

In English we have two uses for the word OR. If a person claims "Either I lost my car keys *or* I locked them in my car," two different conditions are being asserted:

1. The car keys are lost.
2. The keys are locked in the car.

This use of the word "or" is in the *exclusive* sense and means one or the

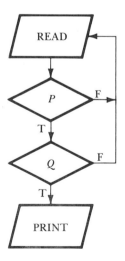

Figure 8.1 *P* AND *Q*

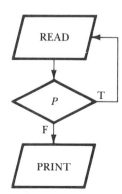

Figure 8.2 ∼*P* OR NOT *P*

other condition, but not both. We will refer to the **exclusive or** as EOR and $\overline{\vee}$. The following table shows the true–false combinations for the exclusive OR.

P	Q	P EOR Q
T	T	F
T	F	T
F	T	T
F	F	F

Note that when both P and Q are true, the exclusive or is false.

The other English use of "or" is in the *inclusive* sense. When the statement, "You must be at least 21 years old *or* have a minimum of 5 years experience" is made, two conditions are being established. However, it is conceivable that both conditions could be true at the same time. The inclusive disjunction actually means *at least one* of the component parts must be true, whereas the exclusive or means *exactly one* must be true. The following table shows the true–false combinations for the **inclusive or**.

P	Q	$P \vee Q$
T	T	T
T	F	T
F	T	T
F	F	F

Note that the inclusive or is false only when both component statements are false.

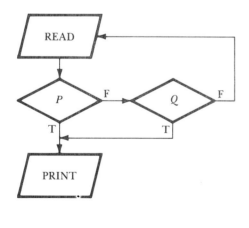

(a)

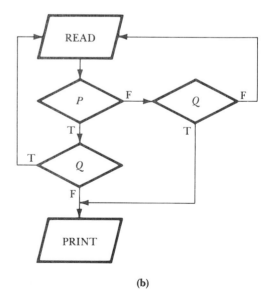

(b)

Figure 8.3   (a) $P \vee Q$; (b) $P \overline{\vee} Q$.

The programmer must be sure which "or" is intended when writing a program. Most computer languages do not provide the direct translation for the exclusive or.

Figure 8.3(a) shows the flowchart for the inclusive case and Fig. 8.3(b) for the exclusive case.

A distinction between OR and EOR is often made in legal documents by the use of and/or for the inclusive case and either/or for the exclusive case.

Exercise Set 8.1

Identify each of the following as either a statement or not a statement.

1. $3 + 7 = 10$

2. $4 \times 3 = 13$

3. Abe Lincoln was the 16th president of the United States.

4. Inflation dropped to 3% in 1982.

5. What time is it?

6. How beautiful you look tonight!

7. London is a city in England.

8. Where have you been?

9. $20 \div 5$

10. Put the key in the lock.

Consider the following statements. Let the variable represent the given statement as indicated.

P: I like to study mathematics.
Q: I study at least 20 hours per week.

Write the English sentence for each of the following.

11. P AND Q

12. Q AND NOT P

13. $\sim P \vee Q$

14. $\sim P \wedge \sim Q$

15. $\sim P \wedge Q$

16. $\sim P \vee \sim Q$

17. NOT (NOT P)

18. NOT P OR Q

Use the appropriate symbols to write logic propositions for each of the following verbal statements.

P: It is snowing.
Q: I am going skiing.

19. It is snowing and I am going skiing.

20. I am going skiing and it is not snowing.

21. It is snowing or I am not going skiing.

22. It is false that it is snowing and I am going skiing.

23. It is not snowing and I am not going skiing.

24. It is false that it is not snowing and I am going skiing.

Construct a truth table for each of the following compound statements.

25. $\sim P \wedge Q$ **26.** $P \vee \sim Q$

27. P AND (NOT Q)   **28.** P EOR (NOT Q)

29. NOT P EOR Q **30.** NOT P EOR (NOT Q)

31. $\sim (P \vee Q)$ **32.** $\sim P \wedge \sim Q$

33. $\sim (P \wedge Q)$ **34.** $\sim P \vee \sim Q$

(Note that the truth table for Exercise 31 is identical to that for Exercise 32 and the truth table for Exercise 33 is identical to the table for Exercise 34. These relationships are logically equivalent and are called DeMorgan's properties.)

Use DeMorgan's properties to write a statement equivalent to each of the following statements. Construct a truth table for the given statement and your statement to verify that they are logically equivalent.

35. $\sim (P \wedge \sim Q)$ **36.** $\sim (\sim P \vee \sim Q)$

37. NOT (NOT P OR Q)   **38.** NOT (NOT P AND (NOT Q))

8.2
Complex Logical Forms and Flowcharts

Not all problems can be stated as simple logical forms such as P OR Q and P AND Q. Fortunately, however, higher-level computer languages such as BASIC, COBOL, FORTRAN, PASCAL, and PL/I are capable of handling more complex logical statements.

■ **EXAMPLE 5**

If we expanded Example 3 to include even more information, a more complex problem could be stated.

1. Employee identification number
2. Employee name
3. Monthly salary
4. Accumulated sick leave in days
5. Age
6. Years of employment

Let's say we need a list of names and identification numbers of those employees who meet the following criteria:

Monthly salary greater than or equal to $2100.00 AND age greater than 35 AND accumulated sick leave less than or equal to 90 days OR years of employment NOT less than 6. (The "or" is interpreted as inclusive.)

Write the problem as a logic statement using the following simple statements.

$$P: \text{Monthly salary} \geq \$2100.00$$
$$Q: \text{Age} > 35$$
$$R: \text{Sick leave} \leq 90$$
$$S: \text{Years of employment} \not< 6$$

Solution P AND Q AND R OR NOT S □

When faced with a complex logical form such as the one in Example 5, we must be cognizant of the hierarchy of operations for logical connectives. Parentheses are interpreted just as they are in algebra. All operations on the same level are executed from left to right, with priority given first to negation, then conjunction, and finally disjunction. We can use parentheses to change the natural order.

■ **EXAMPLE 6**   In each of the following, numbers are used to indicate the order of execution.

a) P AND Q AND R OR NOT S
       2   3   4   1

b) P AND (Q AND R) OR NOT S
       3   1   4   2

c) (P AND Q) AND (R OR NOT S)
        1   4   3   2

d) P AND (Q AND (R OR NOT S))
       4   3   2   1   □

Truth tables for complex logical forms can be constructed, but they must follow the hierarchy of operations. The final column of truth values should be the truth values for the total compound statement.

As we saw earlier, one simple statement has two possible truth values: T or F. A compound statement formed by two simple statements has a total of four truth-value combinations: TT, TF, FT, and FF. A compound statement made up of three simple statements has eight possible truth-value combinations, as we see below:

P	Q	R
T	T	T
T	T	F
T	F	T
T	F	F
F	T	T
F	T	F
F	F	T
F	F	F

The number of truth-value combinations is determined by 2^n, where n is the number of unique simple statements and 2 represents the two-state condition of the statement, that is, either T or F. Remember: $2^1 = 2$, $2^2 = 4$, $2^3 = 8$, $2^4 = 16$, etc.

■ EXAMPLE 7   Construct the truth table for $P \wedge Q \vee {\sim}R$.

Solution

P	Q	R	${\sim}R$	$P \wedge Q$	$P \wedge Q \vee {\sim}R$
T	T	T	F	T	T
T	T	F	T	T	T
T	F	T	F	F	F
T	F	F	T	F	T
F	T	T	F	F	F
F	T	F	T	F	T
F	F	T	F	F	F
F	F	F	T	F	T

Note that the order in which the columns were formed follows the hierarchy of operations.   □

■ EXAMPLE 8   Construct the truth table for $P \wedge (Q \vee {\sim}R)$.

Solution

P	Q	R	${\sim}R$	$Q \vee {\sim}R$	$P \wedge (Q \vee {\sim}R)$
T	T	T	F	T	T
T	T	F	T	T	T
T	F	T	F	F	F
T	F	F	T	T	T
F	T	T	F	T	F
F	T	F	T	T	F
F	F	T	F	F	F
F	F	F	T	T	F

The inclusion of the parentheses, which changed the order of the operations, caused different truth values in the last column.   □

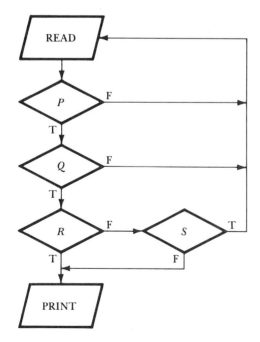

Figure 8.4   *P* AND *Q* AND *R* OR NOT *S*

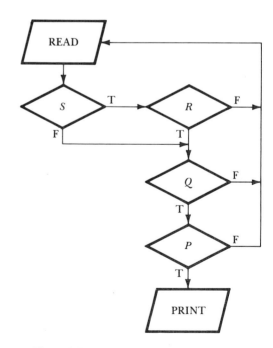

Figure 8.5   *P* AND (*Q* AND (*R* OR NOT *S*))

Flowcharts

We have already seen that the logical connectives AND, OR, EOR, and NOT can be diagrammed in the form of a flowchart. A flowchart is often useful to help visualize the desired flow in a complex logical expression.

Recall that in Example 5 we were looking for a list of the names and identification numbers of the employees whose monthly salary $\geq$ \$2100.00 AND age > 35 AND sick leave $\leq$ 90 OR seniority $\not< 6$. The flowchart is shown in Fig. 8.4. We use the following conditions.

$$P: \text{Monthly salary} \geq \$2100.00$$
$$Q: \text{Age} > 35$$
$$R: \text{Sick leave} \leq 90$$
$$S: \text{Seniority} \not< 6$$

Compare this flowchart to what we would have if we inserted parentheses as we did in Example 6(d). See Fig. 8.5.

Exercise Set 8.2

In each of the following, number the operations to indicate the order of execution.

1. $P \vee \sim (Q \wedge R)$ **2.** $(P \wedge Q) \wedge (P \vee Q)$

3. $P \wedge \sim Q \vee R$ **4.** $(P \wedge Q) \vee (P \wedge R)$

5. P OR Q OR NOT R **6.** NOT P AND Q OR NOT R

7. $(P$ AND $Q)$ OR $(R$ AND NOT $S)$ **8.** $(P$ OR $Q)$ AND NOT $(Q$ AND $R)$

Construct a truth table for each of the following.

9. $P \vee (Q \wedge R)$ **10.** $P \vee (\sim Q \wedge R)$

11. $(\sim P \wedge Q) \vee R$ **12.** $(P \vee \sim Q) \wedge R$

13. P OR NOT $(Q$ AND $R)$ **14.** NOT P OR $(Q$ AND $R)$

15. P AND NOT Q OR NOT R

16. NOT P OR NOT $(Q$ OR NOT $R)$

17. $P \wedge (\sim Q \overline{\vee} R)$

18. P EOR NOT $(Q$ AND $R)$

19. $\sim (P \vee (Q \vee R))$

20. $\sim (P \wedge (Q \vee \sim R))$

Construct a flowchart for each of the following. A name is to be printed for each case in which the logical form is true.

21. $P \vee (Q \wedge R)$ **22.** $P \vee (\sim Q \wedge R)$

23. $P \wedge (\sim Q \vee \sim R)$ **24.** $\sim (P \vee Q) \wedge \sim R$

25. NOT P OR NOT Q OR NOT R AND NOT S

26. P AND (Q OR NOT (R AND S))

27. ((P AND Q) OR R) OR NOT S

28. NOT (P AND (NOT Q OR S))

29. P EOR NOT Q

30. NOT P EOR NOT Q

8.3

Conditional, Biconditional, and Related Statements

The Conditional

Any statement that contains the IF–THEN sequence is called a **conditional** statement (sometimes referred to as an *implication*). The conditional connective is very important in mathematics as the basis for logical proof. We also encounter IF–THEN statements in many higher-level computer languages, as we shall see later. However, there is a difference between the mathematical conditional and the IF–THEN statement as it is used in computer programming.

If P and Q are propositions, then "If P, then Q" is also a proposition. In logic, the implication is written $P \rightarrow Q$ and can be read "P implies Q" or "If P, then Q." The proposition "P" is called the *antecedent,* or *condition,* and Q is the *conclusion,* or *consequent.*

To arrive at the truth table for the conditional, consider the following statement.

> If I step on the brake, then the car will stop.

Case 1: I step on the brake. The car stops.

$$T \rightarrow T \text{ is true.}$$

No one would dispute this outcome for it is exactly what is implied. A condition for the car to stop was established. When it occurred, the conclusion followed.

Case 2: I step on the brake. The car does not stop.

$$T \rightarrow F \text{ is false.}$$

Again, there is no problem with this false outcome. The condition was met, but the conclusion did not follow. Thus the original implication is false.

Case 3: I do not step on the brakes. The car stops.

$$F \rightarrow T \text{ is true.}$$

The outcome here is true since the original implication is not really tested. It does not prove the implication false since nothing is stated, implied, or predicted for what would occur if I do not step on the brake. We have to assume that the original condition is true.

Case 4: I do not step on the brake. The car does not stop.

$$F \rightarrow F \text{ is true.}$$

Again, the outcome is true for I state only what will occur if I step on the brake. There is no condition to be satisfied if I do not step on the brake. Therefore, there is nothing to prove the original implication is false. Therefore, it must be true.

The truth table for the conditional statement is summarized below.

P	Q	$P \rightarrow Q$
T	T	T
T	F	F
F	T	T
F	F	T

The conditional statement is false only when the antecedent is true and the conclusion is false.

The Biconditional

When we make the compound statement (IF P, THEN Q) AND (IF Q, THEN P), we form the **biconditional**, or *equivalence*, statement. The biconditional is a double implication and is written $P \leftrightarrow Q$. The two-directional arrow is read "if and only if." Thus $P \leftrightarrow Q$ is read "P if and only if Q."

The truth table for the biconditional statement is written by forming the conjunction, $P \rightarrow Q \wedge Q \rightarrow P$, as follows.

P	Q	$P \rightarrow Q$	$Q \rightarrow P$	$P \rightarrow Q \wedge Q \rightarrow P$
T	T	T	T	T
T	F	F	T	F
F	T	T	F	F
F	F	T	T	T

The equivalence is true whenever P and Q have the same truth value, both true or both false, and is false otherwise.

Related Conditionals

There are three other forms of the conditional statement that are important in mathematical logic: the converse, the inverse, and the contrapositive.

Given: $P \rightarrow Q$
Converse: $Q \rightarrow P$
Inverse: $\sim P \rightarrow \sim Q$
Contrapositive: $\sim Q \rightarrow \sim P$

To illustrate their usage, consider the following conditional statement and its three related statements:

Conditional: If John lives in Colorado, then John lives in the United States.
Converse: If John lives in the United States, then John lives in Colorado.
Inverse: If John does not live in Colorado, then John does not live in the United States.
Contrapositive: If John does not live in the United States, then John does not live in Colorado.

Examining the truth values of these four conditional forms provides some interesting results. If we accept that Colorado is a state within the United States, then the original conditional statement is true. However, the converse and inverse statements are not necessarily true and are therefore considered false. On the other hand, the contrapositive agrees with the conditional and is true. Is this always the case? Consider the following, remembering the truth values for a conditional statement.

Conditional: If $4 + 5 = 9$, then $4 \times 5 = 20$. $T \rightarrow T$ is true.
Converse: If $4 \times 5 = 20$, then $4 + 5 = 9$. $T \rightarrow T$ is true.
Inverse: If $4 + 5 \neq 9$, then $4 \times 5 \neq 20$. $F \rightarrow F$ is true.
Contrapositive: If $4 \times 5 \neq 20$, then $4 + 5 \neq 9$. $F \rightarrow F$ is true.

This time, all four terms ended up with the same truth value. The truth values for these four conditional forms are summarized below.

P	Q	$P \rightarrow Q$	Converse $Q \rightarrow P$	Inverse $\sim P \rightarrow \sim Q$	Contrapositive $\sim Q \rightarrow \sim P$
T	T	T	T	T	T
T	F	F	T	T	F
F	T	T	F	F	T
F	F	T	T	T	T

■

CONCLUSION
1. The conditional and its contrapositive are logically equivalent (that is, they always have the same truth value).
2. The conditional and its inverse are not logically equivalent.
3. The converse and the inverse are logically equivalent.
4. The four forms may all be true at the same time, but are never all false at the same time.
5. The four forms are all true when P and Q have the same truth value.

Exercise Set 8.3

Rewrite each of the following in the symbolic IF–THEN form.

1. If I work hard, I will succeed.

2. If I earn all A's, I will be a 4.0 student.

3. I will buy a new stereo provided I get the job.

4. I will win the election only if I carry the Ninth District.

5. A is an odd number if and only if A is not divisible by 2.

6. I will win the jackpot if and only if I have 21.

Construct a truth table for each of the following.

7. $P \to \sim Q$
8. $\sim P \to Q$
9. NOT $Q \to P$
10. $Q \to$ NOT P
11. $(P \lor Q) \to P$
12. $\sim (P \lor Q) \to Q$
13. If P AND Q, then NOT Q
14. If P AND NOT Q, then NOT P
15. $P \leftrightarrow$ NOT Q
16. NOT $P \leftrightarrow Q$
17. $\sim (P \leftrightarrow \sim Q)$
18. $(P \lor Q) \leftrightarrow (P \land \sim Q)$
19. $\sim P \leftrightarrow (P \lor \sim Q)$
20. $(P \to Q) \leftrightarrow (\sim P \lor Q)$
21. $[P \lor (Q \land R)] \leftrightarrow [(P \lor Q) \land (P \lor R)]$
22. $[P \land (Q \lor R)] \leftrightarrow [(P \land Q) \lor (P \land R)]$

Write the converse, inverse, and contrapositive of each of the following implications.

23. If $(P$ AND $Q)$, then NOT R
24. $(\sim P \lor Q) \to R$
25. If $x(x - 1) = 0$, then $x = 1$.
26. If $a \cdot b = 0$, then a or $b = 0$.

27. If $3 + 6 = 9$, then $5 \times 4 = 19$ is a conditional that is false. Which one of the following statements must also be false? Why?

a) If $3 + 6 \neq 9$, then $5 \times 4 \neq 19$.

b) If $5 \times 4 \neq 19$, then $3 + 6 \neq 9$.

c) If $5 \times 4 = 19$, then $3 + 6 = 9$.

8.4
Tautologies and Contradictions

So far we have looked at logical connectives and the truth values assigned to the resulting compound statements. Some compound statements turn out to be *always true*, while others turn out to be *always false*.

■ DEFINITION

TAUTOLOGY

A compound statement that is *true* for all possible cases is called a *tautology*.

■ DEFINITION

CONTRADICTION

A compound statement that is *false* for all possible cases is called a *contradiction*.

To determine whether a compound statement is a tautology or a contradiction, we need only construct the truth table for the statement. If the final column contains only T's, then the statement is a tautology. If the final column contains all F's, then the statement is a contradiction.

■ **EXAMPLE 9**   Construct the truth table for $P \rightarrow (P \lor Q)$ and determine whether the statement is a tautology, a contradiction, or neither.

Solution

P	Q	$P \lor Q$	$P \rightarrow (P \lor Q)$
T	T	T	T
T	F	T	T
F	T	T	T
F	F	F	T

This statement is a tautology.   □

■ **EXAMPLE 10**   Construct the truth table for $P \rightarrow (P \wedge Q)$ and determine whether the statement is a tautology, a contradiction, or neither.

Solution

P	Q	$P \wedge Q$	$P \rightarrow (P \wedge Q)$
T	T	T	T
T	F	F	F
F	T	F	T
F	F	F	T

This statement is neither. ☐

■ **EXAMPLE 11**   Construct the truth table for $(P \rightarrow \sim Q) \wedge (P \wedge Q)$ and determine whether the statement is a tautology, a contradiction, or neither.

Solution

P	Q	$\sim Q$	$P \rightarrow \sim Q$	$P \wedge Q$	$(P \rightarrow \sim Q) \wedge (P \wedge Q)$
T	T	F	F	T	F
T	F	T	T	F	F
F	T	F	T	F	F
F	F	T	T	F	F

This statement is a contradiction. ☐

Two statements are said to be *logically equivalent* if and only if the compound biconditional statement between them is a tautology.

■ **EXAMPLE 12**   Show that P and $\sim (\sim P)$ are logically equivalent.

Solution P and $\sim (\sim P)$ are logically equivalent if $P \leftrightarrow \sim (\sim P)$ is a tautology.

P	$\sim P$	$\sim (\sim P)$	$P \leftrightarrow \sim (\sim P)$
T	F	T	T
F	T	F	T

☐

Exercise Set 8.4

Determine whether each of the following propositions is a tautology, a contradiction, or neither.

1. NOT $(P$ AND $Q) \leftrightarrow$ (NOT P) OR (NOT Q)

2. P OR $(Q$ OR NOT $P)$

3. NOT $(P$ OR $Q) \leftrightarrow$ (NOT P) AND (NOT Q)

4. $[(P$ AND $Q)$ AND $P] \rightarrow$ NOT Q

5. $(P \wedge Q) \wedge \sim (P \vee Q)$

6. $[(P \rightarrow Q) \wedge \sim P] \rightarrow \sim Q$

7. $(P \wedge \sim Q) \wedge (\sim P \vee Q)$

8. $[(P \rightarrow Q) \wedge P] \rightarrow Q$

9. $(P \rightarrow Q) \leftrightarrow [(\sim P) \vee Q]$

10. $\sim (P \rightarrow Q) \leftrightarrow (P \wedge \sim Q)$

Determine whether each of the following pairs of compound statements is logically equivalent.

11. $P \leftrightarrow Q$ and $\sim P \leftrightarrow \sim Q$

12. $P \rightarrow Q$ and $\sim P \vee Q$

13. $\sim (P \to \sim Q)$ and $P \wedge Q$

14. $\sim (P \wedge Q)$ and $\sim P \wedge \sim Q$

15. $\sim (P \vee Q)$ and $\sim P \vee \sim Q$

16. $[(P \to Q) \wedge (P \to R)]$ and $[P \to (Q \wedge R)]$

17. P EOR Q and $P \leftrightarrow$ NOT Q

18. $P \leftrightarrow$ NOT Q and NOT $P \leftrightarrow Q$

8.5

Programmed Decision Making

In data processing, the programmer often encounters a problem that requires several tests to be performed before a particular action is taken. Several popular computer languages, such as BASIC, BASIC PLUS, COBOL, FORTRAN, and PASCAL, allow the programmer to compare quantities in a program and select one of two or three branches of the program to follow.

When two quantities are compared, three possible results exist:

1. The two quantities may be equal.
2. The first quantity may be greater than the second.
3. The second quantity may be greater than the first.

Despite these three possible results, a logical expression can be evaluated only in terms of whether the statement is true or false. The IF–THEN statement is used in many programming languages to perform the evaluation of a logical expression.

The IF–THEN statement in programming languages is significantly different from the IF–THEN of symbolic logic. In programming, what follows the IF is a logical expression and may involve a relational operator. What follows the THEN is a dependent or executable step. Only the logical expression has a truth value. The dependent statement *does not* have a truth value. In symbolic logic, both the antecedent (following IF) *and* the conclusion (following THEN) have truth values.

To see how the IF–THEN statement is used in programming, consider the following example.

XYZ Credit Company sends out bills and accepts payments. If the payment is less than the bill, a *balance due* statement must be sent to the customer; if the payment equals the bill, a *thank you* statement is sent; and if the payment is greater than the bill, a *refund* notice is sent.

IF (payment < bill) THEN (send balance due notice).
IF (payment = bill) THEN (send thank you statement).
IF (payment > bill) THEN (send refund notice).

Notice that the action taken is dependent on the result of the relational operator following the IF. The computer first evaluates the relational operator expression and if it is true, executes the dependent statement. Generally, if the relational expression is false, the dependent statement is not executed, and the next line of the program is performed.

Examples of the IF–THEN in several languages follow.

■ **EXAMPLE 13**   BASIC

```
10 . . .
20 IF A >= 0 THEN 80
30 . . .
```

Explanation: If A is greater than or equal to 0 is true, then execute statement 80 next. If A < 0, execute statement 30.   □

■ **EXAMPLE 14**   COBOL

```
. . .
. . .
IF PRICE > 25 THEN
MULTIPLY PRICE BY 0.055
GIVING TAX.
. . .
```

□

■ **EXAMPLE 15**   FORTRAN

a) logical IF (single branch)

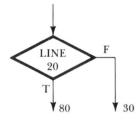

```
10 . . .
20 IF (2 * B .LE. 10) GOTO 80
30 . . .
```

If 2B is less than or equal to 10 is true, control is switched to line 80. If 2B > 10 line 30 is executed next.

b) IF–THEN–ELSE (double branch)

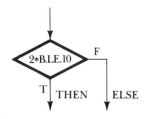

```
. . .
IF (2 * B .LE. 10) THEN
     PRINT 2B < 10
ELSE
     PRINT 2B > 10
ENDIF
. . .
```

This is the TRUE branch.

This is the FALSE branch.

The statement following ENDIF is executed next.

c) Arithmetic IF (triple branch)

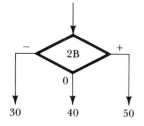

```
10 . . .
20 IF (2 * B) 30, 40, 50
30 . . .
40 . . .
50 . . .
```

Go to line 30 if 2B is negative.
Go to line 40 if 2B is zero.
Go to line 50 if 2B is positive.

d) PASCAL

```
    . . .
IF PROFIT > (0.15 * REVENUE)
THEN TAX: = 0.20 * PROFIT
    . . .
```

If profit is greater than 15% of revenue is true, the tax is computed as 20% of the profit. The next line is executed, even if profit is greater than 15% of revenue is a false statement. No branching is instructed here.

e) BASIC (a second example in BASIC)

```
10 . . .
20 IF A > 10 THEN (3 * A) ELSE (4 * A)
30 . . .
```

If A > 10 is true, A will be multiplied by 3; otherwise, A is multiplied by 4. Line 30 is executed next either way.   □

The logical connectors studied earlier in this section—AND, OR, and NOT—may all be part of the logical expression following the IF in an IF–THEN statement. The established hierarchy of operations is followed. Parentheses may be used to alter the normal sequence.

■ **EXAMPLE 16**   **a)** The word AND means *both* when two or more conditions are specified in an IF statement.

```
10 . . .
20 IF A > 0 AND B > 5 THEN 50
30 . . .
```

If A > 0 and B > 5 are *both* true, then line 50 is executed next. Otherwise, line 30 is executed.

b) The word OR means *either* or *both* when two or more conditions are specified.

```
10 . . .
20 IF A > 0 OR B > 5 THEN 50
30 . . .
```

If either part or both parts are true, then line 50 is executed. Line 30 will be executed only if both statements are false.

Exercise Set 8.5

Identify the logical expression, the relational operator, and the dependent statement.

1. IF A >= 30 THEN 70

2. IF (A ** 2 + 10) < 100 THEN 70

3. IF PRICE > 50 THEN MULTIPLY PRICE BY 0.06 GIVING TAX

4. IF BALANCE < 50 THEN MULTIPLY BALANCE BY 0.015

5. IF (B ** 2 − 4 * A * C) 40, 60, 80

6. IF (A) 40, 60, 80

7. IF A <= 10 THEN (3 + A) ELSE 3 * A

8. IF A <> 10 THEN 40 ELSE 160

9. IF PROFIT > (0.20 * REVENUE) THEN TAX: = 0.15 * PROFIT;

10. IF (PROFIT > 10) AND (PROFIT <= 0.15 * REVENUE)
THEN TAX: = 0.12 * PROFIT;

11. IF (A .EQ. 90) GO TO 100

12. IF (B .LT. A) THEN
    X = B ** 2
    Y = A − X
ELSE
    X = C/A
    Y = X ** 2
ENDIF

13. IF (B .GT. A) THEN
    X = B
ELSE
    X = A
ENDIF

Summary

In this chapter we were introduced to *symbolic logic* and the *logical operators* AND, OR, EOR, and NOT. We used the logical operators to connect simple statements into *compound statements* and learned how to determine truth values for the compound statements using *truth tables*.

We discovered that the *conjunction,* or AND, compound statement is true if and only if both components are true; otherwise, it is false. We saw that the *disjunction,* or *inclusive* OR, is false only when both components are false and is true otherwise. The *exclusive* EOR is true when the components have opposite truth values and is false when the components have the same truth values. We also found that the *negation,* or NOT, of a statement is a statement with the opposite truth value from the original statement.

We found that the logic connective OR closely resembles the *union* operation in set theory, the AND resembles the *intersection,* and the NOT is similar to the *complement* of a set. We examined the flowcharts for AND, OR, EOR, and NOT statements and found a similarity to the flowcharts for intersections, unions, and complements, respectively.

Next we were introduced to *conditional* and *biconditional* statements and learned how to form *converses, inverses,* and *contrapositives* from a conditional statement. We examined truth tables and determined truth values for these forms.

We called logical expressions that have equivalent truth values *logically equivalent.* Statements that are always true are called *tautologies* and statements that are always false are called *contradictions.*

The chapter concluded with a look at logic and computer programming, using examples of conditional (the IF–THEN) statements in several computer languages, including BASIC, COBOL, FORTRAN, PASCAL, and BASIC-PLUS.

Review Exercises

Construct a truth table for each of the following compound statements.

1. $P \wedge \sim Q$
2. $P \vee \sim Q$
3. $\sim (P \vee \sim Q)$
4. $\sim (P \wedge \sim Q)$
5. $(P \wedge Q) \vee \sim R$
6. $(P \vee \sim Q) \wedge R$
7. $\sim P \wedge (Q \wedge \sim R)$
8. $(\sim P \vee Q) \wedge \sim R$
9. $P \rightarrow (Q \wedge R)$
10. $\sim P \rightarrow (Q \overline{\vee} R)$
11. $(P \overline{\vee} \sim Q) \rightarrow Q$
12. $(P \wedge Q) \rightarrow P$

13. $\sim (P \vee \sim Q) \to (P \wedge Q)$

14. $\sim (\sim P \wedge Q) \to (\sim P \vee Q)$

15. $(P \to Q) \to (Q \wedge R)$

16. $(P \to \sim Q) \to (\sim Q \vee R)$

17. $P \leftrightarrow (\sim Q \vee R)$

18. $\sim P \leftrightarrow (Q \wedge \sim R)$

19. $(P \wedge Q) \leftrightarrow (\sim P \vee Q)$

20. $(P \vee Q) \leftrightarrow \sim (P \wedge \sim Q)$

Construct a flowchart for each of the following. Remember, a true result leads to a print.

21. $P \wedge \sim Q$

22. $P \vee \sim Q$

23. $P \wedge \sim (Q \vee R)$

24. $P \vee \sim (Q \wedge R)$

25. $P \vee (\sim Q \wedge R)$

26. $P \wedge (\sim Q \vee R)$

27. $\sim (P \wedge Q) \vee \sim R$

28. $\sim (P \vee Q) \wedge \sim R$

29. $\sim P \overline{\vee} Q$

30. $\sim P \overline{\vee} \sim Q$

Write the converse, the inverse, and the contrapositive for each of the following.

31. If (P OR Q), then NOT R.

32. $(P \vee \sim Q) \to R$

33. $P \to (\sim Q \vee R)$

34. $P \to (Q \wedge \sim R)$

35. If a BOBE is a GADGET, then a WIM is NOT a WIDGET.

36. If $4 \times 5 \neq 19$, then $4 + 3 = 12$.

Determine whether each of the following is a tautology, a contradiction, or neither.

37. $\sim (P \wedge Q) \leftrightarrow (\sim P \wedge \sim Q)$

38. $Q \vee (P \vee \sim Q)$

39. $\sim (\sim P \vee Q) \leftrightarrow (P \wedge \sim Q)$

40. $\sim (P \wedge \sim Q) \leftrightarrow (\sim P \vee Q)$

41. $(P \vee Q) \wedge \sim (P \wedge Q)$

42. $[(\sim P \to Q) \wedge P] \to \sim Q$

43. $(\sim P \wedge Q) \wedge (P \vee \sim Q)$

44. $\sim (P \wedge Q) \wedge (P \vee Q)$

Determine whether each of the following pairs of compound statements is logically equivalent.

45. $\sim (P \vee \sim Q)$ and $\sim P \vee Q$

46. $\sim (P \wedge \sim Q)$ and $\sim P \wedge Q$

47. $\sim (\sim P \wedge \sim Q)$ and $P \vee Q$

48. $\sim (\sim P \vee Q)$ and $P \wedge \sim Q$

49. $P \overline{\vee} Q$ and $\sim P \leftrightarrow Q$

50. $P \leftrightarrow Q$ and $\sim P \leftrightarrow Q$

Boolean Algebra

After completing this chapter, you should be able to:

1. Describe Boolean algebra and its uses.
2. Draw the appropriate parallel-series circuit given a Boolean expression, and vice versa.
3. Represent a given circuit by its correct Boolean expression.
4. Demonstrate the isomorphic properties between the concepts of sets, the concepts from propositional logic, and the concepts from circuits and Boolean algebra.
5. Simplify expressions from propositional logic using the properties of Boolean algebra.
6. Simplify flowcharts using the properties from Boolean algebra.
7. Use fundamental products and sum-of-products form to derive and/or simplify Boolean expressions.
8. Draw and then simplify logic circuits using AND, OR, and NOT gates for a Boolean expression.

9

9.1

Introduction

George Boole
(1815–1864)

The design of today's computers is based on the fundamentals of Boolean algebra. Formulas such as $x + x = x$ and $x \cdot x = x$ are at the heart of this algebra, which is really a study of binary zero–one logic. Although these formulas may not seem correct from your past algebra experience, you will soon see that they are.

In the late 1800s the English mathematician George Boole did quite a bit of work in the field of logic. Noticing the conveniences of algebra, he treated many of the logical propositions as variables. In *logic,* if you recall, the values of a variable are either *true* or *false*. In Boolean algebra, also, a variable can assume only two values. Because each

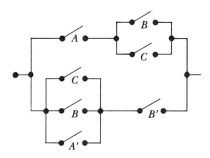

Figure 9.1

variable is "two-valued," Boolean algebra has applications in many other fields. For example, in switching circuits, the switch is either open or closed; and in computer work, signals are either present or absent, and bits are either on or off.

Boolean algebra is often referred to as the algebra of switching circuits (circuits containing switches). With Boolean algebra as a tool, circuits like the one shown in Fig. 9.1 can be analyzed and simplified in a rapid and efficient manner. However, Boolean algebra is more than just a tool to simplify switching circuits—it is much more encompassing. A strong relationship also exists between logic and Boolean algebra; often Boolean algebra is referred to as an algebra of propositional logic. Understanding the elements of logic is fundamental both to the design and to the proper functioning of a digital computer. A programmer uses Boolean algebra to better understand the complex programming in computing. The applications of Boolean algebra range from its use as a tool to aid in the efficient design of the digital system to a "systematic science" to both represent and analyze the seemingly complex circuits, networks, and programs associated with computers.

Undoubtedly, using Boolean algebra in such a broad and varied way would have pleased George S. Boole (1815–1864). The great philosopher and mathematician of the twentieth century, Bertrand Russell, made this tribute to George Boole: "Pure mathematics was discovered by Boole in a work called *The Laws of Thought (1854)*." Boole's work(s) established formal logic and a new algebra known as **Boolean algebra:** the algebra of sets, or the algebra of logic.

What exactly is Boolean algebra? Although a thorough discussion and understanding of Boolean algebra may be a bit premature at this point, the following definition is offered to guide our study throughout this chapter.

■ **DEFINITION**

BOOLEAN ALGEBRA

Boolean algebra is a mathematical structure composed of:

1. a set of elements S containing at least 0 and 1;
2. two binary operations, usually denoted by $+$ and $\cdot$; and
3. a unary operation, denoted by $'$.

In addition, the following axioms must be satisfied: For elements A, B, and C from the set S,

1.	$A + B = B + A$	**Commutative laws**
	$A \cdot B = B \cdot A$	
2.	$A + (B + C) = (A + B) + C$	**Associative laws**
	$A \cdot (B \cdot C) = (A \cdot B) \cdot C$	
3.	$A \cdot (B + C) = (A \cdot B) + (A \cdot C)$	**Distributive laws**
	$A + (B \cdot C) = (A + B) \cdot (A + C)$	
4.	$A + 0 = A$	**Identity elements**
	$A \cdot 1 = A$	**for $+$ and $\cdot$**
5.	$A + A' = 1$	**Complements**
	$A \cdot A' = 0$	

Note: At this point, you may wish to browse through the tables in Section 9.5 to note the similarities between Boolean algebra, sets, and logic propositions.

Exercise Set 9.1

Determine whether the statements in Exercises 1–5 are true or false.

1. Bertrand Russell formulated Boolean algebra in honor of George Boole.

2. Boolean algebra can have only one operation.

3. The commutative laws are valid in Boolean algebra.

4. Multiplication is a unary operation.

5. Boolean algebra has little use in working with computers.

6. A friend asks you, "What is Boolean algebra?" With your limited reading thus far, describe in your own words what Boolean algebra is. Where is it used? How is it used? (After completing this chapter, return and answer this question again. Then compare your two answers!)

7. Match the words on the left with the correct representation on the right. (The symbols $\wedge$, $\vee$, and $'$ were presented in Chapter 8.)

OR	$A \wedge B$
NOT	$A \vee B$
AND	A'

8. (*For the ambitious!*) Locate several other sources that describe Boolean algebra to your satisfaction (try starting in the library). Before studying any further in this chapter, use these resources to construct a good description of Boolean algebra. List all your resources in a bibliography.

9.2
Basic Parallel and Series Switching Circuits

In the past thirty years, the uses and applications of digital equipment and complex digital systems have become extremely widespread. With the advent of the microprocessor, microcomputers and their uses have grown almost exponentially. All operations of a digital computer are ultimately processed by some combination of signals passing through logic elements. In a paper entitled "A Symbolic Analysis of Relay and Switching Circuits," Claude E. Shannon demonstrated how electrical switching networks could be described and analyzed by means of symbolic logic. Boolean algebra had its origins in the algebra of logic. Symbolic logic will be discussed here in relation to simple circuits. Applications will then be shown to switching circuits, which have an isomorphic relationship to electronic gates, so important in the operation of modern high-speed computers. In addition, the basic principles of the logic of circuits may be applied to other logic elements as well, such as flip-flops, transistors, integrated circuits, and other circuit components.

Often referred to as the algebra of computer circuits, Boolean algebra attempts to determine the signal-voltage level on a wire (a 0-volt or a 5-volt signal). A switch in simplest form can be either *open* (no current passing through) or *closed* (current passing through), as in Fig. 9.2.

(a)   **(b)**

Figure 9.2   (a) An open switch; (b) a closed switch.

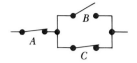

Figure 9.3

An example of a more complex circuit containing several switches is shown in Fig. 9.3. If current enters at the left, through switch *A*, will current reach the point on the extreme right?

Most circuits are combinations of parallel and/or series circuits. Let's begin by examining each of these first-order logic circuits: first a simple parallel (OR) switching circuit, and then a simple series (AND) switching circuit.

A simple first-order **parallel circuit** is diagrammed as shown in Fig. 9.4(a). This simple symbolic circuit is the equivalent of the OR switching circuit in Fig. 9.4(b).

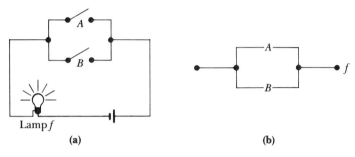

(a)   (b)

Figure 9.4

Questions

When will current pass from left to right through the circuit in Fig. 9.4(a)? Must both switches be closed? Could current pass if only one switch were closed?

Answer

Make a guess after studying the circuit. Then compare your guess to Table 9.1.

By analyzing the logic of the circuit, we reach the conclusion that current *will* pass through the circuit if switch *A* is closed *or* if switch *B* is closed.

If we use a 1 to represent passage of current and a 0 to represent no current, we could simplify Table 9.1 to Table 9.2.

Table 9.1
All possible conditions for switches *A* and *B* and light bulb

A	*B*	CURRENT THROUGH THE PARALLEL CIRCUIT?
Open	Open	*No* current
Open	Closed	Current *yes*
Closed	Open	Current *yes*
Closed	Closed	Current *yes*

Table 9.2
Possible switch settings for an OR (parallel) circuit

A	*B*	*A* OR *B*
0	0	0 (*No* current)
0	1	1 (Current)
1	0	1 (Current)
1	1	1 (Current)

Table 9.3
All possible conditions for switches A and B
and light bulb

A	B	*CURRENT THROUGH THE SERIES CIRCUIT?*
Open	Open	*No* current
Open	Closed	*No* current
Closed	Open	*No* current
Closed	Closed	Current *yes*

Table 9.4
Possible switch settings for an
AND (series) circuit

A	B	A *AND* B
0	0	0 (*No* current)
0	1	0 (*No* current)
1	0	0 (*No* current)
1	1	1 (Current)

In Boolean algebra, a $+$ is used to represent the word OR. With that in mind, we see that

$$0 + 0 = 0 \quad \text{makes sense;}$$
$$0 + 1 = 1 \quad \text{makes sense;}$$
$$1 + 0 = 1 \quad \text{also makes sense;}$$
$$1 + 1 = 1 \quad \text{(What??? Welcome to Boolean algebra!).}$$

A parallel circuit, A OR B, can also be represented by using the Boolean *sum*, $A + B$. A simple **series circuit** is shown in Fig. 9.5(a). The equivalent AND switching circuit is shown in Fig. 9.5(b).

● Question

Refer to Fig. 9.5. Under what conditions of the two switches A and B will current flow through the circuit from left to right?

Give the question some thought, and then compare your answers with those given in Table 9.3.

Again, by analyzing the logic of the circuit, we see that current will pass through the circuit if both switch A *and* switch B are closed.

If we rewrite Table 9.3 by again using the symbols 1 for current passing and 0 for no current passing, we have Table 9.4.

In Boolean algebra, $\cdot$ or $\times$ are the symbols used for the word AND. A series circuit is represented by the Boolean product $A \cdot B$ (or, equivalently, $A \times B$ or AB). With that in mind, we see that the rules of

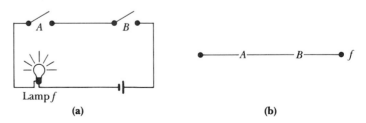

(a)   (b)

Figure 9.5

traditional algebra (or arithmetic) coincide perfectly with the rules of Boolean algebra ($0 \cdot 0 = 0$; $0 \cdot 1 = 0$; $1 \cdot 0 = 0$; $1 \cdot 1 = 1$).

Basic AND circuits and OR circuits constitute first-order logic circuits. Higher-order circuits are developed when several AND circuits supply an OR circuit or when several OR circuits supply an AND circuit.

Some examples of circuits combining both parallel and series circuits are shown below. Could you match each figure on the left with the correct Boolean expression on the right?

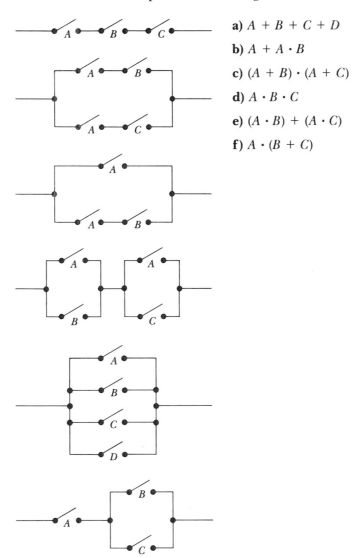

a) $A + B + C + D$

b) $A + A \cdot B$

c) $(A + B) \cdot (A + C)$

d) $A \cdot B \cdot C$

e) $(A \cdot B) + (A \cdot C)$

f) $A \cdot (B + C)$

In the expression $C \cdot E \mid A \cdot B$, an OR circuit (indicated by the +) has as its components two AND circuits (indicated by the ·). This is a higher-order circuit. Could you make a sketch of this circuit?

Exercise Set 9.2

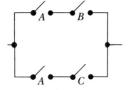

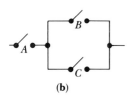

Figure 9.6

In Fig. 9.6, switches A, B, and C can be *open* (0) or *closed* (1). Given each of the possible switch settings below, determine whether current could flow through the circuit.

	A	B	C	FIG. 9.6(a) CURRENT FLOW	FIG. 9.6(b) CURRENT FLOW
1.	0	0	0	yes or no	yes or no
2.	0	0	1	yes or no	yes or no
3.	0	1	0	yes or no	yes or no
4.	0	1	1	yes or no	yes or no
5.	1	0	0	yes or no	yes or no
6.	1	0	1	yes or no	yes or no
7.	1	1	0	yes or no	yes or no
8.	1	1	1	yes or no	yes or no

9. a) Trace all paths through the circuit in Fig. 9.7 that would allow current flow if $A = 1$, $B = 0$, $C = 1$, and $D = 1$. (If there are no paths, state "none.")

   b) Trace all paths that would allow current flow if $A = 0$, $B = 1$, $C = 1$, and $D = 0$. (If there are no paths, state "none.")

   c) (*A real thinker!*) How many possible paths are there through the circuit? How many possible combinations of switch settings are there? How many possible switch settings will allow current through the circuit?

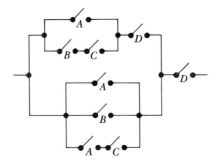

Figure 9.7

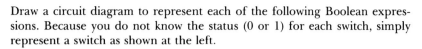

Draw a circuit diagram to represent each of the following Boolean expressions. Because you do not know the status (0 or 1) for each switch, simply represent a switch as shown at the left.

Recall that a *sum* is used to represent parallel circuits and a *product* is used to represent series circuits.

10. $A + B \cdot C$ **11.** $(A + B) \cdot C$

12. $A \cdot C + B \cdot C$ **13.** $(A + B) \cdot (A + C)$

14. $(A + B) + C$ **15.** $A \cdot (B + A) \cdot C$

16. $A + B \cdot C + C \cdot D \cdot E$ **17.** $A \cdot (B + C) + B \cdot (A + C)$

18. $A + A \cdot B \cdot C + A \cdot ((B + C) + D)$

19. $A + A \cdot (B \cdot (C + D) + A \cdot (C + D))$

20. a) Draw a circuit and complete the column in the following table for the Boolean expression $(A + B) \cdot (A + C)$.

All possible conditions for switch settings A, B, and C and two expressions				
A	B	C	$(A + B) \cdot (A + C)$	$A + B \cdot C$
0	0	0		
0	0	1		
0	1	0		
0	1	1		
1	0	0		
1	0	1		
1	1	0		
1	1	1		

b) Draw a circuit and complete the column in the table for the Boolean expression $A + B \cdot C$.

c) Any conclusions?

For each of the following combination parallel–series circuits, write the original Boolean representation.

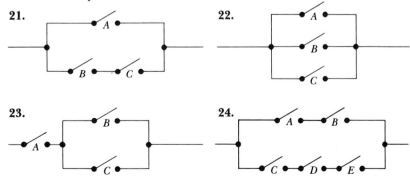

21.

22.

23.

24.

25.

26.

27.

28.

29.

30.

31.

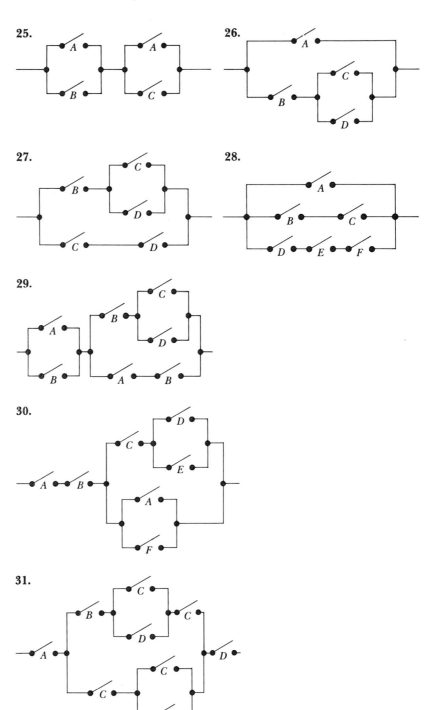

32. a) By constructing a table of all possible values for A and B, compare the Boolean expression $A + A \cdot B$ to the expression A. Are the expressions different or equivalent?

   b) Compare the Boolean expression $A \cdot (A \cdot B + B)$ to the expression $A \cdot B$. Are these two expressions different or equivalent?

33. Current will flow through this series circuit

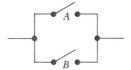

only when A is closed and B is closed. From set theory, we know that an element is in $A \cap B$ only when the element is both in A and in B. From logic, we know that A AND B is true only when A is true and B is true.

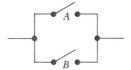 $A \cap B$, A AND B,   and

$A \cdot B$ are "similar" expressions!

What properties from set theory and from logic are "similar" to the following?

34. For the representation $(A + B) \cdot (A + C)$:

   a) Rewrite using set representations.

   b) Rewrite using logic representations.

   c) Draw a Venn diagram.

   d) Draw a parallel–series circuit.

9.3
Boolean Operations and Properties

As you compare Boolean algebra with "ordinary" algebra, you will notice many similarities, as well as a few differences. In ordinary algebra there are five operations: addition, subtraction, multiplication, division, and exponentiation. In Boolean algebra, there are only three operations: the binary operations $+$ and $\times$ or $\cdot$, and the unary operation $'$. (A binary operation is applied to two objects; a unary operation is applied to one object.) Examples of Boolean expressions containing these operations are $A' + B$, $A \cdot C$, and $A(B + C')$. Note that the operation of multiplication, $A \times B$, can be represented as in ordinary algebra by $A \cdot B$, or simply AB, where juxtaposition is understood to mean multiplication. (In most work that follows, the symbol $\cdot$ will be used when multiplication is indicated.)

● Question

If more than one operation occurs in an expression—for example, $A + B \cdot C'$, which operations are to be performed first?

■

> The order (or priority) of operations is as follows:
>
> *First* the operation of complementation;
> *Second,* the operation of multiplication; and
> *Finally,* the operation of addition.
>
> Of course, just as in ordinary algebra, if parentheses are involved, work inside the parentheses has top priority.

■ **EXAMPLE 1**

In the expression $(A + B) \cdot C'$, first, A and B are "summed," then C is complemented, and finally the two results are "multiplied."   □

Boolean Sum: $A + B$, (A OR B)

In Boolean algebra, $A + B$ is interpreted as A OR B and represents the *logical sum,* or *disjunction:*

$$0 + 0 = 0,$$
$$0 + 1 = 1,$$
$$1 + 0 = 1,$$
$$1 + 1 = 1.$$

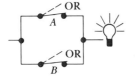

Figure 9.8
A parallel or an
OR switching circuit.

The only case different from traditional algebra (arithmetic) is the case $1 + 1 = 1$. However, if we consider a circuit in which 1 indicates current passing through, we see that $(1 + 1)$, translated as "switch closed" OR "switch closed," definitely produces current at the end of the circuit (see Fig. 9.8).

Boolean Product: $A \times B$, $A \cdot B$, (A AND B)

In Boolean algebra, $A \cdot B$ is interpreted as A AND B, and represents the *logical product,* or *conjunction:*

$$0 \cdot 0 = 0,$$
$$0 \cdot 1 = 0,$$
$$1 \cdot 0 = 0,$$
$$1 \cdot 1 = 1.$$

Here we see that the table of values concides perfectly with the properties from traditional algebra or arithmetic! Remember, at any

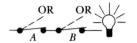

Figure 9.9
A series or an AND
switching circuit.

time you can visualize the corresponding circuit to help understand these concepts from Boolean algebra. Multiplication is made easier by looking at the corresponding series circuit shown in Fig. 9.9.

Boolean Complement: A', (NOT A)

The unary operation $'$ operates on only one object. The complement of A, A', indicates the opposite state of switch A. If A is closed or has a value of 1, then A' will be *open*, or have a value of 0. If A is *open* or has a value of 0, then A' will be *closed*, or have a value of 1. The expression $(A')'$ represents the opposite of the opposite of A; however, with a little thought, you should be able to convince yourself that $(A')'$ is equivalent to A. (The expression A'' is similar to the double negative in English; in algebra, a negative times a negative yields a positive quantity.)

Boolean Algebra Properties

When used on variables, the Boolean operations, $+$, $\cdot$, and $'$, exhibit certain algebraic properties, many of which are similar to those in ordinary algebra. Some examples from ordinary algebra that are identical to those in Boolean algebra are shown below.

COMMUTATIVE LAWS

$$A + B = B + A$$

$$A \cdot B = B \cdot A$$

ASSOCIATIVE LAWS

$$(A + B) + C = A + (B + C)$$

$$(A \cdot B) \cdot C - A \cdot (B \cdot C)$$

DISTRIBUTIVE LAW (FIRST)

$$A \cdot (B + C) = (A \cdot B) + (A \cdot C)$$

Some additional properties (not necessarily having counterparts in ordinary algebra) include the following.

DISTRIBUTIVE LAW (SECOND)

$$A + (B \cdot C) = (A + B) \cdot (A + C)$$ **Try this one by substituting numbers. What happens?**

COMPLEMENTS

$$A + A' = 1,$$
$$A \cdot A' = 0$$

ABSORPTION LAWS

$$A + A \cdot B = A,$$
$$A \cdot (A + B) = A$$

IDENTITIES

$$A + 0 = A,$$
$$A \cdot 1 = A$$

BOUNDEDNESS LAWS

$$A + 1 = 1,$$
$$A \cdot 0 = 0$$

Analysis of Some Properties

Let's examine some of these properties more closely. For example, what is the "sum" of A and its complement?

1. *The property* $A + A' = 1$. By examining Table 9.5 for all possible values for the variables (a truth table), we see that the value of $A + A'$ is always 1. Also, by analyzing the parallel circuit, we find that if switch A is closed (1), current will flow through the circuit. However, if switch A is open (0), then A' will be closed (1), and again current will flow through the circuit. (We could simplify this circuit by eliminating both switches and using a solid piece of wire, since current always passes through!)

2. *The property* $A \cdot A' = 0$. Table 9.6 confirms what we already suspected—anything times zero is zero! Looking at the series circuit, we see that if A is closed (1), then A' will be open (0); hence no current passage! If A is open (0), we need not even worry about A'

Table 9.5
Possible conditions for A, A', and $A + A'$

A	A'	$A + A'$
0	1	1
1	0	1

Table 9.6
Possible conditions for A, A', and $A \cdot A'$

A	A'	$A \cdot A'$
0	1	0
1	0	0

because current will never pass through switch A. (In this case, the easiest way to simplify the circuit is to start with a solid piece of wire and then cut the wire with a tin snip; in other words, to do away with the switches!)

3. *The property $A + 1 = 1$.* Is switch A really needed? In Table 9.7, current always flows independent of the value of A.
4. *The property $A \cdot 0 = 0$.* Regardless of the value of A, no current will pass through this series circuit; 0 represents a switch that is always in the open position. (We'll leave it to you to make the appropriate table and/or draw the circuit.)
5. *The property $A \cdot 1 = A$.* If A is closed, current passes; if A is open, no current passes (see Fig. 9.10). Thus everything is dependent on A. (We leave it to you to construct the table.)
6. *The property $A + 0 = A$.* Note in Table 9.8 that the values of $A + 0$ are exactly the same as the values of A. The circuit, or expression, depends entirely on A. $(A + 0)$ is *equivalent to A*. Again, we leave it to you to draw and analyze the circuit.

Figure 9.10

The Second Distributive Property

Enough of the "easier" properties. Let's now move on to some of the more thought-provoking properties. The second distributive property will take some time to recognize and to use efficiently. Take a look at

Table 9.7
Possible conditions
for A, 1, and $A + 1$

A	1	$A + 1$
0	1	1
1	1	1

Table 9.8
Possible conditions for
A, 0, and $A + 0$

A	0	$A + 0$
0	0	0
1	0	1

Table 9.9
All possible conditions for A, B, C, and the two expressions in the
second distributive property

A	B	C	$A + B \cdot C$	$(A + B) \cdot (A + C)$
0	0	0	0	0
0	0	1	0	0
0	1	0	0	0
0	1	1	1	1
1	0	0	1	1
1	0	1	1	1
1	1	0	1	1
1	1	1	1	1

Table 9.9. (Three switches provide eight possible *open–closed* combinations.) Because the different values for A, B, and C produce identical results, the two Boolean expressions are equivalent:

$$A + B \cdot C = (A + B) \cdot (A + C).$$

You may also want to study the circuits for each Boolean expression as shown in Fig. 9.11; think about all the ways in which current could pass (or could not pass) through both circuits.

DeMorgan's Laws

Two other useful properties in Boolean algebra are *DeMorgan's laws*. Compare columns 7, 8, 9, and 10 in Table 9.10.

Comparing columns 7 and 8, we find that we have equivalent Boolean expressions. You might think that $(A \cdot B)'$ equals $A' \cdot B'$. However, this is *not* true: $A' + B' = (A \cdot B)'$. Columns 9 and 10 are also identical in value. Therefore, $A' \cdot B'$ equals $(A + B)'$.

These two important laws or properties are called DeMorgan's laws:

1. $A' + B' = (A \cdot B)'$
2. $A' \cdot B' = (A + B)'$

Table 9.10
All possible conditions for A, B, A', B', and the expressions involved
in DeMorgan's laws

(1) A	(2) B	(3) A'	(4) B'	(5) $A + B$	(6) $A \cdot B$	(7) $A' + B'$	(8) $(A \cdot B)'$	(9) $A' \cdot B'$	(10) $(A + B)'$
0	0	1	1	0	0	1	1	1	1
0	1	1	0	1	0	1	1	0	0
1	0	0	1	1	0	1	1	0	0
1	1	0	0	1	1	0	0	0	0

Figure 9.11

Absorption Laws

Two other laws that provide for a great deal of efficiency in the simplification process are the *absorption laws:*

1. $A + A \cdot B = A$
2. $A \cdot (A + B) = A$

To verify these, we will not show truth tables, nor will we draw the circuits. (You may, if you wish!) Either of these can be "proved" using the distributive laws.

■ **EXAMPLE 2**

$A + A \cdot B$ becomes $A \cdot (1 + B)$ by the distributive law. Then $A \cdot (1 + B)$ becomes $A \cdot (1)$ because $1 + B = 1$, and $A \cdot (1)$ equals A because of the identity law.   □

You may try your hand at verifying the second absorption property $A \cdot (A + B) = A$ above.

A summary of most of the important and (soon to be) useful properties in Boolean algebra is given in the following box. After studying these properties, we will then show how to use the appropriate laws to simplify Boolean expressions and/or circuits.

PROPERTIES OF BOOLEAN ALGEBRA

COMMUTATIVE	$A + B = B + A$	INVOLUTION	$(A')' = A$
	$A \cdot B = B \cdot A$	COMPLEMENTS	$A + A' = 1$
ASSOCIATIVE	$A + (B + C) = (A + B) + C$		$A \cdot A' = 0$
	$A \cdot (B \cdot C) = (A \cdot B) \cdot C$	BOUNDEDNESS	$A + 1 = 1$
DISTRIBUTIVE	$A \cdot (B + C) = (A \cdot B) + (A \cdot C)$		$A \cdot 0 = 0$
	$A + (B \cdot C) = (A + B) \cdot (A + C)$	DEMORGAN'S	$(A + B)' = A' \cdot B'$
IDENTITIES	$A + 0 = A$		$(A \cdot B)' = A' + B'$
	$A \cdot 1 = A$	ABSORPTION	$A + A \cdot B = A$
IDEMPOTENCY	$A + A = A$		$A \cdot (A + B) = A$
	$A \cdot A = A$		

Now that we have investigated the different properties or laws from Boolean algebra, let's apply these laws to the *simplification* of Boolean expressions.

Several examples will serve to illustrate some of the techniques of simplifying Boolean expressions. Of course, the real learning will take place after you have studied and begun working on your own.

■ EXAMPLE 3   Simplify $A \cdot (A' + B)$.

Solution

$$
\begin{aligned}
A \cdot (A' + B) &= A \cdot A' + A \cdot B &&\text{By the first distributive law} \\
&= 0 + A \cdot B &&\text{Use of complements} \\
&= A \cdot B &&\text{Use of 0 as an identity} \quad \square
\end{aligned}
$$

Let's try another. Keep in mind that many expressions can be simplified in several different ways. You may even find an easier method than the one demonstrated!

■ EXAMPLE 4   Simplify $B + A' + A \cdot B$.

Solution

$$
\begin{aligned}
B + A' + A \cdot B &= A' + B + A \cdot B &&\text{The commutative law for +} \\
&= A' + B \cdot (1 + A) &&\text{The second distributive law} \\
&= A' + B \cdot (1) &&\text{The law of boundedness} \\
&= A' + B &&\text{The identity for } \cdot \quad \square
\end{aligned}
$$

And finally for a tougher one!

■ EXAMPLE 5   Simplify $A(B(A' + B') + (AB' + B))$.

Solution

$$
\begin{aligned}
&A \cdot (B \cdot (A' + B') + (A \cdot B' + B)) \\
&= A \cdot (B \cdot A' + B \cdot B' + (A \cdot B' + B)) &&\text{The first distributive law} \\
&= A \cdot (B \cdot A' + 0 + (A \cdot B' + B)) &&\text{Complements} \\
&= A \cdot (B \cdot A' + (A \cdot B' + B)) &&\text{Identity} \\
&= A \cdot (A' \cdot B + (A \cdot B' + B)) &&\text{The commutative law for multiplication} \\
&= A \cdot (A' \cdot B + (B + A \cdot B')) &&\text{The commutative law for addition} \\
&= A \cdot ((A' \cdot B + B) + A \cdot B') &&\text{The associative law for addition} \\
&= A \cdot ((B + A' \cdot B) + A \cdot B') &&\text{The commutative law for addition} \\
&= A \cdot (B + A \cdot B') &&\text{The law of absorption} \\
&= A \cdot B + A \cdot (A \cdot B') &&\text{The first distributive law} \\
&= A \cdot B + (A \cdot A) \cdot B' &&\text{The associative law for multiplication} \\
&= A \cdot B + A \cdot B' &&\text{Idempotency} \\
&= A \cdot (B + B') &&\text{The first distributive law} \\
&= A \cdot (1) &&\text{Complements} \\
&= A &&\text{The identity for multiplication} \\
&&& \qquad\qquad\qquad \square
\end{aligned}
$$

Example 5 is a good illustration of many of the Boolean properties. Was this method the easiest way? Could the problem have been done in a simpler way? *Try!*

Exercise Set 9.3

Identify which Boolean properties are illustrated in each of the following equations.

1. $A + B = B + A$

2. $A \cdot (B + C) = A \cdot B + A \cdot C$

3. $C + A \cdot B + A = C + A \cdot (B + 1)$

4. $A \cdot A' \cdot B = 0 \cdot B$

5. $A + (C + D) = (A + C) + D$

6. $A + A \cdot B = (A + A) \cdot (A + B)$

7. $A + A + A + A = A$

8. $1 \cdot B = B$

Using the Boolean properties, write an equivalent form for each of the following expressions.

9. $(A + B) + C$

10. $0 + B \cdot C \cdot D$

11. $A + A + A + A$

12. $C \cdot (A + B)$

13. $A + A' \cdot B$

14. $B \cdot B \cdot B \cdot B \cdot B \cdot B$

15. $(A + B) \cdot (A + C)$

16. $A + A \cdot B + A \cdot B \cdot C$

17. $A + B \cdot C$

18. $(A + B \cdot (C + A'))'$

19. $A + A' \cdot B \cdot C \cdot D + 1 + B' \cdot C$

20. $A \cdot B \cdot C \cdot B \cdot A' \cdot B \cdot C$

Determine whether each of the following is true or false in Boolean algebra.

21. $(A + B)' = A' + B'$

22. $A \cdot A \cdot A = A^3$

23. $A + 1 = A$

24. $A + A = 2A$

25. $(A \cdot B \cdot C \cdot D)' = A' \cdot B' \cdot C' \cdot D'$

26. $(A'B')' + (A + B)' = 1$

27. The second distributive property $A + B \cdot C = (A + B) \cdot (A + C)$ will take some studying and work to be able to recognize when to use it. Using this and some of the other Boolean properties, simplify $C + C \cdot D$. (*Hint:* Your final answer should be C. Do not use the absorption law.)

Construct a table to show the equivalence of each of the following pairs of Boolean expressions.

28. $A + A' \cdot B$ and $A + B$

29. $(A + B')'$ and $A' \cdot B$

30. $(A \cdot B)'$ and $A' + B'$

31. a) Draw a simpler circuit that would be equivalent to the circuit shown in Fig. 9.12.

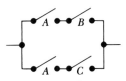

Figure 9.12

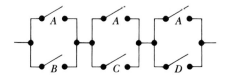

Figure 9.13

    b) Draw a simpler circuit that would be equivalent to the circuit shown in Fig. 9.13.

32. The expression $A \cdot B \cdot C' + A \cdot B \cdot B' + (A' + C)' + (B' + C)'$ is simplified step by step below. However, the reasons have not been given for each of the steps. Supply from the Boolean properties the correct reason and write it in the righthand column. (Could you have simplified the expression in another way? in an easier way?)

    a) $A \cdot B \cdot C' + A \cdot B \cdot B' + (A' + C)' + (B' + C)'$

    b) $A \cdot B \cdot C' + A \cdot B \cdot B' + A'' \cdot C' + B'' \cdot C'$

    c) $A \cdot B \cdot C' + A \cdot B \cdot B' + A \cdot C' + B \cdot C'$

    d) $A \cdot B \cdot C' + A \cdot 0 + A \cdot C' + B \cdot C'$

    e) $A \cdot B \cdot C' + 0 + A \cdot C' + B \cdot C'$

    f) $A \cdot B \cdot C' + A \cdot C' + B \cdot C'$

    g) $A \cdot B \cdot C' + B \cdot C' + A \cdot C'$

    h) $B \cdot C' \cdot (A + 1) + A \cdot C'$

    i) $B \cdot C' \cdot (1) + A \cdot C'$

    j) $B \cdot C' + A \cdot C'$

    k) $C' \cdot (B + A)$

Simplify each of the following Boolean expressions by using the Boolean properties or some combinations thereof.

33. $A \cdot B + A \cdot C$ **34.** $(A + B) \cdot (A + C)$

35. $A \cdot B + A \cdot B' + A$ **36.** $(A' + A' \cdot B) \cdot (A + B)$

37. $(A' \cdot B') + (A + B)$ **38.** $A \cdot (A' \cdot (A + B))$

39. $(A \cdot B + A \cdot B') + (A \cdot (B + B'))$ **40.** $A + B' + (A' \cdot B)$

41. $(A' + B' \cdot C)'$ **42.** $(A' \cdot B')'$

43. $A \cdot B + A' \cdot B + A \cdot B' + A' \cdot B'$

44. $(A \cdot B + A' \cdot B + A \cdot B' + A' \cdot B')'$

45. $A \cdot B \cdot C + A \cdot B' \cdot C + A \cdot B \cdot C' + A \cdot B' \cdot C'$

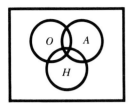

Figure 9.14

Draw a parallel–series circuit for each original Boolean expression below. Then by studying this circuit and/or using the Boolean properties, draw a simpler yet equivalent circuit.

46. $A \cdot (C + B') + B \cdot C'$ **47.** $A \cdot (A' + B)$

48. $(A + B) + A' \cdot B'$ **49.** $A + A' \cdot B + A' \cdot B' \cdot C$

50. IF *"the insured is over 19"* (O), or *"has had no accidents"* (A'), and *"is an honor student"* (H), THEN our company will reduce insurance premiums. Translate the antecedent into a Boolean expression. (The term *antecedent* was defined in Chapter 8 on logic.) Draw a circuit for the Boolean expression. Could you also represent the antecedent by shading the appropriate parts of the Venn diagram in Fig. 9.14?

From your understanding of the Boolean properties and the similarities you have noticed between sets, logic, and circuits, complete each of the following expressions with an equivalent expression from Boolean algebra.

51. $(A \cap B)' =$ **52.** A OR B AND $C =$

53. $(A$ AND $(B$ OR $C))' =$ **54.** A AND B OR A AND C OR A AND $D =$

9.4
The Principle of Duality: Identities and Theorems

A powerful and useful tool in simplifying circuitry—and in Boolean algebra in general—is the principle of **duality.** By referring to the chart of the Boolean properties (see p. 261), you may remember noticing a pattern. (You may want to refer to the chart and review the properties before continuing.) Several properties are listed below. Note the position of the letters: The only things "changing" in each pair are the symbols for operations.

$$\begin{Bmatrix} A + B = B + A \\ A \cdot B = B \cdot A \end{Bmatrix} \qquad \begin{Bmatrix} A \cdot (B + C) = (A \cdot B) + (A \cdot C) \\ A + (B \cdot C) = (A + B) \cdot (A + C) \end{Bmatrix}$$

In each pair of properties, if the operations of $+$ and $\cdot$ are simply interchanged in the first property, a second property is obtained. This newly obtained property is the *dual* of the original. We will state (without proving) that in Boolean algebra if a property is valid, the dual of that property is a statement that is also a valid Boolean property. In addition to interchanging the $+$ and $\cdot$ symbols throughout the property, we should also interchange the corresponding identity elements 0 and 1. Reviewing the list of Boolean properties we cannot help but note the many similarities—almost like a "two-for-one" sale. Awareness of the concept of duality (1) helps us to remember the many properties more efficiently, and (2) is useful in simplifying Boolean expressions.

■

> On the left below are listed some Boolean properties, with their corresponding duals listed on the right.
>
> | | |
> |---|---|
> | $A + A' = 1$ | $A \cdot A' = 0$ |
> | $A + (B + C) = (A + B) + C$ | $A \cdot (B \cdot C) = (A \cdot B) \cdot C$ |
> | $A + 0 = A$ | $A \cdot 1 = A$ |
> | $(A + B)' = A' \cdot B'$ | $(A \cdot B)' = A' + B'$ |
> | $A + 1 = 1$ | $A \cdot 0 = 0$ |
> | $A + A = A$ | $A \cdot A = A$ |
> | $A \cdot (A + B) = A$ | $A + (A \cdot B) = A$ |

● Question

What is the dual property for each of the following Boolean expressions?

a) $(A + B) \cdot B' = A \cdot B'$
b) $A + A \cdot B + A \cdot B \cdot C = A$

The property of duality helps us to remember some of the formulas. For example, you may have no difficulty remembering the distributive law: $A \cdot (B + C) = A \cdot B + A \cdot C$. Often in simplification, however, it is not *this* law that is needed, but rather the second distributive property, which is not as easily remembered. By using the principle of duality, we first write the "familiar" distributive property:

$$A \cdot (B + C) = (A \cdot B) + (A \cdot C);$$

we then rewrite the property by using the principle of duality:

$$A + (B \cdot C) = (A + B) \cdot (A + C).$$

Using the dual of a Boolean function helps greatly in circuit simplification. For a more in-depth discussion of this principle, see the "double-dual technique" for circuit simplification in Angelo Gillie's *Binary Arithmetic and Boolean Algebra* (New York: McGraw-Hill, 1965).

Exercise Set 9.4

Provide the dual property for each of the following properties.

1. $A + B = B + A$
2. $A + 1 = 1$

3. $A \cdot (B + C) = (A \cdot B) + (A \cdot C)$

4. $(A + B)' = A' \cdot B'$

5. $A + (A \cdot B) = A$

6. $A + (A \cdot C)' = A + (A' + C')$

7. $(A \cdot B \cdot C)' = A' + B' + C'$

8. $A \cdot (A' + B) = 0 + A \cdot B$

9. $(A + C) \cdot (B + A) = B \cdot C + A$

Given that the expression $A \cdot (A' + B)$ is equivalent to $A \cdot B$, determine what each of the following expressions is equivalent to.

10. $A + A'B$

11. $(A(A' + B))'$

12. $(A + (A'B))'$

9.5

Isomorphic Relationships Among Concepts from Sets, Switching Circuits, Logic, and Boolean Algebra

The term *isomorphic* means "having the same form or appearance." The purpose of this section is to show the similarities between some familiar concepts. When studying a subject, the time and the opportunity are often too rare to allow one to notice similarities and commonalities among concepts. To illustrate the power and the structure of mathematics, examples showing similar (isomorphic) properties are taken from the areas of sets, propositional logic, and Boolean algebra.

Note how the word *or* is used in the following three sentences.

1. An element is in the set $A \cup B$ if the element is in the set A *or* in the set B (sets).
2. The proposition $P \lor Q$ is true if proposition P is true *or* if proposition Q is true (propositional logic).
3. The Boolean expression $A + B$ is 1 if A is 1 (current passes) *or* B is 1 (current passes) (Boolean algebra).

The word *and* is used in a similar manner in the next three sentences.

1. An element is in the set $A \cap B$ if the element is in set A *and* the element is in set B (sets).
2. The proposition $P \land Q$ is true if proposition P is true *and* if proposition Q is true (propositional logic).

Table 9.11
Properties common to sets, logic, and Boolean algebra

PROPERTY	SETS	LOGIC	BOOLEAN ALGEBRA
Commutative	$A \cup B = B \cup A$	$P \vee Q = Q \vee P$	$A + B = B + A$
	$A \cap B = B \cap A$	$P \wedge Q = Q \wedge P$	$A \cdot B = B \cdot A$
Associative	$A \cup (B \cup C) = (A \cup B) \cup C$	$P \vee (Q \vee R) = (P \vee Q) \vee R$	$A + (B + C) = (A + B) + C$
	$A \cap (B \cap C) = (A \cap B) \cap C$	$P \wedge (Q \wedge R) = (P \wedge Q) \wedge R$	$A \cdot (B \cdot C) = (A \cdot B) \cdot C$
Distributive	$A \cap (B \cup C) =$	$P \wedge (Q \vee R) =$	$A \cdot (B + C) =$
	$(A \cap B) \cup (A \cap C)$	$(P \wedge Q) \vee (P \wedge R)$	$(A \cdot B) + (A \cdot C)$
	$A \cup (B \cap C) =$	$P \vee (Q \wedge R) =$	$A + (B \cdot C) =$
	$(A \cup B) \cap (A \cup C)$	$(P \vee Q) \wedge (P \vee R)$	$(A + B) \cdot (A + C)$
Identities	$A \cup \varnothing = A$	$P \vee$ False $= P$	$A + 0 = A$
	$A \cap U = A$	$P \wedge$ True $= P$	$A \cdot 1 = A$
Idempotency	$A \cup A = A$	$P \vee P = P$	$A + A = A$
	$A \cap A = A$	$P \wedge P = P$	$A \cdot A = A$
Involution	$(A')' = A$	$\sim (\sim P) = P$	$(A')' = A$
Complements	$A \cup A' = U$	$P \vee \sim P =$ True	$A + A' = 1$
	$A \cap A' = \varnothing$	$P \wedge \sim P =$ False	$A \cdot A' = 0$
DeMorgan's laws	$(A \cup B)' = A' \cap B'$	$\sim (P \vee Q) = \sim P \wedge \sim Q$	$(A + B)' = A' \cdot B'$
	$(A \cap B)' = A' \cup B'$	$\sim (P \wedge Q) = \sim P \vee \sim Q$	$(A \cdot B)' = A' + B'$

3. The Boolean expression $A \cdot B$ is 1 if A is 1 (current passes) *and* if B is 1 (current passes) (Boolean algebra).

If you are familiar with sets, for example, you should also be able to work with propositional logic and/or Boolean algebra. Their similarities allow for a great deal of transfer from one area to another. Besides, if you concentrate on learning the common structure, you are not really learning three different topics at all!

Some of the many properties common to these three areas are shown in Table 9.11.

Exercise Set 9.5

Indicate whether or not each of the following is similar in form (isomorphic) to A OR B.

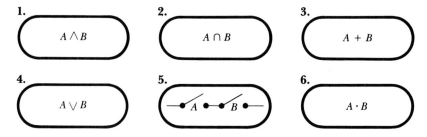

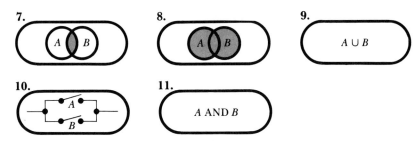

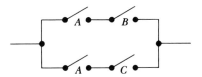

12. Rewrite $(A + B)' = A' \cdot B'$ using unions, intersections, and complements.

13. Write an expression using set notation for the Boolean expression $(A + B) \cdot (A + C) = A + B \cdot C$.

14. Using unions and intersections, write an expression similar to the circuit shown below.

15. Translate $A \cap (B \cup C)' = A \cap (B' \cap C')$ into the notation used in Boolean algebra.

16. a) Verify DeMorgan's law $(A + B)' = A' \cdot B'$ by drawing a Venn diagram to represent each side of the equation.

b) Verify DeMorgan's law $(A + B)' = A' \cdot B'$ by constructing a truth table. (Treat A and B as logical propositions with values of true or false.)

17. The following sentence is taken from a legal document: "NOT OVER 65 AND RECEIVING FINANCIAL SUPPORT OR HAVING NO LIVING RELATIVES AND NOT OVER 65 OR RECEIVING FINANCIAL SUPPORT AND OVER 65." Let A be the phrase "HAVING NO LIVING RELATIVES"; let B be the phrase "RECEIVING FINANCIAL SUPPORT"; and let C be the phrase "OVER 65."

Make the translation to Boolean algebra notation, simplify the expression, and then write the expression in simple English!

18. The flowchart in Fig. 9.15 is given for A AND (B OR C AND NOT B) AND NOT C. Using your knowledge of Boolean algebra, translate the logic to Boolean notation, simplify, and then provide a simpler flowchart.

19. Review the definition of Boolean algebra given in Section 9.3. If the operation $\cdot$ were replaced by intersection ($\cap$), the operation $+$ were replaced by union ($\cup$), and $'$ represented complement:

a) Would the sets form a Boolean algebra?

b) What would replace 0?

c) What would replace 1?

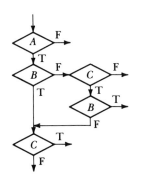

Figure 9.15

9.6
Deriving and Simplifying Boolean Expressions: The Sum-of-products Form

Frequently a logic circuit or a Boolean function is needed, or must be designed, in order to accomplish a particular task. The desired output may be known, but the Boolean function or expression to produce the output is *not* given. The problem then becomes one of finding, or deriving, the Boolean expression (or circuit) and subsequently changing this expression to its simplest form.

For example, suppose that for two switches, X and Y, the outcome desired is as shown in the two cases in Table 9.12.

Table 9.12

X	Y	*DESIRED OUTCOME*	X	Y	*DESIRED OUTCOME*
1	1	0	1	1	1
1	0	1	1	0	0
0	1	1	0	1	0
0	0	0	0	0	0

Fundamental Products

To help us derive the appropriate Boolean expression (or later on, a circuit), we will use the idea of fundamental products. Before we explore this concept, however, a presentation of some needed terminology is in order.

By using X, Y, and Z as variables and the Boolean operations $+$, $\cdot$, and $'$, we can create many Boolean expressions—for example,

$$X \cdot Y' + X \cdot Y \cdot Z + X \cdot Y \cdot Z \cdot X \cdot Y' \cdot Z'.$$

In this expression, $X \cdot Y'$ is a *product* of X and Y'; and X and Y' are literals, a *literal* being a nonnumerical symbol (usually a letter or a complemented letter). In the third product, there are six literals.

■ DEFINITION

FUNDAMENTAL PRODUCT

A product is said to be a *fundamental product* when it is a literal, or a product of two or more literals, in which no two literals involve the same variable.

Table 9.13

	X	Y	$X \cdot Y$	$X \cdot Y'$	$X' \cdot Y$	$X' \cdot Y'$
(1)	1	1	1	0	0	0
(2)	1	0	0	1	0	0
(3)	0	1	0	0	1	0
(4)	0	0	0	0	0	1

The third product in the expression $X \cdot Y' + X \cdot Y \cdot Z + X \cdot Y' \cdot Z'$ is *not* a fundamental product because both the first and the fourth literals involve the same variable, X. The product also contains the variables Y and Y', as well as Z and Z'. The second product in the expression is a fundamental product; there are three literals and three different variables.

Now, exactly how will fundamental products be useful in deriving a desired Boolean expression? By examining the various product terms $X \cdot Y, X \cdot Y', X' \cdot Y$, and $X' \cdot Y'$ in Table 9.13, we see that each of these fundamental products produces a 1 in *exactly one place* in a column.

Relating this to our earlier work with circuits, for example, recall that current can flow through a series circuit *only* when X is closed ($X = 1$) and Y is closed ($Y = 1$). (See Fig. 9.16.) For all other switch settings of X and Y: $X = 1$ and $Y = 0, X = 0$ and $Y = 1, X = 0$ and $Y = 0$, no current will flow; in other words, $X \cdot Y$ has a value of 0.

In a similar manner, we can show that if $X = 1$ and $Y = 0$, then $X \cdot Y'$ is the *product* that will produce a current flow; that is, $X \cdot Y' = 1$ for this setting. For the settings $X = 1$ and $Y = 1, X = 0$ and $Y = 1, X = 0$ and $Y = 0, X \cdot Y'$ has a value of 0.

Think through the similar arguments showing that $X' \cdot Y$ allows current flow (value $= 1$) only when $X = 0$ and $Y = 1$, and that $X' \cdot Y'$ allows current flow (value $= 1$) only when $X = 0$ and $Y = 0$.

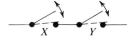

Figure 9.16

■

> In essence, the *fundamental product* to produce a 1 on one given line (lines 1 to 4 in Table 9.13) is obtained by examining the switch settings (0 or 1) for that line. If the variable (or switch) has a value of 1, that variable is part of the product. If the variable has a value of 0, the *complement* of that variable is used in forming the product.

■ **EXAMPLE 6**   If $X = 1$ and $Y = 1$ (line 1), then $X \cdot Y$ will produce a 1 on that line and 0's for all other lines. □

■ **EXAMPLE 7**   If $X = 0$ and $Y = 1$ (line 3), then $X' \cdot Y$ will produce a 1 on that line and 0's for all other lines. □

Table 9.14

	X	Y	DESIRED OUTCOME	$X \cdot Y'$	$X' \cdot Y$	$X \cdot Y' + X' \cdot Y$
(1)	1	1	0	0	0	0
(2)	1	0	1 $\longrightarrow X \cdot Y'$	1	0	1
(3)	0	1	1 $\longrightarrow X' \cdot Y$	0	1	1
(4)	0	0	0	0	0	0

Now let's apply the idea of fundamental products to deriving the original Boolean expression desired in the lefthand portion of Table 9.12.

■ **EXAMPLE 8**   To find a Boolean expression to produce 1's on lines 2 and 3, we first determine what fundamental product produces a 1 on each individual line.

For line 2, $X \cdot Y'$ will be used.
For line 3, $X' \cdot Y$ will be used. (See the lefthand portion of Table 9.14.) By forming the sum of these fundamental products, we obtain the desired Boolean expression:

$$X \cdot Y' + X' \cdot Y.$$

(See the righthand portion of Table 9.14.)   □

■ **EXAMPLE 9**   Derive the Boolean expression to produce the outcome given in Table 9.15. (*Hint:* Extend the same ideas to three variables.)

Solution   We first derive the *individual products* for each line containing a 1. For line 2, $X \cdot Y \cdot Z'$; for line 3, $X \cdot Y' \cdot Z$; and for line 7, $X' \cdot Y' \cdot Z$.

Next, we form the *sum* of these individual fundamental products. Thus the answer for the desired Boolean expression is

$$X \cdot Y \cdot Z' + X \cdot Y' \cdot Z + X' \cdot Y' \cdot Z.$$

We leave it to you to check that this expression produces 1's *only* on lines 2, 3, and 7, and produces 0's on all other lines.   □

Let's take a look at another example with the variables A, B, and C.

■ **EXAMPLE 10**   What Boolean expression represents the values in column R of Table 9.16?

Solution   Remember, first find the fundamental product for each 1 represented in the column; then "add" the different fundamental products together. To provide a check, the answer is given below. Cover the answer, try your best, and then compare!

	X	Y	Z	DESIRED OUTCOME
(1)	1	1	1	0
(2)	1	1	0	1 ←———
(3)	1	0	1	1 ←———
(4)	1	0	0	0
(5)	0	1	1	0
(6)	0	1	0	0
(7)	0	0	1	1 ←———
(8)	0	0	0	0

Table 9.15

A	B	C	R
1	1	1	0
1	1	0	1
1	0	1	0
1	0	0	1
0	1	1	0
0	1	0	0
0	0	1	1
0	0	0	1

Table 9.16

The solution is

$$R = A \cdot B \cdot C' + A \cdot B' \cdot C' + A' \cdot B' \cdot C + A' \cdot B' \cdot C',$$

which simplifies to

$$A \cdot C' + A' \cdot B'.$$

The original Boolean expression could be simplified by using either the Boolean properties we discussed earlier or a technique involving the sum-of-products form. □

Simplifying Boolean Expressions: The Sum-of-products Form

To help us simplify a Boolean expression, we can use a procedure called changing to the *sum-of-products form*.

■ **DEFINITION**

SUM-OF-PRODUCTS

An expression is in *sum-of-products form* if the expression is either (a) a fundamental product or (b) the sum of two or more fundamental products, none of which is contained in another.

■ **EXAMPLE 11**   The expression $X \cdot Y + Z$ *is* in sum-of-products form (that is, it is the sum of two fundamental products $X \cdot Y$ and Z), because $X \cdot Y$ is not contained in Z, or vice versa. □

■ **EXAMPLE 12**   The expression $X \cdot Y + X \cdot Y \cdot Z$ is a sum of fundamental products, but the expression is *not* in sum-of-products form. Why? Study the expression carefully and concentrate on the definition of sum-of-products form. You will probably notice then that although each product is a fundamental product, $X \cdot Y$ is contained in $X \cdot Y \cdot Z$. □

● Question

Because working with the sum-of-products form will take some thinking and work, study the expressions given below. Which Boolean expressions are in sum-of-products form?

a) $X \cdot Y' + X' \cdot Y + X \cdot Y \cdot Z$ yes or no
b) $X \cdot Y' + X' \cdot Y + X \cdot Y' \cdot Z$ yes or no
c) $X \cdot Y' + X + X \cdot Y \cdot Z$ yes or no
d) $X \cdot Y' + X \cdot X' \cdot Y + X' \cdot Y \cdot Z$ yes or no

Answer

The only expression above that is in sum-of-products form is (a). In (b) the first product is contained in the third; in (c) the second product is contained not only in the first but also in the last product. Expression (d) couldn't possibly be a candidate because the second product is not even a fundamental product (it contains X and X', both involving the same variable).

Note: In the remainder of this discussion, Boolean multiplication $A \cdot B$ is represented by AB; two variables "next to each other" (juxtaposition) implies multiplication. (This is similar to traditional algebra in which XY means X times Y.)

● Question

Suppose an expression is not in sum-of-products form. How can the expression be simplified? Several examples will illustrate the procedure.

■ **EXAMPLE 13**

Is the expression $X + XY + Z$ in sum-of-products form? The Boolean expression $X + XY + Z$ is not in sum-of-products form because the product X is contained in the product XY. By using the distributive law, we see that $X + XY + Z$ becomes $X(1 + Y) + Z$, which then simplifies to $X(1) + Z$, or $X + Z$. □

■ **EXAMPLE 14**

The expression $XYZ' + XYY' + (X' + Z)' + (Y' + Z)'$ is worked step by step until it is in sum-of-products form.

$XYZ' + XYY' + (X' + Z)' + (Y' + Z)' =$	**Removing parentheses**
$XYZ' + XYY' + X''Z' + Y''Z' =$	**Using DeMorgan's laws**
$XYZ' + X(0) + X''Z' + Y''Z' =$	**Using complements**
$XYZ' + X''Z' + Y''Z' =$	**Using $X \cdot 0 = 0$**
$XYZ' + XZ' + YZ'$	**By the law of involution**

Now the expression should contain all fundamental products. (This is the first part of the procedure: to remove parentheses and obtain all fundamental products.)

Next, we determine whether the expression is in sum-of-products form. If it is not, then one product is entirely contained in another product(s). We use the distributive law to "factor out" the common factor:

$$XYZ' + XZ' + YZ' = \qquad \text{Note that } XZ' \text{ is contained in } XYZ'.$$
$$XZ'(Y + 1) + YZ' = \qquad \text{Using the distributive law}$$
$$XZ' + YZ'. \qquad \text{Using the identity element}$$

Did you notice that the last two steps could have been consolidated by using the law of absorption? If so, you have a good grasp of Boolean algebra up to this point.

Now, two products remain, both are fundamental products, and neither product is contained in the other. Therefore, the expression is in sum-of-products form. ☐

While it is difficult to list specific steps to put an expression in sum-of-products form, general strategies that will guide you through a simplification process can be given.

To obtain a sum-of-products form for any nonzero Boolean expression:

1. First find the products. If there are any parentheses, remove them. This may require using either a distributive law or perhaps one of DeMorgan's laws. Simplify by using the laws of involution; continue with this process until the complement applies only to single variables, not expressions. The remaining expression should contain only sums and/or products of literals (no parentheses).
2. Use the distributive laws to transform the expression into a sum of products. (However, the expression may not be in sum-of-products form at this point.)
3. Now that you have products, you need to find fundamental products. Use the commutative, associative, idempotent, and complement laws to transform each product into a fundamental product. (Some products may disappear at this step; for example, $BB' = 0$.)
4. Finally, put the expression into sum-of-products form by using the absorption laws.

So, in summary, the progression seems to be from:

1. a Boolean expression to
2. a sum of products to
3. a sum of fundamental products to
4. the sum-of-products form.

Now that you have seen and studied the technique, let's try a more complex expression. You might benefit most by covering the reasons given for each step and trying to supply them yourself. Check at the end to see whether your answers match those listed. (Could the expression have been simplified in any other way?)

■ **EXAMPLE 15**   Simplify the Boolean expression $(A'BC)'(AB' + A')'$.

Solution

$$
\begin{aligned}
(A'BC)'(AB' + A')' &= (A'' + B' + C')((AB')'A'') & \text{DeMorgan's laws} \\
&= (A'' + B' + C')((A' + B'')A'') & \text{DeMorgan's laws} \\
&= (A + B' + C')((A' + B)A) & \text{Involution} \\
&= (A + B' + C')(A'A + BA) & \text{Distributive (first)} \\
&= (A + B' + C')(0 + AB) & \text{Complements} \\
&= (A + B' + C')AB & \text{Identity} \\
&= AAB + B'AB + C'AB & \text{Distributive (first)} \\
&= AB + B'AB + C'AB & \text{Idempotency} \\
&= AB + B'BA + ABC' & \text{Commutative (multi-} \\
& & \text{plication)} \\
&= AB + 0A + ABC' & \text{Complements} \\
&= AB + 0 + ABC' & \text{Boundedness laws} \\
&= AB + ABC' & \text{Identity} \\
&= AB(1 + C') & \text{Distributive (first)} \\
&= AB(1) & \text{Boundedness laws} \\
&= AB & \text{Identity}
\end{aligned}
$$

While this is one method of simplifying the expression, it should be understood that there are several other alternative methods—some perhaps even more efficient. Even in this example, alternate steps could have been taken: From $AB + ABC'$ the answer AB could have been obtained directly by using the absorption law. □

Exercise Set 9.6

1. To add two one-digit binary numbers, we need an expression for the sum column and an expression for the carry column, as shown below.

$$
\begin{array}{cccc}
1 & 1 & 0 & 0 \\
\underline{1} & \underline{0} & \underline{1} & \underline{0} \\
1\,0 & 0\,1 & 0\,1 & 0\,0 \\
\text{C S} & \text{C S} & \text{C S} & \text{C S}
\end{array}
$$

An expression was found for the sum column $(S = X' \cdot Y + X \cdot Y')$ in Example 8 (see also the lefthand portion of Table 9.12). Find a Boolean expression to represent the carry column.

2. For the two variables A and B, use fundamental products and Table 9.13 to find an expression to produce the values in each of the columns in the following table.

A	B	T	W	R	K
0	0	1	1	0	1
0	1	0	0	1	0
1	0	1	1	1	0
1	1	1	0	0	1

List the Boolean expression for each of the following.

a) T _____   **b)** W _____

c) R _____   **d)** K _____

Simplify the expression for each of the following.

e) T _____   **f)** W _____

g) R _____   **h)** K _____

3. Suppose you want to control a hall light by either of two switches: one upstairs, one downstairs. Either switch will turn an ON light OFF, or an OFF light ON. The following table shows the possible positions (1 is ON; 0 is OFF). Assume that the light starts by being ON. Obtain a Boolean expression to show the circuit for this switch problem.

DOWNSTAIRS	UPSTAIRS	LIGHT
Touch	Touch	1
Touch	Don't touch	0
Don't touch	Touch	0
Don't touch	Don't touch	1

4. For the variables A, B, and C, use fundamental products to derive a Boolean expression for each column in the following table.

A	B	C	X	Y	Z	W
0	0	0	0	1	0	1
0	0	1	1	1	1	0
0	1	0	1	0	1	0
0	1	1	0	0	0	0
1	0	0	0	1	1	0
1	0	1	1	1	0	1
1	1	0	1	0	0	0
1	1	1	0	0	1	1

The expressions are:

a) $X = $ _____

b) $Y = $ _____

c) $Z = $ _____

d) $W = $ _____

Simplified, the expressions are:

e) $X = $ _____

f) $Y = $ _____

g) $Z = $ _____

h) $W = $ _____

Construct a Boolean table showing the values for each of the following funda-mental products.

5. $A \cdot B \cdot C \cdot D$ **6.** $A \cdot B \cdot C \cdot D'$

7. $A \cdot B \cdot C' \cdot D$ **8.** $A \cdot B' \cdot C \cdot D$

9. $A \cdot B' \cdot C' \cdot D$ **10.** $A \cdot B' \cdot C \cdot D'$

11. $A \cdot B' \cdot C' \cdot D'$ **12.** $A' \cdot B' \cdot C' \cdot D'$

13. An expression is needed to print a YES on the computer when A is TRUE, B is FALSE, C is TRUE, and D is TRUE. A YES should also be printed when A is FALSE, B is TRUE, C is TRUE, and D is TRUE. For all other values of A, B, C, and D, a NO should be printed. Find the Boolean expression to accomplish this.

Rewrite each expression as the sum of fundamental products. (If the expres-sion is already a sum of fundamental products, so state.)

14. $A(BC')'$ **15.** $(A' + B)'$

16. $XYX'Y + XY'X' + XY'XY$ **17.** $(A + B'C)(A' + B + C)$

18. $A'(C(A + B) + AB)$ **19.** $A(B(A + AB + ABC))$

20. $AB' + ABC' + A'BC'$ **21.** $(A'B)' + (A + B)' + AB$

22. Write each of the Boolean expressions in Exercises 14–21 in sum-of-products form. (If the expression is already in sum-of-products form, so state.)

23. For each of the circuit diagrams in Fig. 9.17:

   a) find the representative Boolean expression;

   b) write the expression as a sum of fundamental products;

   c) write the expression in sum-of-products form; and

   d) draw a new circuit representing the sum-of-products form.

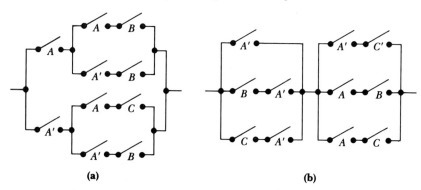

(a)   (b)

Figure 9.17

Simplify each of the following expressions by rewriting it in sum-of-products form.

24. $A(BC + A') + BD'(E'F) + A'DE + CD$

25. $A'(B(A + C) + (C + AB'))C + (A'B)'(A + B)'$

26. $A(B(C' + CD) + BC')D'$

9.7
Applications of Boolean Algebra: Logic Circuits

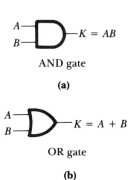

AND gate

(a)

OR gate

(b)

NOT gate

(c)

Figure 9.18

A **logic circuit** is a circuit that contains one or more input devices but has exactly one output device. The elements that make up most digital computer circuits are the AND gate, the OR gate, and a NOT gate, or inverter. Figure 9.18 shows an example of each type of gate. In each case, note that the direction of current is from left to right, and that for each gate, there is one or more inputs, but only one output. The behavior of an AND gate is very similar to the functioning of a *series* circuit; both are represented by the truth table in Fig. 9.19. Likewise, an OR gate has the same output as a *parallel* circuit.

Occasionally, especially when there are many inputs to a gate, you may see the truth tables represented as follows.

For the AND gate, the sequence of values is ($K = A \cdot B$):

$$\begin{array}{lccccc} & A & 0 & 0 & 1 & 1 \\ \text{AND} & B & 0 & 1 & 0 & 1 \\ \hline & K = 0 & 0 & 0 & 1 \end{array}$$

For the OR gate, the sequence of values is ($K = A + B$):

$$\begin{array}{lccccc} & A & 0 & 0 & 1 & 1 \\ \text{OR} & B & 0 & 1 & 0 & 1 \\ \hline & K = 0 & 1 & 1 & 1 \end{array}$$

A	B	$A \cdot B$	⊐D⊢	$A + B$	⊐D⊢
0	0	0	0	0	0
0	1	0	0	1	1
1	0	0	0	1	1
1	1	1	1	1	1

Figure 9.19

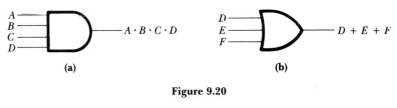

Figure 9.20

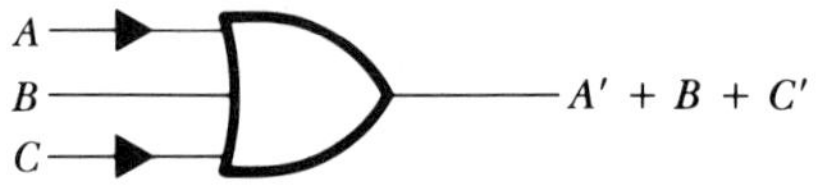

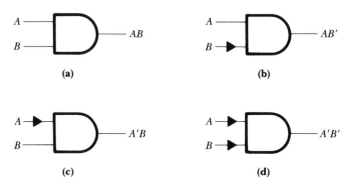

Figure 9.21

Examples of other gates are shown in Fig. 9.20, while Fig. 9.21 illustrates how to incorporate a NOT gate, $A' + B + C'$, into a circuit.

Each of the four fundamental products AB, AB', $A'B$, and $A'B'$ is illustrated in Fig. 9.22. (Again, multiplication is represented by the algebraic terminology AB.)

Can you draw the representations of the following gates?

a) $AB'C'$
b) $A + B' + C'$
c) $ABC'D'E'F$

Thus far we have looked only at simple AND, OR, and NOT gates. After learning each of the simple logic circuits, however, we can combine them into higher-order circuits (Fig. 9.23) in a similar manner to what we did with parallel and series circuits.

By connecting the outputs of several gates to the input terminal of other gates, we can construct more complicated circuits. Several AND gates could supply an OR gate, or several OR gates could supply an AND gate.

Figure 9.22

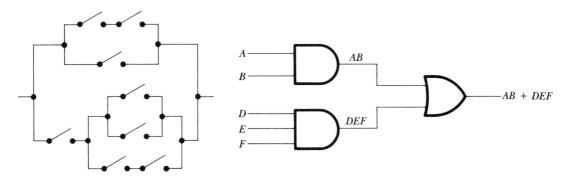

Figure 9.23
A combination of parallel and series circuits.

Figure 9.24

In Fig. 9.24, the output of two AND gates AB (A AND B) and DEF (D AND E AND F) serves as the input for an OR gate (hence, $AB + DEF$).

As another example, study carefully the logic circuit shown in Fig. 9.25. What is the Boolean expression for K?

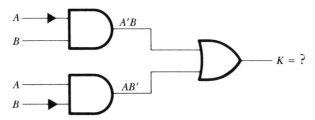

Figure 9.25

After careful study of your truth table for the circuit in Fig. 9.25 and by reviewing Section 9.6 on fundamental products, you may note that the logic circuit in Fig. 9.25 would produce the correct values for the sum column when adding two one-digit binary numbers. (*Check:* We know K is identical to S from our work in fundamental products; that is, $K = S = AB' + A'B$ as shown in Tables 9.17 and 9.18.)

Table 9.17
Values for the sum column
from fundamental products

A	B	S
1	1	0
1	0	1
0	1	1
0	0	0

Table 9.18
Values for the sum
column from logic gates

A	1	1	0	0
B	1	0	1	0
K	0	1	1	0

Because the truth tables for the AND, OR, and NOT gates are similar to those for propositions $P \wedge Q$, $P \vee Q$, and $\sim P$, respectively, the logic circuits satisfy the same laws and properties as those for propositional logic. And although the truth tables for $P \wedge Q$, $P \vee Q$, and $\sim P$ were introduced with propositional logic, they are not an exclusive property of propositional logic, but rather a *property of Boolean expressions in general*.

The AND gate is similar in form and function (isomorphic) to:

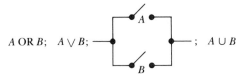

A AND B; $A \wedge B$;   ; $A \cap B$

The OR gate is similar in form and function (isomorphic) to:

A OR B; $A \vee B$;   ; $A \cup B$

Exploring these possibilities of the transfer of ideas permits a great deal of mathematical power because of (a) the properties of a Boolean algebra and (b) the fact that most of these topics, especially logic circuits, are a Boolean algebra. Consider the relationships between sets, propositional logic, and Boolean algebra. Do you think it might be possible to draw a logic circuit (with gates) to correspond to $A \cap B' \cup C \cap D'$? Do you think we could construct a logic circuit given sum-of-products form? Could we analyze—and simplify—propositional logic statements by using logic circuits?

Exercise Set 9.7

Draw a logic circuit for each of the following Boolean expressions by using AND, OR, and NOT gates.

1. $AB' + A'B + AB$ 2. $(A + B)(C + D)$

3. $AB'(A + B')$ 4. $ABC + AB'C' + A'B'C$

5. $(AB + CD)(A'B + C'D)$

6. $(AB + CD)(AB + EF) + (A'BC + A'B')$

Determine the Boolean function for each of the following logic circuits.

7.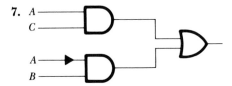

8.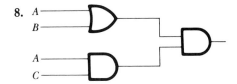

9.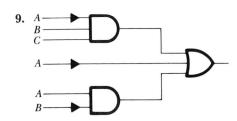

10.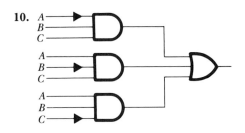

11.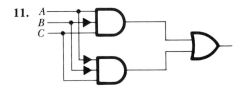

Figure 9.26

12. A NAND gate is formed by using an AND gate followed by a NOT gate, as shown in Fig. 9.26.

a) Compare the truth table for a NAND gate with the truth table for $A' + B'$. Any conclusions?

b) What do you think a NOR gate would be?

c) Draw a NOR gate for $(A + B)'$.

d) Compare the truth table for the NOR gate with the truth table for $A'B'$. Any conclusions?

13. Look carefully at the definition of Boolean algebra in Section 9.1. If an AND gate replaces the operation ·, an OR gate replaces the operation +,

Table 9.19
Obtaining output for the carry column and the sum column for adding binary numbers

0	0	1	1		A	0	0	1	1	A	0	0	0	1
+ 0	+ 1	+ 0	+ 1	or	B	0	1	0	1	B	0	1	0	1
00	01	01	10		C	0	0	0	1	S	0	1	1	0
CS	CS	CS	CS											

and a NOT gate replaces the unary operation ', would logic circuits form a Boolean algebra?

14. After reviewing Section 9.6 on adding binary numbers, draw a logic circuit that would produce two outputs: one for the sum column, S, and another for the carry column, C, as shown in Table 9.19. (The inputs would be A and B.) The term for this logic circuit, when you complete it, is "half-adder."

15. Draw a circuit diagram to correspond to the shaded area in the Venn diagram in Fig. 9.27.

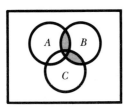

Figure 9.27

Summary

Strong relationships and similarities exist between the concepts developed in (1) sets, (2) logic, (3) circuitry, and (4) Boolean algebra. For example, the word AND is used in a very similar manner in these contexts:

1. An element is in the intersection of two sets ($A \cap B$) if the element is in set A AND in set B.
2. The logical proposition $P \wedge Q$ is true if proposition P is true AND proposition Q is true.
3. Current flows through the series circuit

if switch A is closed AND switch B is closed.
4. The Boolean expression $A \cdot B$ equals 1 if $A = 1$ AND $B = 1$.

An understanding of these four areas and the relationships between them is important to computer programming.

Boolean algebra is a mathematical structure composed of a set of elements S, together with two binary operations, $+$ and $\cdot$, and a unary operation, '. This structure satisfies the commutative laws and the distributive laws; set S contains identity elements and complements.

Boolean algebra is applied in the analysis of the logic of circuits. Formation of circuits is accomplished by using a topic from Boolean algebra called *fundamental products*. Most circuits are developed from the fundamental structures of parallel and/or series circuits. A parallel circuit is analyzed by the use of the logical connective OR, whereas a series circuit is analyzed by the use of the logical connective AND.

By combining parallel and series circuits and using logical connectives, we can represent complex circuits by Boolean expressions; conversely, we can diagram Boolean expressions as circuits. From the mathematical structure of Boolean algebra, many properties other than the commutative and distributive properties are derived. These additional properties (associative laws, idempotency, involution, De-Morgan's laws, boundedness laws, absorption laws) are used in the analysis and simplification of complex circuits by their corresponding Boolean expressions. Techniques such as the use of *sum-of-products form* greatly facilitate both the simplification and the analysis process.

Review Exercises

1. Define what is meant by a *Boolean algebra*.

2. To illustrate your familiarity with the similarities among work with sets, logic, circuits, and Boolean algebra, indicate which of the items listed below has an interpretation similar to A AND B.

   a) $A \wedge B$ **b)** $A \cup B$ **c)** AB

   d) $A \vee B$ **e)** $— A — B —$ **f)**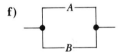

   g) $A + B$ **h)** $A \cap B$

3. Complete Table 9.20 for each diagram below to determine which switch settings will permit current to flow through the circuit.

   a)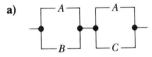

   b)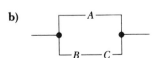

			Table 9.20	
A	B	C	(a)	(b)
0	0	0	y or n	y or n
0	0	1	y or n	y or n
0	1	0	y or n	y or n
0	1	1	y or n	y or n
1	0	0	y or n	y or n
1	0	1	y or n	y or n
1	1	0	y or n	y or n
1	1	1	y or n	y or n

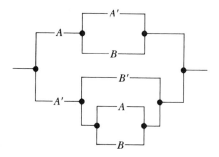

Figure 9.28

4. For what values of the switches *A* and *B* will current flow through the circuit shown in Fig. 9.28? Trace the paths through which current will flow.

5. In a circuit network containing two variables, how many possible results exist? For three variables, how many possible results exist? For *n* variables, how many possible results exist?

For each of the following Boolean expressions, draw the corresponding circuit.

6. $A + AB + ABC$

7. $AB' + A'B$

8. $A(B + C + D) + (B + F)CD$

9. $A(A + B) + (A'B + A')$

10. $A(A + C) + (A + B')BC$

11. $AB + ACD' + A(B' + C')$

12. $A(A + B) + B(A + C)$

13. $A'B'C' + A'B'C$

14. $(A + B')A'B(A + B' + (A'B))$

15. $A(B(A' + B')C + (AB' + B))$

For each of the following switching circuits, write the corresponding Boolean representation (it is not necessary to simplify at this time).

16.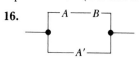

17.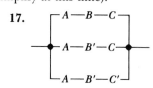

18.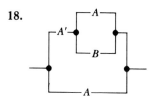

19.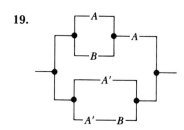

20. 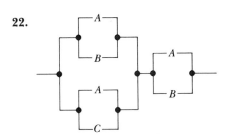   **21.**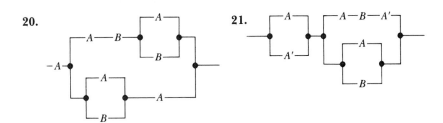

22.

23. To be interviewed by a particular computer firm, an applicant must have either a bachelor's degree in computer science (B) OR a two-year associate degree (A) AND three years of programming experience (P) AND a working familiarity with COBOL (C). Represent these criteria as a Boolean expression.

For each of the expressions in Exercises 24 and 25:

    a) Rewrite the expression using set representation ($\cup$, $\cap$).

    b) Rewrite the expression using logic representation (AND, OR).

    c) Draw a Venn diagram.

    d) Draw a parallel–series circuit.

24. $A + (BC)$ **25.** $(A + B)(A + C)$

Identify each of the Boolean properties illustrated below.

26. $(A + B)' = A'B'$ **27.** $A + BC = (A + B)(A + C)$

28. $(A + B) + C = C + (A + B)$ **29.** $(A')' = A$

30. $(A + B'C)' = A'(B + C')$ **31.** $AB = BA$

Simplify each of the following Boolean expressions using Boolean properties (some expressions may require the use of several properties).

32. $(A' + BA)'$ **33.** $AB(AB' + (A' + B))$

34. $A(B + C) + AC'$ **35.** $A + B' + (A'B) + (A + B')A'B$

36. $A + AB' + AB'C'$ **37.** $(A + C)(A + B) + BC'$

Simplify each of the following circuits by representing it as a Boolean expression and then using the Boolean properties to make a simpler circuit.

38.

39.

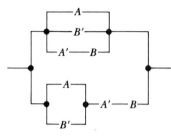

40.

41.

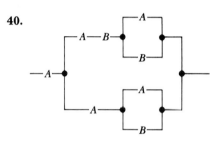

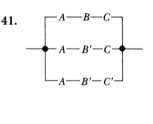

42. The phrase below contains capital letters, each possibly representing a complete sentence or logical condition. Make the phrase simpler, if possible, by using the properties from Boolean algebra.

NOT A AND B AND NOT C OR NOT A AND NOT B AND C OR NOT A AND B AND C OR NOT A AND NOT B AND NOT C

Table 9.21

A	B	C	K
0	0	0	0
0	0	1	1
0	1	0	1
0	1	1	0
1	0	0	0
1	0	1	1
1	1	0	1
1	1	1	0

43. Using fundamental products, find a Boolean expression to produce the output in column K of Table 9.21 for the indicated values of A, B, and C.

Write each of the following expressions as the sum of fundamental products. Then rewrite that expression in sum-of-products form.

44. $ABAC + ABB' + ABCD$

45. $A'B + AB' + AB + ACA$

46. $AB' + ABC + A'B'C'DD' +$
    $A'BC$

47. $AB'C + AB'C' + ABC$

Simplify each of the following Boolean expressions by using sum-of-products form.

48. $AB + A'B$

49. $A'B + AB' + AB$

50. $AB + ABC + (AB)'$

51. $A(BC + BC') + A'C'(B + B')$

52. $B(AC' + C'(A + A') + A'(C + C'))$

By using AND, OR, and NOT gates, draw a logic circuit for each of the following Boolean expressions.

53. $(A + B')(A + B)$

54. $A + AB + ABC$

55. $A'B + AB' + AB$

56. $A'B'C + A'BC' + AB'C' + ABC$

57. $AB' + ABC + A'B'C' + A'BC$ **58.** $A(BC + BC') + A'C'(B + B')$

Determine a Boolean expression to represent each of the following logic circuits.

59.   **60.**

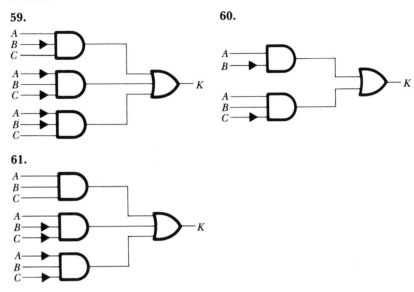

61.

62. Draw a circuit diagram to correspond to the shaded area in the Venn diagram in Fig. 9.29.

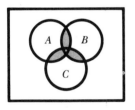

Figure 9.29

63. The Boolean equation $K = A \cdot B$ and the symbolic switch diagram are shown in Table 9.22 with their counterparts from related areas of mathematics. Complete the table for the remaining Boolean equations. Finally, for each equation, replace the multiplication operation with the addition operation, and again complete an appropriate table, showing the similarities among the related mathematical areas.

Table 9.22

EQUATION	SWITCHING DIAGRAM	VENN DIAGRAM	TRUTH TABLE	LOGIC CIRCUIT
$K = A \cdot B$	— A — B — K	(Venn diagram)	$\begin{array}{cc\|c} A & B & K \\ 0 & 0 & 0 \\ 0 & 1 & 0 \\ 1 & 0 & 0 \\ 1 & 1 & 1 \end{array}$	(AND gate: A, B → K)
$K = A'B$	?	?	?	?
$K = AB'$	?	?	?	?
$K = A'B'$	?	?	?	?

Systems of
Equations

CHAPTER 10 OBJECTIVES

After completing this chapter, you should be able to:

1. Identify linear equations in one or more variables.
2. Graph linear equations in two variables.
3. Solve systems of simultaneous equations by graphing.
4. Correctly classify a system of linear equations as independent, dependent, or inconsistent.
5. Solve systems of simultaneous equations by the substitution method.
6. Solve systems of simultaneous equations using linear combinations.
7. Solve systems of simultaneous equations by Gaussian elimination (tabular form).
8. Read, analyze, and solve applied problems.
9. Solve a system of equations by several methods, selecting the most appropriate technique.

10

10.1
Systems of Equations in Two Unknowns

A frequent business application of systems of linear equations is *break-even analysis*. To "break even," total cost should equal total revenue. If the total cost function is given by

$$C = 3.5U + 11{,}250$$

and the total revenue function is given by

$$R = 6U$$

then how many units of merchandise U must be produced in order to achieve an equilibrium, or to break even?

Using methods we will learn about in this chapter, we can find the number of units ($U = 4500$) such that the total *cost* ($\$27{,}000$) will equal the total *revenue* ($\$27{,}000$) (see Fig. 10.1). But, first we should review the graphing of linear equations.

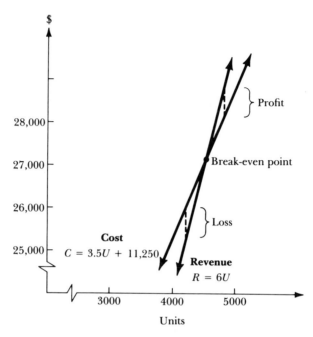

Figure 10.1

Graphing Linear Equations (A Review)

In the discussion of functions in Chapter 7, we worked with equations of the form $2x + 3y = 6$, $x - 4y = 8$, and $7.2x + y = 3.6$. Do you remember that these equations represent straight lines when graphed? In general, an equation of the form

$$ax + by = c$$

represents a *linear equation* in the variables x and y; the equation is called linear because when graphed, it represents a straight line. The equations $x + y = 8$, $3x - 2y = 6$, and $4.28x + 17.892y = 1.7689$ are linear, and each one has an infinite number of solutions, namely all the points (x, y) on the given line. For $x + y = 8$, some solutions are $(0, 8)$, $(1, 7)$, $(2, 6)$, $(3, 5)$, $(4, 4)$, $(5, 3)$, $(6, 2)$, $(7, 1)$ and $(8, 0)$. Could you name at least five more points that are solutions? These points and others can be used to plot this particular linear equation.

Systems of Simultaneous Equations

Often we must plot not only one linear equation, but several *simultaneously*. By reviewing the example of break-even analysis above, we can see that a solution is often needed that will solve several equations at the same time. The point (4500, 27,000) satisfies both equations

$$C = 3.5U + 11250 \quad \text{and} \quad R = 6U$$

in that example.

A simpler example is that the point (1, 7) satisfies both $x + y = 8$ and $9x - y = 2$ at the same time. A simultaneous solution to the set of equations

$$x + y = 8,$$
$$9x - y = 2,$$

then, is the *common* solution (1, 7).

Two or more equations that impose conditions simultaneously on the same variables form a *system of simultaneous equations.*

■

A fact that will be stated without proof is:

In order for a unique solution to occur in a system of simultaneous linear equations, there must be as many different equations as there are different variables.

The system of equations

$$3x - 2y + 4z = 5,$$
$$7x + y - 5z = 6$$

has three variables but only two equations. A unique solution is not possible.

The system of equations

$$x + y = 5,$$
$$x - y = 3$$

contains two variables and has two equations. A unique solution may be possible. Furthermore, the system imposes the *simultaneous conditions* on the two variables x and y that (1) the *sum* of the two values for x and y should equal 5, and (2) the *difference* of these same values for x and y should equal 3. Can you guess the solution to this system of equations? (In other words, two numbers when added yield 5, yet when subtracted yield 3.)

Geometrically, $x + y = 5$ represents a straight line when graphed (see Fig. 10.2a), as does the equation $x - y = 3$ (Fig. 10.2b). When both equations are drawn *simultaneously* on the same set of axes, the two different straight lines can intersect in at most one common point. This common point, then, is the solution to the set of the two linear equations. We would say that this point is the *simultaneous solution* for the system. (This point of intersection is shown in Fig. 10.2c.)

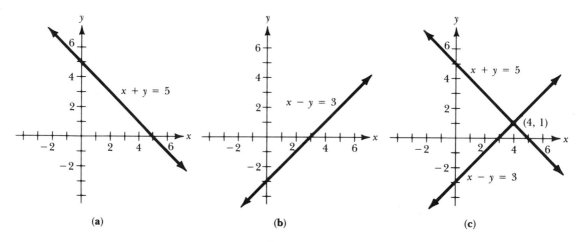

(a)   (b)   (c)

Figure 10.2

Although the system

$$x + y = 5,$$
$$x - y = 3$$

could be solved by guessing, it is not that easy to find the simultaneous solution of every system. For example, would you have guessed that the solution to the simultaneous system

$$1.3x - 1.68\ y = -0.24408,$$
$$2.0x + 0.006y =\quad 2.502666$$

is the common point $(1.248, 1.111)$? There are, however, several techniques to help us solve more complex systems.

As the systems become more complex, you will probably begin to ask yourself: "What are some techniques I could use to solve systems of equations?"

The following are some methods we will learn to help us find solutions to simultaneous systems of equations.

■

Solution of simultaneous systems may be accomplished by:

a) guessing (it works on the easier ones);
b) graphing;
c) elimination by substitution;
d) elimination by linear combinations;
e) Gaussian elimination;
f) Cramer's rule (use of determinants);
g) matrix methods.

These methods will not only make our work much more efficient, but also, because of their repetitive nature, lend themselves well to solution by today's high-speed computers.

As an illustration of a more complicated system, consider the following business application.

■ **EXAMPLE 1**   A clothing manufacturer wishes to explore the relationship of clothing expenses to annual net income. To do this, the following system must be solved for E and I:

$$8.2E + 4,750I = 3,878,$$
$$47,500E + 32,179,000I = 26,630,200. \quad \square$$

Systems may also be not only more complicated, but also larger, as illustrated in Example 2.

■ **EXAMPLE 2**   Jane must use the following system of equations to decide whether to put her money in bonds (B), mortgages (M), and stocks (S):

$$B + M + S = 13,000,$$
$$9.2B + 8.41M + 10.26S = 1230.50,$$
$$1.6(B + 2.8M) = 9.4S. \quad \square$$

While thus far our discussion has been limited to systems of linear equations, the systems need not necessarily be linear.

■ **EXAMPLE 3**   A solution to the simultaneous system

$$x^2 = y,$$
$$x + y = 6$$

is the point (2, 4). Another solution to this system of equations, however, is the point $(-3, 9)$, as we can see in Fig. 10.3. (You can verify each of these points as solutions to the system by substituting them into the equations.) $\square$

■ **EXAMPLE 4**   From the field of economics, a pair of supply and demand equations, each of which is a function of price per unit, produces this system of equations:

$$\text{for supply,} \quad Q = 100P - 10,$$
$$\text{for demand,} \quad Q = 4.8/P^2.$$

By examining the graph in Fig. 10.4, can you estimate the common (simultaneous) solution? The point at which these two functions intersect (the simultaneous solution) produces market equilibrium. (A commodity will tend to be sold at its equilibrium price by the law of supply and demand.) $\square$

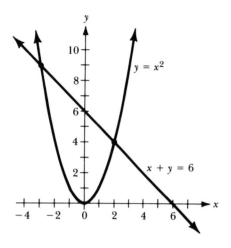

Figure 10.3

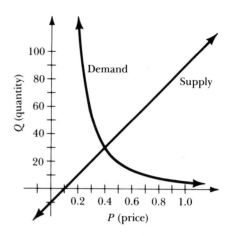

Figure 10.4

Exercise Set 10.1

A linear equation in standard form is represented by an equation of the form $ax + by = c$. In Exercises 1–6, write the equation in standard form, and then identify a, b, and c.

1. $3x = y - 6$

2. $x + y = 8$

3. $2x - y - 8 = 0$

4. $6x + 2y = 0$

5. $y - 7 = 0$

6. $y = 3x - 2$

7. Graph each of the linear equations in Exercises 1–6.

8. For each of the linear equations in Exercises 1–6, identify (i) the slope, (ii) the y-intercept, and (iii) the x-intercept.

Study each of the following systems of linear equations, and then guess the solution.

9. $x - y = 9,$
$\quad x + y = 11$

10. $3x = y,$
$\quad x + y = 4$

11. $2x - y = 7,$
$\quad x + y = 5$

12. By guessing, find the common solution to this system:

$$x + y + z = 6,$$
$$2x - y + 3z = 9,$$
$$-x + 2y - z = 0.$$

(*Hint:* If $z = 3$, find the values for x and y.) Is there only one solution to this system?

13. By inspection and then guessing, try to find the solution(s) to this system:

$$y = x^2 - 4,$$
$$2x - y = 4.$$

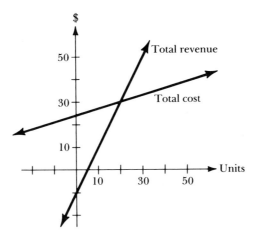

Figure 10.5

14. By examining the graph in Fig. 10.5, estimate at what point the total revenue (TR) will equal the total cost (TC), if

$$TC = 0.3U + 24,$$
$$TR = 2U - 10.$$

Determine whether each of the following is true or false.

a) When $U = 30$ units, the revenue will exceed the cost.
b) When $U = 10$ units, the revenue will exceed the cost.
c) More than one break-even point is possible.

10.2
Solution of Systems by Graphing

The system of equations

$$x + y = 5,$$
$$x - y = 3$$

represents two straight lines when graphed. Some points for each equation are shown in Fig. 10.6(a) and (b).

By inspecting the graph of each linear equation, shown in Fig. 10.6(c), we see that the point that is common to both equations is (4, 1). Graphing in this manner provides a valuable aid for solving simultaneous systems of equations and an intuitive illustration of the goal of the solution process. So, when we solve systems of equations, we are really seeking this "common point of intersection."

As a further review of graphing, let's try another example.

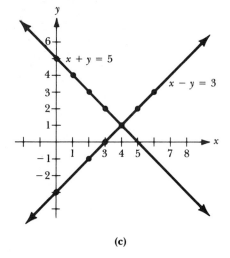

$x + y = 5$ (a)

x	y
0	5
1	4
2	3
3	2
4	1

$x - y = 3$ (b)

x	y
6	3
5	2
4	1
3	0
2	-1

Figure 10.6

■ EXAMPLE 5

Find the point common to this system of equations:

$$y = (1/2)x - 4,$$
$$2x + y = 6.$$

Solution

Each equation can be drawn by:

a) finding the slope and y-intercept;
b) finding the x-intercept and y-intercept;
c) "plugging in" values for the x-coordinate and then finding the corresponding y-value; or
d) any combination of the above.

The linear equation $y = (1/2)x - 4$ is already in "slope, y-intercept form" ($y = mx + b$). The slope is $1/2$ and the y-intercept ($b = -4$) is $(0, -4)$. The graph of this line is shown in Fig. 10.7.

The equation $2x + y = 6$ is in "standard form." Let's try finding the x- and y-intercepts. For the x-intercept (the point on the x-axis), the y-coordinate must be zero. Solving the equation with $y = 0$, we find that $x = 3$.

$$2x + y = 6$$
$$2x + 0 = 6$$
$$2x = 6$$
$$x = 3$$

x	y
?	0
3	0

For the y-intercept (the point on the y-axis), the x-coordinate must be zero. Solving for y, we find that $y = 6$.

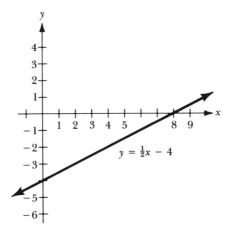

Figure 10.7

$$
\begin{array}{ll}
2x + y = 6 & \quad \begin{array}{c|c} x & y \\ \hline 0 & ? \\ \\ 0 & 6 \end{array} \\
\rule{3cm}{0.4pt} & \\
2(0) + y = 6 & \\
0 + y = 6 & \\
y = 6 &
\end{array}
$$

Plotting the two intercepts (3, 0) and (0, 6), we get the line shown in Fig. 10.8.

Plotting the two lines simultaneously (see Fig. 10.9), we see that the point of intersection is (4, −2). By substituting these values into each equation in the system, we can check our solution.

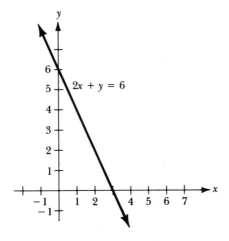

Figure 10.8

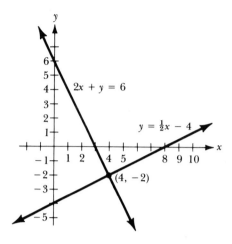

Figure 10.9

Check: $2x + y = 6$ Check: $y = (1/2)x - 4$
        $2(4) + (-2) = 6$ $-2 = (1/2)(4) - 4$
        $8 - 2 = 6$ $-2 = 2 - 4$ □

Each of the methods in Example 5 required *two items* to graph the linear equation: either (a) a slope and the *y*-intercept or (b) two intercepts. What if a mistake is made on either part? We were fortunate that $(4, -2)$ checked. What if the point did *not* check?

To eliminate potential errors and decrease the probability of graphing the linear equations incorrectly, use of three items will help. For example, when using the slope (item one) and the *y*-intercept (item two), why not also substitute some value for *x* and compute the corresponding *y*-coordinate? This additional point (item three) will serve to check our work from the other two items. It is especially advantageous when using the intercept method to have at least three points. If all three points are in a straight line, your chances of correctly graphing the linear equations are very good!

You may now wish to try some of the graphing exercises in Exercise Set 10.2 to review sketching linear equations. When you have had some practice, try solving the system in Example 6 by graphing.

■ **EXAMPLE 6**   Find the point of intersection for the system

$$3x - y = 6,$$
$$6x - 2y = 4.$$

Solution   We see from Fig. 10.10 that the lines *do not* intersect; they are parallel. (When viewed as two sets of points, the lines have an intersection that is the null or empty set.)   □

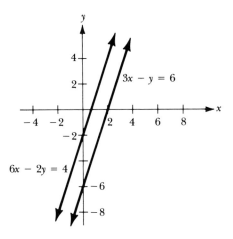

Figure 10.10

■ **EXAMPLE 7**   Solve this system graphically:

$$2x + y = 4,$$
$$4x + 2y = 8.$$

Solution   Here, the lines are not really different. Both linear equations represent the *same* straight line (see Fig. 10.11). (Try dividing the second equation by 2. Are the equations really two different equations?)   □

In general, the solution of simultaneous systems of linear equations includes three cases.

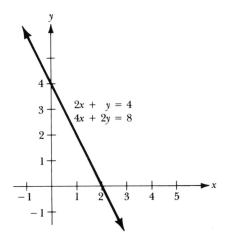

Figure 10.11

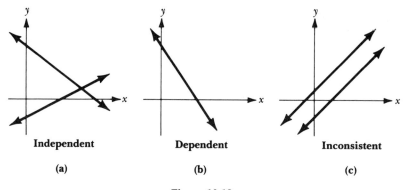

Independent   **Dependent**   **Inconsistent**

(a)   (b)   (c)

Figure 10.12

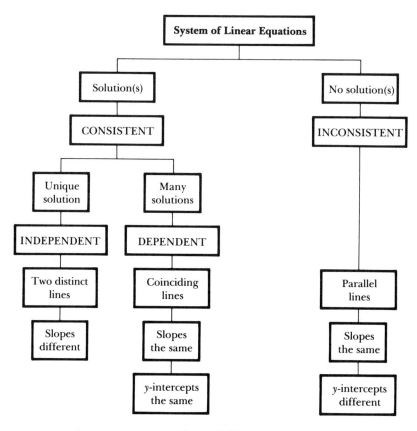

Figure 10.13

1. The system is *independent.*

   The system has exactly *one solution.* Geometrically, the linear equations represent two different lines intersecting at one common point. (See Fig. 10.12a.)

2. The system is *dependent.*

   The system has *infinitely many solutions.* Geometrically, the linear equations represent the same line. The graphs of the linear equations coincide, hence all the points on the line are possible solutions. (See Fig. 10.12b.)

3. The system is *inconsistent.*

   The system has *no common solution.* Geometrically, the linear equations represent two parallel lines. (See Fig. 10.12c.)

By graphing, we can often identify a system as one of these three cases. However, consideration should also be given to methods other than graphing to determine the type of system. Think about the role of the slopes and the y-intercepts.

If the system is independent, the slopes must be different.

If the system is dependent, the slopes must be the same. A dependent system represents lines that coincide—in other words, the same line! Therefore, not only the slopes, but also the y-intercepts must be the same.

If the system is inconsistent, the slopes must be the same. And if the lines are parallel, the y-intercepts must be different, since parallel lines cross the y-axis at different points.

The above discussion is summarized in Fig. 10.13.

Exercise Set 10.2

For each of the following linear equations, review graphing by finding the slope, the x-intercept, and the y-intercept. Also, find the corresponding y-value when $x = 1$, when $x = 5$, and when $x =$ any value of your choice.

	Slope	*x-intercept*	*y-intercept*	*Additional Points*
1. $3x - 2y = 6$	_____	(__, __)	(__, __)	(1, __) (5, __) (__, __)
2. $y = 4x - 8$	_____	(__, __)	(__, __)	(1, __) (5, __) (__, __)
3. $x + y = 12$	_____	(__, __)	(__, __)	(1, __) (5, __) (__, __)
4. $y = (2/3)x - 6$	_____	(__, __)	(__, __)	(1, __) (5, __) (__, __)
5. $1.2x - 3.6y = 7.2$	_____	(__, __)	(__, __)	(1, __) (5, __) (__, __)

How could you check to determine whether your (x, y) coordinates are correct?

Solve each of the following systems of linear equations by graphing.

6. $x + y = 8$,
$2x - y = 4$

7. $x + 2y = 8$,
$2x + y = 7$

8. $x = 2$,
$x + 2y = 10$

9. $-3x + 2y = 6$,
$y = 2x + 3$

10. $4x + y = 9$,
$3x - 5y = 1$

11. $y = x$,
$x - 2y = 2$

12. $4x + 3y = 15$,
$y = 4$

13. $0.5x - y = 8$,
$0.25x - 0.5y = 4$

14. $y = 4$,
$x = -2$

15. $6x - 5y = 30$,
$5x - 4y = 20$

16. $(x + y)/2 = 15$,
$1.5x + 1.5y = 6$

17. $2(3x - 2y) = 10 - 2x$,
$x = 10 - 2y$

Identify each of the following systems as independent (different lines intersecting in a common point), inconsistent (parallel lines), or dependent (coinciding lines). To do this you may either (1) graph each system or (2) study the slopes and the y-intercepts.

18. $x + y = 6$,
$x - y = 2$

19. $y = 5x$,
$y = 5x - 2$

20. $2x - y = 4$,
$4x + 2y = 8$

21. $2x - y = 4$,
$4x - 2y = 8$

22. $2x - y = 4$,
$4x - 2y = -8$

23. $ax + by = 1$,
$bx + ay = ab$

24. For the system

$$3x - 5y = 2,$$
$$6x - 10y = K,$$

find a value for K (if possible) so that:

a) the graphs coincide;
b) the graphs are parallel;
c) the graphs are distinct intersecting lines.

(*Hint:* Study the slopes and y-intercepts.)

10.3

Solution of Systems by Substitution

Substitution is a method of solution that is valuable for systems of linear equations, but it is even more valuable—almost indispensable—

for nonlinear systems such as

$$x^2 + y^2 = 5,$$
$$y = 3x^2 - 5.$$

The technique will be demonstrated first on a system of linear equations:

$$3x - y = 5,$$
$$x + 2y = 4.$$

In Section 10.1 we learned that in order for a system to have a unique solution, the number of different equations must equal the number of unknowns. In this case we have two equations in two unknowns, or a 2-by-2 system. We will try to obtain a 1-by-1 system, or one equation in one unknown, by eliminating one of the unknowns. That we can then solve!

Remember that each equation imposes a different "condition" on x and y. In the equation $3x - y = 5$, the condition imposed on x and y is that "three times the x-value, minus the y-value, must equal five." The equation $x + 2y = 4$ imposes a different condition on x and y. In Example 8 we will attempt to combine both conditions into one equation.

■ **EXAMPLE 8**

We have the following system

$$\text{Equation (1)} \quad 3x - y = 5,$$
$$\text{Equation (2)} \quad x + 2y = 4.$$

If we choose to solve Eq. (2) for x, we will then *substitute* this condition upon x and y into Eq. (1). Thus $x + 2y = 4$ becomes $x = 4 - 2y$ when we are solving for x.

Eq. (1) $3x - y = 5$ becomes $3\underline{(4 - 2y)} - y = 5$

               ↑   ↑

Eq. (2) $x = \boxed{4 - 2y}$ x

The transformed equation (1) is now one equation with one variable: $3(4 - 2y) - y = 5$. Yet this equation contains both of the original *conditions* on x and y: "three times x minus y equals five," and "x is equal to four minus two times y."

$$3(4 - 2y) - y = 5 \longrightarrow 12 - 6y - y = 5$$
$$12 - 7y = 5$$
$$12 - 5 = 7y$$
$$7 = 7y$$
$$1 = y$$

Now that we know that $y = 1$, how do we find the x-coordinate? By substituting $y = 1$ into one of the original equations, we can then find the x-coordinate of the simultaneous solution. If we substitute $y = 1$ into Eq. (1), we find that $3x - (1) = 5$ yields an x-value of 2. The simultaneous solution to the system is then (2, 1). □

■ **EXAMPLE 9**   In Example 8 we could also have solved the system by using substitution for y, as follows:

$$\text{Eq. (1)} \quad 3x - y = 5,$$
$$\text{Eq. (2)} \quad x + 2y = 4.$$

We solve Eq. (1) for y, and then substitute this value into Eq. (2):

Eq. (1) $3x - y = 5$ could be changed to yield:

$$\boxed{3x - 5} = y$$
$$\downarrow$$

Eq. (2) $x + 2y = 4$ would then become:
$$x + 2(\underline{3x - 5}) = 4.$$

Now this equation in one unknown can be solved for x:

$$x + 2(3x - 5) = 4 \longrightarrow x + 6x - 10 = 4$$
$$7x - 10 = 4$$
$$7x = 14$$
$$x = 2$$

By substituting back into the equation, we find that y is equal to 1. The solution (x, y) is (2, 1). □

Although the work would have been more involved, we could also have (a) solved for x in Eq. (1), and then *substituted* into Eq. (2), or (b) solved for y in Eq. (2), and then substituted into Eq. (1).

Would solving for x in Eq. (1) and then substituting into Eq. (1) work? Why or why not?

In general, substitution is a desirable method *if one of the coefficients is either a +1 or a −1*. (In Eq. (1) above, the coefficient of y is −1. In Eq. (2), the coefficient of x is 1.)

Substitution works on larger systems as well (however, be prepared for a little more work!). The technique of substitution is useful in linear programming as well as in higher mathematics involving simultaneous solutions of nonlinear systems. An example of each follows.

■ **EXAMPLE 10**   Solve the following system for (x, y, z):

$$x + 2y - z = 2, \tag{1}$$
$$x - y + 2z = 5, \tag{2}$$
$$2x + 2y + z = 9. \tag{3}$$

Solution   Solving Eq. (1) for x yields

$$x = \boxed{2 - 2y + z}.$$

This quantity could then be substituted into either Eq. (2) or (3). We'll substitute $(2 - 2y + z)$ for x in Eq. (2):

$$x - y + 2z = 5 \tag{2}$$

then becomes

$$(2 - 2y + z) - y + 2z = 5,$$

which simplifies to

$$-3y + 3z = 3;$$

and dividing by -3 yields

$$y - z = -1.$$

However, this yields one equation with two unknowns! We need another equation containing the same two unknowns. Thus far we have combined Eqs. (1) and (2). Let's substitute $x = (2 - 2y + z)$ into Eq. (3) for x:

$$2x + 2y + z = 9 \tag{3}$$

becomes

$$2(2 - 2y + z) + 2y + z = 9,$$

which simplifies to

$$-2y + 3z = 5.$$

The condition imposed on the coordinates x, y, and z in Eq. (1) has now been combined with those in Eqs. (2) and (3). We now have two equations in two unknowns:

$$\begin{aligned} y - z &= -1, \\ -2y + 3z &= 5. \end{aligned}$$

We now solve the first equation for y, and substitute the expression for y into the second equation:

$$-2y + 3z = 5$$

becomes

$$-2(z - 1) + 3z = 5,$$

which when simplified produces the value $z = 3$.

Now, by substituting back into the equations, we find the values for y and x. Substituting $z = 3$ into $y = z - 1$ yields $y = 2$. Substituting $z = 3$ and $y = 2$ into $x = 2 - 2y + z$ yields $x = 1$. Therefore, we have $x = 1$, $y = 2$, and $z = 3$.

A check should be made to verify that $(1, 2, 3)$ does satisfy all three original equations. □

Substitution is most useful in work with nonlinear systems. For example, let's solve this system containing one linear equation and one quadratic equation.

■ **EXAMPLE 11**   Solve the system

$$y = x^2, \tag{1}$$
$$2x - y = -3. \tag{2}$$

Solution   Because in Eq. (1) y has a coefficient of 1, we substitute this expression for y into Eq. (2):

$$2x - y = -3 \tag{2}$$

then becomes

$$2x - (x^2) = -3$$
$$2x - x^2 = -3$$
$$-x^2 + 2x + 3 = 0 \quad \text{or} \quad x^2 - 2x - 3 = 0.$$

This quadratic equation factors as

$$(x - 3)(x + 1) = 0,$$

yielding

$$x = 3 \quad \text{or} \quad x = -1.$$

(For a review of quadratic functions, see Chapter 7.)

Note that we now have two values for x. Substituting $x = 3$ in $y = x^2$ yields $y = 9$; and substituting $x = -1$ into $y = x^2$ yields $y = 1$. The simultaneous solutions are then $(-1, 1)$ and $(3, 9)$.

If we graph the linear equation $2x - y = -3$ and the quadratic function $y = x^2$ on the same set of axes, we can verify the solutions as the points of intersection, as shown in Fig. 10.14. □

■ **EXAMPLE 12**   (*A business application.*) Find the equilibrium price, P, and the corresponding number of units supplied, Q, if the supply function is given by $P^2 + P - 60 = Q$, and the demand function is given by $420 - 3P = Q$.

Solution   Because $420 - 3P$ equals Q, we can substitute this expression for Q into the equation for the supply function:

$$P^2 + P - 60 = Q$$
$$P^2 + P - 60 = 420 - 3P$$
$$P^2 + 4P - 480 = 0.$$

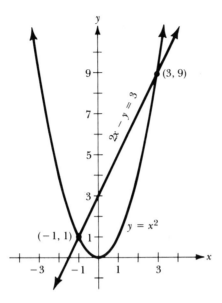

Figure 10.14

This is a quadratic that can be factored,

$$(P + 24)(P - 20) = 0,$$

yielding the solutions

$$P = -24 \quad \text{or} \quad P = 20.$$

Since negative price is not meaningful here, we choose \$20 as the solution. By substituting \$20 back into either of the original equations, we find that the number of units Q is 360. At \$20 per item, supply and demand will both equal 360 units. Supply will equal demand and equilibrium has been reached. □

Exercise Set 10.3

Solve each of the following systems of equations by using substitution. (If any of the systems is inconsistent or dependent, so state.)

1. $y = 2$,
   $x + y = 3$

2. $y = 2x - 5$,
   $x + 3y = 6$

3. $2x - 3y = 7$,
   $4x + y = 0$

4. $3x - 1.5y = 6$,
   $-2x + y = 2$

5. $7x - 2y = 6$,
   $3y = 4x + 1$

6. $(2/3)x + (3/5)y = 1/15$,
   $(1/2)x + y = -1/2$

7. $0.67x - 2.1y = 5.605,$
$\quad x + 1.34y = 5.1468$

8. $x = 5,$
$\quad y = 3,$
$\quad x + y - z = 6$

9. $x = -3,$
$\quad x + 2y = -7,$
$\quad x + y + z = -6$

10. $x + 2y + z = 8,$
$\quad 2x + y - z = 13,$
$\quad 3x - 2y + z = 4$

11. $2x + y - z = 1,$
$\quad 3x - 2y + 3z = 8,$
$\quad 8x - 6y + 4z = 8$

12. $3x + 2y + 17z = 1,$
$\quad 2x + 3y + 8z = 4,$
$\quad 2x + 4y + 6z = 6$

Solve each of the following systems (some nonlinear) by substitution.

13. $y = 3x,$
$\quad x^2 - y + 2 = 0$

14. $x^2 + y^2 = 5,$
$\quad y = 3x - 5$

15. $y = x^2 - 4,$
$\quad x - y = 2$

16. $x^2 + y^2 = 16,$
$\quad x^2 - y^2 = 16$

17. $x^2 = y,$
$\quad y = x$

18. $xy = 12,$
$\quad 2x - 3y = 1$

19. a) This system is an *inconsistent* system:

$$3x - y = 8,$$
$$6x - 2y = 5.$$

Attempt to solve by substitution; explain the outcome of your work.

b) This system is a *dependent* system:

$$3x - y = 8,$$
$$6x - 2y = 16.$$

Attempt to solve by substitution; explain the outcome of your work.

20. In an investment analysis problem, a business associate must solve the following system of equations for R and N:

$$N = 1000/R,$$
$$NR + 10R - N = 1050.$$

Find all solutions for R and N.

21. Supply is represented by the function $S = U^2 + 100$. Demand is represented by the function $D = 10U + 700$. How many units U are needed in order that the supply will exceed the demand?

10.4

Solution of Systems by Linear Combinations

Using linear combinations is probably the most popular method for solving systems of linear equations. (This method is also referred to as linear combination, elimination, elimination by addition or subtraction, addition/subtraction, multiplication/addition.) It is most often used for smaller systems: 2 equations in 2 unknowns or 3 equations in 3 unknowns. More efficient methods are generally needed for larger systems. With the advent of the microcomputer, other methods lend themselves to programming a little better than does the method of linear combinations.

First we'll try to capture the "idea" of linear combinations, and then we'll illustrate the technique. If you recall from earlier sections, in order for a system to have a unique simultaneous solution, the number of equations should equal the number of unknowns. For example, if there are three unknowns A, B and C, then for a unique solution, three equations should be present in the system. *The objective of linear combinations is to reduce the "dimension" of the system.* If you have a 3-by-3 system (three equations in three unknowns), you should try to reduce it to a 2-by-2 system (two equations in two unknowns). Then try to reduce the 2-by-2 system to a 1-by-1 system, which can then be solved! The general goal is to reduce the size of the system until a 1-by-1 system has been obtained.

Let's begin with some examples of 2-by-2 systems.

■ **EXAMPLE 13**   Solve

$$x - y = 6,$$
$$x + y = 8$$

for (x, y).

Solution   We would like to combine the equations (hence the use of linear combinations) and simultaneously *eliminate* one of the variables. (Read that statement one more time!) Thus

$$
\begin{array}{r}
x - y = 6 \\
x + y = 8 \\
\hline
2x = 14
\end{array}
$$

Here we note that by "adding" the two equations, we can form a linear combination and, at the same time, *eliminate* y. Thus we have a 1-by-1 system (one equation in one unknown). Solving for x, we find

$$x = 7.$$

How do we find the value for y? By substituting $x = 7$ into *either* original equation. (This is similar to "back substitution" in the substitution method.) Thus

$$
\begin{array}{r}
x - y = 6 \\
7 - y = 6 \\
y = 1.
\end{array}
$$

To verify the solution (7, 1), we check $x = 7$ and $y = 1$ in *both* original equations. □

The key to success in solving this example rested in the fact that the coefficients of y were *inverses* of each other: in the first equation a $+1$ and in the second a -1. For the variable you wish to eliminate, the coefficients should be inverses of each other.

Remember, geometrically a system represents straight lines, and two distinct lines can intersect in at most one point. Therefore, if the point checks in both equations, the point must be the correct solution to the simultaneous system.

Let's try another example.

■ EXAMPLE 14

Solve the system

$$2x + y = 11,$$
$$x + y = 7.$$

Solution

If we try to add these equations, we get another equation—but the resulting equation has *two* variables. We would have formed a linear combination, but we would *not* have eliminated one of the variables!

$$
\begin{array}{r}
2x + y = 11 \\
x + y = 7 \\
\hline
3x + 2y = 18
\end{array}
$$

We would have a 1-by-2 system (one equation in two unknowns), but we need a 1-by-1.

We remember from our previous work in algebra that we can multiply both sides of an equation by any desired constant. The resulting equation will be *equivalent* to the original equation; they will both have the same solutions. To eliminate y, for example, we would like the coefficients to have opposite signs. So we multiply the second equation by -1.

We will use the symbol M(-1) to indicate that we have multiplied that equation by a (-1):

$$
\begin{array}{l}
2x + y = 11 \\
x + y = 7 \quad \text{M}(-1)
\end{array}
$$

Now by adding the two equations, we get $x = 4$:

$$
\begin{array}{r}
2x + y = 11 \\
-x - y = -7 \\
\hline
x = 4
\end{array}
$$

And then by substituting back, we find that y has a value of 3. The common solution to the system is (4, 3). How could you check this solution? □

On the system just solved, we could also have formed a linear combination by subtracting the second equation from the first. The work then would be similar to what was just completed.

Let's try a harder one!

■ **EXAMPLE 15**   Solve

$$3x - 2y = 8,$$
$$x + y = 1.$$

Solution   Remember, we wish to form a *linear combination*, but we must also eliminate one of the variables. Suppose we decide to eliminate y. What should the second equation be multiplied by?

$$3x - 2y = 8,$$
$$x + y = 1 \quad \text{M(2)}$$

Multiplying by 2 transforms the system into the equivalent system:

$$3x - 2y = 8,$$
$$2x + 2y = 2.$$

And then by *adding* the two equations, we have

$$5x = 10$$
$$x = 2$$

Substituting back gives us

$$y = -1.$$

Thus the common solution to this simultaneous linear system is $(2, -1)$. □

Looking at the same system, could we have eliminated x rather than y? What would be needed to have both the x-coefficients similar but with opposite signs? We have

$$3x - 2y = 8,$$
$$x + y = 1.$$

Multiplying the second equation by -3 gives us the equivalent system

$$
\begin{array}{rcr}
3x - 2y = & 8 \\
-3x - 3y = & -3 \\
\hline
-5y = & 5 \\
y = & -1.
\end{array}
$$

We obtain the linear combination by adding, and by substituting $y = -1$ back into either original equation, we find that $x = 2$.

Could we also have eliminated x by multiplying the first equation by $-1/3$ and then adding the resulting equations?

To form a linear combination, remember that you choose which variable you wish to eliminate. You will also want to make sure that the coefficients of this variable are inverses of each other.

■ **EXAMPLE 16**   Solve

$$5x - 7y = 19,$$
$$2x + 3y = -4$$

by using linear combinations.

Solution   Let's choose to eliminate y. Again, we need to have the coefficients identical (in absolute value) but with differing signs. Rather than multiply the second equation by 7/3 (which would work), let's make life a little easier. (What's the least common multiple of 3 and 7?) There is no reason why we couldn't "adjust" both equations rather than just one. Multiplying the first equation by 7 and the second equation by 3 gives us the following equivalent system:

$$5x - 7y = \ \ 19 \quad M(3) \longrightarrow 15x - 21y = \ \ \ 57,$$
$$2x + 3y = -4 \quad M(7) \longrightarrow 14x + 21y = -28.$$

Adding gives us the linear combination, $29x = 29$. Solving for x, we get $x = 1$ and substituting back into $5x - 7y = 19$ yields $y = -2$, for a common solution of $(1, -2)$. (We could then check this answer by substituting into $2x + 3y = -4$.)   □

● **Question**   In Example 16, could we have eliminated the variable x rather than y? How?

Answer   By multiplying the first equation by 2 and the second equation by -5, and then adding the resulting equations, we could have eliminated x.

Try this technique of using the least common multiple on the following system:

$$4x - 6y = \ \ 5,$$
$$6x + 8y = 16.$$

(*Hint:* To eliminate y, the least common multiple is 24. Therefore, the first equation would have to be multiplied by 4 and the second by 3.)

The technique of linear combinations can be generalized to find the solution of a system of any size. Remember the objective for this technique:

■

To reduce the dimensions of a system, linear combinations are formed by eliminating a variable from the system of equations.

We now illustrate the effectiveness of the method of linear combinations on a 3-by-3 system (three equations in three unknowns).

■ EXAMPLE 17   Solve the system

$$x - 2y + z = 0, \tag{1}$$
$$2x + y - z = 1,$$
$$x - 3y + 2z = 1. \tag{3}$$

Solution   We can choose any two equations and eliminate one of the variables. Here, to illustrate we will choose to eliminate z from the first two equations:

$$x - 2y + z = 0$$
$$\underline{2x + y - z = 1}$$
$$3x - y \qquad = 1$$

By simply adding the first two equations, we can obtain a linear combination of one equation in two unknowns. However, if we have two unknowns, how many equations will we need?

To get a second equation also containing the variables x and y, we must eliminate z from two other equations. Although we could also work on Eqs. (1) and (3), for this example, we will choose to form a linear combination of Eqs. (2) and (3):

$2x + y - z = 1$ M(2)	To eliminate the variable z, we mul-
$x - 3y + 2z = 1.$	tiply the first equation by 2 to get the equivalent system:

$4x + 2y - 2z = 2$	Now, by adding we obtain another
$\underline{x - 3y + 2z = 1}$	linear combination in the variables
$5x - y \qquad = 3$	x and y.

So, from the 3-by-3 system we have obtained a smaller 2-by-2 system:

$$\left.\begin{array}{l} x - 2y + z = 0 \\ 2x + y - z = 1 \\ x - 3y + 2z = 1 \end{array}\right\} \longrightarrow \begin{array}{l} 3x - y = 1 \\ 5x - y = 3 \end{array}$$

Again, we now wish to reduce the dimensions of the 2-by-2 system, as follows:

$3x - y = 1$ M(−1)	To eliminate y (which would be
$5x - y = 3$	easier!) we multiply the first equation by −1.

$-3x + y = -1$	Now by adding two equations, we
$\underline{5x - y = 3}$	get $x = 1$.
$2x = 2$	
$x = 1$	

In summary, the progression has gone from:

$$\begin{array}{ccccc} \text{a 3 by 3} & & \text{to} & \text{a 2 by 2} & \text{to} & \text{a 1 by 1} \end{array}$$

$$\left.\begin{array}{rcl} x - 2y + z & = & 0 \\ 2x + y - z & = & 1 \\ x - 3y + 2z & = & 1 \end{array}\right\} \longrightarrow \left.\begin{array}{rcl} 3x - y & = & 1 \\ 5x - y & = & 3 \end{array}\right\} \longrightarrow x = 1$$

By substituting back, we can find the values for y and z. Going back to either $3x - y = 1$ or $5x - y = 3$ and substituting $x = 1$, we find that $y = 2$:

$$\begin{array}{rcl} x = 1 \text{ in} & \longrightarrow & 3x - y = 1 \\ \text{or in} & \longrightarrow & 5x - y = 3 \quad \text{yields } y = 2. \end{array}$$

Then, going back to either $x - 2y + z = 0$ or $2x + y - z = 1$ or $x - 3y + 2z = 1$, and substituting $x = 1$ and $y = 2$, we find that $z = 3$:

$$\begin{array}{rcl} x = 1 \text{ and } y = 2 \text{ in} & \longrightarrow & x - 2y + z = 0, \\ \text{or in} & \longrightarrow & 2x + y - z = 1, \\ \text{or in} & \longrightarrow & x - 3y + 2z = 1 \quad \text{yields } z = 3. \end{array}$$

Thus the common solution to the simultaneous system

$$\begin{array}{rcl} x - 2y + z & = & 0, \\ 2x + y - z & = & 1, \\ x - 3y + 2z & = & 1 \end{array}$$

by the technique of linear combinations is $(x, y, z) = (1, 2, 3)$. □

● Questions

How would you verify this solution? Would checking $(1, 2, 3)$ in the first equation guarantee the correctness of the solution? Would checking $(1, 2, 3)$ in all three original equations guarantee that the solution is correct?

Exercise Set 10.4

Solve each of the following systems of linear equations by linear combinations (elimination by addition). Some of the systems may be inconsistent, some may be dependent; to check these systems, Fig. 10.13 will help.

1. $x + y = 9,$
   $x - y = 7$

2. $2x + y = 2,$
   $3x + y = 0$

3. $7x - 4y = 5,$
   $3x + 2y = 4$

4. $x = 2,$
   $2x + y = 19$

5. $(1/2)x + 3y = -2,$
   $(1/4)x + 2y = -2$

6. $2x + 1.5y = -8,$
   $1.3x - 2.05y = 6.9$

7. $ax + by = c,$
   $dx + ey = f$

8. $3x - 2y = 8,$
   $4.5x - 3y = 5$

9. $1.0R + 1.25S = 20,$
$0.433R - 0.4S = 10$

10. $4.22W - 6.38T = 20.48,$
$6.33W - 9.57T = 30.72$

Solve each of the following "larger" systems of equations, again by forming linear combinations by elimination.

11. $x + y - z = 4,$
$x + y + z = 6,$
$x + 2y - z = 5$

12. $4x + 2y - 2z = 2,$
$4x - 3y + 2z = 4,$
$3x - 2y + 3z = 8$

13. $3A = 6,$
$A + B = 3,$
$2A - B + 4C = 23$

14. $2x - y - z = 30,$
$y + z = 30,$
$x + y + z = 70$

15. $2x + 3y + z = 5,$
$3x + y - 4z = -6,$
$5x + 4y - 3z = -2$

16. $(1/2)x - y - z = 3/4,$
$2x + 3y - (1/2)z = -4,$
$3x + 8y + 2z = -11$

17. $2R - 3S + 5T = 100,$
$2.5S + 5T = 175,$
$0.2T = 2$

18. $2x - y + 10 = 0,$
$3y + z + 40 = 0,$
$3x + 2z + 50 = 0$

19. $0.6A + 0.5B + 0.3C = 40,$
$0.3A + 0.3B = 15,$
$0.1A + 0.2B + 0.7C = 45$

20. $x + 2y - 3z = 18,$
$3x + 6y - 9z = 18,$
$0.5x + y - 1.5z = 9$

21. a) Identify each system as either dependent, inconsistent, or independent.

(i) $3x - 4y = 5,$
$6x - 8y = 2$

(ii) $3x - 4y = 5,$
$6x + 8y = 2$

(iii) $3x - 4y = 5,$
$6x - 8y = 10$

b) Now, try to solve each system by using linear combinations.

c) What result is obtained when attempting to solve a dependent system?

d) What result is obtained when attempting to solve an inconsistent system?

22. The sum of two numbers x and y is 84. The difference of the two numbers is 6.

a) Write the two equations—one to represent the sum, the other to represent the difference.

b) Solve for x and y.

c) Check your solutions.

23. The equation of a straight line in slope–y-intercept form is $y = mx + b$. If a straight line passes through the points (4, 1) and (2, 3), then these points can be substituted in the equation $y = mx + b$. The system of equations that results from this work is

$$1 = m \cdot 4 + b,$$
$$3 = m \cdot 2 + b.$$

Find the slope, m, and the y-intercept, b.

24. Two amounts of money are invested, some in real estate (R) and some in growth stocks (G). The total amount invested is \$18,000. Three times as much money is invested in real estate as in the growth stocks. By finding the appropriate system of equations and solving, determine how much of the \$18,000 is invested in real estate and how much is invested in growth stocks.

10.5
Solution of Systems by Gaussian Elimination

With larger systems of linear equations, finding a common solution "by hand" can become extremely time-consuming, very tedious, and probably very prone to errors. Computers, however, allow us to find the solutions of very complex systems, and avoid these problems at the same time.

A typical personal microcomputer can run a business model that incorporates a maximum of 40 or 50 equations to solve for such economic variables as interest rates, unemployment levels, and rates of inflation. The larger computers often handle systems of more than 1000 equations! (Would you care to try solving that system "by hand"?) Therefore, we need methods of solution that lend themselves to being programmable and that are more systematic than those discussed in earlier chapters.

One of the oldest methods of solving systems of linear equations is called *Gaussian elimination,* in honor of the German mathematician Karl Friedrich Gauss, who developed the process. Understanding this process will be very useful not only for solving systems of equations, but also as the basis for understanding the *simplex method* in our later work in linear programming. Gaussian elimination is also very systematic, a desirable trait if a program is to be written to allow the computer to solve systems of equations! You will find that certain operations (multiplication and addition) are performed over and over again in Gaussian elimination. This ability to handle repetitive computation rapidly is one of the biggest advantages of using a computer.

Gaussian Elimination Explained

Gaussian elimination can be explained best by first illustrating the procedure in outline form. The objective is to transform a system of equations such as

$$\begin{aligned} x + 4y - 2z &= 3, \\ 5x - 2y + z &= 4, \\ 2x + 4y + z &= 13 \end{aligned}$$

into an equivalent system such as

$$x + 4y - 2z = 3,$$
$$2y - z = 1,$$
$$z = 3.$$

If a diagonal were drawn through x, $2y$, and z, as shown below, you may notice that all terms below this diagonal are now missing in this equivalent system. This system of equations is said to be in *triangular* form:

$$x + 4y - 2z = 3,$$
$$2y - z = 1,$$
$$z = 3.$$

The objective of Gaussian elimination is to transform the original system of equations into an equivalent system in triangular form. This is accomplished by forming linear combinations.

Then, by substituting back $z = 3$, we find a y-value of 2 by solving the middle equation for y: $(y = 2)$.

$$x + 4y - 2z = 3,$$
$$2y - (3) = 1$$
$$\uparrow$$
$$z = 3$$

Again, by substituting back with $y = 2$ and $z = 3$, we find an x-value of 1 by solving the first equation for x:

$$x + 4(2) - 2(3) = 3$$
$$\uparrow \qquad \uparrow$$
$$y = 2 \qquad |$$
$$z = 3$$

The common solution is $(1, 2, 3)$. Thus the solution $(1, 2, 3)$ to the *equivalent* system of equations above is the solution to the *original* system of equations.

   With the general idea of the objective in mind, let's now concentrate on the *algorithm* (the step-by-step procedure) to change the original given system into the simpler equivalent system. To facilitate our work, we will label our equations A1, B1, and C1, with A being the top (or first) equation, B the middle equation, and C the bottom (or last) equation. Subsequent changes in any one of these equations will be denoted by changing the number after the letter (for example, A1 to A2, B3 to B4, etc.).

■ EXAMPLE 18

Solve the following system using Gaussian elimination:

$$
\begin{array}{lrcl}
\text{A1} & x + 4y - 2z & = & 3, \\
\text{B1} & 5x - 2y + z & = & 4, \\
\text{C1} & 2x + 4y + z & = & 13.
\end{array}
$$

The first step is to eliminate the variable x from equations B1 and C1. By using the coefficient of x in equation A1, we'll use a linear combination of A1 and B1 to yield a new equation B2. (This leading coefficient of x will later be called the pivot in linear programming.)

$$
\begin{array}{lrcl}
\text{A1} & \textcircled{x} + 4y - 2z & = & 3, \\
\text{B1} & 5x - 2y + z & = & 4, \\
\text{C1} & 2x + 4y + z & = & 13.
\end{array}
$$

By using a multiple of A1 ($-5 \times$ A1), and adding this to equation B1, we'll get a transformed middle equation:

$$
\begin{array}{lrcl}
(-5) \times (\text{A1}) & -5x - 20y + 10z = -15 \\
\text{B1} & \underline{5x - 2y + z = 4} \\
\text{B2} & - 22y + 11z = -11.
\end{array}
\qquad (-5)(\text{A1}) + \text{B1} = \text{B2}
$$

An equivalent system to the original system is now

$$
\begin{array}{lrcl}
\text{A1} & x + 4y - 2z & = & 3, \\
\text{B2} & - 22y + 11z & = & -11, \\
\text{C1} & 2x + 4y + z & = & 13.
\end{array}
$$

Note that only the middle equation has been changed. Keeping in mind to simplify as we go, let's divide the entire second equation by the common factor of -11. The system now becomes

$$
\begin{array}{lrcl}
\text{A1} & x + 4y - 2z & = & 3, \\
\text{B3} & 2y - z & = & 1, \\
\text{C1} & 2x + 4y + z & = & 13.
\end{array}
\qquad \text{B3} = \text{B2}/(-11)
$$

Next, we'll use a similar process to eliminate the variable x from equation C1:

$$
\begin{array}{lrcl}
\text{A1} & \textcircled{x} + 4y - 2z & = & 3, \\
\text{B3} & 2y - z & = & 1, \\
\text{C1} & 2x + 4y + z & = & 13.
\end{array}
$$

A linear combination of A1 and C1 can be formed. *How?* By multiplying equation A1 by -2, and adding this to equation C1, we'll form a new bottom equation:

$$
\begin{array}{lll}
(-2) \times A1 & -2x - 8y + 4z = -6 & (-2)(A1) + C1 = C2 \\
C1 & \underline{2x + 4y + z = 13} & \\
C2 & - 4y + 5z = 7.
\end{array}
$$

The equivalent system now becomes

$$
\begin{array}{lll}
A1 & x + 4y - 2z = 3, \\
B3 & 2y - z = 1, \\
C2 & - 4y + 5z = 7,
\end{array}
$$

where C2 was obtained by the linear combination $(-2) \times$ (A1) + C1. Now the variable x has been eliminated from all the equations below the first one. The first "stage" of work has been completed!

Next, onward to the middle equation. We wish to use the coefficient of y in the second equation to eliminate the y-variable from any equations below the second. (Here, we wish to eliminate $(-4y)$ from equation C2.) By using a linear combination of B3 and C2, we will form a new bottom equation C3. How will this linear combination be formed?

$$
\begin{array}{ll}
A1 & x + 4y - 2z = 3, \\
B3 & \boxed{2y} - z = 1, \\
C2 & - 4y + 5z = 7
\end{array}
$$

The coefficient of y in equation B3 is now considered the pivot.

By multiplying B3 by 2 and adding the result to C2, we will form C3 (with no y-term present!):

$$
\begin{array}{lll}
(2)(B3) & 4y - 2z = 2 \\
C2 & \underline{-4y + 5z = 7} \\
C3 & 3z = 9.
\end{array}
$$

The equivalent system becomes

$$
\begin{array}{lll}
A1 & x + 4y - 2z = 3, \\
B3 & 2y - z = 1, \\
C3 & 3z = 9
\end{array}
\quad \text{or} \quad
\begin{array}{lll}
A1 & x + 4y - 2z = 3, \\
B3 & 2y - z = 1, \\
C4 & z = 3.
\end{array}
$$

by simplifying equation C3.

And now we are finished with the process of transforming the original system of equations into a system in triangular form. The *common solution* (x, y, z) to the original system can be found by substituting back into the derived equivalent system. To find the value of y, we have

$$
\begin{array}{ll}
A1 & x + 4y - 2z = 3 \\
B3 & 2y - z = 1 \longrightarrow 2y - 3 = 1 \longrightarrow \boxed{y = 2} \\
C4 & \boxed{z = 3}
\end{array}
$$

And to find the value of x, we have

A1 $x + 4y - 2z = 3 \longrightarrow x + 4(2) - 2(3) = 3 \longrightarrow \boxed{x = 1}$

B3 $2y - z = 1 \longrightarrow \boxed{y = 2}$

C4 $\boxed{z = 3}$

The common solution (x, y, z) is $(1, 2, 3)$. □

In summary, the entire process of Gaussian elimination can be represented by the following schematics:

$$\begin{array}{l} x + 4y - 2z = 3, \\ 5x - 2y + z = 4, \\ 2x + 4y + z = 13 \end{array}$$ becomes $$\begin{array}{l} x + 4y - 2z = 3, \\ 2y - z = 1, \\ z = 3. \end{array}$$

By forming linear combinations, we can eliminate terms until the original system of equations has been transformed into an equivalent system of equations in triangular form. Solutions are then determined by substitution back into the original system.

Although this technique requires some thought and diligent work, the investment of time is rewarding. Gaussian elimination is useful in solving systems of equations, but also serves as the basis for our future work with matrices in Chapter 11 (finding inverses) as well as most of the solution techniques in linear programming in Chapter 12.

Because of the importance of the technique of Gaussian elimination, let's try a few more examples.

■ **EXAMPLE 19**   Solve

$$\begin{array}{rcl} 600x - 500y &=& 1900, \\ x + 2y &=& 6. \end{array}$$

Solution   Before starting, we simplify the work as much as possible, by dividing the first equation by the common factor 100. In this case, switching the two equations also helps.

$$\begin{array}{rcl} x + 2y &=& 6, \\ 6x - 5y &=& 19 \end{array}$$

Hint: Linear combinations are more easily formed if the beginning coefficient, or pivot, is a 1.

$$\text{A1} \quad x + 2y = 6,$$
$$\text{B1} \quad 6x - 5y = 19$$

In this much simpler example, there is only one entry ($6x$) below the diagonal.

$$
\begin{array}{rl}
(-6) \times \text{A1} & -6x - 12y = -36 \\
\text{B1} & 6x - 5y = 19 \\
\hline
\text{B2} & - 17y = -17
\end{array}
$$

The following linear combination of A1 and B1 will replace the bottom row.

$$\text{A1} \quad x + 2y = 6,$$
$$\text{B2} \quad - 17y = -17$$

The equivalent system is in triangular form. Simplifying B2 by dividing by the common factor -17 yields the equivalent system in final form:

$$\text{A1} \quad x + 2y = 6,$$
$$\text{B3} \quad y = 1.$$

Substituting $y = 1$ back into A1, we have $x + 2(1) = 6$, or $x = 4$. The common solution, then, to the original system is $(x, y) = (4, 1)$. $\square$

Let's examine another system, this time a 3-by-3 system.

■ EXAMPLE 20   Solve

$$
\begin{aligned}
-2R + 2S + T &= 0, \\
R + 4S - 3T &= 5, \\
R - 6S - 2T &= -1.
\end{aligned}
$$

Solution   To obtain a leading coefficient of 1, we interchange either the first and second or the first and third equations. (Here, the first and second were interchanged.) We use the coefficient of R in A1 to eliminate first $-2R$ in B1, then R in C1:

$$\text{A1} \quad R + 4S - 3T = 5,$$
$$\text{B1} \quad -2R + 2S + T = 0,$$
$$\text{C1} \quad R - 6S - 2T = -1.$$

When added to B1, the linear combination $(2) \times$ (A1) will eliminate $-2R$ in B1:

$$
\begin{array}{lrcrcrcr}
2 \times \text{A1} & 2R & + & 8S & - & 6T & = & 10 \\
\text{B1} & -2R & + & 2S & + & T & = & 0 \\
\hline
\text{B2} & & & 10S & - & 5T & = & 10;
\end{array}
$$

and, dividing by a common factor of 5 gives us

$$
\begin{array}{lrcrcr}
\text{B3} & 2S & - & T & = & 2
\end{array}
$$

for the following equivalent system:

$$
\begin{array}{lrcrcrcr}
\text{A1} & R & + & 4S & - & 3T & = & 5, \\
\text{B3} & & & 2S & - & T & = & 2, \\
\text{C1} & R & - & 6S & - & 2T & = & -1.
\end{array}
$$

The linear combination $(-1) \times$ (A1) + C1 will eliminate R from C1:

$$
\begin{array}{lrcrcrcr}
(-1)\text{A1} & -R & - & 4S & + & 3T & = & -5, \\
\text{C1} & R & - & 6S & - & 2T & = & -1, \\
\hline
\text{C2} & & - & 10S & + & T & = & -6.
\end{array}
$$

The equivalent system is now

$$
\begin{array}{lrcrcrcr}
\text{A1} & R & + & 4S & - & 3T & = & 5, \\
\text{B3} & & & 2S & - & T & = & 2, \\
\text{C2} & & - & 10S & + & T & = & -6.
\end{array}
$$

Next, we use the coefficient of S in the equation B3 to eliminate $-10S$ in C2. The linear combination $(5) \times$ (B3) added to C2 will eliminate $-10S$:

$$
\begin{array}{lrcrcr}
5 \times \text{B3} & 10S & - & 5T & = & 10 \\
\text{C2} & -10S & + & T & = & -6 \\
\hline
\text{C3} & & - & 4T & = & 4,
\end{array}
$$

and after simplifying C3 by dividing by -4, we obtain the desired equivalent system in triangular form:

$$
\begin{array}{lrcrcrcr}
\text{A1} & R & + & 4S & - & 3T & = & 5, \\
\text{B3} & & & 2S & - & T & = & 2, \\
\text{C4} & & & & & T & = & -1.
\end{array}
$$

Substituting $T = -1$ back into equation B3 yields $S = 0.5$. And if $T = -1$ and $S = 0.5$, then by substituting into equation A1, we find $R = 0$. The common solution then is $(R, S, T) = (0, 0.5, -1)$.   □

Gaussian Elimination: Tabular Form

The method of Gaussian elimination can be considerably condensed if work is performed on the *coefficients only.* The 3-by-3 from Example 18 will again be illustrated here; however, only the coefficients will be shown in our work.

$$\begin{aligned} x + 4y - 2z &= 3, \\ 5x - 2y + z &= 4, \\ 2x + 4y + z &= 13. \end{aligned}$$

We can transform this system into the one shown below:

$$\begin{pmatrix} 1 \\ 5 \\ 2 \end{pmatrix} \begin{array}{ccc|c} 4 & -2 & 3 \\ -2 & 1 & 4 \\ 4 & 1 & 13 \end{array}$$

In this "new" system, it is understood that the *first column* represents the coefficients of x, the *second column* represents the coefficients of y, the *third column* represents the coefficients of z, and the *fourth column*— separated by a line—represents the constants.

A1	①	4	−2	3	All terms below the diagonal are to be eliminated. We begin by using 1 in A1 as the pivot.
B1	5	−2	1	4	
C1	2	4	1	13	

A1	1	4	−2	3	B2 is formed by the linear combination (-5)A1 $+$ B1; and C2 is formed by the linear combination (-2)A1 $+$ C1.
B2	0	−22	11	−11	
C2	0	−4	5	7	

A1	1	4	−2	3	B3 is formed by dividing B2 by -22.
B3	0	2	−1	1	
C2	0	−4	5	7	

A1	1	4	−2	3	We use 2 in equation B2 as the pivot to eliminate −4 in C2.
B2	0	②	−1	1	
C2	0	−4	5	7	

A1	1	4	−2	3	C3 is formed by the linear combination (2)B2 $+$ C2.
B2	0	2	−1	1	
C3	0	0	3	9	

A1	1	4	−2	3
B2	0	2	−1	1
C4	0	0	1	3

Because C3 has a common factor of 3, we divide C3 by 3 to form C4. The system is now in triangular form.

The system in triangular form can be rewritten in equation form:

$$1x + 4y - 2z = 3,$$
$$2y - 1z = 1,$$
$$1z = 3.$$

We can find solutions for x, y, and z by substitution. Thus $(x, y, z) = (1, 2, 3)$.

Larger Systems, a Flowchart, Additional Methods

Larger systems of equations can be handled in a manner similar to that shown in the preceding examples. Let's examine the general outline of Gaussian elimination, denoting the coefficients by lower-case letters with subscripts. A subscript of $_{31}$ indicates the coefficient in the *third* equation, column *one*. The first number indicates the row or equation, and the second number indicates the column or the variable. The variables are denoted by subscripts after the x-terms (x_1, x_2, etc.). The constants are denoted by subscripts after the c-terms (c_1, c_2, etc.). A general system would appear as follows:

$$a_{11}x_1 + a_{12}x_2 + a_{13}x_3 + \cdots + a_{1n}x_n = c_1,$$
$$a_{21}x_1 + a_{22}x_2 + a_{23}x_3 + \cdots + a_{2n}x_n = c_2,$$
$$a_{31}x_1 + a_{32}x_2 + a_{33}x_3 + \cdots + a_{3n}x_n = c_3,$$
$$a_{41}x_1 + a_{42}x_2 + a_{43}x_3 + \cdots + a_{4n}x_n = c_4,$$
$$\vdots \qquad\qquad\qquad \vdots$$
$$a_{n1}x_1 + a_{n2}x_2 + a_{n3}x_3 + \cdots + a_{nn}x_n = c_n.$$

The coefficient of x_1 in the first equation (a_{11}) will be used to form linear combinations with each equation below to successively eliminate the values in the positions occupied by $a_{21}, a_{31}, a_{41}, \ldots, a_{n1}$.

Then the coefficient of x_2 in the second equation (a_{22}) will be used to form linear combinations with each equation below to successively eliminate the numbers currently occupied by $a_{32}, a_{42}, a_{52}, \ldots, a_{n2}$.

Then the coefficient of x_3 in the third equation (a_{33}) will be used to form linear combinations with each equation below to successively eliminate the numbers currently occupied by $a_{43}, a_{53}, a_{63}, \ldots, a_{n3}$.

This procedure continues, working with each coefficient (pivot) on the diagonal, until all terms are eliminated below the diagonal. At that

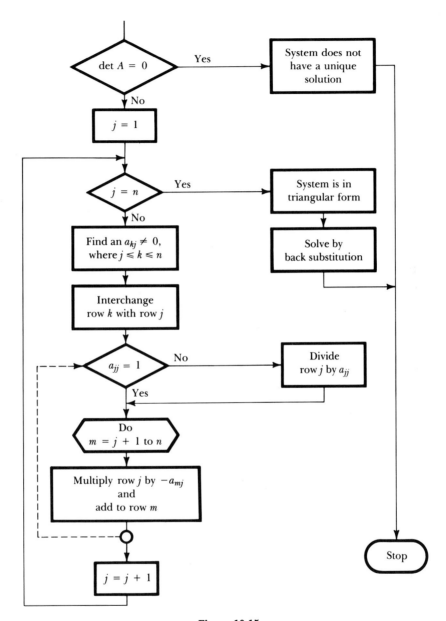

Figure 10.15

point, the system should be in triangular form. The common solution to the system of equations is then found by back substitution.

The flowchart in Fig. 10.15, using the generalized method just described, succinctly illustrates the technique of Gaussian elimination.

$$
\begin{array}{cccccc|c}
x_{11} & 0 & 0 & 0 & \ldots & 0 & c_1 \\
0 & x_{22} & 0 & 0 & \ldots & 0 & c_2 \\
\cdot & \cdot & \cdot & \cdot & & \cdot & \cdot \\
\cdot & \cdot & \cdot & \cdot & & \cdot & \cdot \\
\cdot & \cdot & \cdot & \cdot & & \cdot & \cdot \\
0 & 0 & 0 & 0 & \ldots & x_{nn} & c_n
\end{array}
$$

Figure 10.16

Use the general system of equations we discussed earlier while following the flow of the technique.

Gaussian elimination can effectively handle (or be programmed to handle) any system in which the number of equations equals the number of unknowns. While our work has been primarily with 2-by-2 or 3-by-3 systems, keep in mind that in business applications, the "order" of the systems can be quite large—50-by-50 or even 600-by-600 systems are not uncommon. However, remember that although we concentrated on smaller systems, once you have mastered the technique and acquired some programming ability, you should be able to write a program to handle the larger systems.

An extension of Gaussian elimination is obtaining "diagonal form," which is useful for larger systems. Whereas by Gaussian elimination we obtained all 0's below the diagonal, we can also obtain 0's above the diagonal with relatively little extra work, as shown in Fig. 10.16. The solutions are readily given by $x_{11} = c_1$, $x_{22} = c_2$, . . . , $x_{nn} = c_n$. For larger systems, other methods besides Gaussian elimination may effectively solve for the simultaneous solution. An understanding of matrices and matrix algebra will allow the use of matrices in solving larger n-by-n systems of equations. Cramer's rule allows the use of determinants in solving smaller systems of equations. Additional methods will be presented in Chapter 11.

Exercise Set 10.5

■

OBJECTIVE OF GAUSSIAN ELIMINATION

By forming linear combinations, transform the original system of equations into an equivalent system in triangular form. The simultaneous solution is then found by back substitution.

Each of the following systems of simultaneous equations is already in triangular form. Find the common solution by back substitution.

1. $x + y = 5,$
$\qquad y = 3$

2. $6x - 2y = 9,$
$\qquad 4.5y = 18$

3. $x + y + z = \quad 1,$
$\qquad y + z = -3,$
$\qquad\qquad z = \quad 2$

4. $6x - \quad 2y - \quad 8z = \quad 14,$
$\qquad 100y - 300z = 100,$
$\qquad\qquad 55z = 110$

5. $x + y + z + w = \quad 5,$
$\qquad y + z + w = \quad 7,$
$\qquad\qquad z + w = \quad 2,$
$\qquad\qquad\qquad w = -4$

6. $4R - 2S + 5T - \quad U = \quad 9,$
$\qquad 5S - 2T + \quad U = \quad 6,$
$\qquad\qquad 5T - 4U = \quad 8,$
$\qquad\qquad\qquad 2U = 10$

7. $68x - 2y + 4z = 9,$
$\qquad 3y - 4z = 5,$
$\qquad\qquad 8z = 1$

8. $R + 3S - 7T + \quad U + 2V - \quad W = \quad -1,$
$\qquad 2S + \quad T - 2U - 2V + 3W = -28,$
$\qquad\qquad T + 6U - 3V + \quad W = -39,$
$\qquad\qquad\qquad 5U - \quad V - 2W = \quad -4,$
$\qquad\qquad\qquad\qquad V - 4W = \quad 37,$
$\qquad\qquad\qquad\qquad\qquad W = \quad -8$

Use Gaussian elimination to find the common solution for each of the following systems of simultaneous equations.

9. $x + y + \quad z = \quad 9,$
$\qquad y - 2z = -1,$
$\quad 3x + y + 2z = \quad 14$

10. $x + 2y - \quad z = -3,$
$\quad 2x - \quad y + 2z = \quad 6,$
$\quad 3x + 2y - 2z = -3$

11. $x - \quad 7y = \quad 4,$
$\quad 2x - 11y = 17$

12. $6x - 2y + 7z = \quad 45.7,$
$\qquad 5y - 2z = -19,$
$\qquad\qquad 4z = \quad 6$

13. $2x + 6y - 2z = -12,$
$\quad 2x \qquad + \quad z = \quad 6,$
$\quad -x + 2y + 3z = \quad 11$

14. $2x - \quad y + 3z = -9,$
$\quad x + 3y - \quad z = \quad 10,$
$\quad 3x + \quad y - \quad z = \quad 8$

15. $2x - \quad y + 6z = \quad 5,$
$\quad 3x - 5y - 9z = -7,$
$\quad 3x \qquad + 3z = \quad 0$

16. $\quad x - 2y + 3z = \quad -8,$
$\quad -4x + 5y - 6z = \quad 20,$
$\quad 7x - 8y + 9z = -32$

17. $\quad 3x + \quad y \qquad = 100,$
$\qquad 5y + 3z = \quad 20,$
$\quad 18x \qquad + 8z = 460$

18. $\quad x + y - z = 19.5,$
$\quad x - y + z = 19.5,$
$\quad -x + y + z = 19.5$

19. $1400R - 100S + 33T = \quad 16,800,$
$\quad 20R - \quad 14S - \quad 5T = \quad -370,$
$\quad 18R + 200S - 40T = -2,820$

20. $7.2B - 3.1M + 1.5S = 132,$
$\quad 2.8B + \quad M + 0.6S = \quad 52,$
$\quad 0.8B - 2.2M - 1.2S = \quad 56$

21. Each of the following systems represents a *dependent* system. Attempt to solve each system by Gaussian elimination. Summarize, in words, the outcome when a dependent system is solved (or attempted) by Gaussian elimination.

a) $3x - 2y = 7,$
$\quad\; 6x - 4y = 14$

b) $x + 4y - 3z = 5,$
$\qquad x - 6y - 2z = 1,$
$\qquad 2x + 8y - 6z = 10$

22. Each of the following systems represents an *inconsistent* system. Attempt to solve each system by Gaussian elimination. Summarize, in words, the outcome when an inconsistent system is solved (or attempted) by Gaussian elimination.

a) $3x - 2y = 7,$
$\quad\; 6x - 4y = 15$

b) $x + 4y - 3z = 5,$
$\qquad x - 6y - 2z = 1,$
$\qquad 2x + 8y - 6z = 13$

"Realistic" examples are usually larger systems: 3-by-3, 4-by-4, even 50-by-50. Demonstrate your expertise by solving these "larger" systems using Gaussian elimination.

23. $x - 2y + 3z + 4w = 18,$
$\quad\; 2x + y - z + 6w = 17,$
$\quad\; 3x - y + 2z + w = 11,$
$\quad\; 8x - 2y + z - w = 9$

24. $4A - 2B - 3C + D = 20,$
$\quad\; 2A + B - C - D = 9,$
$\quad\;\; A + 2B - C - 2D = 3,$
$\quad\;\; A + 3B + C + 3D = 14$

25. $R + S - T + U + V + W = 15,$
$\quad R - S \quad\;\; - U + 2V + W = 15,$
$\quad 2R + 3S \qquad\quad - 2V - W = -8,$
$\qquad\quad 2S + T - U + V + 3W = 26,$
$\quad R - S + T + U - V - 3W = -17,$
$\qquad\qquad\quad T + U - V - W = -4,$

26. Translate this business problem into a system of equations and find the simultaneous solution by using Gaussian elimination.

A broker invests some money in three stocks, A, B, and C. The total amount invested is $6000. The percent return on the three stocks is 12%, 8%, and 11%, respectively. The combined income on stocks A and C is $635. The combined income on all three stocks is $675.

Find how much money was invested in each of the three stocks.

10.6

Applications of Simultaneous Systems

Solving systems of equations simultaneously finds applications in many diverse areas. To successfully apply mathematics, (1) the given problem must be thoroughly understood, (2) the problem must be "translated" to mathematical formulas and/or equations, and (3) the problem must be solved (quite often the easiest part). Several illustrations using practical applications from mathematics and business follow.

■ **EXAMPLE 21**

The sum of two unknown numbers is 84. The difference of the two numbers is 6. Find the numbers.

Solution

After you have read the problem several times, the most important step is to *identify the unknown(s)*. In this case we will be trying to find two numbers. We let N_1 represent the first number, and N_2 represent the second number.

Now try reading the problem several more times. Look for key words; in this case, the words *sum* and *difference*. What information is known about the two numbers? "The *sum* of the numbers is 84" *translates* to

$$N_1 + N_2 = 84.$$

"The *difference* of the numbers is 6" *translates* to

$$N_1 - N_2 = 6.$$

Solving the system will often be the easiest part of the application problem; as a matter of fact, once the problem is *translated* into the appropriate equation or equations, the problem could then be solved by a computer! We have

$$\begin{aligned} N_1 + N_2 &= 84 \\ \underline{N_1 - N_2} &= \underline{6} \\ 2N_1 &= 90 \quad \text{or} \quad N_1 = 45. \end{aligned}$$ **Adding the two equations**

By back substitution, we find that $N_2 = 39$. The solutions $N_1 = 45$ and $N_2 = 39$ should now be checked in the original equations. In practical applications, the "correctness" of the solution is of ultimate importance!   □

■ **EXAMPLE 22**

An investment counselor invests a total of $8000. Part of the investment is in stocks, yielding a 10% return. The remainder of the investment is in bonds, yielding a 12% return. If the combined investment returns $860, how much money is invested in stocks and how much in bonds?

Solution

The unknowns in this case are the *amount* of money invested in *stocks* and the *amount* of money invested in *bonds*. (Often the unknowns are what are being asked for in the final sentence of the problem!)

We let S denote the amount of money in stocks and B the amount of money in bonds. Having two unknowns, we will need at least *two equations*. Let's reread the problem very carefully. What relationships exist (or what can we find) about the money invested in stocks and bonds?

Using the key words "a total of $8000," we find that one of the equations is

$$S + B = \$8000.$$

Again, by reading and reflecting, we know that the return on the money is $860. How was this return obtained? The stocks returned 10% and the bonds returned 12%. Combined, the total yield is $860. The second equation is then

$$10\%S + 12\%B = \$860.$$

Therefore, the system to be solved is

$$S + \quad B = \$8000,$$
$$0.10S + 0.12B = \quad \$860.$$

For variety, as well as review, let's solve this system by substitution:

Eq. (1) $S + \quad B = 8000$
Eq. (2) $0.10S + 0.12B = \quad 860.$

Solving Eq. (1) for S gives us $S = 8000 - B$. If we then substitute $8000 - B$ for S in Eq. (2), we have one equation in one unknown. And then solving, we find that $3000 is invested in bonds.

Eq. (2) $\qquad\qquad\qquad\qquad 0.10S + 0.12B = \quad 860$
$$0.10(8000 - B) + 0.12B = \quad 860$$
$$800 - 0.10B + 0.12B = \quad 860$$
$$800 + 0.02B = \quad 860$$
$$0.02B = \quad 60$$
$$B = 3000$$

By substituting $3000 for B into $S = 8000 - B$, we find that $8000 - 3000$, or $5000 is invested in stocks.

Let's now check the solution. We know that $3000 + $5000 does equal $8000. But does 10%($5000) + 12%($3000) = $860? Solving gives us

$$\$500 + \$360 = \$860.$$

The correct answers: The amount of money invested in stocks is $5000. The amount of money invested in bonds is $3000.

■ **EXAMPLE 23**   An application of interest in business economics concerns the law of supply and demand. In most cases as the price of a product increases, the manufacturer's supply increases and the consumer's demand decreases. Both supply and demand are functions of the price.

Suppose the supply function of a certain product is given by $S(p) = 30p$ and the demand function by $D(p) = 240/p^2$. Find the point at which supply equals demand, or, as it is referred to in business, the point of market equilibrium.

Solution   In order for the supply to equal the demand, the two functions should be equal to each other, that is,

$$30p = 240/p^2.$$

Solving for p gives us

$$30p^3 = 240$$
$$p^3 = 8$$
$$p = 2.$$

This is the equilibrium price, $2.00 per item. At $2.00 per item the supply, 30(2) or 60, does equal the demand, $240/(2)^2$ or 60.

A sketch of this business application is shown in Fig. 10.17.

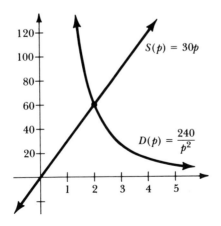

Figure 10.17

(In a situation of pure competition, a product will tend to be sold at its equilibrium price.) □

Exercise Set 10.6

1. The sum of two numbers is 132. The difference of the same two numbers is 100. Find the two numbers by first obtaining a system of equations and then solving.

2. The sum of two numbers is 132. Twice the first number minus the second is 6. Find the two numbers.

3. The sum of two numbers is 132. The product of the two numbers is 387. Find the two numbers by obtaining a system of equations and then solving by substitution.

4. Find values for a and b such that the line $ax + by = 15$ will pass through the points $(4, 1)$ and $(3, -3)$.

5. In business, the point at which total cost equals total revenue is called the break-even point. Suppose the cost function is given by $y = C(x) = 30x + 2800$, and the revenue function is given by $y = R(x) = 100x$, where x is the number of units made and sold. Determine each of the following.

   a) How many units must be sold to break even?

   b) What is the manufacturer's profit or loss if 40 units are sold?

   c) What is the manufacturer's profit or loss if 20 units are sold? if 100 units are sold?

   d) How many units must be sold in order to realize a profit of $2100?

6. By graphing, determine where the cost function

$$C(x) = (x^2 + 3200)/x$$

   intersects the profit function

$$P(x) = 16x + 40.$$

7. Three loans were assumed by the Stampen Printing Company. The loans totalled $18,000. Money was borrowed at rates of 8.5%, 9%, and 10%. The annual interest charge was $1565. The amount borrowed at the 8.5% rate was four times as much as the amount borrowed at the 9% rate. How much money was loaned at each of the three different rates? (*Hint:* Form a system of equations.)

8. To "fit a line" through some points, m and b must be determined for the linear equation $y = mx + b$, in slope y-intercept form. If you know that the line passes through the two points $(7, 2)$ and $(3, -6)$, find the equation of the line by determining the values for m and b. (Use a system of equations to solve the problem.)

9. To fit a cost function (parabola) through the points $(10, 20)$, $(30, 40)$, and $(70, 10)$, the coordinates of each point are separately substituted into the general formula for a parabola:

$$y = ax^2 + bx + c.$$

A system of equations results:

$$20 = 100a + 10b + c,$$
$$40 = 900a + 30b + c,$$
$$10 = 4900a + 70b + c.$$

Find the equation of the cost function by solving the system for a, b, and c.

10. The manufacturing of two products, standard and deluxe, requires some material costs and some labor costs. Each standard requires $1.76 in material costs and $4.20 in labor costs. Each deluxe requires $1.88 in material costs and $8.40 in labor costs.

   The total amount spent for material is $270.
   The total amount spent for labor is $840.

   Find how many of *each* product can be manufactured.

11. A pair of supply and demand functions are given as follows.

   Manufacturer's supply: $S(p) = p^2 + 3p - 10$,
   Consumer's demand: $C(p) = 290 - 2p$;

   where p represents the price per unit. In general, when the price increases, supply increases and demand decreases. Find the point of intersection of these two functions. (The point at which supply equals demand is called the point of market equilibrium.)

Summary

A system of equations is a set of two or more equations for which is sought a *common solution*, a solution that will *simultaneously* satisfy all the equations in the system. The majority of our work (and for a wide variety of applications) involves systems of simultaneous linear equations. When graphed, a linear equation represents a straight line (hence the term "linear.") By graphing a system of linear equations simultaneously on the same set of axes, we find that lines (1) may intersect in a unique solution (an *independent* system), (2) may be parallel (an *inconsistent* system), or (3) may be coinciding lines (a *dependent* system).

Assuming a system represents an independent system, several methods were developed to allow you to find the simultaneous solution of a system of equations. A common solution may be found by using:

a) graphing,
b) elimination by substitution,
c) elimination by use of linear combinations,
d) Gaussian elimination (tabular form),
e) matrix methods,
f) Cramer's rule (determinants).

Although discussion of matrix methods and Cramer's rule is deferred until Chapter 11, they are mentioned here because the development of these methods is very closely related to an intuitive understanding of the methods of linear combinations and Gaussian elimination.

Although the solution of systems is most important and received most of our attention, nonlinear systems were not overlooked. Business applications using supply-and-demand curves and cost-and-revenue functions, for example, require a working knowledge of solving nonlinear systems.

In conclusion, many methods for solving systems of simultaneous equations were developed. Most of you will undoubtedly encounter computer programs—or you may write programs—to handle the complex computations. Understanding the mechanics of a technique is important. Equally important, though, is the ability to read, analyze, and appropriately apply the mathematics you have learned to practical situations. Toward that end, we concluded the chapter with a section on applications of simultaneous systems of equations (both linear and nonlinear).

Review Exercises

1. The point $(2, 3)$ is a simultaneous solution to which of the following systems of equations?

a) $4x - 2y = 7,$
$\qquad x + y = 5$

b) $2x - y = 1,$
$\qquad x + y = 5$

c) $x + y = 5,$
$\qquad x - y = 1$

d) $1.5x - 4y = -9,$
$\qquad 4.5x + 2y = 15$

For each of the following equations, identify the slope, the y-intercept, and the x-intercept.

2. $3x - 2y = 6$

3. $2x + 3y = -6$

4. $7x + 4y = 14$

5. $4x - 3y = -24$

Solve each of the following systems of simultaneous equations by graphing.

6. $2x + y = 8,$
$\quad x - y = 4$

7. $x + 2y = 7,$
$\quad x - 2y = 3$

8. $y = x,$
$\quad x + y = 6$

9. $y = 2x - 3,$
$\quad 3x + 2y = 22$

Identify each of the following systems as independent, inconsistent, or dependent. (If the system is independent, find the common solution!)

10. $3x - 2y = 12,$
$\quad 6x - 4y = 8$

11. $3x - 2y = 12,$
$\quad 6x - 4y = 24$

12. $3x - 2y = 12,$
$\quad 6x + 4y = 8$

13. $3x - 2y = 12,$
$\quad 6x + 4y = 12$

For each of these systems of simultaneous equations, find the common solution by use of any (or all) of the methods presented in this chapter: (1) substitution, (2) linear combinations, and/or (3) Gaussian elimination.

14. $x + y = 6,$
$\quad x - y = 2$

15. $3x - y = 20,$
$\quad x + 2y = 9$

16. $y = x - 6,$
$\quad 3x - 2y = 20$

17. $x + 4y = -5,$
$\quad 5x = 1 - 7y$

18. $3x = 9,$
$\quad 2x - y = 5,$
$\quad 4x + 3y + 2z = 1$

19. $x - y + 4z = 7,$
$\quad 2x - y - 5z = -2,$
$\quad 9x - 4y - z = 8$

20. $x + y + z = 6,$
$\quad x + 2y + 2z = 11,$
$\quad -x - 4y + 3z = 0$

21. $x + 4y - 5z = -10,$
$\quad 2x + 7y - z = -9,$
$\quad x + 3y + 4z = 1$

22. $x + y - z = 2,$
$\quad x - y + z = 12,$
$\quad x - y - z = 8$

23. $x + y + z + w = 17,$
$\quad 2x - 4y - z - 3w = -35,$
$\quad x - 2y + 3z + 2w = 7,$
$\quad x + 2y - 3z + w = -3$

(*Applications of systems.*) After reading and analyzing each of the following problems, formulate an appropriate system of equations to solve the problem; then solve by a method of your choice.

24. The sum of two numbers is -1. Twice the larger number minus five times the smaller number is 26. Find the numbers.

25. Two investments total $8000. One pays interest at the rate of 10% yearly; the other pays interest at the rate of 5% yearly. If the income from the 10% investment is six times more than the income from the 5% investment, how much money is invested at each rate?

26. A pair of supply and demand functions is given by

$$S(p) = p^2 - 2p - 8,$$
$$D(p) = 4p - 8,$$

where each is a function of the price per unit, p. Find the equilibrium point ($p > 2$).

Arrays, Matrices, and Determinants

CHAPTER 11 OBJECTIVES

After completing this chapter, you should be able to:

1. Define a matrix by identifying and using such terms as array, element, row, column, order, and dimension.
2. Differentiate between a matrix and a determinant.
3. Identify these terms associated with determinants: element, row, column, principal and secondary diagonals, minor, and cofactor.
4. By evaluating determinants, solve systems of simultaneous linear equations using Cramer's rule.
5. Identify these terms associated with matrix algebra: subscript, square matrix, vector, scalar multiplication, adjoint, row transformations, invertible, and nonsingular.
6. Perform the operations of addition, scalar multiplication, and multiplication in matrix algebra.
7. Find the inverse of a matrix using either (a) the adjoint or (b) a variation of Gaussian elimination.
8. Solve systems of linear equations with matrix algebra by (a) exploiting the idea of elementary row operations or (b) using the inverse of a matrix.

11

■ CHAPTER OUTLINE

11.1
The Notions of Arrays, Matrices, and Vectors

Many applications in data processing require manipulating an entire set of numbers, or quantities, as a single entity. Some examples are updating production charts, providing an inventory of parts, totalling sales figures, and evaluating investment portfolios.

Let's use the information given in Table 11.1 as an example. To compute the average daily sales for each department, all individual sales figures would be added for each row (department); then that total would be divided by six (for 6 days). This same sequence of operations would then be repeated for *each* department. To compute the average departmental sales for each day, the sales figures in each column would be added; then this total would be divided by five (for 5 departments). This same sequence of operations would then be repeated for each day of the week.

339

Table 11.1
Daily Sales by
Department—Week of
August 7

	MON.	TUES.	WED.	THURS.	FRI.	SAT.
DEPT. SP	480	760	350	400	502	380
C	1000	1200	430	875	902	603
LG	510	620	405	382	190	635
DT	85	40	75	125	94	56
HA	800	1250	130	800	720	984

Using Table 11.2, in order to project a 5.2% increase in quarterly production figures for the forthcoming year, how many individual calculations would we need to make? How many individual computations would we need if rather than only 3 factories, there were 118 factories?

Table 11.2
Quarterly
Productivity—Present
Year

	Q1	Q2	Q3	Q4
FACTORY 1	12,580	17,645	15,829	19,080
FACTORY 2	32,450	35,498	37,645	38,580
FACTORY 3	18,345	19,895	20,400	21,990

When we are working with a group of numbers such as these sales figures or productivity figures and often performing repetitive computations, we can be far more efficient by treating each group of numbers as a single quantity. Most computer languages provide techniques for easily manipulating and/or operating on such groups of numbers, called **arrays.**

■ DEFINITION

ARRAY

An *array* is a collection of elements, usually in rectangular form. (Generally, the elements have some common attribute.)

The entire collection of sales figures in Table 11.1 is an example of an array. A listing or collection of all inventory part numbers with quantity on hand at the beginning of each month is yet another example. If you wished to alphabetize a list of all business clients, computer programming and the use of arrays would greatly facilitate the task!

■ EXAMPLE 1

Some examples of arrays are as follows:

$$A = \begin{bmatrix} 2 & 3 & 8 \\ 7 & 1 & 5 \end{bmatrix};$$

$$B = \begin{bmatrix} 4 & 5046 \\ 1 & 1034 \\ 3 & 6052 \\ 8 & 3150 \\ 7 & 8004 \end{bmatrix};$$

$$C = \begin{bmatrix} 3 & 2 \\ 5 & 1 \end{bmatrix};$$

$$D = [7.85 \quad 3.95 \quad 4.10 \quad 8.80 \quad 10.25]. \quad \square$$

Often an array is referred to by a single capital letter. Array A in Example 1 contains six numbers. Each entry in an array is called an *element.* In array A, then, there are six elements; and it is made up of two *rows,*

$$\begin{bmatrix} \boxed{2 \quad 3 \quad 8} \\ \boxed{7 \quad 1 \quad 5} \end{bmatrix},$$

and three columns,

$$\begin{bmatrix} \boxed{\begin{matrix}2\\7\end{matrix}} & \boxed{\begin{matrix}3\\1\end{matrix}} & \boxed{\begin{matrix}8\\5\end{matrix}} \end{bmatrix}.$$

When we refer to the size of an array, we first describe how many rows, and then how many columns the array contains. Array A is a 2-by-3 array (2 rows, 3 columns).

What size array is array B in Example 1? What about array C? Array D has only one row with five columns; thus it is a 1-by-5 array.

■ **DEFINITION**

MATRIX
A *matrix* is a rectangular array of elements arranged in rows and columns.

A **matrix,** being an array, is also denoted by a capital letter and the elements of the array are enclosed either in brackets or parentheses. The *order,* or dimension, of a matrix is given by (1) the number of rows and (2) the number of columns.

■ **EXAMPLE 2**   The following matrix has dimension 2 by 7 (generally written as 2×7).

$$M = \begin{bmatrix} 1 & 3 & 6 & 5 & 7 & 3 & 9 \\ 2 & 0 & 4 & 5 & 5 & 8 & 6 \end{bmatrix} \quad \square$$

Each of the arrays illustrated in Example 1 is also an example of a matrix.

Since the following array is *not* a rectangular array, it is not an example of a matrix:

$$G = \begin{bmatrix} 1 & 3 & 6 & 5 & 9 & 3 \\ & & 8 & 2 & & \\ & 1 & 6 & 4 & 5 & 3 & 5 \end{bmatrix}.$$

Two matrices having the same number of rows and columns are said to have the same *dimension:*

$$R = \begin{bmatrix} 1 & 3 & 6 \\ 2 & 3 & 9 \end{bmatrix} \quad \text{and} \quad S = \begin{bmatrix} -105 & -36.8 & 48.22 \\ 19.1 & 0.1 & 48.21 \end{bmatrix}.$$

Matrices R and S have the same dimension; both are 2-by-3 matrices.

If the number of rows in a matrix is equal to the number of columns, then the matrix is called a *square matrix.* The following is an example:

$$J = \begin{bmatrix} 2.33 & 4.65 & 5.44 \\ 1.45 & 9.33 & 5.46 \\ 3.44 & 2.65 & 5.58 \end{bmatrix}.$$

J is a 3-by-3 matrix (three rows, three columns).

■ DEFINITION

VECTOR
A matrix consisting of either only one row or only one column is called a *vector*.

Matrix K is an example of a *column vector,* whereas matrix D is called a *row vector:*

$$K = \begin{bmatrix} 2 \\ 9 \\ 5 \end{bmatrix}; \qquad D = [1.2 \quad -3.4 \quad 1.9 \quad -3.6].$$

Vectors, and matrices for that matter, are often used in business applications containing several hundred elements. When vast quantities of data must be processed and analyzed rapidly, a good working knowledge of matrices and computers go hand in hand!

Before we can apply matrices, or operate on and manipulate the elements in the matrices, we must first become more familiar with the position of elements in a matrix. If a matrix is identified by a capital letter, then often the elements in the matrix are referred to by the same letter in lower case:

$$A = \begin{bmatrix} a & a & a & a \\ a & a & a & a \end{bmatrix}.$$

But wait! How can we differentiate between the elements in the different positions if each element is represented by the same lower-case letter?

Two methods are used to represent matrix elements—both are illustrated here. The more common method is to use a lower-case letter for every element in the matrix. But to aid in differentiating among elements, two subscripts are attached to each lower-case letter. This method is called the *double-subscript* method. For example, in the matrix

$$A = \begin{bmatrix} a_{11} & a_{12} & a_{13} & a_{14} \\ a_{21} & a_{22} & a_{23} & a_{24} \end{bmatrix},$$

the first number of the subscript indicates the *row* the element is in, and the second number indicates the *column* the element is in. (Note that no comma is used to separate the two numbers.) The double subscript on a_{23} positions the element in the *second row, third column*. Is a_{23} the same as a_{32}? Definitely not! While a_{23} is located in matrix A, the element a_{32} cannot be found; matrix A does not contain three rows!

Another method of referring to the elements in a matrix uses a single subscript. In D, a 2-by-3 matrix,

$$D = \begin{bmatrix} a_1 & b_1 & c_1 \\ a_2 & b_2 & c_2 \end{bmatrix},$$

each element in the same row has the same subscript. In row 1, each letter has a subscript 1. Each entry in a column then uses the same lower-case letter. In column 2, for example, every entry begins with a b. Single-subscripted elements are used most frequently when working with a row or column matrix (vector), such as

$$C = \begin{bmatrix} c_1 \\ c_2 \\ c_3 \end{bmatrix}.$$

Although the single-subscript method is used, and should be mentioned, most of our work will make use of the double-subscript format. Much of data manipulation and processing involves matrices with more than just one row or column.

In solving systems of equations, for example, matrices will prove most useful. For a 3-by-3 system of simultaneous equations,

$$3x - 2y + z = 15,$$
$$5x + 4y - 8z = 13,$$
$$7x - 6y + 9z = 14,$$

the entire set (array) of coefficients could—and will—form a special matrix:

$$\begin{bmatrix} 3 & -2 & 1 \\ 5 & 4 & -8 \\ 7 & -6 & 9 \end{bmatrix}.$$

Would matrices have been helpful when we were using the tabular form of Gaussian elimination? Do you see any similarities?

In general, the coefficients of a system of simultaneous equations of any size can be set into array form; this array is called the *coefficient matrix*. Using the general notation we discovered in Chapter 10, we can write a system of simultaneous equations with n equations and n unknowns with a matrix of coefficients, as follows:

$$\begin{bmatrix} a_{11} & a_{12} & a_{13} & a_{14} & a_{15} & a_{16} & \cdots & a_{1n} \\ a_{21} & a_{22} & a_{23} & a_{24} & a_{25} & a_{26} & \cdots & a_{2n} \\ a_{31} & a_{32} & a_{33} & a_{34} & a_{35} & a_{36} & \cdots & a_{3n} \\ a_{41} & a_{42} & a_{43} & a_{44} & a_{45} & a_{46} & \cdots & a_{4n} \\ \vdots & \vdots & \vdots & \vdots & \vdots & \vdots & & \vdots \\ a_{n1} & a_{n2} & a_{n3} & a_{n4} & a_{n5} & a_{n6} & \cdots & a_{nn} \end{bmatrix}.$$

Matrices, then, are helpful in better describing and handling larger systems of equations! To efficiently solve larger systems of equations found in data processing situations, we will pursue matrix notation further. Using the terminology associated with matrices given in this section, we will define operations to be performed on the matrices and their elements. After learning some matrix algebra, we will be able to manipulate matrices in order to solve these larger systems of simultaneous equations.

Exercise Set 11.1

1. How many elements are in each matrix?

   a) $\begin{bmatrix} 1 & 3 & 19 \\ 7 & 0 & 5 \end{bmatrix}$

   b) $\begin{bmatrix} 1 & 3 & 9 \\ 6 & 2 & 4 \\ 5 & 2 & 1 \\ 1 & 1 & 2 \end{bmatrix}$

   c) $[1.3 \quad 2.6 \quad 9.4 \quad 1.3 \quad 3.4]$

2. What is the dimension of each matrix in Exercise 1?

Determine whether each of the following statements is true or false.

3. A 5-by-4 matrix is the same as a 4-by-5 matrix.

4. Every matrix is an array.

5. [1.3 2.4 3.8] is an example of a column vector.

6. A blank in a matrix is treated as if it contains a zero.

Using matrices A and B, list each of the following elements from their respective matrices.

$$A = \begin{bmatrix} 1 & 3 & 2 \\ -4 & 7 & -5 \\ 8 & 5 & 0 \end{bmatrix}, \quad B = \begin{bmatrix} 11 & 13 & 17 & 14 & 15 \\ 1.7 & 2.3 & 8.9 & 1.5 & 6.2 \end{bmatrix}$$

7. a_{11} **8.** a_{23} **9.** a_{32}

10. a_{31} **11.** a_{12} **12.** a_{21}

13. b_{13} **14.** b_{21} **15.** b_{24}

State whether each of these arrays is a row vector, a column vector, a square matrix, or none of these.

16. $\begin{bmatrix} 3 & 1 & 2 \\ 4 & 7 & 5 \end{bmatrix}$ **17.** $\begin{bmatrix} 1 & 3 \\ 2 & 0 \end{bmatrix}$ **18.** $\begin{bmatrix} 1 \\ 7 \\ 3 \end{bmatrix}$

19. $[1 \quad 7 \quad 3]$ **20.** $\begin{bmatrix} m & n \\ r & t \end{bmatrix}$ **21.** $\begin{bmatrix} 1 & 3 & 5 \\ 2 & 6 & 4 \\ 1 & 0 & 0 \end{bmatrix}$

22. $\begin{bmatrix} 1 & -7 & 0 & 4 \\ 2 & 5 & 0 & 3 \end{bmatrix}$

23. Construct a matrix with the following elements:

$$\begin{array}{llll} a_{22} = 4, & a_{24} = -6, & a_{12} = 1, & a_{13} = 3, \\ a_{23} = -2, & a_{21} = 0, & a_{11} = 7.2, & a_{14} = 5. \end{array}$$

24. If a matrix has six elements, how many possible matrices with different dimensions can be constructed?

25. Construct a 5-by-5 matrix M where the elements in M are given by the formula $m_{ij} = i + j$.

11.2
Determinants

Cramer's rule is yet another technique for solving systems of linear equations. When the system is relatively small, this method has a particular utility for computers. Cramer's rule makes use of *determi-*

nants. As you will see, determinants are invaluable in the theoretical analysis of systems of equations. First we must learn about determinants and some of their applications, and then examine their role in Cramer's rule.

What is a determinant? Remember that a matrix is a rectangular array of elements, arranged in rows and columns. Since a matrix is an array, it does not have a quantitative value. Associated with every given square matrix is an assigned number, called the **determinant** of the matrix. We will investigate determinants of order 2 and order 3.

Determinants of Order 2

■ **DEFINITION**

DETERMINANT

A *determinant* of the 2-by-2 square matrix

$$A = \begin{bmatrix} a & b \\ c & d \end{bmatrix}$$

is the single number associated with this matrix. The determinant of A, denoted by

$$|A| = \begin{vmatrix} a & b \\ c & d \end{vmatrix},$$

is defined by the equation $ad - bc$.

The *notation* differentiating a matrix from a determinant should be emphasized! The *parentheses,* or *brackets,* used for matrices indicate an array of numbers; the *vertical lines* used for a determinant denote the number that the determinant assigns to the enclosed array of numbers.

■ **EXAMPLE 3**

For matrix

$$C = \begin{bmatrix} 3 & 2 \\ 4 & 5 \end{bmatrix},$$

the determinant is

$$\begin{vmatrix} 3 & 2 \\ 4 & 5 \end{vmatrix} \qquad \begin{vmatrix} a & b \\ c & d \end{vmatrix} = ad - bc$$

$$= (3)(5) - (4)(2) = 7. \quad \square$$

■ EXAMPLE 4

For matrix

$$R = \begin{bmatrix} 3.2 & -2 \\ 10 & 50 \end{bmatrix},$$

the determinant is $(3.2)(50) - (-2)(10) = 180.$ □

Because determinants represent the *numerical* value of a square matrix, much of the terminology associated with matrices also applies to determinants. A *determinant*

$$\begin{vmatrix} a & b \\ c & d \end{vmatrix}$$

is made up of

elements: the numbers represented by a, b, c, d;
rows: the elements in a horizontal line;
columns: the elements in a vertical line.

A determinant with two rows and two columns is said to be of the second order, or *order two*.
    Two other terms associated with determinants are (1) the *principal, or main diagonal* and (2) the *secondary*, or *"off" diagonal:*

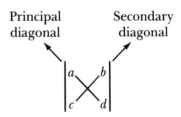

They are useful for providing a schematic for finding the value of the determinant. Using this schematic, we can summarize the process of finding the value of a determinant as follows:

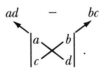

The value of a 2-by-2 determinant is the product of the elements on the principal diagonal *minus* the product of the elements on the secondary diagonal.

■ EXAMPLE 5

Some examples of evaluating 2-by-2 determinants follow:

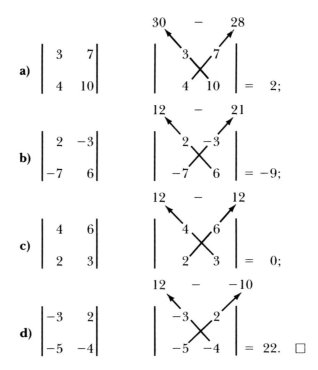

a)
$$\begin{vmatrix} 3 & 7 \\ 4 & 10 \end{vmatrix} \quad = \quad 2;$$

b)
$$\begin{vmatrix} 2 & -3 \\ -7 & 6 \end{vmatrix} \quad = \quad -9;$$

c)
$$\begin{vmatrix} 4 & 6 \\ 2 & 3 \end{vmatrix} \quad = \quad 0;$$

d)
$$\begin{vmatrix} -3 & 2 \\ -5 & -4 \end{vmatrix} \quad = \quad 22. \quad \square$$

Determinants of Order 3

A third-order determinant is one having three rows and three columns, for a total of nine elements.

■ DEFINITION

A *determinant of order three* is defined by the equation

$$\begin{vmatrix} a_1 & b_1 & c_1 \\ a_2 & b_2 & c_2 \\ a_3 & b_3 & c_3 \end{vmatrix} = \begin{aligned} &(a_1)(b_2)(c_3) + (a_3)(b_1)(c_2) + (a_2)(b_3)(c_1) \\ &- (a_3)(b_2)(c_1) - (a_1)(b_3)(c_2) - (a_2)(b_1)(c_3). \end{aligned}$$

As you can see, the expansion of a 3-by-3 determinant is slightly more complex, or at least more difficult to remember! With some work—just as with 2-by-2 determinants—we will arrive at some schematics that provide more convenient and easier ways to remember this definition.

The expansion of a third-order determinant is defined by the expression:

$$a_1b_2c_3 + a_3b_1c_2 + a_2b_3c_1 - a_3b_2c_1 - a_1b_3c_2 - a_2b_1c_3$$
$$= a_1b_2c_3 - a_1b_3c_2 - a_2b_1c_3 + a_3b_1c_2 + a_2b_3c_1 - a_3b_2c_1$$
$$= a_1(b_2c_3 - b_3c_2) - b_1(a_2c_3 - a_3c_2) + c_1(a_2b_3 - a_3b_2)$$

or

$$a_1 \begin{vmatrix} b_2 & c_2 \\ b_3 & c_3 \end{vmatrix} - b_1 \begin{vmatrix} a_2 & c_2 \\ a_3 & c_3 \end{vmatrix} + c_1 \begin{vmatrix} a_2 & b_2 \\ a_3 & b_3 \end{vmatrix}.$$

Now try to find each of these 2-by-2 determinants in the original 3-by-3 determinant:

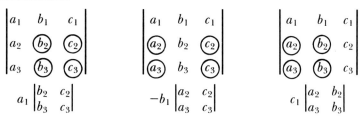

We obtain each 2-by-2 determinant by *deleting* from the original matrix the row and the column the coefficient is in.

An example for clarification is needed, before we continue with more terminology.

■ **EXAMPLE 6**

Find the value of

$$\begin{vmatrix} 1 & 3 & 2 \\ 4 & 7 & 0 \\ 6 & 8 & 1 \end{vmatrix}.$$

Solution

By using the numbers in positions a_1, b_1, and c_1, we have

$$\begin{vmatrix} \cancel{1} & 3 & 2 \\ 4 & 7 & 0 \\ \cancel{6} & 8 & 1 \end{vmatrix} \qquad \begin{vmatrix} 1 & \cancel{3} & 2 \\ 4 & \cancel{7} & 0 \\ 6 & \cancel{8} & 1 \end{vmatrix} \qquad \begin{vmatrix} 1 & 3 & \cancel{2} \\ 4 & 7 & \cancel{0} \\ 6 & 8 & \cancel{1} \end{vmatrix}$$

$$1 \begin{vmatrix} 7 & 0 \\ 8 & 1 \end{vmatrix} \quad + \quad -3 \begin{vmatrix} 4 & 0 \\ 6 & 1 \end{vmatrix} \quad + \quad 2 \begin{vmatrix} 4 & 7 \\ 6 & 8 \end{vmatrix}.$$

Then, evaluating each of the 2-by-2 determinants, we get

$$\begin{aligned} &= 1(7 - 0) - 3(4 - 0) + 2(32 - 42) \\ &= 7 - 12 - 20 \\ &= -25. \quad \square \end{aligned}$$

Expansion of a third-order determinant in terms of the constituent second-order determinants is called *expansion by cofactors* or *expansion by minors*.

■ **DEFINITION**

MINOR

The *minor* of an element is the remaining determinant obtained after deleting the row and the column containing that particular element.

■ **EXAMPLE 7**

a)
$$\begin{vmatrix} \text{\textcircled{a_1}} & b_1 & c_1 \\ a_2 & b_2 & c_2 \\ a_3 & b_3 & c_3 \end{vmatrix}$$

The minor of the element a_1 is
$$\begin{vmatrix} b_2 & c_2 \\ b_3 & c_3 \end{vmatrix}$$

b)
$$\begin{vmatrix} a_1 & b_1 & \text{\textcircled{c_1}} \\ a_2 & b_2 & c_2 \\ a_3 & b_3 & c_3 \end{vmatrix}$$

The minor of the element c_1 is
$$\begin{vmatrix} a_2 & b_2 \\ a_3 & b_3 \end{vmatrix}$$

c)
$$\begin{vmatrix} a_1 & b_1 & c_1 \\ a_2 & b_2 & c_2 \\ a_3 & \text{\textcircled{b_3}} & c_3 \end{vmatrix}$$

The minor of the element b_3 is
$$\begin{vmatrix} a_1 & c_1 \\ a_2 & c_2 \end{vmatrix} \quad \square$$

Third-order determinants contain nine elements. For each element, there is an associated **minor.** Thus a third-order determinant has a total of nine possible minors.

In the expansion of the 3-by-3 determinant earlier, we used three of the possible minors, each minor multiplied by its associated element— for example,

$$a_1 \begin{vmatrix} b_2 & c_2 \\ b_3 & c_3 \end{vmatrix} \quad - b_1 \begin{vmatrix} a_2 & c_2 \\ a_3 & c_3 \end{vmatrix} \quad c_1 \begin{vmatrix} a_2 & b_2 \\ a_3 & b_3 \end{vmatrix}$$

Now the question arises, "Why is the negative sign preceding the element b_1?"

Associated with each position in a determinant is a *sign,* positive or negative. *The sign for a given position is determined as follows:*

a) The sign is *positive* if the sum of the element's row number and column number is even.
b) The sign is *negative* if the sum of the element's row number and column number is odd.

For example, in the matrix

$$\begin{vmatrix} a_1 & b_1 & c_1 \\ a_2 & b_2 & c_2 \\ a_3 & b_3 & c_3 \end{vmatrix}$$

the sign for the position c_2 is *negative* (c_2 is in row 2, column 3; $2 + 3$ is odd). The sign for the position a_1 is *positive* (a_1 is in row 1, column 1; $1 + 1$ is even). The sign for the position b_1 is *negative* (b_1 is in row 1, column 2; $1 + 2$ is odd).

■ **DEFINITION**

COFACTOR

The *cofactor* of an element is the minor of that element preceded by the appropriate sign for that particular element.

■ **EXAMPLE 8**

Find the cofactor of the element 3 in

$$\begin{vmatrix} 1 & 3 & 2 \\ 4 & 7 & 5 \\ 6 & -2 & 8 \end{vmatrix}$$

Solution

a) The sign for the position (row 1, column 2) is *negative*.

b) The minor for the element 3 is

$$\begin{vmatrix} 4 & 5 \\ 6 & 8 \end{vmatrix}$$

Thus the cofactor of the element 3 is

$$-\begin{vmatrix} 4 & 5 \\ 6 & 8 \end{vmatrix} \qquad \square$$

● Questions

In Example 8, what is the cofactor of 1? What is the cofactor of 8?

Answers

Evaluated, the cofactor of 1 is 66, and the cofactor of 8 is −5.

Finally, we are in a position to put all the pieces together. It can be proved (by referring to several good textbooks on matrix algebra) that a determinant can be expanded by using the cofactors of *any row or* the cofactors of *any column.*

As an example, we will return to a 3-by-3 determinant used earlier and compute the value of the determinant $|A|$ by using several of the many possible ways.

■ **EXAMPLE 9**

Find the value of the determinant

$$A = \begin{vmatrix} 1 & 3 & 2 \\ 4 & 7 & 0 \\ 6 & 8 & 1 \end{vmatrix}$$

Solution

Expanding by using all elements in the first row

$$\begin{vmatrix} 1 & 3 & 2 \\ 4 & 7 & 0 \\ 6 & 8 & 1 \end{vmatrix}$$

we have

(1) × (cofactor of 1) + (3) × (cofactor of 3) + (2) × (cofactor of 2)

or

$$(1)(+)\begin{vmatrix} 7 & 0 \\ 8 & 1 \end{vmatrix} + (3)(-)\begin{vmatrix} 4 & 0 \\ 6 & 1 \end{vmatrix} + (2)(+)\begin{vmatrix} 4 & 7 \\ 6 & 8 \end{vmatrix}$$

$$\underset{\text{Sign for position}}{\overline{\underset{\uparrow}{\text{Element}} \quad \underset{\uparrow}{\text{Minor}}}} \Big\} \text{ Cofactor}$$

Evaluating further gives us

$$1(7 - 0) - 3(4 - 0) + 2(32 - 42) = -25. \quad \square$$

Let's try another!

■ EXAMPLE 10

Expanding by using all elements in the third column

$$\begin{vmatrix} 1 & 3 & 2 \\ 4 & 7 & 0 \\ 6 & 8 & 1 \end{vmatrix}$$

we have

$$(2)(+)\begin{vmatrix} 4 & 7 \\ 6 & 8 \end{vmatrix} + (0)(-)\begin{vmatrix} 1 & 3 \\ 6 & 8 \end{vmatrix} + (1)(+)\begin{vmatrix} 1 & 3 \\ 4 & 7 \end{vmatrix}.$$

Evaluating further gives us

$$2(32 - 42) - 0(8 - 18) + 1(7 - 12) = -25. \quad \square$$

You can now try expanding $|A|$ by using any other row *or* any other column; the answer you get should be -25! (Because multiplication of a value by zero yields zero, often the easiest row or column to select is the one containing the most zeros!)

Expansion of determinants by cofactors is a method that could be used on larger-order determinants. For example, a fourth-order determinant can be expanded in terms of third-order determinants (again, using any row or any column). These third-order determinants would then be expanded in terms of second-order determinants. As you can see, the amount of work increases dramatically. (For expanding a fourth-order determinant, it is possible to finish by evaluating 12 2-by-2 determinants! Why?)

Because the majority of work will involve primarily second- or third-order determinants, a shortcut for evaluating third-order determinants will be shown. However, keep in mind that this shortcut does not work, in general, for evaluating all high-order determinants.

■ EXAMPLE 11

Solution

Evaluate determinant $|A|$ from Example 9 using a shortcut.

Again, with

$$|A| = \begin{vmatrix} 1 & 3 & 2 \\ 4 & 7 & 0 \\ 6 & 8 & 1 \end{vmatrix},$$

we use the concept of diagonals. We first recopy the first two columns to the right of $|A|$:

$$\begin{vmatrix} 1 & 3 & 2 \\ 4 & 7 & 0 \\ 6 & 8 & 1 \end{vmatrix} \begin{matrix} 1 & 3 \\ 4 & 7 \\ 6 & 8 \end{matrix}$$

7 0 64

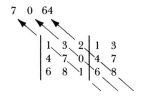

84 0 12

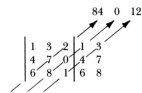

$(7 + 0 + 64) - (84 + 0 + 12) = -25$

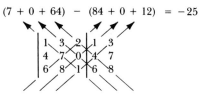

Figure 11.1　　　　　　　Figure 11.2　　　　　　　Figure 11.3

We then draw a principal diagonal and two parallel diagonals, as shown in Fig. 11.1. We then find the *product* of all the elements on *each* of these diagonals.

Now we draw a secondary diagonal and two parallel diagonals, as shown in Fig. 11.2. For each diagonal, we find the product of all elements on that diagonal.

From the sum of the first products $(7 + 0 + 64)$ on the main diagonal, we subtract the sum of the second products $(84 + 0 + 12)$ from the secondary diagonal, which gives us the value of the determinant, -25. This shortcut is summarized in Fig. 11.3.　□

Exercise Set 11.2

Find the value of each of the following determinants.

1. $\begin{vmatrix} 7 & 2 \\ 4 & 3 \end{vmatrix}$

2. $\begin{vmatrix} 7 & 4 \\ 2 & 3 \end{vmatrix}$

3. $\begin{vmatrix} 4 & 3 \\ 7 & 2 \end{vmatrix}$

4. $\begin{vmatrix} 3 & 4 \\ 2 & 7 \end{vmatrix}$

5. $\begin{vmatrix} -3 & -4 \\ -2 & -7 \end{vmatrix}$

6. $\begin{vmatrix} -2 & 8 \\ -1 & 4 \end{vmatrix}$

7. $\begin{vmatrix} 1.3 & -5.1 \\ 8 & 2.6 \end{vmatrix}$

8. $\begin{vmatrix} r - s & -s \\ s & r + s \end{vmatrix}$

9. $\begin{vmatrix} 0 & -6 \\ 0 & -2 \end{vmatrix}$

10. $\begin{vmatrix} 0.02 & 200 \\ 0.01 & 100 \end{vmatrix}$

Find those values of k for which each of the following determinants will equal zero.

11. $\begin{vmatrix} 2k & -k \\ 4 & 3 \end{vmatrix}$

12. $\begin{vmatrix} 3k & -2 \\ 9 & 2 \end{vmatrix}$

13. $\begin{vmatrix} k & k \\ 4 & -2k \end{vmatrix}$

14. Given the determinant

$$\begin{vmatrix} 4 & -2 & 3 \\ 2 & 5 & -7 \\ -3 & 8 & 6 \end{vmatrix}.$$

a) What is the *minor* for the element 4? for the element 5? for the element 6? for the element 8?

b) What is the *cofactor* of each of the elements in part (a)?

Evaluate each of the following determinants. (Do several by using the method of expansion by cofactors.)

15. $\begin{vmatrix} 0 & 1 & 2 \\ 3 & 4 & 5 \\ 6 & 7 & 8 \end{vmatrix}$

16. $\begin{vmatrix} 4 & 2 & 3 \\ 5 & 7 & 6 \\ 9 & 8 & 1 \end{vmatrix}$

17. $\begin{vmatrix} 1 & 0 & 3 \\ 0 & 1 & 2 \\ 2 & -1 & 0 \end{vmatrix}$

18. $\begin{vmatrix} 2 & 1 & -1 \\ -3 & 2 & -2 \\ 4 & -3 & 5 \end{vmatrix}$

19. $\begin{vmatrix} 4 & -3 & 7 \\ 0 & 2 & -1 \\ 0 & 0 & 5 \end{vmatrix}$

20. $\begin{vmatrix} 6 & -3 & 15 \\ -1 & 1 & -8 \\ -2 & 1 & -5 \end{vmatrix}$

21. $\begin{vmatrix} 3 & 1 & 4 \\ 2 & 0 & 9 \\ 5 & -2 & 7 \end{vmatrix}$

22. $\begin{vmatrix} 0 & 17.2 & 19.1 \\ 0 & 3.8 & -6.4 \\ 0 & -5.2 & 4.11 \end{vmatrix}$

23. $\begin{vmatrix} r & t & t \\ t & r & r \\ r & r & t \end{vmatrix}$

24. Find the value of the determinant

$$\begin{vmatrix} 1 & -1 & 2 \\ 6 & 1 & 5 \\ 2 & 4 & -3 \end{vmatrix}.$$

Then recompute the value of the determinant after:

a) each element in row 1 has been multiplied by 2;

b) each element in row 1 has been multiplied by -6;

c) row 1 has been added to row 3;

d) any two rows have been interchanged.

Formulate some observations about your work.

Solve each of the following for x.

25. $\begin{vmatrix} x & 3 \\ -2 & 1 \end{vmatrix} = 3$

26. $\begin{vmatrix} 2 & 0 & x \\ -1 & x & -1 \\ 0 & 1 & 1 \end{vmatrix} = 17$

27. $\begin{vmatrix} x^2 & 4 & 9 \\ x & 2 & 3 \\ 1 & 1 & 1 \end{vmatrix} = 0$

28. Show that the area of the triangle formed by the points (2, 1), (6, 10), and (8, 1) can be found by the formula $A = (1/2)bh$ as well as by the following formula:

$$\text{The absolute value of } \frac{1}{2} \times \begin{vmatrix} 1 & x_1 & y_1 \\ 1 & x_2 & y_2 \\ 1 & x_3 & y_3 \end{vmatrix}$$

where (x_1, y_1), (x_2, y_2), and (x_3, y_3) represent the coordinates of the three given points.

Using the method of expansion by cofactors, find the value of each of the following.

29. $\begin{vmatrix} 3 & 2 & 6 & 8 \\ 0 & -1 & 5 & -2 \\ 0 & 0 & 4 & 5 \\ 0 & 0 & 0 & 2 \end{vmatrix}$
30. $\begin{vmatrix} 3 & 4 & 2 & 1 \\ 0 & 7 & 0 & 5 \\ 8 & 0 & 9 & 1 \\ 0 & 6 & 0 & 0 \end{vmatrix}$

11.3
Solving Systems of Equations Using Cramer's Rule

Determinants provide a convenient shorthand notation for writing and computing the solution of systems of linear equations. The following process of solving systems for equations was named in honor of a Swiss mathematician of the eighteenth century, Gabriel Cramer.

Cramer's Rule for Two Equations in Two Unknowns

The solution set to the system of equations

$$a_1x + b_1y = k_1,$$
$$a_2x + b_2y = k_2$$

is $x = D_x/D_c$, and $y = D_y/D_c$, where the D's refer to determinants. D_c represents the determinant of the coefficients:

$$D_c = \begin{vmatrix} a_1 & b_1 \\ a_2 & b_2 \end{vmatrix}.$$

D_x refers to D_c when the coefficients of x are replaced by the constant terms:

$$D_x = \begin{vmatrix} k_1 & b_1 \\ k_2 & b_2 \end{vmatrix}.$$

D_y refers to D_c when the coefficients of y are replaced by the constant terms:

$$D_y = \begin{vmatrix} a_1 & k_1 \\ a_2 & k_2 \end{vmatrix}.$$

Let's take a look at an example.

■ **EXAMPLE 10**   Solve

$$3x + 2y = 13$$
$$4x + y = 14$$

using Cramer's rule.

Solution   We know that

$$x = \frac{D_x}{D_c} \quad \text{and} \quad y = \frac{D_y}{D_c}.$$

Thus

$$D_c = \begin{vmatrix} 3 & 2 \\ 4 & 1 \end{vmatrix} = -5,$$

$$D_x = \begin{vmatrix} 13 & 2 \\ 14 & 1 \end{vmatrix} = -15,$$

$$D_y = \begin{vmatrix} 3 & 13 \\ 4 & 14 \end{vmatrix} = -10,$$

and

$$x = \frac{D_x}{D_c} = \frac{-15}{-5}, \quad y = \frac{D_y}{D_c} = \frac{-10}{-5}.$$

The solution is $x = 3, y = 2$. □

Cramer's Rule for Three Equations in Three Unknowns

The solution set to the system of equations

$$a_1x + b_1y + c_1z = k_1,$$
$$a_2x + b_2y + c_2z = k_2,$$
$$a_3x + b_3y + c_3z = k_3$$

is given by

$$x = \frac{D_x}{D_c}, \quad y = \frac{D_y}{D_c}, \quad z = \frac{D_z}{D_c}.$$

D_c again refers to the determinant of the coefficients:

$$D_c = \begin{vmatrix} a_1 & b_1 & c_1 \\ a_2 & b_2 & c_2 \\ a_3 & b_3 & c_3 \end{vmatrix}.$$

D_x refers to D_c when the coefficients of x are replaced by the constants:

$$D_x = \begin{vmatrix} k_1 & b_1 & c_1 \\ k_2 & b_2 & c_2 \\ k_3 & b_3 & c_3 \end{vmatrix}.$$

D_y refers to D_c when the coefficients of y are replaced by the constants:

$$D_y = \begin{vmatrix} a_1 & k_1 & c_1 \\ a_2 & k_2 & c_2 \\ a_3 & k_3 & c_3 \end{vmatrix}.$$

D_z refers to D_c when the coefficients of z are replaced by the constants:

$$D_z = \begin{vmatrix} a_1 & b_1 & k_1 \\ a_2 & b_2 & k_2 \\ a_3 & b_3 & k_3 \end{vmatrix}.$$

As an example of solving a system of three equations in three unknowns by Cramer's rule, let's solve the following problem.

■ **EXAMPLE 12**   Solve

$$\begin{aligned} 2x + 3y + z &= 15, \\ 4x - y + 2z &= 9, \\ 5x + y + z &= 12. \end{aligned}$$

Solution   Since we know that

$$x = \frac{D_x}{D_c}, \qquad y = \frac{D_y}{D_c}, \qquad z = \frac{D_z}{D_c},$$

then

$$x = \frac{D_x}{D_c} = \frac{\begin{vmatrix} 15 & 3 & 1 \\ 9 & -1 & 2 \\ 12 & 1 & 1 \end{vmatrix}}{\begin{vmatrix} 2 & 3 & 1 \\ 4 & -1 & 2 \\ 5 & 1 & 1 \end{vmatrix}}$$

$$= \frac{15 \begin{vmatrix} -1 & 2 \\ 1 & 1 \end{vmatrix} - 3 \begin{vmatrix} 9 & 2 \\ 12 & 1 \end{vmatrix} + 1 \begin{vmatrix} 9 & -1 \\ 12 & 1 \end{vmatrix}}{2 \begin{vmatrix} -1 & 2 \\ 1 & 1 \end{vmatrix} - 3 \begin{vmatrix} 4 & 2 \\ 5 & 1 \end{vmatrix} + 1 \begin{vmatrix} 4 & -1 \\ 5 & 1 \end{vmatrix}}$$

$$= \frac{15(-1 - 2) - 3(9 - 24) + 1(9 + 12)}{2(-1 - 2) - 3(4 - 10) + 1(4 + 5)}$$

$$= \frac{21}{21} = 1.$$

Similarly,

$$y = \frac{D_y}{D_c} = \frac{\begin{vmatrix} 2 & 15 & 1 \\ 4 & 9 & 2 \\ 5 & 12 & 1 \end{vmatrix}}{\begin{vmatrix} 2 & 3 & 1 \\ 4 & -1 & 2 \\ 5 & 1 & 1 \end{vmatrix}} = \frac{63}{21} = 3$$

and

$$z = \frac{D_z}{D_c} = \frac{\begin{vmatrix} 2 & 3 & 15 \\ 4 & -1 & 9 \\ 5 & 1 & 12 \end{vmatrix}}{\begin{vmatrix} 2 & 3 & 1 \\ 4 & -1 & 2 \\ 5 & 1 & 1 \end{vmatrix}} = \frac{84}{21} = 4.$$

Thus the common solution to the system is (1, 3, 4).   □

As you can see, with a working knowledge of determinants, applying Cramer's rule to solve systems of simultaneous equations is a rather convenient mechanical procedure.

Deriving Cramer's Rule

Before we show some other applications of Cramer's rule, a brief presentation should be made showing how Cramer's rule was developed. Our illustration will focus on the use of Cramer's rule for two equations in two unknowns. You will then be asked to derive Cramer's rule for three equations in three unknowns.

In addition to providing a more thorough understanding of Cramer's rule, the derivation will also review some important work on solving systems of equations.

First, we suggest that you go back to Chapter 10 and review a numerical example of solving systems of equations using the method of linear combinations.

To solve the system

$$\begin{cases} a_1x + b_1y = k_1, \\ a_2x + b_2y = k_2 \end{cases}$$

by the method of linear combinations, or elimination, how could we eliminate y from the system? (Remember that the coefficients of y should have coefficients that will be inverses of each other.)

We multiply Eq. (1) by b_2 and Eq. (2) by $-b_1$:

$$\text{Eq. (1)} \quad a_1x + b_1y = k_1 \quad \text{M}(b_2)$$
$$\text{Eq. (2)} \quad a_2x + b_2y = k_2 \quad \text{M}(-b_1)$$

to obtain the equivalent system

$$\left\{ \begin{array}{l} a_1b_2x + b_1b_2y = b_2k_1 \\ -a_2b_1x - b_1b_2y = -b_1k_2 \end{array} \right\}$$

Adding these two equations, we have an expression containing only the variable x. Thus y was eliminated:

$$a_1b_2x - a_2b_1x = b_2k_1 - b_1k_2.$$

We now solve for x:

$$x(a_1b_2 - a_2b_1) = b_2k_1 - b_1k_2$$

or

$$x = \frac{b_2k_1 - b_1k_2}{a_1b_2 - a_2b_1}.$$

By comparing this expression to the *original* system above and by using the more efficient notation of determinants, we have

$$x = \frac{b_2k_1 - b_1k_2}{a_1b_2 - a_2b_1} = \frac{\begin{vmatrix} k_1 & b_1 \\ k_2 & b_2 \end{vmatrix}}{\begin{vmatrix} a_1 & b_1 \\ a_2 & b_2 \end{vmatrix}} = \frac{D_x}{D_c}.$$

Hence the evolution of Cramer's rule: a shortcut developed from the procedure of elimination!

To solve for y, we must eliminate x from the system. We multiply Eq. (1) by $-a_2$ and Eq. (2) by a_1

$$\text{Eq. (1)} \quad a_1x + b_1y = k_1 \quad \text{M}(-a_2)$$
$$\text{Eq. (2)} \quad a_2x + b_2y = k_2 \quad \text{M}(a_1)$$

to obtain the following equivalent system:

$$\left\{ \begin{array}{l} -a_1a_2x - b_1a_2y = -a_2k_1 \\ a_1a_2x + a_1b_2y = a_1k_2 \end{array} \right\}.$$

By adding the two equations, we obtain the expression for y:

$$a_1b_2y - b_1a_2y = a_1k_2 - a_2k_1$$

or

$$y(a_1b_2 - b_1a_2) = a_1k_2 - a_2k_1.$$

Now we solve for y:

$$y = \frac{a_1k_2 - a_2k_1}{a_1b_2 - a_2b_1}.$$

By comparing this expression to the original system above and again using the more efficient notation of determinants, we have

$$y = \frac{a_1k_2 - a_2k_1}{a_1b_2 - a_2b_1} = \frac{\begin{vmatrix} a_1 & k_1 \\ a_2 & k_2 \end{vmatrix}}{\begin{vmatrix} a_1 & b_1 \\ a_2 & b_2 \end{vmatrix}} = \frac{D_y}{D_c}.$$

(Wasn't that Mr. Cramer clever to relate determinants to solving systems of equations by linear combinations!)

In a similar manner, you could derive Cramer's rule by solving the system of three equations in three unknowns for x, then y, then z:

$$a_1x + b_1y + c_1z = k_1,$$
$$a_2x + b_2y + c_2z = k_2,$$
$$a_3x + b_3y + c_3z = k_3.$$

Related Applications of Determinants—Analysis of Systems

Although Cramer's rule and the use of determinants is economical and practical for solving systems of equations in two or three unknowns, they have a more powerful application in analyzing the type of system a set of equations may be. Figure 11.4 shows the different types of systems of simultaneous linear equations.

By combining Cramer's rule and the properties of zero (discussed in Chapter 0), an easy means is developed for classifying systems of simultaneous equations. For example, we know that

$$\frac{0}{6} = 0, \qquad \frac{6}{0} = \text{undefined}, \quad \text{and} \quad \frac{0}{0} = \text{undefined};$$

and from Cramer's rule:

$$x = \frac{D_x}{D_c}, \qquad y = \frac{D_y}{D_c}, \qquad \text{and} \quad z = \frac{D_z}{D_c}.$$

If $D_c \neq 0$, the system of simultaneous equations has a *unique solution* (that is, it is an independent system). If $D_c = 0$, two possibilities exist:

1. If *all* the numerators $= 0$, the system is dependent.
2. If *at least one* of the numerators $\neq 0$, the system is inconsistent.

These relationships are summarized in Fig. 11.5.

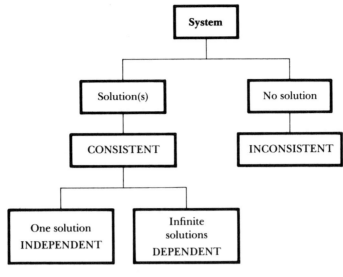

Figure 11.4

A Practical Application

The analysis of systems of equations is very important in the following respect. In most cases, a system of equations must be solved by programming a computer. The first question to ask is, "Is the system solvable?" One of the first checks in a computer program to solve systems of equations—regardless of the method used (elimination,

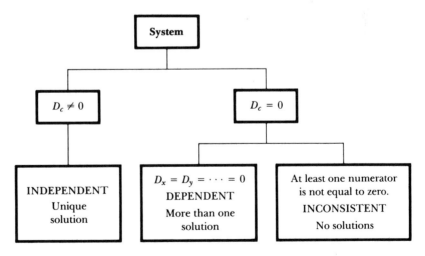

Figure 11.5

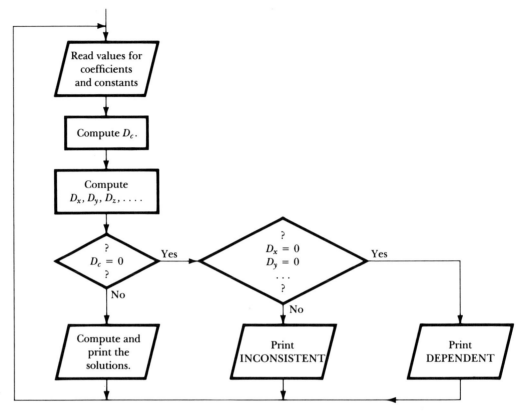

Figure 11.6

substitution, matrices, etc.)—*should* be a check to determine whether the system is independent (or at least not inconsistent). A simple check to accomplish this is to calculate the determinant of the coefficients, D_c; this *cannot* equal 0.

A flowchart showing the steps in solving a system of simultaneous equations by Cramer's rule is shown in Fig. 11.6.

Exercise Set 11.3

Solve each of the following systems using Cramer's rule (determinants).

1. $x - 3y = 5,$
$\quad 4x + 6y = 7$

2. $2x + y = 6,$
$\quad 5x - 7y = 15$

3. $7x - 5y = 3,$
$\quad 4x + 2y = -8$

4. $y = 3x - 2,$
$\quad y + 2 = x + 200$

5. $2(x + 3) = 3y - 2,$
$\quad 3x + 5y = -12$

6. $\quad r + \quad s = 770,$
$\quad 0.6r + 0.3s = 381$

7. $\quad 3x - 7y = -1,$
$\quad -6x + 5y = 20$

8. $ax + by = e,$
$\quad cx + dy = f$

9. $2x + y = \quad 4,$
$\quad x - y = -0.7$

10. $3x - 4y = 8,$
$\quad 6x - 8y = 9$

11. How many solutions does the system of equations

$$x + y = h,$$
$$x + (h - 1)y = 2$$

have if

a) $h = 2?$

b) $h = 1?$

Solve each of the following systems using Cramer's rule.

12. $2x + 9y + 3z = 14,$
$\quad 4x + 5y + 6z = 15,$
$\quad 7x + \quad y + 8z = 13$

13. $2x - \quad y + \quad z = 2,$
$\quad 3x - \quad y - 3z = 1,$
$\quad 2x + 2y - \quad z = 3$

14. $2x - \quad y + \quad z = 8,$
$\quad 3x - \quad y - 3z = 8,$
$\quad 2x + 2y - \quad z = 9$

15. $2x - \quad y + \quad z = 11,$
$\quad 3x - \quad y - 3z = -4,$
$\quad 2x + 2y - \quad z = -5$

16. $\quad x + 2y - 3z = 1,$
$\quad 5x + 3y - 6z = 0,$
$\quad 3x - \quad y \quad\quad = 2$

17. $3a - \quad 2b + c = 26,$
$\quad 4a - \quad 2b \quad\quad = 30,$
$\quad 3a + 12b \quad\quad = 3c$

18. $a + b = 11,$
$\quad a + c = 12,$
$\quad b + c = 13$

19. $0.2x - 0.3y + 0.3z = 0.2,$
$\quad 1.5x + \quad y + 0.5z = 3,$
$\quad 0.1x + 0.1y - 0.1z = 0.1$

20. $3x + 2y - \quad z = 0,$
$\quad 4x + 7y - 3z = 0,$
$\quad x - 7y + \quad z = 0$

21. $4r - 3s + 2t = \quad 4,$
$\quad 5s - 7t = -11,$
$\quad 8t = \quad 24$

22. Solve the following simultaneous equations for z. What value of k will create an independent system?

$$2x - \quad y + \quad z = 1,$$
$$2x + \quad y - kz = 1,$$
$$2x + 3y - 3z = 5$$

23. Using determinants and Cramer's rule, find the value of x in the following system. Then obtain values for y, z, and w by substitution.

$$x + y + w = \quad 7,$$
$$x + y + z = \quad 8,$$
$$y + z + w = \quad 9,$$
$$x + z + w = 10$$

24. For the system

$$a_1x + b_1y + c_1z = k_1,$$
$$a_2x + b_2y + c_2z = k_2,$$
$$a_3x + b_3y + c_3z = k_3,$$

algebraically derive the x-component of Cramer's rule:

$$x = \frac{D_x}{D_c} = ?$$

(*Hint:* Study the algebraic procedure for deriving Cramer's rule for two equations in two unknowns!)

Note: For additional work with systems of equations, other systems are available: see Exercise Set 10.2, Exercises 6–17; Exercise Set 10.4, Exercises 1–20; Exercise Set 10.5, Exercises 1–20 and 25–27.

11.4
Mathematical Operations: The Algebra of Matrices

In preparation for applying matrices to solving systems of simultaneous equations, we first need to learn the basics of matrix algebra. Operations between matrices are similar to operations between real numbers. Specifically, sums, differences, products, and inverses of matrices exist. While matrices were first introduced in 1858, matrix algebra was devised by the English mathematician Arthur Cayley in the nineteenth century. Matrix algebra, a logical deductive system, was built on several basic definitions.

■ DEFINITION

Two matrices are *equal* if and only if the elements in all the corresponding positions are equal.

Matrix A and matrix B below are *not* equal:

$$A = \begin{bmatrix} 17.2 & 8 \\ 9.4 & 5.1 \end{bmatrix}, \qquad B = \begin{bmatrix} 17.2 & 8 \\ 9.4 & 4.3 \end{bmatrix}.$$

Although most corresponding positions are the same, the element in the second row, second column is different for the two matrices.

● Question

If the two matrices

$$D = \begin{bmatrix} 1 & m & 3 \\ n & 2 & 5 \\ 4 & 8 & r \end{bmatrix}, \qquad E = \begin{bmatrix} 1 & 17 & 3 \\ 7 & 2 & 5 \\ 4 & 8 & 6 \end{bmatrix},$$

are equal, then by the definition, what must be the value for m? for n? for r?

Answer

Because the matrices are equal, elements in corresponding positions must be equal. Thus m in the first row, second column of matrix D must equal the element in the first row, second column of matrix E ($m = 17$). Correspondingly, n in matrix D must equal 7 in matrix E, and r must equal 6.

● Question

If two matrices A and B are equal, what can be said about their dimensions?

Answer

If the dimension of matrix A is m-by-n, then by the definition, matrix B must have corresponding elements. Thus matrix B must also have dimension m-by-n.

Using the general notation of double subscripts, we can summarize equality of matrices as follows:

■

For matrices A and B, $A = B$ if and only if $a_{ij} = b_{ij}$ for each of the values for i and j.

■ **DEFINITION**

The *sum* of two matrices of the same dimension is the matrix whose elements are the sums of the corresponding elements of the given matrices.

■ **EXAMPLE 13**

$$\begin{bmatrix} a & 3 \\ 7 & 4 \end{bmatrix} + \begin{bmatrix} b & 5 \\ 2 & f \end{bmatrix} = \begin{bmatrix} a+b & 3+5 \\ 7+2 & 4+f \end{bmatrix} \quad \square$$

■ **EXAMPLE 14**

$$\begin{bmatrix} 2 & 4 \\ 7 & -3 \\ 5 & 6 \end{bmatrix} + \begin{bmatrix} 3 & 1 \\ 2 & 5 \\ -4 & -3 \end{bmatrix} = \begin{bmatrix} 2+3 & 4+1 \\ 7+2 & -3+5 \\ 5-4 & 6-3 \end{bmatrix} \quad \text{or} \quad \begin{bmatrix} 5 & 5 \\ 9 & 2 \\ 1 & 3 \end{bmatrix} \quad \square$$

Again, using the general notation with double subscripts to indicate the matrix elements, we summarize the sum of two matrices as follows:

■

$C = A + B$ if and only if $c_{ij} = a_{ij} + b_{ij}$ for all the values for i and j.

■ **EXAMPLE 15**

A microcomputer dealer's inventory consists of CRT's, disc drives, printers, and keyboards. Suppose the inventory on hand, for three

Table 11.3

	CRT	DD	PR	KB
EXSON 12	48	32	41	40
BDP-PC	50	48	49	51
ARS 80B	8	6	14	10

different brands of computers, is as shown in Table 11.3. This table could be represented by matrix S

$$S = \begin{bmatrix} 48 & 32 & 41 & 40 \\ 50 & 48 & 49 & 51 \\ 8 & 6 & 14 & 10 \end{bmatrix}.$$

Now suppose a new shipment of microcomputer components arrived, denoted in matrix form by:

$$R = \begin{bmatrix} 20 & 30 & 20 & 20 \\ 10 & 12 & 11 & 9 \\ 7 & 9 & 1 & 5 \end{bmatrix}.$$

How many components, of each brand are *now* in stock?

Solution

$$S + R = \begin{bmatrix} 48+20 & 32+30 & 41+20 & 40+20 \\ 50+10 & 48+12 & 49+11 & 51+9 \\ 8+7 & 6+9 & 14+1 & 10+5 \end{bmatrix}$$

$$= \begin{bmatrix} 68 & 62 & 61 & 60 \\ 60 & 60 & 60 & 60 \\ 15 & 15 & 15 & 15 \end{bmatrix} \quad \square$$

In Example 15, how would we indicate the components in stock after a widely publicized week-long sale? The following matrix indicates the quantities of each component for each brand sold during the sale:

$$T = \begin{bmatrix} 20 & 10 & 13 & 10 \\ 15 & 16 & 15 & 15 \\ 8 & 8 & 20 & 8 \end{bmatrix}.$$

As you may have guessed, subtraction of matrices must be defined. And, as you may also have guessed, subtraction of matrices is obtained by subtracting elements in the corresponding positions.

Let's be a little more thorough. In matrix algebra, "ordinary" numbers are called *scalars*. Multiplying a matrix by a scalar is done in this manner.

If

$$A = \begin{bmatrix} 1 & 3 \\ -2 & 7 \end{bmatrix},$$

then

$$2A = \begin{bmatrix} (2)(1) & (2)(3) \\ (2)(-2) & (2)(7) \end{bmatrix} = \begin{bmatrix} 2 & 6 \\ -4 & 14 \end{bmatrix},$$

where 2 is a scalar. Thus $2A$ is an example of *scalar multiplication*.

Using matrix A above, find $3A$ and $-5A$.

If M represents a matrix, then $(-1)M$ or $-M$ will represent its additive inverse. Further, $M + (-M) = 0$, where 0 represents the zero matrix (every element $= 0$) of the same order as matrix M.

If

$$M = \begin{bmatrix} 2 & 3 \\ 4 & -1 \end{bmatrix},$$

then

$$-M = \begin{bmatrix} -2 & -3 \\ -4 & 1 \end{bmatrix}$$

by scalar multiplication and

$$M + (-M) = \begin{bmatrix} 2 - 2 & 3 - 3 \\ 4 - 4 & -1 + 1 \end{bmatrix} = \begin{bmatrix} 0 & 0 \\ 0 & 0 \end{bmatrix}.$$

Similar to the definition of subtraction of real numbers (see Chapter 0), we have the following.

■ DEFINITION

Subtraction of two matrices A and B is equivalent to adding to matrix A the additive inverse of matrix B:

$$A - B = A + (-B).$$

If

$$A = \begin{bmatrix} 3 & 2 \\ 1 & 7 \end{bmatrix} \quad \text{and} \quad B = \begin{bmatrix} 4 & -1 \\ 2 & -5 \end{bmatrix},$$

then

$$A - B = A + (-B) = \begin{bmatrix} 3 & 2 \\ 1 & 7 \end{bmatrix} + (-1)\begin{bmatrix} 4 & -1 \\ 2 & -5 \end{bmatrix}$$

$$= \begin{bmatrix} 3 - 4 & 2 + 1 \\ 1 - 2 & 7 + 5 \end{bmatrix} = \begin{bmatrix} -1 & 3 \\ -1 & 12 \end{bmatrix}.$$

■ EXAMPLE 15 (continued)

Now returning to Example 15, we see that matrix S represents micro-computer components in stock originally, R represents components recently received in a shipment, and T represents components sold during the week-long sale:

$$S = \begin{bmatrix} 48 & 32 & 41 & 40 \\ 50 & 48 & 49 & 51 \\ 8 & 6 & 14 & 10 \end{bmatrix}, \qquad R = \begin{bmatrix} 20 & 30 & 20 & 10 \\ 10 & 12 & 12 & 9 \\ 7 & 9 & 1 & 5 \end{bmatrix},$$

$$T = \begin{bmatrix} 20 & 10 & 13 & 10 \\ 15 & 16 & 15 & 15 \\ 8 & 8 & 10 & 8 \end{bmatrix}.$$

How many components, for each brand, are in stock *after* the sale?

Solution

Using matrix algebra notation, we can represent the amount currently in stock by the matrix expression $S + R - T$. You may verify the "arithmetic" that gives us

$$S + R - T = \begin{bmatrix} 48 & 52 & 48 & 40 \\ 45 & 44 & 46 & 45 \\ 7 & 7 & 5 & 7 \end{bmatrix}. \quad \square$$

Can you imagine the usefulness of matrices if, rather than 3 computer brands, there were 14, or instead of 4 components, there were 9? Think of how useful matrices are if in a factory inventory there are 1045 separate items, each with 328 components! Most computers are equipped to handle arrays of data in the form of matrices; using matrices, then, becomes a very efficient way of processing such quantities of data.

We have thus far defined the operations of addition, scalar multiplication, and subtraction in matrix algebra. Addition of real numbers is both commutative and associative. Is matrix addition commutative? Is it associative? Try some examples of your own to either verify your guesses or disprove them! (Answers are deferred to the exercise set.)

Sums and differences of matrices are determined by corresponding positions. Are matrices multiplied by multiplying the elements in corresponding positions? For example, if

$$A = \begin{bmatrix} 7 & 3 \\ 2 & 8 \end{bmatrix} \quad \text{and} \quad B = \begin{bmatrix} 2 & 5 \\ 6 & 1 \end{bmatrix},$$

does

$$AB = \begin{bmatrix} (7)(2) & (3)(5) \\ (2)(6) & (8)(1) \end{bmatrix} = \begin{bmatrix} 14 & 15 \\ 12 & 8 \end{bmatrix}?$$

The answer is no! Although matrices *could* have been defined in this manner, matrix multiplication finds many more applications if the product is defined in a different manner.

■ **DEFINITION**

The *product* $C = AB$ of an m-by-p matrix A and a p-by-n matrix B is an m-by-n matrix. An element in the product in row r and column c is obtained by multiplying row r of matrix A by column c of matrix B.

Let's look at the following example:

$$\begin{bmatrix} a_{11} & a_{12} & \cdots & a_{1p} \\ a_{21} & a_{22} & \cdots & a_{2p} \\ \vdots & \vdots & \ddots & \vdots \\ a_{r1} & a_{r2} & a_{r3} & \cdots & a_{rp} \\ \vdots & \vdots & & \vdots \\ a_{m1} & a_{m2} & \cdots & a_{mp} \end{bmatrix} \times \begin{bmatrix} b_{11} & b_{12} & \cdots & b_{1c} & \cdots & b_{1n} \\ b_{21} & b_{22} & \cdots & b_{2c} & \cdots & b_{2n} \\ \vdots & \vdots & & b_{3c} & \cdots & b_{3n} \\ \vdots & \vdots & & \vdots & & \vdots \\ b_{p1} & b_{p2} & \cdots & b_{pc} & \cdots & b_{pn} \end{bmatrix} = \begin{bmatrix} c_{11} & c_{12} & \cdots & c_{1n} \\ c_{21} & c_{22} & \cdots & c_{2n} \\ \vdots & \vdots & & \vdots \\ \vdots & \vdots & c_{rc} & \vdots \\ \vdots & \vdots & & \vdots \\ c_{m1} & c_{m2} & \cdots & c_{mn} \end{bmatrix}.$$

To obtain element c_{rc} in the product, *row r of matrix A* (a row vector) is multiplied by *column c of matrix B* (a column vector).

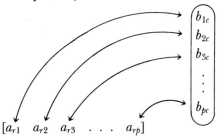

Note: The number of columns of matrix A *must equal* the number of rows of matrix B or matrix A cannot be multiplied by matrix B.

Multiplication is performed as follows: The first position of the row vector times the first position of the column vector, the second position of the row times the second position of the column, etc., until the last position of the row times the last position of the column. All of these products are then added.

Using summation notation, we determine the element c_{rc} in the product AB by the following formula:

$$c_{rc} = (a_{r1})(b_{1c}) + (a_{r2})(b_{2c}) + (a_{r3})(b_{3c}) + \cdots + (a_{rp})(b_{pc})$$

$$= \sum_{k=1}^{p} (a_{rk})(b_{kc}).$$

Before giving a complete example of the entire matrix multiplications, let's illustrate more clearly with a numerical example exactly how a row vector is multiplied by a column vector. To multiply a 1-by-4 row vector by a 4-by-1 column vector, we have

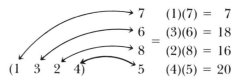

$$\begin{array}{ll} 7 & (1)(7) = 7 \\ 6 & (3)(6) = 18 \\ 8 & (2)(8) = 16 \\ 5 & (4)(5) = 20 \end{array}$$

Elements in the row are multiplied position by position with the elements in the column. These products are then added. The multiplication of this particular row vector times this column vector is equal to 7 + 18 + 16 + 20, or 61.

We can see from this multiplication that each of the four positions in the row vector needs a corresponding position in the column vector. The number of positions in the row vector is really determined by how many columns the row vector has. The number of positions in the column vector is actually the number of rows the column vector has.

To assure proper "matching" of positions for multiplication, then, the number of columns in the first matrix must equal the number of rows in the second matrix. In general,

$$\begin{array}{cc} \text{Matrix } A & \text{Matrix } B \\ m \times p & p \times n \end{array}$$

If matrix A has p columns and matrix B has p rows, the matrices are said to be **conformable.** The product AB then can be computed. In addition,

$$\begin{array}{cc} \text{Matrix } A & \text{Matrix } B \\ m \times p & p \times n \end{array}$$

The dimension of the resulting product is $m \times n$. Again, each row in matrix A (m rows) is multiplied by each column in matrix B (n columns); thus the dimension of the product AB is $m \times n$.

A numerical illustration is needed to clarify these ideas. The example will be a fairly large example; pay careful attention to all the intricate details!

■ **EXAMPLE 16**   Given

$$A = \begin{bmatrix} 1 & 3 & 2 \\ 4 & -5 & 0 \end{bmatrix} \quad \text{and} \quad B = \begin{bmatrix} 7 & 1 & 6 & 3 \\ 5 & 2 & 4 & 0 \\ 8 & -3 & 5 & 9 \end{bmatrix},$$

find AB.

Solution

Because the number of columns in A equals the number of rows in B, the matrices are conformable and the product *can* be determined.

$$\text{Matrix } A \qquad \text{Matrix } B$$
$$2 \times 3 \qquad 3 \times 4$$

The *dimension* of the product matrix AB will be 2×4:

$$\text{Matrix } A \qquad \text{Matrix } B$$
$$2 \times 3 \qquad 3 \times 4$$

The product $C = AB$ will appear as the 2×4 matrix:

$$C = \begin{bmatrix} c_{11} & c_{12} & c_{13} & c_{14} \\ c_{21} & c_{22} & c_{23} & c_{24} \end{bmatrix}.$$

Now using the definition for multiplying matrices, we find the value of each element in matrix C.

$c_{11} \longrightarrow$ *first* row of A times *first* column of B

$$\begin{bmatrix} 1 & 3 & 2 \\ 4 & -5 & 0 \end{bmatrix} \quad \begin{bmatrix} 7 & 1 & 6 & 3 \\ 5 & 2 & 4 & 0 \\ 8 & -3 & 5 & 9 \end{bmatrix}$$

$c_{11} = (1)(7) + (3)(5) + (2)(8) = 38$

$c_{12} \longrightarrow$ *first* row of A times *second* column of B

$$\begin{bmatrix} 1 & 3 & 2 \\ 4 & -5 & 0 \end{bmatrix} \quad \begin{bmatrix} 7 & 1 & 6 & 3 \\ 5 & 2 & 4 & 0 \\ 8 & -3 & 5 & 9 \end{bmatrix}$$

$c_{12} = (1)(1) + (3)(2) + (2)(-3) = 1$

Thus far,

$$C = \begin{bmatrix} 38 & 1 & c_{13} & c_{14} \\ c_{21} & c_{22} & c_{23} & c_{24} \end{bmatrix}.$$

Continuing in a similar manner, we have

$c_{13} \longrightarrow$ *first* row of A times *third* column of B
$c_{13} = (1)(6) + (3)(4) + (2)(5) = 28$

$c_{14} \longrightarrow$ *first* row of A times *fourth* column of B
$c_{14} = (1)(3) + (3)(0) + (2)(9) = 21$

$c_{21} \longrightarrow$ *second* row of A times *first* column of B
$c_{21} = (4)(7) + (-5)(5) + (0)(8) = 3$

$c_{22} \longrightarrow$ *second* row of A times *second* column of B
$c_{22} = (4)(1) + (-5)(2) + (0)(-3) = -6$

$c_{23} \longrightarrow$ *second* row of A times *third* column of B

$c_{23} = (4)(6) + (-5)(4) + (0)(5) = 4$

$c_{24} \longrightarrow$ *second* row of A times *fourth* column of B

$c_{24} = (4)(3) + (-5)(0) + (0)(9) = 12$

The product of matrix A and matrix B is

$$C = \begin{bmatrix} 38 & 1 & 28 & 21 \\ 3 & -6 & 4 & 12 \end{bmatrix}. \quad \square$$

The complexity of matrix multiplication together with attention to all the details of the arithmetic computations will require some extra studying. Review the example several times and perhaps try several exercises at the end of the section before proceeding.

• Question

Addition of matrices is commutative and associative. Is multiplication of matrices commutative? Does $A \times B = B \times A$?

Answer

In general, matrix multiplication is *not* commutative. In Example 15, matrix A has order 2×3; B has order 3×4. These matrices were conformable; the product was defined. However, to compute the product BA,

Matrix B Matrix A

$$3 \times 4 \qquad\qquad 2 \times 3$$

the matrices are not even comformable. The product BA is *not* defined. Here, $A \times B \neq B \times A$. Can you think of other "noncommutative" examples?

An Application of Matrix Multiplication

Let's look at an application. The number of shares of each stock held by a client are given in the row matrix form below, taken from Table 11.4:

$$A = [110 \quad 148 \quad 80 \quad 95 \quad 125].$$

Stock prices for each quarter (see Table 11.5) are given in matrix B:

$$B = \begin{bmatrix} 52.25 & 53.00 & 54.75 & 55.00 \\ 38.25 & 38.75 & 39.00 & 42.25 \\ 24.00 & 25.50 & 26.25 & 26.75 \\ 19.75 & 21.00 & 22.25 & 25.75 \\ 82.00 & 83.25 & 84.25 & 85.00 \end{bmatrix}.$$

If we multiply matrix A times matrix B, the product would indicate, in one table, the total value of stocks for each quarter of this particular

Table 11.4

CB	DRF	R1	UAR	XYZ
110	148	80	95	125

Table 11.5

	Q_1	Q_2	Q_3	Q_4
CB	52.25	53.00	54.75	55.00
DRF	38.25	38.75	39.00	42.25
RT	24.00	25.50	26.25	26.75
UAR	19.75	21.00	22.25	25.75
XYZ	82.00	83.25	84.25	85.00

investment year:

$$\begin{matrix} Q_1 & Q_2 & Q_3 & Q_4 \\ AB = [25454.75 & 26006.25 & 26539.50 & 27514.25]. \end{matrix}$$

The ease with which these calculations are performed on today's computers is amazing. Illustrations of multiplication by use of the computer will be shown in Section 11.6.

Exercise Set 11.4

1. Select two matrices that are equal from the following group.

a) $\begin{bmatrix} 1 & 3 & 0 \\ 2 & 4 & 1 \end{bmatrix}$
   b) $\begin{bmatrix} 1 & 3 \\ 2 & 4 \end{bmatrix}$
   c) $\begin{bmatrix} 1 & 3 \\ 4 & 2 \end{bmatrix}$

d) $\begin{bmatrix} 1 & 3 \\ 2 & 4 \\ 0 & 1 \end{bmatrix}$
   e) $\begin{bmatrix} 2 & 4 \\ 1 & 3 \end{bmatrix}$
   f) $\begin{bmatrix} 1 & 3 & 0 \\ 4 & 2 & 0 \\ 0 & 0 & 0 \end{bmatrix}$

g) $\begin{bmatrix} 1 & 3 \\ 2 & 4 \end{bmatrix}$

Add each of the following.

2. $\begin{bmatrix} 3 & -2 \\ 7 & 4 \end{bmatrix} + \begin{bmatrix} 1 & 5 \\ -2 & 1 \end{bmatrix}$
   3. $\begin{bmatrix} 3 & 2 \\ 1 & -5 \end{bmatrix} + \begin{bmatrix} -2 & -2 \\ -1 & 6 \end{bmatrix}$

4. $\begin{bmatrix} 1 & 3 & -2 \\ 0 & 7 & 4 \\ 1 & 5 & 4 \end{bmatrix} + \begin{bmatrix} 0 & 0 & 0 \\ 1 & 3 & 2 \\ -1 & -5 & -7 \end{bmatrix}$
   5. $\begin{bmatrix} 8 & 3 & 0 & 2 \\ 1 & 7 & 4 & 5 \end{bmatrix} + \begin{bmatrix} 1 & 3 & 0 & 4 \\ -1 & 6 & 2 & 4 \end{bmatrix}$

6. $\begin{bmatrix} 2 & 1 & 3 \\ 4 & 5 & 7 \\ 6 & 8 & 9 \end{bmatrix} + \begin{bmatrix} -2 & -1 & -3 \\ -4 & -5 & -7 \\ -6 & -8 & -9 \end{bmatrix}$

7. What matrix is equal to

$$\begin{bmatrix} 3+2 & 10/20 & 2(5) \\ 0 & 7-4 & 6 \end{bmatrix}?$$

8. Given that

$$\begin{bmatrix} 1 & 3 \\ a & b \end{bmatrix} = \begin{bmatrix} c & 3 \\ 2 & 5 \end{bmatrix},$$

find the values of a, b, and c.

9. Given that

$$\begin{bmatrix} 3 & a & b & 4 & 5 \\ c & 2 & 5 & d & -2 \end{bmatrix} = \begin{bmatrix} e & 9 & -6 & 4 & f \\ 1 & 2 & g & 6 & h \end{bmatrix},$$

find the values of a, b, c, d, e, f, g, and h.

Find the additive inverse of each matrix.

10.
$$\begin{bmatrix} 1 & 4 & -3 & 2 \\ 2 & 1.6 & -2 & -3 \\ -2 & 4 & 5 & 0 \end{bmatrix}$$

11.
$$\begin{bmatrix} -2.3 & 3.1 \\ 4.1 & 5.2 \\ 1.2 & -7.3 \end{bmatrix}$$

Given that

$$B = \begin{bmatrix} 4 & 2 & 10 \\ -2 & 8 & 6 \end{bmatrix} \quad \text{and} \quad C = \begin{bmatrix} 1 & 3 & 2 \\ 4 & 1 & 5 \end{bmatrix},$$

find each of the following.

12. $2B$

13. $3C$

14. $2B + 3C$

15. $2B - 3C$

16. $3B$

17. $-2C$

18. $3B + 2C$

19. $3B - 2C$

20. Given that

$$A = \begin{bmatrix} 1 & -2 \\ 3 & 4 \end{bmatrix}, \quad B = \begin{bmatrix} 4 & -1 \\ 3 & 2 \end{bmatrix}, \quad C = \begin{bmatrix} 0 & -2 \\ 2 & -3 \end{bmatrix}.$$

a) Verify that addition of matrices is commutative by showing that $A + B = B + A$.

b) Verify that addition of matrices is associative by showing that $(A + B) + C = A + (B + C)$.

c) Does $A - B = B - A$?

d) Does $(A - B) - C = A - (B - C)$?

21. If

$$\begin{bmatrix} 4 \\ 3 \\ -2 \\ 5 \end{bmatrix} + \begin{bmatrix} X \\ Y \\ Z \\ W \end{bmatrix} = \begin{bmatrix} 7 \\ 6 \\ 2 \\ -3 \end{bmatrix},$$

find values for X, Y, Z, W.

22. If

$$3 \begin{bmatrix} X \\ 2Y \end{bmatrix} + \begin{bmatrix} -2 \\ 4 \end{bmatrix} = \begin{bmatrix} 12 \\ 10 \end{bmatrix},$$

find X and Y.

23. If

$$\begin{bmatrix} 5 & -3 \\ 4 & 1 \end{bmatrix} \begin{bmatrix} X \\ Y \end{bmatrix} = \begin{bmatrix} 7 \\ 9 \end{bmatrix},$$

find X and Y.

Multiply each of the following.

24. $2 \begin{bmatrix} 6 & 3 \\ -2 & 4 \end{bmatrix}$

25. $(-1/3) \begin{bmatrix} 6 & -9 & 12 \\ 3 & 15 & 18 \\ 12 & 0 & -3 \end{bmatrix}$

26. $0.01 \ [180 \quad 1400 \quad 130]$

27. $\begin{bmatrix} 3 & 2 \\ 1 & 7 \end{bmatrix} \begin{bmatrix} 5 & 2 \\ 0 & 4 \end{bmatrix}$

28. $\begin{bmatrix} 3 & -2 & 5 \\ 1 & 7 & 6 \end{bmatrix} \begin{bmatrix} 1 & 3 \\ 0 & -2 \\ 7 & 5 \end{bmatrix}$

29. $[0.5 \quad 0.2 \quad 0.3] \begin{bmatrix} 60 & 80 & 50 \\ 90 & 20 & 80 \\ 10 & 30 & 70 \end{bmatrix}$

30. $\begin{bmatrix} r & s & t \\ u & v & w \\ x & y & z \end{bmatrix} \begin{bmatrix} 1 & 0 & 0 \\ 0 & 1 & 0 \\ 0 & 0 & 1 \end{bmatrix}$

31. $\begin{bmatrix} 1 & 0 & 0 \\ 0 & 1 & 0 \\ 0 & 0 & 1 \end{bmatrix} \begin{bmatrix} r & s & t \\ u & v & w \\ x & y & z \end{bmatrix}$

32. $\begin{bmatrix} 4 & 3 & -5 \\ 2 & 6 & 0 \\ 5 & 0 & 2 \end{bmatrix} \begin{bmatrix} 1 \\ 3 \\ -8 \end{bmatrix}$

33. $\begin{bmatrix} 1 & 3 & -2 \\ 7 & 6 & 5 \\ 4 & 8 & 0 \end{bmatrix} \begin{bmatrix} 1 & 13 & 2 \\ -4 & -2 & -6 \\ 5 & -3 & 0 \end{bmatrix}$

34. $\begin{bmatrix} 1 & 1 & 0 \\ 0 & 1 & 0 \\ 0 & 0 & 1 \end{bmatrix} \begin{bmatrix} a & b & c \\ d & e & f \\ g & h & i \end{bmatrix}$

35. $\begin{bmatrix} 1 & -2 & 4 & 5 \\ 3 & 0 & 0 & 2 \\ 0 & 1 & 7 & 0 \\ 0 & 5 & 0 & 2 \end{bmatrix} \begin{bmatrix} r \\ s \\ t \\ u \end{bmatrix}$

36. If

$$A = \begin{bmatrix} 1 & 3 \\ 2 & 6 \end{bmatrix} \quad \text{and} \quad B = \begin{bmatrix} 4 & 7 \\ 5 & 0 \end{bmatrix},$$

determine

a) $A + B$;

b) A^2;

c) B^2;

d) $(A + B)^2$;

e) $A^2 + B^2$.

37. If

$$A = \begin{bmatrix} 9 & 7 \\ 5 & 4 \end{bmatrix} \quad \text{and} \quad B = \begin{bmatrix} 4 & -7 \\ -5 & 9 \end{bmatrix},$$

  a) find AB;
  b) find BA.
  c) Any conclusions?

Find and describe the matrix product in each of the following.

38. $\begin{bmatrix} 2 & 3 \\ 4 & -7 \end{bmatrix}\begin{bmatrix} x \\ y \end{bmatrix}$

39. $\begin{bmatrix} 1 & 0 \\ 0 & 1 \end{bmatrix}\begin{bmatrix} r \\ s \end{bmatrix}$

40. $\begin{bmatrix} 1 & 2 & 3 \\ 4 & 5 & 6 \\ 7 & 8 & 9 \end{bmatrix}\begin{bmatrix} x \\ y \\ z \end{bmatrix}$

41. $\begin{bmatrix} 1 & 0 & 0 \\ 0 & 1 & 0 \\ 0 & 0 & 1 \end{bmatrix}\begin{bmatrix} x \\ y \\ z \end{bmatrix}$

42. $\begin{bmatrix} r & 0 & 0 \\ 0 & s & 0 \\ 0 & 0 & t \end{bmatrix}\begin{bmatrix} x \\ y \\ z \end{bmatrix}$

43. $\begin{bmatrix} 1 & 0 & 0 \\ 0 & 3 & 0 \\ 0 & 0 & 1 \end{bmatrix}\begin{bmatrix} x \\ y \\ z \end{bmatrix}$

44. $\begin{bmatrix} r & 0 & 0 \\ 0 & 3s & 0 \\ 0 & 0 & t \end{bmatrix}\begin{bmatrix} x \\ y \\ z \end{bmatrix}$

45. $\begin{bmatrix} 0 & 1 & 0 \\ 1 & 1 & 0 \\ 0 & 0 & 1 \end{bmatrix}\begin{bmatrix} x \\ y \\ z \end{bmatrix}$

46. $\begin{bmatrix} 0 & s & 0 \\ r & s & 0 \\ 0 & 0 & t \end{bmatrix}\begin{bmatrix} x \\ y \\ z \end{bmatrix}$

47. $\begin{bmatrix} 0 & 0 & 1 \\ 0 & 1 & 0 \\ 1 & 0 & 0 \end{bmatrix}\begin{bmatrix} x \\ y \\ z \end{bmatrix}$

48. Find values for a, b, c, and d:

$$\begin{bmatrix} 2 & 5 \\ 3 & 8 \end{bmatrix}\begin{bmatrix} a & b \\ c & d \end{bmatrix} = \begin{bmatrix} 1 & 0 \\ 0 & 1 \end{bmatrix}.$$

(*Hint:* First find the product on the left side; then form two systems of equations, one in a and c, the other in b and d; then solve.)

	Model type		
	A B C D		
I			
II	Brands		
III			

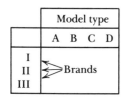

Figure 11.7

49. Inventory for a certain product is given in the form of a matrix by brand and model (see Fig. 11.7):

$$R = \begin{bmatrix} 7 & 3 & 2 & 4 \\ 5 & 8 & 2 & 1 \\ 9 & 3 & 7 & 6 \end{bmatrix},$$

$$S = \begin{bmatrix} 3 & 1 & 1 & 3 \\ 4 & 8 & 2 & 0 \\ 3 & 2 & 7 & 5 \end{bmatrix},$$

where R represents the beginning inventory, and S represents the number of items sold during a sale.

a) Represent by a matrix expression the inventory *after* the sale.

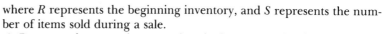

b) How many additional model types should be ordered so the updated inventory after the sale will contain:

14 of each model type for brand I,
9 of each model type for brand II, and
6 of each model type for brand III.

(Represent the quantities to order in matrix form.)

50. Quantities of individual stocks are represented by

$$Q = [17 \quad 30 \quad 400 \quad 25 \quad 140]$$

and costs per share for each of the stocks are represented by

$$P = \begin{bmatrix} 24.625 \\ 37.50 \\ 14.25 \\ 64.125 \\ 58.375 \end{bmatrix}.$$

a) Compute QP.

b) Explain the significance of QP. What does QP represent?

11.5
Solving Simultaneous Systems of Equations Using Matrix Algebra

For large systems of equations with many unknowns, matrix methods are particularly suitable for solution with computers. We will discuss two common methods in this section.

Method 1 makes use of elementary row transformations. This method closely parallels the procedure of solving systems of linear equations by Gaussian elimination, as discussed in Section 10.5.

Method 2 uses the inverse of a matrix. We will demonstrate three separate procedures for finding the inverse of a matrix.

In the following discussion we will first present an illustration with two equations in two unknowns and then three equations in three unknowns. Generalizations for solving n equations with n unknowns ($n > 3$) will then also be made.

Method 1

When we are solving systems of simultaneous linear equations by Gaussian elimination, most of our work is centered on elimination by forming linear combinations.

Part of a solution to a system is illustrated below:

$$\text{Eq. (1)} \quad 2x - 3y = 11,$$
$$\text{Eq. (2)} \quad 4x + 5y = 33$$

Equation (1) will be multiplied by -2 and then added to Eq. (2). This linear combination would eliminate the x-term from Eq. (2). Work continues until the system is in triangular form.

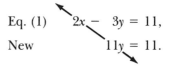

$$\text{Eq. (1)} \quad 2x - 3y = 11,$$
$$\text{New} \quad\quad 11y = 11.$$

Multiplying an equation by some constant and adding two equations (forming linear combinations) are two examples of **elementary transformations.** Our earlier work with solving systems of simultaneous equations can be summarized with the following three "transformations":

1. Two equations may be interchanged.
2. An equation may be multiplied by a nonzero constant.
3. One equation may be added to another.

These transformations have the effect of producing other equivalent systems—systems that have the same solutions as the original system of equations!

A system of equations can also be represented in matrix notation as

$$
\begin{matrix}
2x - 3y = 11, \\
4x + 5y = 33
\end{matrix}
\qquad
\left[
\begin{array}{cc|c}
2 & -3 & 11 \\
4 & 5 & 33
\end{array}
\right].
$$

Joining the column of constants with the coefficient matrix forms the *augmented matrix* on the right above. This augmented matrix is actually a shortened form of the system of equations on the left; each row represents an equation! Analogous to the elementary transformations listed above are the following *row operations* that can be performed on the augmented matrix:

1. Two rows of a matrix may be interchanged.
2. The elements of a row may be multiplied by a nonzero constant.
3. Each element of a row (or multiple of a row) may be added to the corresponding elements of another row.

Transformations on a matrix such as these are called *elementary row operations.*

To solve a system of equations, elementary row operations are performed on the augmented coefficient matrix. The process continues until the augmented matrix has been transformed or reduced to a form called *diagonal form,* as shown below:

$$\begin{bmatrix} 1 & 0 & | & ? \\ 0 & 1 & | & ? \end{bmatrix} \text{ or } \begin{bmatrix} 1 & 0 & 0 & | & ? \\ 0 & 1 & 0 & | & ? \\ 0 & 0 & 1 & | & ? \end{bmatrix} \text{ or } \begin{bmatrix} 1 & 0 & 0 & 0 & | & ? \\ 0 & 1 & 0 & 0 & | & ? \\ 0 & 0 & 1 & 0 & | & ? \\ 0 & 0 & 0 & 1 & | & ? \end{bmatrix}.$$

The coefficient matrix to the left of the vertical bar has been transformed into an **identity matrix.** (An identity matrix has 1's on the principal diagonal and 0's in every other position.)

The elements to the right of the vertical bar—upon completion of the process—will be the solutions to the original system of equations!

Let's demonstrate the process in the following example.

■ **EXAMPLE 17**   The system

$$x - 3y = 8,$$
$$2x + 4y = 6$$

would be written

$$\begin{bmatrix} 1 & -3 & | & 8 \\ 2 & 4 & | & 6 \end{bmatrix}$$

as the augmented coefficient matrix.

$\begin{bmatrix} 1 & -3 & | & 8 \\ 2 & 4 & | & 6 \end{bmatrix}$ Elementary row operations are performed in a manner similar to that used in Gaussian elimination.

$\begin{bmatrix} 1 & -3 & | & 8 \\ 2 & 4 & | & 6 \end{bmatrix}$ The element in the first row, first column is 1, the pivot. We use the pivot to eliminate the 2 in row two, column one.

Multiplying row one by -2 and adding this to row two is the linear combination to use for eliminating the 2. Symbolically, this will be shown as: (-2) R1 + R2. The matrix is transformed into the following:

$\begin{bmatrix} 1 & -3 & | & 8 \\ 0 & 10 & | & -10 \end{bmatrix}$ Now, moving down the principal diagonal, we see that a 1 is needed in row two, column two.

$\begin{bmatrix} 1 & -3 & | & 8 \\ 0 & 1 & | & -1 \end{bmatrix}$ We multiply this row two by the constant 1/10.

$\begin{bmatrix} 1 & -3 & | & 8 \\ 0 & 1 & | & -1 \end{bmatrix}$ Next, we use this newly created 1 as the pivot to eliminate the -3. To do this, we use the linear combination (3) R2 + R1 (that is, three times row two added to row one).

$\begin{bmatrix} 1 & 0 & | & 5 \\ 0 & 1 & | & -1 \end{bmatrix}$

"Translating" these results into the form of equations, we have

$$1x \quad = \quad 5,$$
$$1y = -1.$$

Thus the solution to the original system is $x = 5, y = -1$. $\square$

■ **EXAMPLE 18** Solve

$$3x + y + z = 9,$$
$$3x + y \quad = 7,$$
$$2x \quad + 2z = 12.$$

Solution The augmented matrix is

$$\begin{bmatrix} 3 & 1 & 1 & | & 9 \\ 3 & 1 & 0 & | & 7 \\ 2 & 0 & 2 & | & 12 \end{bmatrix}.$$

$$\begin{bmatrix} 3 & 1 & 1 & | & 9 \\ 3 & 1 & 0 & | & 7 \\ 2 & 0 & 2 & | & 12 \end{bmatrix}$$
A 1 is needed in row one, column one. Noticing that row three has a common factor, let's first divide row three by 2, and then interchange rows one and three.

$$\begin{bmatrix} 1 & 0 & 1 & | & 6 \\ 3 & 1 & 0 & | & 7 \\ 3 & 1 & 1 & | & 9 \end{bmatrix}$$
We then use the 1 as the pivot to eliminate the 3's.

$$\begin{bmatrix} 1 & 0 & 1 & | & 6 \\ 3 & 1 & 0 & | & 7 \\ 3 & 1 & 1 & | & 9 \end{bmatrix}$$
To eliminate the 3 in row two, we use the linear combination (-3) R1 + R2. To eliminate the 3 in row three, we use (-3) R1 + R3.

$$\begin{bmatrix} 1 & 0 & 1 & | & 6 \\ 0 & 1 & -3 & | & -11 \\ 0 & 1 & -2 & | & -9 \end{bmatrix}$$
Now we use the 1 in row two, column two as the new pivot.

Here we are fortunate: The pivot already is a 1. What would we have done had the 1 been something different, say a 5?

$$\begin{bmatrix} 1 & 0 & 1 & | & 6 \\ 0 & 1 & -3 & | & -11 \\ 0 & 1 & -2 & | & -9 \end{bmatrix}$$
We eliminate the 1 in row three, column two by using the linear combination (-1) R2 + R3.

$$\begin{bmatrix} 1 & 0 & 1 & | & 6 \\ 0 & 1 & -3 & | & -11 \\ 0 & 0 & 1 & | & 2 \end{bmatrix}$$
Note that the first two columns are in "finished" form. On to column three. . . .

$$\begin{bmatrix} 1 & 0 & 1 & \vline & 6 \\ 0 & 1 & -3 & \vline & -11 \\ 0 & 0 & \boxed{1} & \vline & 2 \end{bmatrix}$$

Using the last element on the principal diagonal, we see that our new pivot again is already a 1.

Indeed we are lucky to have such a "neat" example. What would we have done had that 1 been perhaps -2?

$$\begin{bmatrix} 1 & 0 & 1 & \vline & 6 \\ 0 & 1 & -3 & \vline & -11 \\ 0 & 0 & \boxed{1} & \vline & 2 \end{bmatrix}$$

Using the 1 as the pivot, we see that the linear combination to eliminate the other 1 in column three is (-1) R3 + R1. To eliminate the -3, we use the combination (3) R3 + R1.

$$\begin{bmatrix} 1 & 0 & 0 & \vline & 4 \\ 0 & 1 & 0 & \vline & -5 \\ 0 & 0 & 1 & \vline & 2 \end{bmatrix}$$

The finished matrix!

The transformed system is then

$$\begin{aligned} 1x & & & = & 4, \\ & 1y & & = & -5, \\ & & 1z & = & 2. \end{aligned}$$

Thus $x = 4$, $y = -5$, and $z = 2$ represent the solutions to the original system of equations. $\square$

● Question

Does this method *always* work?

Answer

Usually, yes. If, however, during the process a row to the left of the bar contains all 0's, then the system does *not* have a unique solution.

Method 2

To begin with, we know that a system of two equations in two unknowns

$$\begin{aligned} 3x + 5y &= 14, \\ 4x + 7y &= 19 \end{aligned}$$

can be rewritten in the following form by using matrices:

$$\begin{bmatrix} 3 & 5 \\ 4 & 7 \end{bmatrix} \begin{bmatrix} x \\ y \end{bmatrix} = \begin{bmatrix} 14 \\ 19 \end{bmatrix}.$$

To verify that this matrix expression is equivalent to the original system of equations, we use the following:

1. By multiplication of matrices,

$$\begin{bmatrix} 3 & 5 \\ 4 & 7 \end{bmatrix} \begin{bmatrix} x \\ y \end{bmatrix} = \begin{bmatrix} 14 \\ 19 \end{bmatrix} \quad \text{becomes} \quad \begin{bmatrix} 3x + 5y \\ 4x + 7y \end{bmatrix} = \begin{bmatrix} 14 \\ 19 \end{bmatrix}.$$

2. By the definition of equality of matrices,

$$\begin{bmatrix} 3x + 5y \\ 4x + 7y \end{bmatrix} = \begin{bmatrix} 14 \\ 19 \end{bmatrix} \quad \text{becomes} \quad \begin{matrix} 3x + 5y = 14, \\ 4x + 7y = 19. \end{matrix}$$

The matrix form for this system of equations

$$\begin{bmatrix} 3 & 5 \\ 4 & 7 \end{bmatrix} \begin{bmatrix} x \\ y \end{bmatrix} = \begin{bmatrix} 14 \\ 19 \end{bmatrix}$$

can be more succinctly stated by using symbolic notation $AX = B$, where

$$A = \begin{bmatrix} 3 & 5 \\ 4 & 7 \end{bmatrix}, \quad X = \begin{bmatrix} x \\ y \end{bmatrix}, \quad \text{and} \quad B = \begin{bmatrix} 14 \\ 19 \end{bmatrix}.$$

Remember, these capital letters *represent matrices!*

The solution to $AX = B$ can then be found by dividing both sides of the equation by A. Yet B/A has no meaning if A and B are matrices, which they are in this case! However, the notion of inverses does exist in matrix algebra.

Let's relate our work to real numbers for a moment. If a is a non-zero real number, then a has a multiplicative inverse a^{-1} with the property that $a^{-1}a = 1$.

A corresponding example from matrix algebra is as follows.

■ **EXAMPLE 19**   For matrix

$$A = \begin{bmatrix} 3 & 5 \\ 4 & 7 \end{bmatrix},$$

the inverse is

$$A^{-1} = \begin{bmatrix} 7 & -5 \\ -4 & 3 \end{bmatrix}.$$

(The procedures for actually finding the inverse will be given later in this section; for now, do not worry about *how* to find the inverse.)

Further,

$$A^{-1}A = \begin{bmatrix} 7 & -5 \\ -4 & 3 \end{bmatrix} \begin{bmatrix} 3 & 5 \\ 4 & 7 \end{bmatrix}$$

$$= \begin{bmatrix} 1 & 0 \\ 0 & 1 \end{bmatrix} = I.$$

This matrix is called the *identity matrix.* It is the matrix counterpart to the identity 1 in the real-number system.

■ **DEFINITION**

IDENTITY MATRIX

A square matrix of n rows and n columns with 1's on the principal diagonal and 0's everywhere else is called an *identity matrix* of order n. This matrix is denoted by I.

● Question

If

$$C = \begin{bmatrix} 1 & 2 \\ 3 & 4 \end{bmatrix},$$

what does IC equal?

Answer

$$IC = \begin{bmatrix} 1 & 0 \\ 0 & 1 \end{bmatrix}\begin{bmatrix} 1 & 2 \\ 3 & 4 \end{bmatrix} = \begin{bmatrix} 1 & 2 \\ 3 & 4 \end{bmatrix}$$

IC does equal C; thus matrix I is indeed the identity matrix for multiplication.

Let's return now to our system of equations in matrix form. To solve for X, then, we use the inverse of matrix A as follows:

$$AX = B$$
$$A^{-1}(AX) = A^{-1}B$$
$$(A^{-1}A)X = A^{-1}B$$
$$IX = A^{-1}B$$
$$X = A^{-1}B.$$

■

The solution set for a system of equations can be found by multiplying the constant matrix B by the inverse of the coefficient matrix, A^{-1}.

We illustrate this technique in the following numerical example.

■ **EXAMPLE 20**

Using $AX = B$ and A^{-1} from Example 19, solve

$$\begin{bmatrix} 3 & 5 \\ 4 & 7 \end{bmatrix}\begin{bmatrix} x \\ y \end{bmatrix} = \begin{bmatrix} 14 \\ 19 \end{bmatrix}.$$

Solution

$$A^{-1}AX = A^{-1}B \qquad \begin{bmatrix} 7 & -5 \\ -4 & 3 \end{bmatrix}\begin{bmatrix} 3 & 5 \\ 4 & 7 \end{bmatrix}\begin{bmatrix} x \\ y \end{bmatrix} = \begin{bmatrix} 7 & -5 \\ -4 & 3 \end{bmatrix}\begin{bmatrix} 14 \\ 19 \end{bmatrix}$$

$$IX = A^{-1}B \qquad \begin{bmatrix} 1 & 0 \\ 0 & 1 \end{bmatrix}\begin{bmatrix} x \\ y \end{bmatrix} = \begin{bmatrix} 7 & -5 \\ -4 & 3 \end{bmatrix}\begin{bmatrix} 14 \\ 19 \end{bmatrix}$$

$$X = A^{-1}B \qquad \begin{bmatrix} x \\ y \end{bmatrix} = \begin{bmatrix} 3 \\ 1 \end{bmatrix}$$

and, by the definition of equality of matrices, $x = 3$ and $y = 1$. □

Got the technique? After reviewing the examples several times and practicing you'll feel like an expert at this procedure. All that remains now is to fill in one important detail: *How do you find the inverse of a matrix?*

Finding the Inverse of a Matrix

Finding the inverse of a matrix is not quite as easy as finding the inverse of a real number. Many sophisticated techniques have been devised, some of which are very involved and complex. For finding matrix inverses of higher orders ($n = 9$, or 80, or 1000), a computer is mandatory. (On some computing systems in which operations with matrices are defined, if X represents a matrix in the computer, the simple program direction MAT Y = INV(X) will produce the inverse of matrix X!) We will show several of these techniques, ranging from a simple shortcut to some of the more elaborate and generalized techniques.

A note of caution before we begin: Not all matrices have inverses. We know that every *nonzero* real number has an inverse. The word *nonzero* has a counterpart in matrix inversion.

■ DEFINITION

A matrix is *invertible* (has an inverse) if and only if its determinant is *not* zero. (A matrix for which an inverse does exist is said to be nonsingular; the determinant of a singular matrix is equal to zero.)

■ **EXAMPLE 21**   Matrix

$$A = \begin{bmatrix} 3 & 4 \\ 5 & 2 \end{bmatrix}$$

does have an inverse; $|A| = -14$.

Matrix

$$B = \begin{bmatrix} 6 & 12 \\ 4 & 8 \end{bmatrix}$$

does *not* have an inverse because the value of the determinant is 0.   □

A Shortcut for Finding the Inverse of a 2-by-2 Matrix
For the matrix

$$\begin{bmatrix} a & b \\ c & d \end{bmatrix}$$

(provided the determinant is not equal to zero), the inverse is obtained by:

1. *interchanging* the elements of the principal diagonal,
2. taking the *negative* of the elements on the secondary diagonal, and
3. *dividing each element* by the determinant of the original matrix.

■ **EXAMPLE 22**

If possible, find the inverse for

$$A = \begin{bmatrix} 8 & -3 \\ 4 & 2 \end{bmatrix}.$$

Solution

We first compute $|A|$. Because $|A| = 28$ and $|A| \neq 0$, an inverse for A *can* be determined.

$$\begin{bmatrix} 2 & \\ & 8 \end{bmatrix}$$

1. Change the main diagonal elements.

$$\begin{bmatrix} 2 & 3 \\ -4 & 2 \end{bmatrix}$$

2. Negate the elements on the secondary diagonal.

$$A^{-1} = \begin{bmatrix} 2/28 & 3/28 \\ -4/28 & 2/28 \end{bmatrix}$$

3. Divide each element by $|A|$.

Check: A times A^{-1} should equal I.

$$\begin{bmatrix} 8 & -3 \\ 4 & 2 \end{bmatrix}\begin{bmatrix} 2/28 & 3/28 \\ -4/28 & 8/28 \end{bmatrix} = \begin{bmatrix} 1 & 0 \\ 0 & 1 \end{bmatrix} \quad \square$$

Now you have all the information you need to solve the system

$$3x + 5y = 14,$$
$$4x + 7y = 19$$

by matrix methods!

Finding the Inverse of a Matrix Using the Adjoint
A formal development of this technique can be found in a more complete textbook on matrix algebra. For our purposes, we will illustrate only the "mechanics" of the procedure.

The inverse of matrix A can be obtained by dividing each element of the adjoint of A by the determinant of A; symbolically,

$$A^{-1} = \text{INV } A = \frac{\text{ADJ } A}{|A|}.$$

But what is the adjoint of a matrix?

■ DEFINITION

> ADJOINT
>
> The *adjoint* of a matrix A (ADJ A) is found by replacing each element of A by its cofactor and then transposing the results.

■ EXAMPLE 23

Let

$$A = \begin{bmatrix} 3 & 1 & 1 \\ 3 & 1 & 0 \\ 2 & 0 & 2 \end{bmatrix}$$

and then replace *each* element by its cofactor:

$$\begin{bmatrix} \begin{vmatrix} 1 & 0 \\ 0 & 2 \end{vmatrix} & -\begin{vmatrix} 3 & 0 \\ 2 & 2 \end{vmatrix} & \begin{vmatrix} 3 & 1 \\ 2 & 0 \end{vmatrix} \\[2mm] -\begin{vmatrix} 1 & 1 \\ 0 & 2 \end{vmatrix} & \begin{vmatrix} 3 & 1 \\ 2 & 2 \end{vmatrix} & -\begin{vmatrix} 3 & 1 \\ 2 & 0 \end{vmatrix} \\[2mm] \begin{vmatrix} 1 & 1 \\ 1 & 0 \end{vmatrix} & -\begin{vmatrix} 3 & 1 \\ 3 & 0 \end{vmatrix} & \begin{vmatrix} 3 & 1 \\ 3 & 1 \end{vmatrix} \end{bmatrix}.$$

Evaluating each cofactor, we obtain:

$$\begin{bmatrix} 2 & -6 & -2 \\ -2 & 4 & 2 \\ -1 & 3 & 0 \end{bmatrix}.$$

Transposing means to *interchange* each row with its column. Row one will now be written as column one, row two will be written as column two, etc.

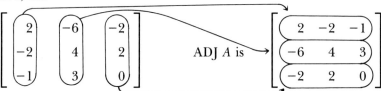

$$\begin{bmatrix} 2 & -6 & -2 \\ -2 & 4 & 2 \\ -1 & 3 & 0 \end{bmatrix} \quad \text{ADJ } A \text{ is} \rightarrow \begin{bmatrix} 2 & -2 & -1 \\ -6 & 4 & 3 \\ -2 & 2 & 0 \end{bmatrix}$$

Finally, in order to find the inverse of A, we must divide each of the elements of the adjoint by the determinant of A.

Using the shortcut for evaluating this 3-by-3 determinant, we have

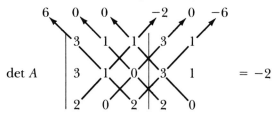

$$\det A = -2.$$

The inverse of A, A^{-1}, is

$$\begin{bmatrix} 2/-2 & -2/-2 & -1/-2 \\ -6/-2 & 4/-2 & 3/-2 \\ -2/-2 & 2/-2 & 0/-2 \end{bmatrix}$$

or

$$\begin{bmatrix} -1 & 1 & 0.5 \\ 3 & -2 & -1.5 \\ 1 & -1 & 0 \end{bmatrix}.$$

Check:

$$A^{-1}A = \begin{bmatrix} -1 & 1 & 0.5 \\ 3 & -2 & -1.5 \\ 1 & -1 & 0 \end{bmatrix} \begin{bmatrix} 3 & 1 & 1 \\ 3 & 1 & 0 \\ 2 & 0 & 2 \end{bmatrix} = \begin{bmatrix} 1 & 0 & 0 \\ 0 & 1 & 0 \\ 0 & 0 & 1 \end{bmatrix}. \quad \square$$

Suggestion. Try this method of using the adjoint to find the inverse of the 2-by-2 example shown on p. 385. Any interesting results?

Finding the Inverse of a Matrix by Gaussian Elimination

This method is a variation of the method discussed in Chapter 10 for solving systems of equations. Let's use the following example as an explanation.

■ **EXAMPLE 24**   Suppose we wish to find the inverse for a matrix A. We write down the matrix A, and adjacent to the matrix, write the identity matrix.

$$\begin{matrix} A & \quad & I \end{matrix}$$
$$\begin{bmatrix} 3 & 0 & 1 \\ 3 & 1 & 0 \\ 2 & 0 & 2 \end{bmatrix} \begin{bmatrix} 1 & 0 & 0 \\ 0 & 1 & 0 \\ 0 & 0 & 1 \end{bmatrix}$$

Now we operate on the rows of matrix A and reduce matrix A to the identity matrix. (This is somewhat similar to the work done by using linear combinations to change a system of equations into triangular form.)

Simultaneously as row operations are performed on matrix A, we perform the *same* operations on the rows of the identity matrix I. After an investment of some time and work, the result shows that by transforming matrix A into the identity matrix I, matrix I is simultaneously transformed into the inverse matrix A^{-1}.

$$\begin{matrix} A & \quad & I \\ \downarrow & & \downarrow \\ I & & A^{-1} \end{matrix}$$

You may want to review Gaussian elimination first, but then let's get started!

$$\begin{bmatrix} 3 & 0 & 1 \\ 3 & 1 & 0 \\ 2 & 0 & 2 \end{bmatrix} \begin{bmatrix} 1 & 0 & 0 \\ 0 & 1 & 0 \\ 0 & 0 & 1 \end{bmatrix}$$

We need a 1 in row one, column one. We divide row three by 2 (the common factor) and then interchange rows one and three.

$$\begin{bmatrix} 1 & 0 & 1 \\ 3 & 1 & 0 \\ 3 & 1 & 1 \end{bmatrix} \begin{bmatrix} 0 & 0 & 1/2 \\ 0 & 1 & 0 \\ 1 & 0 & 0 \end{bmatrix}$$

We use the new 1 in row one, column one as the pivot to eliminate the 3's in positions a_{21} and a_{31} by forming linear combinations.

$$\begin{bmatrix} 1 & 0 & 1 \\ 0 & 1 & -3 \\ 0 & 1 & -2 \end{bmatrix} \begin{bmatrix} 0 & 0 & 1/2 \\ 0 & 1 & -3/2 \\ 1 & 0 & -3/2 \end{bmatrix}$$

We multiply row one by (-3) and add to row two: (-3) R1 + R2. We then multiply row one by (-3) and add to row two: (-3) R1 + R3.

Note: Hereafter, only the symbolic notation will be used to represent the linear combinations.

$$\begin{bmatrix} 1 & 0 & 1 \\ 0 & 1 & -3 \\ 0 & 1 & -2 \end{bmatrix} \begin{bmatrix} 0 & 0 & 1/2 \\ 0 & 1 & -3/2 \\ 1 & 0 & -3/2 \end{bmatrix}$$

Working down the main diagonal, we see that the element in row two, column two already is a 1; we're lucky! We use that 1 as the pivot to eliminate the 1 in position a_{32}. The correct linear combination to accomplish this task is (-1) R2 + R3.

$$\begin{bmatrix} 1 & 0 & 1 \\ 0 & 1 & -3 \\ 0 & 0 & 1 \end{bmatrix} \begin{bmatrix} 0 & 0 & 1/2 \\ 0 & 1 & -3/2 \\ 1 & -1 & 0 \end{bmatrix}$$

The first two columns are already as we wish them to be.

$$\begin{bmatrix} 1 & 0 & 1 \\ 0 & 1 & -3 \\ 0 & 0 & 1 \end{bmatrix} \begin{bmatrix} 0 & 0 & 1/2 \\ 0 & 1 & -3/2 \\ 1 & -1 & 0 \end{bmatrix}$$

On to element a_{33} on the main diagonal. Again, we are lucky that this element is a 1.

$$\begin{bmatrix} 1 & 0 & 1 \\ 0 & 1 & -3 \\ 0 & 0 & 1 \end{bmatrix} \begin{bmatrix} 0 & 0 & 1/2 \\ 0 & 1 & -3/2 \\ 1 & -1 & 0 \end{bmatrix}$$

We use this 1 as the pivot to eliminate both other numbers in column three.

$$\begin{bmatrix} 1 & 0 & 0 \\ 0 & 1 & 0 \\ 0 & 0 & 1 \end{bmatrix} \begin{bmatrix} -1 & 1 & 1/2 \\ 3 & -2 & -3/2 \\ 1 & -1 & 0 \end{bmatrix}$$
$$\quad\uparrow\qquad\qquad\uparrow$$
$$\quad I\qquad\qquad A^{-1}$$

To remove the 1, we use (-1) R3 + R1. To remove the -3, we use (3) R3 + R2.

□

This inverse could be checked to verify that indeed $A^{-1}A = I$. We know, however, that the inverse is correct because this same inverse was also found through the method of using the adjoint (see Example 23).

■

In summary, to solve the matrix equation

$$AX = B,$$

A^{-1} can be found by either

1. the shortcut if A is 2-by-2;
2. use of the adjoint; or
3. a variation of Gaussian elimination.

Then X can be found as follows:

$$AX = B \longrightarrow A^{-1}AX = A^{-1}B$$
$$IX = A^{-1}B$$
$$X = A^{-1}B.$$

Exercise Set 11.5

1. What is the identity matrix for:

a) all 2-by-2 matrices?

b) all 3-by-3 matrices?

c) all 4-by-4 matrices?

d) all n-by-n matrices?

Rewrite each of the following systems of equations in matrix form.

2. $2x + y = 1,$
   $3x + 2y = 0$

3. $3x - 2y = 13,$
   $10x - 7y = 43$

4. $8x + 9y = 36,$
   $7x + 8y = 32$

5. $14x + y = 0,$
   $8x - 2y = 18$

6. $25x - 11y = 600,$
   $-9x + 4y = -212$

7. $1.2x + 1.7y = -11.6,$
   $0.8x + 0.3y = -4.4$

8. $10x + 20y = -36,$
   $30x - 20y = -252$

9. $ax + by = e,$
   $cx + dy = f$

10. For each system in Exercises 2–9:

a) Write the coefficient matrix.

b) Find the inverse matrix for the coefficient matrix.

c) Check your answer by multiplying A^{-1} times A ($= I$?).

d) Use the inverse to solve the system.

Write the transpose for each of the following matrices.

11. $\begin{bmatrix} 1 & 3 \\ 5 & 7 \end{bmatrix}$

12. $\begin{bmatrix} a & b \\ c & d \end{bmatrix}$

13. $\begin{bmatrix} 1 & 3 & 4 \\ -2 & 6 & 5 \\ 8 & 7 & 2 \end{bmatrix}$

14. $\begin{bmatrix} 1 & 2 & 3 & 4 \\ 5 & 6 & 7 & 8 \\ 9 & 10 & 11 & 12 \\ 13 & 14 & 15 & 16 \end{bmatrix}$

15. For matrix

$$M = \begin{bmatrix} 1 & -1 & 1 \\ 4 & 5 & 6 \\ 3 & 2 & -7 \end{bmatrix}:$$

a) Find the determinant of M.

b) Find the cofactor for each of the nine elements.

c) Find the adjoint.

d) Find the inverse of M, M^{-1}.

Find inverses for each of these matrices by the three different methods: (1) shortcut; (2) adjoint; and (3) Gaussian elimination. (*Caution:* Some matrices may be singular!)

16. $\begin{bmatrix} 15 & 4 \\ 11 & 3 \end{bmatrix}$

17. $\begin{bmatrix} 7 & 3 \\ 2 & 1 \end{bmatrix}$

18. $\begin{bmatrix} 7 & -3 \\ -2 & 1 \end{bmatrix}$

19. $\begin{bmatrix} 4 & 7 \\ 6 & 8 \end{bmatrix}$

20. $\begin{bmatrix} 1 & 3 \\ -2 & 4 \end{bmatrix}$

21. $\begin{bmatrix} 6 & 4 \\ 12 & 8 \end{bmatrix}$

22. $\begin{bmatrix} 1 & 6 & -7 \\ -1 & 5 & 2 \\ 1 & 4 & 3 \end{bmatrix}$

23. $\begin{bmatrix} 6 & -1 & 2 \\ -9 & -5 & 3 \\ 3 & 0 & 3 \end{bmatrix}$

24. $\begin{bmatrix} 1 & 1 & 2 \\ 2 & 3 & 1 \\ -1 & -2 & 3 \end{bmatrix}$

25. $\begin{bmatrix} 2 & -2 & 1 \\ -3 & 4 & -2 \\ 11 & -13 & 7 \end{bmatrix}$

26. $\begin{bmatrix} 3 & 3 & 3 & 2 \\ 3 & 2 & 2 & 2 \\ 3 & 2 & 1 & 1 \\ 2 & 2 & 1 & 0 \end{bmatrix}$

27. $\begin{bmatrix} 1 & 2 & 3 & 4 \\ 5 & 6 & 7 & 8 \\ 8 & 7 & 6 & 5 \\ 4 & 3 & 2 & 1 \end{bmatrix}$

Using some of the inverses obtained in Exercises 16–27, solve these equations by means of writing the system in matrix form $AX = B$, and solving for X using A^{-1}.

28.
$$x + 6y - 7z = 10,$$
$$-x + 5y + 2z = 7,$$
$$x + 4y + 3z = 14$$

29.
$$2x - 2y + z = 4,$$
$$-3x + 4y - 2z = -4,$$
$$11x - 13y + 7z = 16.5$$

30.
$$2x + 3y + z = 49,$$
$$-x - 2y + 3z = -43,$$
$$x + y + 2z = 8$$

11.6
Operations with Matrices on the Computer

As collections or tables of elements with a common attribute, *arrays* are very useful in data processing. Most commonly used computer languages, such as BASIC, COBOL, PL/1, and FORTRAN, have techniques for storing and manipulating arrays. Although each programming language has its own syntax for handling arrays, there are many common characteristics in *how* an array is handled by a computer, as we show in the following discussion.

First, when working with arrays in most computer languages, the computer must know—before encountering an array in a program—the *name of the array* and the *number of elements in the array*. The number of storage positions needed must be known in advance.

In BASIC and in FORTRAN, a dimension statement is used to create an array. In the few lines of the BASIC program below, an array called *IQ* with 12 locations, 2 rows of 6 columns, is created; and an array *R* with 4 locations is created. (In BASIC, dimension is abbreviated DIM.)

```
 5 . . . . . . . . .
10 DIM IQ(2,6), R(4)
15 . . . . . . . . . .
```

In PL/1 the equivalent statement to create and reserve storage positions is

```
DECLARE IQ(2,6);
DECLARE R(4);
```

In COBOL, the language used most often in business data processing environments, the OCCUR clause is used to create arrays.

For the array

$$A = \begin{bmatrix} a_{11} & a_{12} & a_{13} \\ a_{21} & a_{22} & a_{23} \end{bmatrix} = \begin{bmatrix} 1 & 3 & 7 \\ 9 & 2 & 4 \end{bmatrix},$$

the elements of the array are accessed by the capital letter representing the matrix, A, followed by the subscripts written as an ordered pair. For example, to refer to the element 4 in matrix A, computer reference would be made by using $A(2,3)$, the element in the second row, third column.

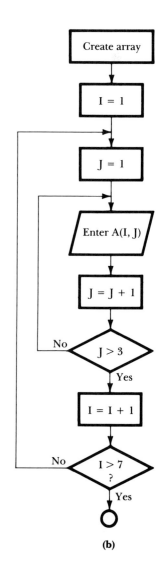

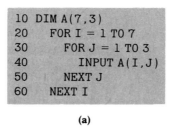

```
10  DIM A(7,3)
20     FOR I = 1 TO 7
30        FOR J = 1 TO 3
40           INPUT A(I,J)
50        NEXT J
60     NEXT I
```

(a)   (b)

Figure 11.8

A program in BASIC and its accompanying flowchart to illustrate placement of elements into an array is shown in Fig. 11.8.

In the pseudocode program shown in Fig. 11.9(a), used to illustrate how to average daily sales, the values would be handled in an array called SALES and would appear as shown in Fig. 11.9.

A small BASIC program for the addition of two matrices is shown in Fig. 11.10.

One final example, written in pseudocode, illustrates the process of multiplying matrices (Fig. 11.11). To multiply two matrices ($C = AB$), the formula presented earlier to compute each element c_{ij} in the product C was

$$c_{ij} = \sum_{k=1}^{n} (a_{ik})(b_{kj}).$$

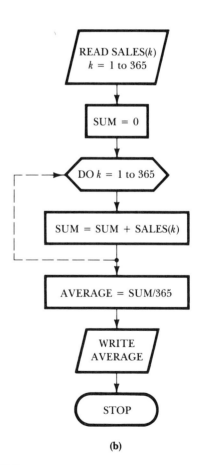

```
read SALES(k), k = 1 TO 365
SUM = 0
DO  k = 1 TO 365
    SUM = SUM + SALES(k)
ENDDO
    AVERAGE = SUM/365
write AVERAGE
END
```

(a)   (b)

Figure 11.9

You may wish to follow the procedure in the flowchart (Fig. 11.11b) by working on the following two matrices:

$$A = \begin{bmatrix} 3 & 2 & 5 \\ 4 & 5 & 1 \end{bmatrix}, \quad B = \begin{bmatrix} 1 & 7 & 8 & 9 \\ 3 & 9 & 6 & 0 \\ 5 & 4 & 2 & 7 \end{bmatrix}.$$

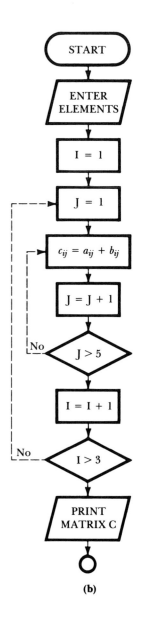

```
10 DIM A(3,5), B(3,5), C(3,5)
15   . . .⎫
     . . .⎬ enter elements
65   . . .⎭
70 FOR I = 1 TO 3
75   FOR J = 1 TO 5
80     C(I,J) = A(I,J) + B(I,J)
85   NEXT J
90 NEXT I
120   . . .⎫
      . . .⎬ print matrix C
150   . . .⎭
```

(a)

(b)

Figure 11.10

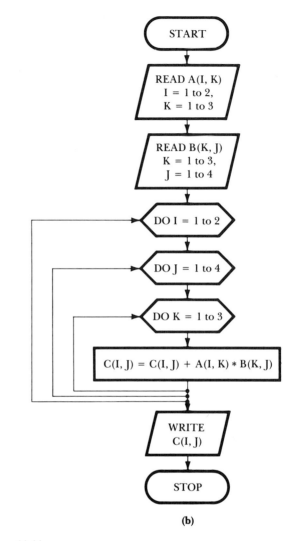

```
read A(I,K), I = 1 to 2, K = 1 to 3
read B(K,J), K = 1 to 3, J = 1 to 4
DO I = 1 to 2
 DO J = 1 to 4
  DO K = 1 to 3
  C(I,J) = C(I,J) + A(I,K) * B(K,J)
  ENDDO
  ENDDO
ENDDO
write C(I,J)
stop
```

(a)   (b)

Figure 11.11

Additional computer programs dealing with matrices could include ordering elements in an array, finding the value of a determinant, and finding the inverse of a matrix. Programs to solve systems of simultaneous equations can be found in most beginning computer language textbooks.

Summary

In order to facilitate the processing of mass quantities of data, the individual items can be stored or manipulated as parts of a single quantity called an *array*. An example might be an investment grid or table, where each line represents a stock followed by the 12 monthly price increases.

The elements in an array are arranged in rows and columns. A *matrix* is a rectangular array. The size and type of a matrix are given by denoting the number of rows and the number of columns: A 2-by-3 matrix contains 2 rows and 3 columns. If the number of rows equals the number of columns, then the matrix is a *square matrix*. A matrix with only one row is called a *row vector;* a matrix with only one column is called a *column vector*.

Generally, matrices are denoted by capital letters and elements within a matrix are denoted by subscripted letters:

$$A = \begin{bmatrix} a_{11} & a_{12} & a_{13} \\ a_{21} & a_{22} & a_{23} \\ a_{31} & a_{32} & a_{33} \end{bmatrix}.$$

Since we can regard a matrix as a set, the operations of addition, scalar multiplication, and subtraction may be performed on matrices. Addition and subtraction are performed by adding, or subtracting, elements in corresponding positions. Matrix multiplication is more complex and is defined differently. In essence, an element in the multiplication of two matrices $(C = AB)$ can be defined by the formula

$$c_{ij} = \sum_{k=1}^{p} (a_{ik})(b_{kj}).$$

In order for the product of two matrices to be defined, the matrices must be *conformable*.

For smaller systems of equations, an extremely useful method is Cramer's rule. Cramer's rule uses determinants, a *determinant* being an assigned number associated with each square matrix. In addition to their use in solving systems of simultaneous equations, determinants are invaluable in the analysis of the type of system. If the determinant of the coefficients of a system of equations is not zero, then and only then is the system *independent* (that is, a unique solution is possible).

Matrices are important in solving larger systems of simultaneous equations. Two methods were shown in this chapter to augment the number of methods given in Chapter 10. The first method relies heavily on the concept of *elementary row transformations* to solve systems. With this method the augmented coefficient matrix is transformed into the identity matrix together with a column representing the solutions to the original system of equations.

The second method uses the *inverse* of the coefficient matrix. The system to be solved can be represented in the form $AX = B$, where each letter denotes a matrix: A the coefficient matrix, X the variable matrix, and B the matrix of constants. The matrix expression $AX = B$ is then solved for X as follows:

$$AX = B$$
$$A^{-1}AX = A^{-1}B$$
$$X = A^{-1}B.$$

Several procedures were shown for finding the inverse of a matrix. By using the *adjoint* or by using a variation of Gaussian elimination, inverses can be found for quite large matrices. However, as the dimension of the matrix increases, the use of computers becomes much more desirable—if not totally necessary!

Review Exercises

1. Given

$$A = \begin{bmatrix} 1.3 & -2 & 7 & 5.1 \\ 0 & -3 & 8 & 1 \\ 4.2 & 6 & -9 & 4 \end{bmatrix}$$

identify these elements.

a) $a_{13} =$ _____ b) $a_{21} =$ _____

c) $a_{24} =$ _____ d) $a_{32} =$ _____

e) $a_{23} =$ _____ f) $a_{31} =$ _____

For each of the following matrices, (a) give the order of the matrix, and (b) identify the matrix as either a row vector, a column vector, a square matrix, or none of these.

2.
$$A = \begin{bmatrix} 7 & 4 \\ 6 & 8 \\ 2 & 1 \\ 3 & 9 \end{bmatrix}$$

3. $B = \begin{bmatrix} 1 & -3 & 2 & -5 \end{bmatrix}$

4. $C = \begin{bmatrix} 1 & 3 \\ 2 & -4 \end{bmatrix}$

5. $D = \begin{bmatrix} 1 & -3 & 2 & 7 \\ 4 & 6 & -5 & 0 \end{bmatrix}$

6.
$$E = \begin{bmatrix} 1 & 0 & 0 \\ 0 & 1 & 0 \\ 0 & 0 & 1 \end{bmatrix}$$

7. a) Using a, b, c, d, e, f, g, and h to represent the matrix elements, construct a matrix with dimension 4-by-2.

   b) How many possible matrices *with different dimensions* can be constructed using these eight elements?

8. Complete each of the following, given that A and B represent two matrices of the same order.

   a) A equals B if _____

   _____

   b) $A + B$ is obtained by _____

   _____

   c) $A - B$ is obtained by _____

   _____

Solve for the unknowns in each of the following.

9. $\begin{bmatrix} 3 & r & t \\ 6 & 7 & u \end{bmatrix} = \begin{bmatrix} s & 9 & -2 \\ 6 & v & 8 \end{bmatrix}$

10. $\begin{bmatrix} x+2 & 4 \\ 5 & 7 \\ x-2 & x-1 \end{bmatrix} = \begin{bmatrix} 9 & 4 \\ 5 & 7 \\ 5 & 6 \end{bmatrix}$

Perform the indicated matrix operations.

11. $\begin{bmatrix} 3 & -2 \\ 6 & 4 \end{bmatrix} + \begin{bmatrix} -2 & 2 \\ -6 & -3 \end{bmatrix}$

12. $\begin{bmatrix} 6 & 3 \\ -2 & 4 \end{bmatrix} + \begin{bmatrix} 0 & 0 \\ 0 & 0 \end{bmatrix}$

13. $2\begin{bmatrix} 3 & 2 & 5 \\ -1 & 6 & 0 \end{bmatrix} - 3\begin{bmatrix} 4 & -1 & 2 \\ 3 & -9 & 8 \end{bmatrix}$

14. $\begin{bmatrix} 1 & 8 & 2 \\ 7 & 6 & 4 \\ 5 & 2 & 9 \end{bmatrix} + \begin{bmatrix} 0 & 0 & 0 \\ 0 & 0 & 0 \\ 0 & 0 & 0 \end{bmatrix} - \begin{bmatrix} 3 & -2 & -4 \\ -5 & -6 & 8 \\ -5 & -2 & -9 \end{bmatrix}$

15. Given that

$$A = \begin{bmatrix} 3 & -2 \\ 4 & -5 \end{bmatrix}, \quad B = \begin{bmatrix} 4 & 2 \\ 6 & -4 \end{bmatrix}, \quad \text{and} \quad C = \begin{bmatrix} 6 & -2 \\ 3 & -4 \end{bmatrix}.$$

   a) Find $(A + B) + C$.

   b) Find $(B + A) + C$.

   c) Find $B + (A + C)$.

   d) Does $(A + B) + C$ equal $B + (A + C)$?

Perform the indicated multiplications, if the matrices are *conformable*.

16. $\begin{bmatrix} 3 & 2 \\ 6 & 4 \end{bmatrix}\begin{bmatrix} 1 \\ 1 \end{bmatrix}$

17. $\begin{bmatrix} 2 & 3 & -1 \\ 4 & 0 & 5 \end{bmatrix}\begin{bmatrix} -3 \\ 2 \\ 6 \end{bmatrix}$

18. $\begin{bmatrix} 2 & 3 & -1 \\ 4 & 0 & 5 \end{bmatrix}\begin{bmatrix} 2 & 6 \\ 1 & 0 \end{bmatrix}$

19. $\begin{bmatrix} -3 & 2 & 5 \\ 4 & 6 & -1 \\ 7 & 0 & 1 \end{bmatrix}\begin{bmatrix} 3 & -2 \\ 6 & 5 \\ 8 & 9 \end{bmatrix}$

20. $\begin{bmatrix} 1 & 1 \\ 0 & 1 \end{bmatrix}\begin{bmatrix} x \\ y \end{bmatrix}$

21. $\begin{bmatrix} 1 & 0 & 0 \\ 0 & 2 & 0 \\ 0 & 0 & 3 \end{bmatrix}\begin{bmatrix} x \\ y \\ z \end{bmatrix}$

22. $\begin{bmatrix} 1 \\ 2 \\ 3 \\ 4 \end{bmatrix}[5 \quad 6 \quad 7 \quad 8]$

23. $[1 \quad 2 \quad 3 \quad 4]\begin{bmatrix} 5 \\ 6 \\ 7 \\ 8 \end{bmatrix}$

24. $\begin{bmatrix} 1 & 0 \\ 0 & 1 \end{bmatrix}\begin{bmatrix} x \\ y \end{bmatrix}$

25. $\begin{bmatrix} 0 & 1 \\ 1 & 0 \end{bmatrix}\begin{bmatrix} x \\ y \end{bmatrix}$

26. $\begin{bmatrix} 1 & 1 \\ 1 & 1 \end{bmatrix}\begin{bmatrix} x \\ y \end{bmatrix}$

27. $\begin{bmatrix} 1 & 2 & -3 & 1 \\ 2 & -4 & -1 & -3 \\ -1 & -1 & -1 & -1 \\ 1 & -2 & 3 & 2 \end{bmatrix}\begin{bmatrix} x \\ y \\ z \\ w \end{bmatrix}$

Find the value of each of the following determinants.

28. $\begin{vmatrix} 2 & 7 \\ 6 & 4 \end{vmatrix}$

29. $\begin{vmatrix} 3 & -2 \\ 8 & -4 \end{vmatrix}$

30. $\begin{vmatrix} 1 & 3 & -2 \\ 6 & 4 & 5 \\ 1 & 0 & 1 \end{vmatrix}$

31. $\begin{vmatrix} 2 & 0 & 0 \\ 0 & 4 & 0 \\ 0 & 0 & 3 \end{vmatrix}$

32. $\begin{vmatrix} 2 & -6 & 5 \\ 0 & 4 & 1.8 \\ 0 & 0 & 3 \end{vmatrix}$

33. $\begin{vmatrix} 4 & 1 & 3 \\ 9 & 0 & 2 \\ 7 & -2 & 5 \end{vmatrix}$

34. $\begin{vmatrix} 2 & 0 & 0 & 0 \\ 5 & 4 & 0 & 0 \\ -2 & 5 & -1 & 0 \\ 8 & 6 & 2 & 3 \end{vmatrix}$

35. $\begin{vmatrix} 1 & 4 & 3 \\ 0 & 9 & 2 \\ -2 & 7 & 5 \end{vmatrix}$

36. Solve each system of simultaneous equations in Exercises 46–52 using Cramer's rule (*determinants*).

Derive a system of equations from each of the following matrix expressions.

37. $\begin{bmatrix} 3 & -2 \\ 7 & 6 \end{bmatrix}\begin{bmatrix} x \\ y \end{bmatrix} = \begin{bmatrix} 10 \\ 22 \end{bmatrix}$

38. $\begin{bmatrix} 0.1 & 0.3 \\ 3 & -1 \end{bmatrix}\begin{bmatrix} x \\ y \end{bmatrix} = \begin{bmatrix} 5 \\ 0 \end{bmatrix}$

39. $\begin{bmatrix} 1 & -1 & 2 \\ 1 & -3 & 3 \\ 1 & 2 & -3 \end{bmatrix}\begin{bmatrix} r \\ s \\ t \end{bmatrix} = \begin{bmatrix} 7 \\ -1 \\ 5 \end{bmatrix}$

40. $\begin{bmatrix} 2 & -2 & 3 \\ 1 & 0 & -2 \\ 1 & 0 & 2 \end{bmatrix}\begin{bmatrix} x \\ y \\ z \end{bmatrix} = \begin{bmatrix} -5 \\ 3 \\ 1 \end{bmatrix}$

41. For the matrix

$$M = \begin{vmatrix} 1 & 3 & -2 \\ 1 & 2 & -1 \\ 2 & 1 & 3 \end{vmatrix}:$$

  a) Find the determinant $|M|$.

  b) Find the adjoint.

  c) Find the inverse, M^{-1}.

Using the variation of Gaussian elimination, find the inverse of each of the following matrices.

42. $\begin{bmatrix} 1 & 8 \\ 7 & 35 \end{bmatrix}$

43. $\begin{bmatrix} 1 & 3 & -2 \\ 1 & 2 & -1 \\ 2 & 1 & 3 \end{bmatrix}$

44. $\begin{bmatrix} 1 & 2 & -3 & 1 \\ 1 & -2 & 3 & 2 \\ 2 & -4 & -1 & -3 \\ 1 & 1 & 1 & 1 \end{bmatrix}$

45. $\begin{bmatrix} 8 & 7 \\ 2 & -3 \end{bmatrix}$

For each of the following systems, write the coefficient matrix, find the inverse of the coefficient matrix, and then solve the system by using the inverse.

46. $x - 2y = 3,$
$5x - 12y = 5$

47. $23x + 9y = 40,$
$5x + 2y = 10$

48. $x + 4y - 5z = -10,$
$2x + 7y - z = -9,$
$x + 3y + 4z = 1$

49. $x - y + 2z = 2,$
$3x + y - z = -1,$
$x - 2y - 2z = -3$

Using the inverses of the matrices obtained in Exercises 42–45, solve these systems using matrix algebra.

50. $x + 8y = 10,$
$7x + 35y = 69$

51. $x + 3y - 2z = 2,$
$x + 2y - z = 2,$
$2x + y + 3z = 6$

52. $r + 2s - 3t + u = -3,$
$r - 2s + 3t + 2u = 7,$
$2r - 4s - t - 3u = -35,$
$r + s + t + u = 17$

Linear Programming

CHAPTER 12 OBJECTIVES

After completing this chapter, you should be able to:

1. Find and graph the common solution set to a system of linear inequalities.
2. Give a good description of what linear programming is, and how it may be used.
3. Define and use correctly the terms related to linear programming, such as objective function, constraints, feasible region, optimal solution, half-plane, tableau, key variable, pivotal equation, pivot, and slack variable.
4. Read a linear programming problem and formulate the objective function and the necessary constraints.
5. Solve a linear programming problem by each of these methods: (a) the graphical method (when possible), (b) the algebraic method, (c) the algebraic method in matrix form, and (d) the simplex method.

12

■ CHAPTER OUTLINE

12.1

Inequalities

The concepts on graphing presented in Chapter 10 on systems of equations can be applied to solve some elementary problems in a fairly modern subject called *linear programming*. A good understanding of graphing systems of equations and/or inequalities is a prerequisite for learning about linear programming. The graphing of systems also provides an intuitive underpinning for the more complex methods of linear programming such as the simplex method, presented later in the chapter. You may first wish to review Chapter 10 on graphing systems of equations; the following discussion concerns graphing inequalities and systems of inequalities.

An equation in the two variables x and y represents a straight line. This *linear equation*, $ax + by = c$ (where both constants a and b cannot be zero at the same time), divides, or partitions, the plane into three subsets:

1. the set of all points (x, y) whose coordinates satisfy $ax + by = c$;
2. the set of all points (x, y) whose coordinates satisfy the inequality $ax + by < c$, and
3. the set of all points (x, y) whose coordinates satisfy the inequality $ax + by > c$.

Although subset (1) represents a straight line, the subsets of pairs representing $ax + by < c$ and $ax + by > c$ represent *open half-planes*.

Let's look at the following example.

■ **EXAMPLE 1**   The equation $x + y = 5$ represents a straight line. Each of the points on the line satisfies the condition that the sum of the x-coordinate and the y-coordinate equals 5. The graph of $x + y = 5$ is shown in Fig. 12.1. (One way to graph the line is to use the x- and y-intercepts, (5, 0) and (0, 5).) □

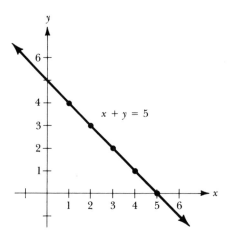

Figure 12.1

What about points that are *not* on the line? Figure 12.2 shows several points that *do not* lie on the line represented by $x + y = 5$. These points each have the common property that the sum of their x-coordinate and y-coordinate is *less than* 5. As a matter of fact, it can be shown that all points in the shaded region satisfy the *linear inequality* $x + y < 5$. (*Note:* The points on the line do not satisfy the inequality $x + y < 5$; a dashed line is used to indicate that these points are *not* included with the shaded area.)

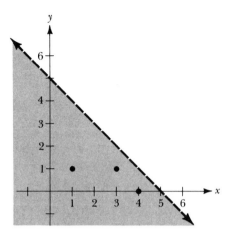

Figure 12.2

What mathematical inequality would describe all the points in the shaded area of Fig. 12.3? (Try picking several points and adding their x-coordinate and y-coordinate.) The two regions on either side of the line $x + y = 5$ represent the two open half-planes, $x + y < 5$ and $x + y > 5$ (see Fig. 12.4).

■

> In general, a linear equation, when graphed, divides the xy-plane into three subsets: (1) all the points *on* the line, (2) all the points *below* the line, and (3) all the points *above* the line.

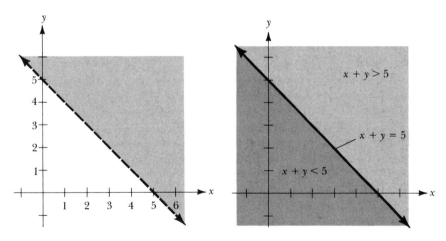

Figure 12.3   **Figure 12.4**

To correctly associate each half-plane with the correct inequality, the procedure is simple. Let's arbitrarily choose a point that does not belong on the line $x + y = 5$, say, $(0, 0)$. We check to see which inequality is satisfied by $(0, 0)$. The point $(0, 0)$ does satisfy $x + y < 5$. Thus, this region containing the chosen point represents the inequality $x + y < 5$. And as you have guessed, by process of elimination, the inequality $x + y > 5$ is represented by the other open half-plane.

While $ax + by < c$ represents an open half-plane, combining $ax + by = c$ and $ax + by < c$ to form $ax + by \leq c$ (or $ax + by = c$ and $ax + by > c$ to form $ax + by \geq c$) represents a *closed half-plane*. To distinguish between open and closed half-planes, the following convention is used. For open half-planes, the dividing line ($ax + by = c$) is a dotted line, whereas to represent closed half-planes, the dividing line is a solid line to indicate that the line is part of the solution set.

■ EXAMPLE 2

Graph the solution set to the system of inequalities

$$x - 4y \leq 4,$$
$$y \leq 2 - \tfrac{2}{3}x,$$
$$x \geq 0.$$

Solution

We begin by graphing each inequality separately. The line $x - 4y = 4$ can be plotted using the intercept method. If $x = 0$, then $y = -1$, or $(0, -1)$. If $y = 0$, then $x = 4$, or $(4, 0)$. The graph of this equation is a solid line, as shown in Fig. 12.5.

If we test a point, say $(0, 0)$, we see that $(0, 0)$ satisfies the inequality $x - 4y \leq 4$. Thus the region above the line is shaded in Fig. 12.6.

You may have noticed that the inequality $y \leq 2 - \tfrac{2}{3}x$ is close to the form $y = mx + b$, where m represents the slope and b the y-intercept. If

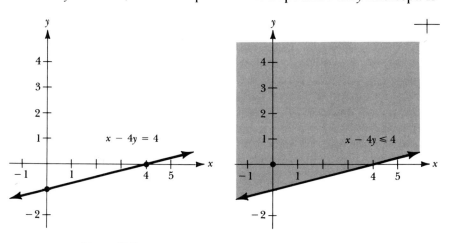

Figure 12.5   **Figure 12.6**

we rewrite the equation as $y = -\frac{2}{3}x + 2$, we know that the coefficient of x is the slope $-\frac{2}{3}$ and the constant term is the y-intercept (2).

Beginning at the y-intercept (0, 2) and using the slope $-2/+3$, we can plot additional points to graph the line. Testing the point (0, 0), we have

$$y < -\tfrac{2}{3}x + 2$$
$$0 \;?\; -\tfrac{2}{3}(0) + 2$$
$$0 \;?\; 0 + 2$$
$$0 < 2.$$

Thus the corresponding region is shaded below the line, as shown in Fig. 12.7.

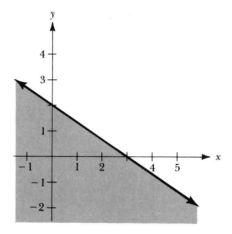

Figure 12.7

For $x \geq 0$, all points are shaded to the right of the y-axis, which is also included (see Fig. 12.8).

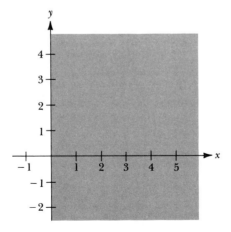

Figure 12.8

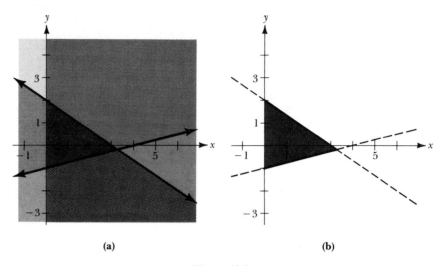

(a)   (b)

Figure 12.9

The solution set to the system of inequalities

$$x - 4y \le 4,$$
$$y \le 2 - \tfrac{2}{3}x,$$
$$x \ge 0$$

is the *intersection* of all three shaded regions. The three separate regions are superimposed on the same set of axes in Fig. 12.9(a), and the final solution set is shaded in Fig. 12.9(b). □

This common intersection in linear programming problems is referred to as the *feasible region*. Any point in the feasible region has the property that the coordinates of the point simultaneously satisfy each inequality in the given set.

■ **EXAMPLE 3**   (*A business example.*) To manufacture two items A and B, the cost for materials per item is $30 and $20, respectively. Total material costs cannot exceed $12,000. Labor costs are $80 for each A item and $40 for each B item. These labor costs must be less than or equal to $16,000 per union contract. What are some possible points (A, B) that meet these requirements?

Solution   For material costs the inequality is

$$30A + 20B \le 12,000.$$

For labor costs the inequality is

$$20A + 80B \le 16,000.$$

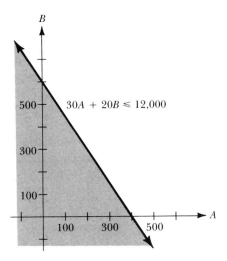

Figure 12.10

We will sketch each inequality separately, and then for the final solution, we will superimpose the two regions on the same set of axes.

For material costs, $30A + 20B \leq 12,000$, we first sketch the line $30A + 20B = 12,000$. Using intercepts, we see that if $A = 0$, then $B = 600$, and if $B = 0$, then $A = 400$. The line will be solid because material costs can equal but not exceed 12,000. Testing a point on one side of the line, say $(0, 0)$, we see that $30(0) + 20(0) < 12,000$. Therefore, the solution set for this inequality is shaded in Fig. 12.10.

Next we will work with the inequality for labor costs, $20A + 80B \leq 16,000$. The intercepts for plotting the line are $(800, 0)$ and $(0, 200)$. The solution set for the inequality is the shaded region in Fig. 12.11. (Do you see how this region was obtained?)

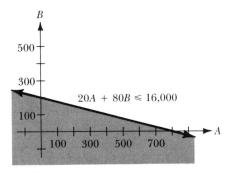

Figure 12.11

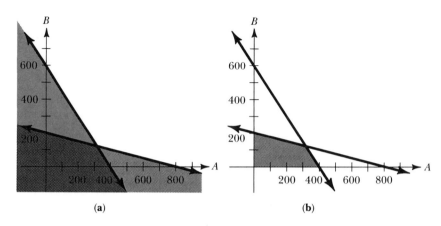

(a)   (b)

Figure 12.12

To satisfy *both* inequalities, or the system of inequalities,

$$30A + 20B \leq 12{,}000,$$
$$20A + 80B \leq 16{,}000,$$

we must find the *intersection* of the two different solution sets. This common solution set is shown simultaneously in Fig. 12.12(a) by the two shaded regions.

Because the problem requires that $A \geq 0$ and $B \geq 0$, we restrict our attention to the points in quadrant I. The solution set to the system of inequalities

$$30A + 20B \leq 12{,}000,$$
$$20A + 80B \leq 16{,}000,$$
$$A \geq 0, B \geq 0,$$

is illustrated in Fig. 12.12(b). Any of the points in the darkest shaded region will simultaneously satisfy the system of inequalities. (Later in the chapter, we will show how to select the "best" point from this shaded area.)   □

Exercise Set 12.1

Graph each of the following linear inequalities by shading the appropriate region.

1. $x + y > 4$ **2.** $x - y \geq 5$

3. $y \geq 2x - 1$ **4.** $x \leq 3$

5. $y \geq 2$ **6.** $2x - 4 < y$

7. $2x - 3y > 12$ **8.** $y > x$

9. $x - 2y - 4 \leq 0$ 10. $3x - 2y < 2x + 4y - 12$

11. $0.16x - 0.24y > 4.8$ 12. $3y < 6x - 12$

Graph the half-planes that are associated with each of the following linear inequalities.

13. $x + y \leq 6$ 14. $y < 2x + 3$ 15. $y \geq 2x + 3$

16. $3x - 2y \leq -6$ 17. $x \leq -2$ 18. $3x - 2y \geq 18$

19. $y > 2$ 20. $x + 2y \leq 7$ 21. $1.3x - 2.6y \geq 3.9$

Determine graphically the region that represents the solution set described by each of the following sets of inequalities.

22. $x + y < 6,$
$\quad x - 3y > -6$

23. $x - y \leq 4,$
$\quad 2x + y > 2$

24. $x - 2y < -1,$
$\quad 3x + y > 4$

25. $y < 3x + 2,$
$\quad x > 0,$
$\quad y > 2$

26. $x + y \leq 6,$
$\quad x - y \leq 6$

27. $3x + 5y \leq 15,$
$\quad x + y \leq 8$

28. $y < 4,$
$\quad x > -2,$
$\quad x - y < 0$

29. $x + y \leq 6,$
$\quad y \geq x + 1$

30. $x + y \leq 6,$
$\quad x - y \leq 6,$
$\quad x \geq 0, y \geq 0$

31. $x \geq 0, y \geq 0,$
$\quad 2x + 3y \geq 12,$
$\quad 2x + y \geq 8$

32. $2x - 3y \leq 6,$
$\quad x + y \leq 4,$
$\quad y \geq -x$

33. $x \geq 0, y \geq 0$
$\quad 2x + y \leq 14,$
$\quad x + 6y \geq 12,$
$\quad 3x + 2y \leq 18$

34. $30x + 20y \leq 12{,}000,$
$\quad 20x + 80y \leq 16{,}000,$
$\quad x \geq 0, y \geq 0$

35. $4x + 3y \leq 48,$
$\quad 12x + 16y \geq 48,$
$\quad x \geq 0, y \geq 0$

36. $x \geq 0, y \geq 0,$
$\quad x + 2y \leq 50,$
$\quad x - 3y \geq -25$

37. $x \geq 0, x \leq 100,$
$\quad y \geq 0, y \leq 400,$
$\quad 2x + 5y \geq 300,$
$\quad x + y \leq 400,$
$\quad 5x - y \geq -200$

38. What would be the solution set to the intersection of all the points whose coordinates satisfy simultaneously $ax + by < c$ and $ax + by > c$?

39. Sketch the common intersection of this set of inequalities:

$$4x + y \geq 4,$$
$$x - y \geq -1,$$
$$x + 3y \geq 4.$$

How do the graph representing this set of inequalities and the graph in Exercise 31 differ from the graphs in Exercises 22–37?

40. Find the intersection of the individual solution sets for each of the following inequalities.

a) $9A + 3B \geq 18$ **b)** $8A + 8B \geq 20$

c) $A + 10B \geq 10$ **d)** $A \geq 0, B \geq 0$

41. Find the solution set for the following problem:

Two items are produced, A and B. The total number of items must be less than 50 per week. The amount of material needed to produce A is always at least twice the amount as that to produce B.

Find the inequalities, graph each, and shade the appropriate region.

42. For the set of inequalities

$$4x + y \geq 4,$$
$$x - y \geq -4,$$
$$x + 3y \geq 4,$$
$$x + 2y \leq 12,$$
$$y \geq 0,$$

an enclosed five-sided region is formed. Make a sketch of this region, and then determine the coordinates of each vertex of this closed region.

12.2
Linear Programming: An Introduction and Practical Examples

In contrast to most other developments in mathematics, linear programming has gained an increased importance and application in a relatively short period of time. The early history of linear programming can be traced from the development of economic theories in the 1930s by John von Neumann and W. Leontief. In the 1940s many problems arose concerning the best allocation of resources to meet certain objectives. Specifically, how to allocate supplies for the armed services in the least expensive way was a problem facing George B. Dantzig. In 1947 Dantzig formulated the general linear programming problem and developed a method called the *simplex method* as one means of solving linear programming problems. Since then, linear programming has become one of the most useful topics in mathematics, finding applications in such diverse areas as economics, engineering, agriculture, management, transportation, nutrition, business, and industry. Saul Gass' *Linear Programming* contains many illustrations of the wide uses of linear programming.

Many mathematical models for practical application involve either *maximizing* or *minimizing* a given function. Using profit as an example, why couldn't we just continue increasing production, thereby maximizing profit? Often there are restrictions or constraints on time and money; supplies and materials are usually limited; unions may limit total hours of production per week. These restrictions introduce an extra concern as we attempt to maximize our profit. Often in production, a point of diminishing returns is reached. Further investments of time and/or materials may actually cause a decline in profit.

■

> Linear programming supplies the solution of finding the best or optimal solution possible subject to certain given restrictions or constraints.

Linear programming serves to either maximize or minimize a given linear function (called the objective function) subject to a set of linear constraints (which are expressed as linear equations or inequalities). *Maximizing profit* subject to constraints on production materials and *minimizing advertising costs* subject to constraints on time are examples of linear programming problems.

A Sample Linear Programming Problem (An Overview)

To illustrate the need for linear programming as a method of solution, the following simple problem is presented—simple, but common to many businesses today.

■ **EXAMPLE 4**   A manufacturing firm makes two models, A and B, of a particular item. The number of hours (hr) required for each model for each of the three manufacturing processes is shown presented in Table 12.1.

For each of the A models, the manufacturing firm realizes a $72 profit; for each B model, the profit is $35. The problem is to determine how many of each model should be manufactured in order to

Table 12.1

	ASSEMBLY	PAINT AND FINISH	QUALITY CHECK AND TEST
MODEL A	50 hr	15 hr	3 hr
MODEL B	20 hr	12 hr	7 hr

realize the greatest profit; hence, we wish to maximize:

$$\text{Profit} = \$72A + \$35B.$$

Solution

At first glance it might appear feasible to manufacture many more of the A models (because of the greater profit); however, note that each A model also requires more hours of production. In addition, there is usually *more* to a typical linear programming problem. For each production run, the number of available hours is limited!

For assembly, there are no more than 9000 hours possible. For paint and finishing, 3000 hours are the most allowable, and the number of hours for quality control cannot exceed 1470.

With some analysis and study—as we shall learn in this chapter—the above problem translates, in more mathematical terms, to the following linear programming problem.

The company wishes to find A and B (how many models of each should be produced) such that:

1. the *profit function* $P = \$72A + \$35B$ is maximized, and
2. the combination of A and B satisfies these *constraints:*

$$50A + 20B \leq 9000 \quad \text{(for assembly)},$$
$$15A + 12B \leq 3000 \quad \text{(for paint and finish)},$$
$$3A + 7B \leq 1470 \quad \text{(for testing)}.$$

Partial solution
While attempts at guessing would prove most bewildering in practical applications, several combinations of A and B are given below to demonstrate some different possible profits.

		Profit	Assembly	Paint and finish	Quality tests
A	B	$72A + 35B$	$50A + 20B \leq 9000$	$15A + 12B \leq 3000$	$3A + 7B \leq 1470$
10	10	\$ 1080	700 < 9000	270 < 3000	100 < 1470

(*Comment:* Too few items. Manufacture more!)

200	100	\$18,000	12,000 $\not\leq$ 9000	4200 $\not\leq$ 3000	1300 < 1470

(*Comment:* Profit is great, good increase—however, constraints exceeded!)

150	75	\$13,500	9000 $\leq$ 9000	3150 $\not\leq$ 3000	975 < 1470

(*Comment:* Improving. Now only one constraint exceeded.)

| 150 | 100 | $14,400 | 9500 $\neq$ 9000 | 3450 $\neq$ 3000 | 1150 < 1470 |

(*Comment:* Two constraints are still exceeded. Guess again!)

| 140 | 75 | $12,780 | 8500 $\leq$ 9000 | 3000 $\leq$ 3000 | 945 $\leq$ 1470 |

(*Comment:* Good profit . . . all constraints satisfied . . . can we do better?)

As you are probably realizing, the eventual solution may take years (?) of this type of guessing to find the combination that maximizes profit. By the graphical method, and a more general algebraic method developed later in the chapter, the solution for this problem is found to be: Produce 160 of model A and 50 of model B. For comparative purposes, figures are given for this production level.

| 160 | 50 | $13,320 | 9000 $\leq$ 9000 | 3000 $\leq$ 3000 | 830 < 1470 |

(At this point, do not be concerned with how this solution was determined. We'll soon learn.)

Practical Examples

A discussion of several other examples of the successful use of linear programming is presented to help demonstrate the wide variety of possible applications.

Production Control

Some products are in greater demand in a particular season (skis, for example). Does this mean that the production facilities should operate only during the peak season? On the contrary, production continues year round to allow better utilization of facilities and to avoid depleting the stock during the peak season. Often seasonal goods are stored until peak demand. With knowledge of shipping and storage costs and sales forecasts, linear programming can be used to minimize the total cost of shipping, storage, and production.

Nutritional Foods

Many foods require certain minimal amounts of vitamins and/or minerals. To obtain the required nutritional values, different ingredients may be used, each of which contributes a certain number of units of the required nutritional contents. For example, each ounce of ingredient A may provide 10 mg of calcium and 20 mg of iron. Each ounce of

Table 12.2

	CALCIUM	IRON
A	10 mg	20 mg
B	15 mg	5 mg

ingredient B may provide 15 mg of calcium and 5 mg of iron. If the minimum federal requirements for calcium and iron are 60 mg and 30 mg, respectively, how many ounces of each ingredient (A and B) should be used (see Table 12.2)?

● Question

Would one ounce of ingredient A and two ounces of ingredient B satisfy the requirements?

Answer

For calcium: $1(10 \text{ mg}) + 2(15 \text{ mg}) = 40 \text{ mg}$
For iron: $1(20 \text{ mg}) + 2(5 \text{ mg}) = 30 \text{ mg}$

While the requirements for iron are met (at least 30 mg is needed), the requirements for calcium are not met (60 mg of calcium are required).

Different combinations of the two ingredients would be tried until *both* requirements are met. Then to add to the complexity of such problems, the ingredients usually cost different amounts. Each ounce of ingredient A in our example costs $0.04 and each ounce of ingredient B costs $0.06. Not only must the minimum federal requirements be met for the nutritional requirements, but hopefully these requirements will be met with a *minimum cost!* Problems of this type are rather common to blending industries (foodstuffs, pharmaceuticals, and even oil).

Transportation costs
Linear programming permits the efficient scheduling of airplanes to minimize cost and also to maximize utilization of different aircraft to service the widest geographical area. By combining information on passenger revenues, cargo revenues, flight operating costs, expected demands, and so on, a linear programming model can be formulated to yield a maximum profit.

Finally, to set the stage for our further work, we present the following in-depth example of a linear programming problem. Attention, at this time, should not be focused too much on all the details. Try to concentrate instead on understanding the overall framework.

■ EXAMPLE 5

For advertising on television in prime time during a recent Super-bowl, a company paid over $900,000 per minute. The more viewers a company can reach, the greater the probability that its sales will increase. (Even if not a better product, the sales may still increase because more people know of the product and consequently more people will be tempted to try it.)

The company is faced with a dilemma. Too much advertising and no entertainment—the number of viewers will decline. No advertising and only entertainment—how will the TV audience know of the product? By sampling some TV audiences, the company found that during afternoon programs, for every 20-minute segment of entertainment provided, the company has one minute of advertising. For evening programs, for every 10-minute segment of entertainment, the company has one minute of advertising.

At least 8 minutes must be devoted to advertising, but the company can afford to sponsor no more than 120 minutes of entertainment. If an afternoon program attracts 10,000 viewers and an evening program attracts an average of 40,000 viewers, how many segments of TV time should be purchased in the afternoon and how many segments in the evening? The company obviously wishes to attract the maximum number of viewers.

Solution

The variables to be determined are how many afternoon segments (A) and how many evening segments (B) to purchase. For each afternoon segment, 10,000 viewers are attracted; for each evening segment, 40,000 viewers are attracted. The total number of viewers is given by

$$10,000A + 40,000B = N.$$

However, the problem poses several *constraints,* or restrictions. At least 8 minutes is needed for advertising. For each afternoon segment, one minute is devoted to advertising; for each evening segment, the same. The resulting inequality is

$$1A + 1B \geq 8.$$

Advertising is costly and no more than 120 minutes of entertainment can be purchased. For each segment in the afternoon, 20 minutes is entertainment; for each evening segment, 10 minutes is entertainment. The inequality here is

$$20A + 10B \leq 120.$$

Finally, the number of segments cannot be negative quantities, so two additional constraints are

$$A \geq 0 \quad \text{and} \quad B \geq 0.$$

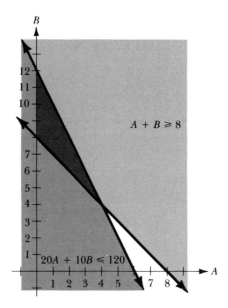

Figure 12.13

The linear programming problem, then, is

Maximize: $10{,}000A + 40{,}000B = N,$
subject to the $A + B \geq 8,$
constraints: $20A + 10B \leq 120,$
                                    $A \geq 0,$
                                    $B \geq 0.$

First, borrowing from our work in Section 12.1, we graph the set of inequalities, as shown in Fig. 12.13. The common intersection of these inequalities (constraints) is called the feasible region. Any point in this convex region is a feasible solution (in relation to the constraints). (A convex region has the property that if any two points on the figure are connected with a line segment, that segment lies in the region.)

However, from this set of feasible points, we must determine which point will give the company the most exposure. We wish to maximize the number of viewers. The expression to be maximized is called the *objective function;* here, $10{,}000A + 40{,}000B$.

A fundamental theorem of linear programming states:

■

> If an optimal value exists for the objective function, this optimal value will occur at one (possibly more) of the vertex points of the feasible region.

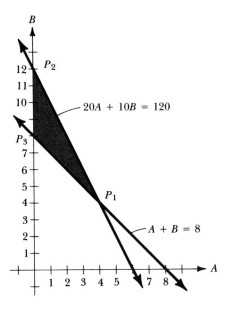

Figure 12.14

Although the proof of the theorem is beyond the scope of this text, its usefulness cannot be overestimated. For the feasible region in Fig. 12.13, there are three vertices. By simultaneously solving the appropriate system of equations, we can find each vertex. These points are then tested in the objective function to find which point produces the optimal value (here, the maximum number of viewers).

Using Fig. 12.14, we see that point P_1 is the intersection of $A + B = 8$ and $20A + 10B = 120$. Solved simultaneously, $A = 4$ and $B = 4$.

Point P_2 is the intersection of $A = 0$ and $20A + 10B = 120$. Solved simultaneously, $(A, B) = (0, 12)$.

Point P_3 is the intersection of $A = 0$ and $A + B = 8$. Solved simultaneously, P_3 is $(0, 8)$.

These three points and their respective values, when substituted into the objective function $(10,000A + 40,000B)$, are as follows:

P_1: 200,000 viewers;
P_2: 480,000 viewers; and
P_3: 320,000 viewers.

Thus the optimal solution for this simplistic problem is to purchase no afternoon segments and 12 evening segments.

Check: $A = 0$ and $B = 12$ will attract 480,000 viewers

$$[10,000(0) + 40,000(12) = 480,000],$$

will exceed the 8-minute minimum requirement

$$[(0) + (12) > 8],$$

and will purchase exactly 120 minutes of entertainment

$$[20(0) + 10(12) \geq 120]. \quad \square$$

The procedure for solving most linear programming problems is very similar to that illustrated here. Perhaps the only item that may change is the complexity of the problem.

Exercise Set 12.2

Determine the profit for each of the ordered pairs, while at the same time making sure all constraints are satisfied.

1. $7x + 2y = P$ subject to:

$$5x + 3y \leq 300,$$
$$x + 3y \leq 108.$$

a) (10, 30)   **b)** (30, 10)   **c)** (20, 20)   **d)** (30, 30)

e) (50, 20)   **f)** (20, 50)   **g)** (52, 17)   **h)** (53, 17)

i) (52, 18)   **j)** (53, 18)

2. $P = x + 2y$ subject to:

$$x \geq 0, y \geq 0,$$
$$x \leq 40, y \leq 30,$$
$$x + y \leq 45.$$

a) (10, 30)   **b)** (10, 40)   **c)** (30, 10)

d) (15, 30)   **e)** (30, 15)   **f)** (40, 10)

3. $P = x + 3y$ subject to:

$$x + y \geq 2,$$
$$x + y \leq 5,$$
$$0 \leq x \leq 4,$$
$$0 \leq y \leq 3.$$

a) (0, 4)   **b)** (1, 4)   **c)** (2, 4)   **d)** (3, 4)

e) (4, 4)   **f)** (4, 3)   **g)** (4, 2)   **h)** (4, 1)

i) (4, 0)   **j)** (0, 3)   **k)** (1, 3)   **l)** (2, 3)

m) (3, 3)   **n)** (3, 2)   **o)** (3, 1)   **p)** (3, 0)

By evaluating the objective function for each of the points, determine which of the *possible* ordered pairs will yield a minimum cost, yet satisfy the given constraints. (*Note:* These ordered pairs may not yield *the* minimum cost.)

4. Cost $= 11x + 9y$ constraints: $7x + 8y \geq 112$,
$$x + 10y \geq 50,$$
$$6x + y \geq 30.$$

(9, 9), (15, 9), (9, 15), (15, 0), (0, 15), (8, 8), (8, 9), (9, 8)

5. Cost $= 3x + 2y$ constraints: $x + 8y \geq 80$,
$$4x + 5y \geq 200,$$
$$70x + 10y \geq 1400.$$

(19, 30), (80, 0), (0, 140), (30, 20), (19, 29), (45, 18), (20, 30), (50, 50)

Translate each of the following problems into linear programming format: (a) state the objective function and (b) determine the constraints.

6. A furniture company makes tables and chairs. A table requires 1/2 hr on the drill and 1 hr on the finisher. A chair requires 1/2 hr on each machine. There is a maximum of 4 hr available on the drill and a maximum of 6 hr available on the finisher per day. Each chair yields a profit of $3, and each table yields a profit of $5. How many chairs and how many tables should be manufactured?

7. An investment firm has up to $15,000 available for speculating. Two growth stocks are available, C yielding a 14% return and D yielding an 18% return. The amount invested in C stock should be at least $1000 more than the amount invested in D stock. A maximum of $7500 should be invested in C stock and no more than $5000 should be invested in D stock. The firm wishes to maximize its return on these stocks. How much money should be invested in each stock?

In the following exercise, (a) write the constraints only and then (b) check for feasible points.

8. A business firm's peak sale season is in the winter. To accumulate stock, some items will be made and held in storage until the sale season. A maximum of 3600 ft^3 of storage space is available for two sales items. The first sales item is packaged in 5-ft^3 containers, and the second item in 3-ft^3 containers. The cost of storing each of the first sales item is $20, and that of the second item $50. No more than $19,000 of the budget is allocated for storage expenses.

After writing the constraints, check to see which (if any) of these coordinates of sales items satisfies both constraints:

(200, 400), (300, 400), (400, 300), (275, 390), (270, 385), (280, 390).

Read the original problem again. Try posing a possible objective function for this linear programming problem.

12.3
A Graphical Method

The graphical method is an excellent technique to illustrate the theory and to help you visualize the workings of linear programming problems.

■

> A general linear programming problem involves *finding* the maximum (or minimum) *optimal value* of a linear function called the *objective function* subject to a set of *constraints*.

Illustrations of the Graphical Method

■ EXAMPLE 6

Find values for x and y that will make $2x + y = P$ as large as possible, where x and y are restricted by the following constraints:

$$4x + y \leq 8,$$
$$x + 2y \leq 9,$$
$$x \geq 0,$$
$$y \geq 0.$$

Solution (Graphical)

First we graph the permissible values of x and y by individually graphing each constraint and finding the common intersection. (This follows exactly the method illustrated in Section 12.1 on systems of inequalities.) The shaded area in Fig. 12.15 represents all points common to each constraint; thus this shaded area, including the boundary, is the set of all permissible points. Any point from this region is a feasible solution. Hence, in linear programming this area is referred to as the *feasible region*.

$$\begin{aligned} \text{Maximize:} \quad & 2x + y = P, \\ \text{Subject to:} \quad & 4x + y \leq 8, \\ & x + 2y \leq 9. \end{aligned}$$

From the feasible region, we choose the point (or points) that will provide the maximum (or minimum) value of the objective function. A feasible point that maximizes (or minimizes) the objective function is called *optimal*.

● Question

How is the optimal point located within the feasible region?

Answer

Several procedures exist; most are based on one of the most fundamental theorems of linear programming. We repeat the theorem,

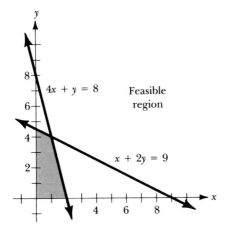

Figure 12.15

again without proof, because it is the key behind most all work with linear programming.

■

> If an optimal value exists for the objective function, it will occur at one (maybe more than one) of the vertex points of the feasible region.

One method of locating the optimal point is to methodically find the coordinates of each vertex, and substitute the coordinates into the objective function. Then continue in this manner until the optimal value is obtained.

Continuing with Example 6, we see in Fig. 12.16 that the vertices are labeled A, B, C, and D. For vertex A, the system of equations $y = 0$ and $4x + y = 8$ solved simultaneously yields $x = 2, y = 0$ (the x-intercept). For vertex B, $4x + y = 8$ and $x + 2y = 9$ when solved simultaneously yield $(1, 4)$. For vertex C, $x + 2y = 9$ and $x = 0$ when solved simultaneously yield $(0, 4.5)$ (the y-intercept). Vertex D is the origin.

Vertex	Substituted into $2x + y = P$
$(2, 0)$	$2(2) + 0 = 4$
$(1, 4)$	$2(1) + 4 = 6$
$(0, 4.5)$	$2(0) + 4.5 = 4.5$
$(0, 0)$	$2(0) + 0 = 0$

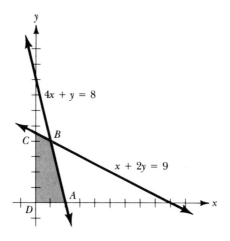

Figure 12.16

The optimal point is $(1, 4)$; $x = 1$ and $y = 4$ and the optimal value is 6. □

■ **EXAMPLE 7**

Using the graphical method,

$$\begin{aligned}
\text{Maximize:} \quad & x + 3y = P \\
\text{Subject to:} \quad & x + y \geq 2, \\
& x + y \leq 5, \\
& x \leq 4, \\
& x - 2y \geq -4, \\
& y \geq 0.
\end{aligned}$$

Solution

After graphing each of the inequalities, we find that the feasible region is as shown in Fig. 12.17. By simultaneously solving each pair of

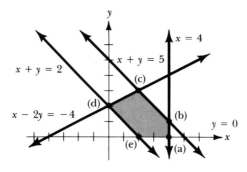

Figure 12.17

equations, we find the five vertices:

a) $y = 0$ and $x = 4 \longrightarrow (4, 0)$
b) $x = 4$ and $x + y = 5 \longrightarrow (4, 1)$
c) $x + y = 5$ and $x - 2y = -4 \longrightarrow (2, 3)$
d) $x - 2y = -4$ and $x + y = 2 \longrightarrow (0, 2)$
e) $x + y = 2$ and $y = 0 \longrightarrow (2, 0)$

Vertex point	Profit
(4, 1)	$ 7.00
(2, 3)	$11.00
(0, 2)	$ 6.00
(2, 0)	$ 2.00
(4, 0)	$ 4.00

We then substitute these five vertices into the objective function, $x + 3y = P$, and examine the corresponding values to find the optimal solution (see the chart on the left).

The optimal point (2, 3) yields the optimal—in this case, maximum—value of $11.00.   □

A comment on our work thus far. Examining Examples 6 and 7, we see that as the number of constraints increases, the number of vertex points also increases. If we assume all constraints form intersection points, how many vertex points will there be for a linear programming problem with 5 constraints? with 80 constraints?

For finding the coordinates of each vertex, a separate system of equations must be solved. In the last example, five systems of equations had to be solved simultaneously! The work increases dramatically. Even with much of the computation being done on computers (almost every type of computer presently being operated has some linear programming codes written for it), the solution to a linear programming problem requires a great deal of time and work. A bit of "refinement in procedure" is in order.

A method is suggested to find the optimal point from all the possible vertex points by using an additional graphical technique. The procedure is most useful when using the graphical method for solving linear programming problems.

If we use Example 7 for illustrating the additional refinement, the feasible region is as shown in Fig. 12.18:

$$
\begin{aligned}
\text{Maximize:} \quad & x + 3y = P \\
\text{Subject to:} \quad & x + y \geq 2, \\
& x + y \leq 5, \\
& x \leq 4, \\
& x - 2y \geq -4, \\
& y \geq 0.
\end{aligned}
$$

For a given profit, say $P = 3$, all points (x, y) yielding $P = 3$ lie on a straight line (see Fig. 12.19). When substituted into the objective function, any point on this line yields a P-value of 3.

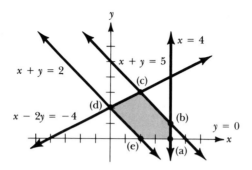

Figure 12.18

For a larger value of P, say $P = 12$, the solution points lie on a parallel line, but with a higher y-intercept. For $x + 3y = 3$, or $y = -\frac{1}{3}x + 1$, the slope is $-\frac{1}{3}$ and the y-intercept is $(0, 1)$.

For $x + 3y = 12$, or $y = -\frac{1}{3}x + 4$, the slope is $-\frac{1}{3}$ and the y-intercept is $(0, 4)$.

The greater the value of P, the higher the y-intercept. Continuing with this line of reasoning, we see that the parallel line with the highest y-intercept that still contains a point in the feasible region will yield the optimal value for P (in this case, the maximum value).

Let's return to our illustration. By drawing lines parallel to our objective function (lines with a slope of $-\frac{1}{3}$) (Fig. 12.20), we note that the last line to intersect the feasible region passes through point A.

To find the coordinates of A, we solve the one system of equations that represents the intersecting lines at that point. (Note that with our

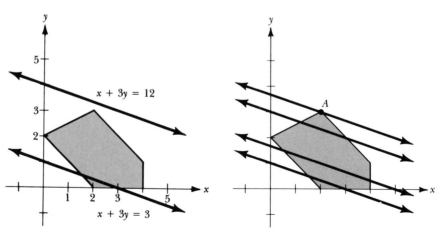

Figure 12.19   **Figure 12.20**

refinement, our work has been greatly reduced from the five systems of simultaneous equations we worked with earlier.)

Suggestion. Try this technique on Example 6 to find the optimal point.

An Application of the Graphical Method

With a technique for solving linear programming problems in hand, we now need to concentrate our efforts on one more topic: *formulating the problem.* When we are solving linear programming problems, often the variables, the objective function, and the constraints are not explicitly given. On computer systems, rental time and/or run time can be costly if errors are made in setting up the problem. The following application shows the solution of a linear programming problem *from the beginning* (problem formulation) *to the end* (finding the optimal value).

Keep in mind that for learning purposes, our linear programming examples tend to be more simplistic than real-life applications. Large-scale linear programming problems in business usually involve more than solving problems with only 3 or 4 constraints; problems with over 500 or 600 constraints are not at all uncommon!

Comment. At this point you are encouraged to return to Section 12.2 and reread, very carefully, Example 5.

An area of business that frequently utilizes linear programming is transportation and the storage of goods. Often a manufacturing firm will distribute finished products or goods from the manufacturing plant to perhaps even hundreds of warehouses and/or retail stores. Our example will be concerned not with hundreds, but only two warehouses.

■ **EXAMPLE 8**

To assure a building supply company of maintaining adequate goods on hand to meet consumer demand, the company will transport and store bags of concrete in two locations. Transportation costs to ship and store one bag of concrete at the southern location are $0.15 and at the eastern location $0.20. The southern location can store a maximum of 8000 bags, whereas the eastern location can store a maximum of 6000 bags. The company must ship at least 9000 bags. How many bags should be shipped to each location in order to minimize the transportation and storage costs? What will be this minimum cost?

Solution

The *variables* to determine are: How many bags of concrete should be shipped to the southern location, S, and how many bags to the eastern location, E.

Transportation costs are to be minimized so the *objective function* is:

$0.15S + 0.20E = $ Cost ($0.15 to store each bag in the southern location and $0.20 in the eastern location).

We must minimize $0.15S + 0.20E$, but subject to what constraints? The *constraints* will also contain the variables S and E. From our reading, we know that no more than 8000 bags can be transported and stored in the south:

$$S \leq 8000.$$

No more than 6000 bags can be transported and stored in the east:

$$E \leq 6000.$$

At least 9000 bags, total, must be transported and stored:

$$S + E \geq 9000.$$

The bags represent nonnegative quantities, so the additional constraints $S \geq 0$ and $E \geq 0$ are needed. The problem then becomes

Minimize: $0.15S + 0.20E$
Subject to: $S \leq 8000,$
  $E \leq 6000,$
  $S + E \geq 9000,$
  $S \geq 0, E \geq 0.$

A graphical sketch of each constraint and the feasible region are shown in Fig. 12.21. Several costs are given and their respective parallel lines are drawn to represent different cost functions.

The line given by $2400 = 0.15S + 0.20E$ does intersect the feasible region at the point (8000, 6000); however, the cost $2400 could be less. In maximization problems, we learned that the greater the value of the objective function, the greater the y-intercept. In maximization problems the goal is to increase the objective function's value as much as possible. In minimization problems, just the opposite strategy is needed. We need to lower the y-intercept in order to decrease the value of the objective function, again only as much as possible so the line still intersects the feasible region. The cost function given by $600 = 0.15S + 0.20E$ does represent a lower total cost than $2400, but it intersects no point in the feasible region. The line represented by $1800 = 0.15S + 0.20E$ is a better choice than that for 2400. By moving parallel lines "downward," we find the optimal point at which a cost function still intersects the feasible region at point A.

To find the coordinates of point A, we solve the system of equations

$$S = 8000 \quad \text{and} \quad S + E = 9000$$

simultaneously to obtain

$$S = 8000 \quad \text{and} \quad E = 1000.$$

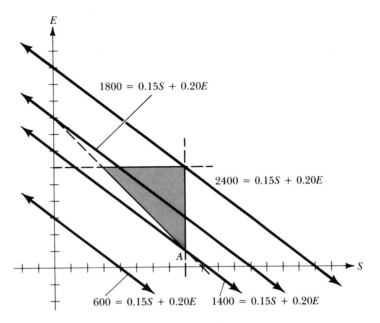

Figure 12.21

By substituting $S = 8000$ and $E = 1000$ into the objective function, we can then find the minimum cost:

$$0.15S + 0.20E = C$$
$$0.15(8000) + 0.20(1000) = \$1400. \quad \square$$

Note: Keep in mind that these examples are given to illustrate the ideas of linear programming and consequently are rather simplistic. Practice on the easier one; more difficult ones appear later in the chapter. As a matter of fact, for the problem above on minimum cost (unlike most practical applications), could the solution have been found by using no linear programming at all? How? (Think!)

SUMMARY

To solve a linear programming problem by the graphical method:

1. Formulate the problem mathematically; identify the variables, the objective function, and then the constraints.
2. Graph each of the constraints.
3. Shade the feasible region.
4. Plot an objective function and sketch parallel lines.
5. Evaluate the corner points (vertices) of the feasible region.
6. The smallest of these values determines the minimum and the largest determines the maximum for the objective function.

Exercise Set 12.3

Graph the feasible region for each of the following linear programming problems.

1. $x + 5y \leq 10$,
   $3x + y \leq 6$,
   $x \geq 0, y \geq 0$

2. $3x + 2y \leq 150$,
   $x + 2y \geq 40$,
   $x \geq 20, y \geq 0$

3. $5x + y \geq 10$,
   $x + 4y \geq 8$

4. $6x + y \leq 6$,
   $x + 6y \leq 6$

5. $x \geq 0, y \geq 0$,
   $x \leq 4, y \leq 2$,
   $x + y \leq 4$

6. $2x - 5y \leq 10$,
   $x + 2y \geq 8$,
   $x \geq 0, y \geq 0$

7. $x + y \geq 25$,
   $6x + 2y \geq 60$,
   $x + 4y \geq 40$

8. $3x + 4y \geq 12$,
   $2x - 3y \leq 6$,
   $2x + 3y \geq 18$,
   $x \geq 0, y \geq 0$

9. Show the feasible region for these constraints and then answer the following questions.

$$2x - 5y \leq 10,$$
$$14x + 9y \leq 126,$$
$$3x + 7y \leq 42,$$
$$x - y \geq 0$$

   a) If an equal profit is made from x and y, circle the point that will yield the maximum profit.
   b) If x earns twice as much profit as y, circle the point that will yield the maximum profit.
   c) If y earns twice as much profit as x, circle the point that will yield the maximum profit.

Shown below are several regions of feasible solutions. Use these regions and parallel lines for the given objective function to find both maximum and minimum values for each objective function.

10.

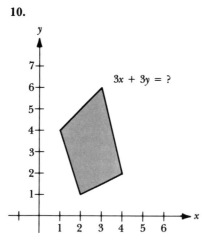

$3x + 3y = ?$

11.

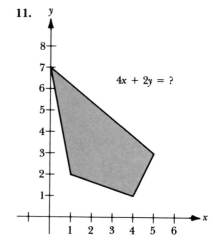

$4x + 2y = ?$

12. Use the figure in Exercise 10 but now use the function

$$4x + y = ?$$

13. Use the figure in Exercise 11 but now use the function

$$2x + y = ?$$

(Notice anything here?)

14.

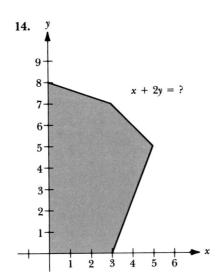

$x + 2y = ?$

15.

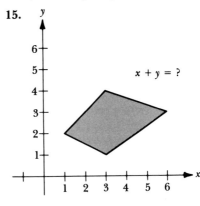

$x + y = ?$

Graphically, solve each of the following linear programming problems.

16. Maximize: $x + 2y = P$
     Subject to: $x + 5y \leq 10,$
                 $3x + y \leq 6,$
                 $x \geq 0, y \geq 0$

17. Maximize: $5x + 3y = P$
     Subject to: $3x + 2y \leq 150,$
                 $x + 2y \geq 40,$
                 $x \geq 20, y \geq 0$

18. Maximize: $3x - 11y = P$
     Subject to: $x - 4y \leq 8,$
                 $-x + 4y \leq 8,$
                 $x + 3y \geq 6,$
                 $x \leq 12$

19. Maximize: $11x - 3y = P$
     Subject to: $x - 4y \leq 8,$
                 $-x - 4y \leq 8,$
                 $x + 3y \geq 6,$
                 $x \leq 12$

20. Minimize: $3x - 11y = C$
Subject to the constraints in Exercise 18.

21. Minimize: $11x - 3y = C$
Subject to the constraints in Exercise 19.

22. Maximize: $x + 3y = P$
Subject to the constraints in Exercise 18.

23. Minimize: $4.5x + 3.0y = C$
     Subject to: $x + y \geq 25,$
                 $6x + 2y \geq 60,$
                 $x + 4y \geq 40$

24. Maximize: $x + 3y = P$
     Subject to: $4x + 3y \leq 12,$
                 $x - 2y \leq 2,$
                 $3x - y \geq -3,$
                 $x + y \geq 1$

24. Maximize: $x + 3y = P$
   Subject to: $4x + 3y \leq 12,$
               $x - 2y \leq 2,$
               $3x - y \geq -3,$
               $x + y \geq 1$

25. Maximize: $3x + y = P$
   Subject to: $4x + 3y \leq 12,$
               $2x - y \geq -1,$
               $10x - 7y \leq 1,$
               $x \geq 0, y \geq 0$

26. Minimize: $20A + 20B = C$
   Subject to: $A + 2B \geq 8,$
               $4A + 2B \geq 16,$
               $5A + 2B \geq 20,$
               $A \geq 0, B \geq 0$

27. Minimize: $4A + 2B = C$
   Subject to: $12A + 4B \geq 48,$
               $2A + 16B \geq 32,$
               $14A + 21B \geq 140,$
               $A \geq 0, B \geq 0$

28. An agriculture group has a maximum of $240,000 to invest in growing two cash crops, corn and soybeans. To grow corn, the approximate cost to the group is $240 per acre; to grow soybeans, the approximate cost is $150 per acre. The total cost may be less than or equal to $240,000; the total cost cannot exceed their total capital for investment. The group has a ready market for at least 400 acres of corn and 300 acres of soybeans. The approximate profit on each crop per acre is $270 for corn and $315 for soybean. How much acreage of each crop should be planted in order to realize the largest profit?

After having read the above paragraph, reread it and then answer these questions:

a) What are the two variables of interest?

b) What is the objective function?

c) What are the constraints?

d) How many acres of corn should be planted? of soybeans?

e) What will be the profit?

29. To manufacture two models of calculators, A and B, an electronics firm uses three machines, R, S, and T. The time requirements for each model calculator on each machine are listed in the following table.

MACHINE	CALCULATOR MODEL		MAXIMUM TIME AVAILABLE
	A	B	
R	30	50	200
S	40	20	160
T	60	20	240

If the profit on model A is $4.00 apiece and on model B is $3.00 apiece, find the production of each model that best maximizes the profit.

The objective function is _____

The constraints are _____

_____

_____

$A = $ ___ ; $B = $ ___ ; Profit is _____ .

12.4
An Algebraic Method

Of the several methods of solving linear programming problems, the graphical method best illustrates the basic ideas of linear programming. However, for problems with more than two variables, or with two variables and many constraints, the graphical method is either impossible or simply not suitable. For example, a linear programming problem involving 6 variables and satisfying 5 other linear constraints may have a feasible region with as many as 462 vertices! And how do we graph an equation in more than 2 variables? For three-variable linear programming problems, three-dimensional configurations are possible, but extremely difficult to picture. For four or more variables the graphical method cannot be used. (Aren't we limited to a three-dimensional world?) Thus there is a need for an alternative method of solving linear programming problems.

The method most frequently used is called the **simplex method,** due to G. B. Dantzig. Although the method is rather involved and intricate for programming on a computer, it is very systematic. To help in understanding how the simplex method works, we will first introduce the *algebraic method.* The simplex method can be regarded as a refinement and a streamlining of the algebraic method. The progression of our developing an understanding of the simplex method will be to present the *algebraic method,* to make refinements in the algebraic method by using *matrix notation,* and finally to simplify the matrix notation to the systems of *tableaus* of the simplex method. The algebraic method will provide an intuitive understanding of the basic terminology and processes of the computationally more efficient simplex method.

At the basis of all these methods for solving linear programming problems is the elimination procedure for solving systems of simultaneous equations, discussed in Chapter 10. Before beginning this chapter you are strongly encouraged to become proficient in dealing with systems of equations!

The algebraic method will be presented by analyzing a linear programming problem. Before beginning, let's look at the overall outline of this method, which will serve as a guide.

■

The algebraic method is rather algorithmic in nature.

1. A feasible solution is chosen as a starting point.
2. The objective function is evaluated for this point.
3. By examining the objective function, determination is made whether an improved solution exists.
4. If an improved solution is possible, return to step (2); if an improved solution is not possible, the linear program has found the optimal solution.

The pseudocode (abbreviated statements) above appears as a flowchart in Fig. 12.22.

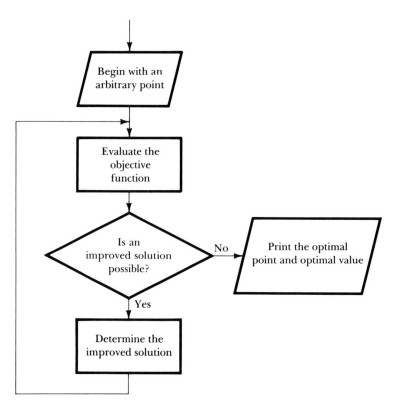

Figure 12.22

Want to buy a gadget?

■ **EXAMPLE 9**   Suppose a manufacturer makes two items: gadgets (G) and widgets (W). The profits earned on the sale of gadgets and widgets are \$20 and \$30, respectively. Each item requires some time for assembly and for finishing. Each gadget requires 4 hr for assembly and 2 hr for finishing; each widget requires 12 hr for assembly and 2 hr for finishing. Total time for assembly may not exceed 120 hr, while total time for finishing may not exceed 40 hr.

How many gadgets and how many widgets should be manufactured so that the greatest profit can be realized?

Solution   First the linear program is translated into mathematical formulas. The variables of interest are G and W. The objective function to be maximized is \$20G + \$30W.

The constraints are related to assembly and to finishing time. For *assembly,* the total time is represented by the expression $4G + 12W$ and this cannot exceed 120 hr. For *finishing,* the total time constraint is $2G + 2W \leq 40$. Because G and W cannot be negative items, the two additional constraints $G \geq 0$ and $W \geq 0$ are assumed.

The problem is:

$$\begin{aligned} \text{Maximize:} \quad & 20G + 30W = P \\ \text{Subject to:} \quad & 4G + 12W \leq 120, \\ & 2G + 2W \leq 40, \\ & G \geq 0, W \geq 0. \end{aligned}$$

We can use Fig. 12.23 as a guide.

In preparation for the algebraic method, each inequality must be converted to an equation, using slack variables. Because of the possibility that a given solution may *not* use all the available time either for assembly or for finishing, a variable is added to "take up the slack."

For example, if 10 gadgets and 5 widgets were made, the total

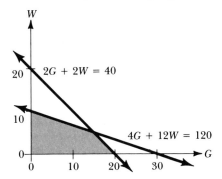

Figure 12.23

assembly time would be $4(10) + 12(5) - 100$ and the total finishing time would be $2(10) + 2(5) = 30$.

For assembly work, there would then be 20 hr of *slack* (unused) time $(120 - 100)$; and for finishing work, there would be 10 hr of *slack* time $(40 - 30)$. Upon completion of the linear programming problem, both assembly and finishing should be utilized to full capacity and the slack variables will then have a value of zero, although this is not always the case.

Adding the slack variables S_1 for assembly and S_2 for finishing, our linear programming problem becomes:

$$\begin{aligned} \text{Maximize:} \quad & 20G + 30W + 0S_1 + 0S_2 = P \\ \text{Subject to:} \quad & 4G + 12W + S_1 + 0S_2 = 120, \\ & 2G + 2W + 0S_1 + S_2 = 40. \end{aligned}$$

(*Note:* For a complete mathematical formulation, any unknown occurring in one equation should occur in the other equations. By assigning coefficients of zero, the equality relationships are not affected.)

Because the number of variables is greater than the number of equations, more than one solution exists. Solutions are determined by arbitrarily assigning values to two of the four variables, and then solving the resulting system of two equations in two unknowns.

Instead of considering all possible vertices (as in the graphical method), the strategy of the algebraic method is to proceed step by step from the initial basic feasible solution to other basic feasible solutions. At each step, one of the variables with a value of zero is changed to a nonzero variable and vice versa. From one basic solution to the next solution, the value of the objective function is usually increased; it will always be at least as large as the preceding solution.

An initial solution. In the algebraic method a systematic progression is made, starting with the worst possible solution (no gadgets, no widgets) and proceeding to the optimal solution. Initially with $G = 0$ and $W = 0$, the profit is zero and both the assembly and finishing departments are completely idle. Presently, G and W are the zero variables, and S_1 and S_2 are the nonzero variables.

In our example,

$$\begin{aligned} \text{Maximize:} \quad & 20G + 30W + 0S_1 + 0S_2 = P \\ \text{Subject to:} \quad & 4G + 12W + S_1 + 0S_2 = 120, \\ & 2G + 2W + 0S_1 + S_2 = 40 \end{aligned}$$

with $G = 0$ and $W = 0$, the values of the variables are computed:

$$\begin{aligned} 20(0) + 30(0) + 0(S_1) + 0(S_2) = 0 &\longrightarrow P = 0; \\ 4(0) + 12(0) + S_1 + 0(S_2) = 120 &\longrightarrow S_1 = 120; \\ 2(0) + 2(0) + 0(S_1) + S_2 = 40 &\longrightarrow S_2 = 40; \end{aligned}$$

or we have that the profit = 0 with $G = 0$, $W = 0$, $S_1 = 120$, $S_2 = 40$. (After several more steps, the optimal solution will be reached: that is, profit = $450 with $G = 15$, $W = 5$, $S_1 = 0$, $S_2 = 0$.)

An improved solution (?). To determine whether an improved solution is possible, we examine the objective function. By manufacturing either gadgets or widgets, the profit will increase (from $0.00) and the idle or slack time should decrease. To proceed toward an optimal solution, the company should manufacture the item that increases profit the most—in this case, widgets. The coefficient for widgets ($30) is larger than the coefficient for gadgets ($20).

■

> In general, the largest positive coefficient in the objective function is called the **key variable.**

Next we must determine the maximum number of widgets that we are allowed to make (without exceeding the constraints). Suppose all allowable time is spent making widgets.

$$4G + 12W + S_1 = 120 \quad \text{(for assembly)}$$
$$2G + 2W + S_2 = 40 \quad \text{(for finishing)}$$

Now, in the assembly department, with $G = 0$ and $S = 0$, a maximum of $12W = 120$, or 10 widgets, can be made. In the finishing department, with $G = 0$ and $S = 0$, a maximum of $2W = 40$, or 20 widgets, can be made.

We choose to make 10 widgets, not 20. Why? Had we chosen to make 20 widgets, the finishing department would have had no slack, or idle, time—that's good! To assemble the same 20 widgets, with *each* widget requiring 12 hr, the total time would be 240 hr. But with only 120 hr available for assembly, manufacturing this number of widgets is not possible. The constraint equation for assembly becomes what is called the pivotal equation.

■

> In general, the *pivotal equation* is determined by the smaller quotient, resulting from division of the constant term in each constraint by the nonzero coefficient of the key variable.

Correct identification and selection of the *key variables* and the *pivotal equations* are the most important components in understanding and working with the algebraic method. Our linear programming problem up to this point is summarized:

$$\text{(1) } W \text{ is the key variable (largest coefficient)}$$

(2) Pivotal equation $\longrightarrow$ identified by the smaller quotient

$$20G + 30W + 0S_1 + 0S_2 = P,$$
$$4G + 12W + S_1 + 0S_2 = 120, \qquad (120 \div 12 = 10)$$
$$2G + 2W + 0S_1 + S_2 = 40. \qquad (40 \div 2 = 20)$$

$$G \qquad W \qquad S_1 \qquad S_2$$

Entering variable $\uparrow$ $\downarrow$ Departing variable

Initially, G and W were the zero variables, which gave us $S_1 = 120$ and $S_2 = 40$. We then determined that making 10 widgets would increase our profit. With $G = 0$ and W now 10, S_1 will become 0 (no idle time for assembly). From

G	W	S_1	S_2
0	0	120	40

we have proceeded to

G	W	S_1	S_2
0	10	0	20

The amount of slack time in the assembly department has been reduced to zero, and W replaces S_1 for an improved solution. Now W and S_2 are the nonzero variables, and G and S_1 are the zero variables. To reflect this fact, the constraint equations must be changed. Because W is going to replace S_1, we solve the pivotal equation for W. This will be used to eliminate W in the other constraint equations, reflecting the change in manufacturing. Changes will also be brought about in the objective function. We use substitution to eliminate W in all equations other than the pivotal equation.

$$20G + 30W + 0S_1 + 0S_2 = P,$$
$$4G + 12W + S_1 + 0S_2 = 120,$$
$$\text{or} \quad W = 10 - \tfrac{1}{3}G - \tfrac{1}{12}S_1, \text{ when solved for } W$$
$$2G + 2W + 0S_1 + S_2 = 40$$

When this expression is substituted for W in the objective function, and also in the bottom constraint, a "transformed system" of equations is obtained:

$$10G + 0W - \tfrac{5}{2}S_1 + 0S_2 + 300 = P,$$
$$\tfrac{1}{3}G + W + \tfrac{1}{12}S_1 + 0S_2 = 10,$$
$$\tfrac{4}{3}G + 0W - \tfrac{1}{6}S_1 + S_2 = 20.$$

With $G = 0$ and $S_1 = 0$, the values of the remaining variables are:

$$10(0) + 0W - \tfrac{5}{2}(0) + 0(S_2) + 300 = P \longrightarrow P = 300,$$
$$\tfrac{1}{3}(0) + W + \tfrac{1}{12}(0) + 0(S_2) = 10 \longrightarrow W = 10,$$
$$\tfrac{4}{3}(0) + 0W - \tfrac{1}{6}(0) + S_2 = 20 \longrightarrow S_2 = 20,$$

or in other words, profit = 300 when $G = 0$, $W = 10$, $S_1 = 0$, $S_2 = 20$. The solution has improved from

G	W	S_1	S_2	P
0	0	120	40	0

to

G	W	S_1	S_2	P
0	10	0	20	$300

An improved solution, then, is to manufacture 10 widgets rather than no gadgets and no widgets. Ten widgets can be made at a profit of $300 with no idle time in the assembly department. In the finishing department, 20 hr of finishing are required for the 10 widgets. The remaining $(40 - 20 = 20)$ hours are idle or slack time. The original zero variables G and W have been changed to G and S_1.

Can we further improve the solution? By examining the "revised" profit function (reflecting the net effect of changes in manufacturing), we see that a positive coefficient indicates the amount of increase in total profit if one additional unit of this variable were produced:

$$10G + 0W - \tfrac{5}{2}S_1 + 0S_2 + 300 = P,$$
$$\tfrac{1}{3}G + W + \tfrac{1}{12}S_1 + 0S_2 = 10,$$
$$\tfrac{4}{3}G + 0W - \tfrac{1}{6}S_1 + S_2 = 20.$$

The "new" key variable is G; by manufacturing gadgets, an additional $10 will be added to total profit for each gadget manufactured. How many gadgets can be made without exceeding the "new" revised constraints? In

$$\tfrac{1}{3}G + W + \tfrac{1}{12}S_1 + 0S_2 = 10,$$

if W and S_1 were 0, G would be 30. In

$$\tfrac{4}{3}G + 0W - \tfrac{1}{6}S_1 + S_2 = 20,$$

if S_1 and S_2 were 0, G would be 15. We choose to manufacture 15 gadgets, not 30! Why? If 30 gadgets were made, the revised constraint equation for finishing would be exceeded. Let's see why.

$$\tfrac{4}{3}(30) + 0W - \tfrac{1}{6}S_1 + S_2 = 20, \quad \text{or}$$
$$40 - \tfrac{1}{6}S_1 + S_2 = 20.$$

If 30 gadgets were made, W and S_1 would have to be 0. But then S_2 would have to be a negative value. Impossible!

Recall that the smaller quotient determines the pivotal equation.

An improved solution can be found with $G = 15$. (Note that S_1 and S_2 will then be the zero variables.) From

G	W	S_1	S_2
0	10	0	20

we proceed to

G	W	S_1	S_2
15	5	0	0

From the system

$$
\begin{aligned}
10G &+ 0W - \tfrac{5}{2}S_1 + 0S_2 + 300 = P \\
\tfrac{1}{3}G &+ W + \tfrac{1}{12}S_1 + 0S_2 \quad\quad = 10 \quad (10 \div \tfrac{1}{3} = 30) \\
\tfrac{4}{3}G &+ 0W - \tfrac{1}{6}S_1 + S_2 \quad\quad = 20 \quad (20 \div \tfrac{4}{3} = 15)
\end{aligned}
$$

$$
\begin{array}{cccc}
\uparrow & & & \downarrow \\
G & W & S_1 & S_2 \\
\end{array}
$$

Entering   Departing   (variables)

with G the "new" key variable, the pivotal equation is now the one for finishing. Because G is going to replace S_2 in the improved solution, the system should be revised to reflect the newest changes in manufacturing. The pivotal equation is solved for G:

$$\tfrac{4}{3}G + 0W - \tfrac{1}{6}S_1 + S_2 = 20$$

yields

$$G = 15 - 0W + \tfrac{1}{8}S_1 - \tfrac{3}{4}S_2.$$

We will then use this to eliminate the variable G in the other two equations. When substituted into the first two equations, the "new transformed system" of equations becomes

$$0G + 0W - \tfrac{5}{4}S_1 - \tfrac{3}{4}S_2 + 450 = P,$$
$$0G + W + \tfrac{1}{8}S_1 - \tfrac{1}{4}S_2 \qquad\quad = 5,$$
$$G + 0W - \tfrac{1}{8}S_1 + \tfrac{3}{4}S_2 \qquad\quad = 15.$$

(The third equation was obtained from the equation solved for G.)

With S_1 and S_2 now the zero variables, the values of the remaining variables are

$$0G + 0W - \tfrac{5}{4}(0) - \tfrac{3}{4}(0) + 450 = P \longrightarrow P = 450,$$
$$0G + W + \tfrac{1}{8}(0) - \tfrac{1}{4}(0) \qquad = 5 \longrightarrow W = 5,$$
$$G + 0W - \tfrac{1}{8}(0) + \tfrac{3}{4}(0) \qquad = 15 \longrightarrow G = 15,$$

or, in other words, profit = \$450 when $G = 15$, $W = 5$, $S_1 = 0$, $S_2 = 0$.

In summary, the solution(s) have proceeded from beginning to end in the following sequence:

	G	W	S_1	S_2	P
	0	0	120	40	\$ 0
to	0	10	0	20	\$300
to	15	5	0	0	\$450

By manufacturing 15 gadgets and 5 widgets, the company will realize an increase of \$150 in profit, and both slack variables are now zero (no idle time)!

● Question

Could the solution $(G, W) = (15, 5)$ be further improved?

Answer

No, by examining the latest revised objective function, we see that there is no positive coefficient to increase profit. All negative coefficients in the objective function are the signal that the process is finished. An optimal solution has been reached. □

We will illustrate with another example after reviewing and slightly modifying the algebraic method to better conform—eventually—to the simplex method.

■

AN OUTLINE OF THE ALGEBRAIC METHOD

1. Determine the key variable by using the variable with the largest positive coefficient in the objective function.
2. In each constraint, divide the constant term by the nonzero coefficient of the key variable. The smallest quotient determines the pivotal equation.
3. The pivotal equation is solved for the key variable.
4. This expression is substituted into the other equations to eliminate the key variable.
5. The value of the nonzero variables in the new solution will be the constant terms on the right side of each equation.

To further simplify the algebra, steps (3) and (4) (using substitution to eliminate the key variable) will be easier handled by forming linear combinations. They will be replaced with the statements below; then we will use another example to illustrate all five steps.

(new step 3)   The coefficient of the key variable in the pivotal equation is changed to 1.

(new step 4)   This transformed pivotal equation is used to eliminate the key variable in all other equations by forming linear combinations.

■ **EXAMPLE 10**

Maximize: $2x + y = P$
Subject to: $3x + 2y \leq 12,$
                    $x + 2y \leq 8$

Solution

From the inequalities, a system of equations is formed by introducing the slack variables S_1 and S_2:

$2x + y + 0S_1 + 0S_2 = P,$	**Objective function**
$3x + 2y + S_1 + 0S_2 = 12,$	**Constraint (1)**
$x + 2y + 0S_1 + S_2 = 8.$	**Constraint (2)**

An initial solution. With $x = 0$ and $y = 0$, substitution in the system yields $P = 0$ with $x = 0, y = 0, S_1 = 12, S_2 = 8$.

An improved solution. The variable in the objective function with the largest positive coefficient increases the profit the most. The key variable, then, is x. The pivotal equation is the constraint equation with the smaller resulting quotient.

$$\downarrow \text{Key variable}$$

$$2x + \ y + 0S_1 + 0S_2 = P,$$

Pivotal $\longrightarrow$ $3x + 2y + \ \ S_1 + 0S_2 = 12,$ $(12 \div 3 = 4)$

equation $\quad\quad\quad x + 2y + 0S_1 + \ \ S_2 = 8.$ $(8 \div 1 = 8)$

For the maximum number of x's, y and S_1 must equal 0 in constraint equation (1). With y a zero variable, S_1 now becomes a zero variable, and x becomes the other nonzero variable with S_2.

The coefficient of x in the pivotal equation is changed to 1:

$$2x + \ y + 0S_1 + 0S_2 = P,$$

$$\textcircled{x} + \tfrac{2}{3}y + \tfrac{1}{3}S_1 + 0S_2 = 4, \quad\quad (\text{row } 2 \div 3)$$

$$x + 2y + 0S_1 + \ \ S_2 = 8.$$

$$\uparrow \quad\quad\quad\quad\quad \downarrow$$

$$x \quad\quad y \quad\quad S_1 \quad\quad S_2$$

Entering   Departing   (variables)

This transformed pivotal equation is used to eliminate the key variable in all other equations. (You may need to review the work in Chapter 10 on the use of row operations to form linear combinations in systems of equations.) The basis of the algebraic method is the elimination procedure for solving systems of simultaneous equations.

$$2x + \ y + 0S_1 + 0S_2 = P$$

$$\left(x + \tfrac{2}{3}y + \tfrac{1}{3}S_1 + 0S_2 = 4 \right)$$

$$x + 2y + 0S_1 + 1S_2 = 8$$

$$0x - \tfrac{1}{3}y - \tfrac{2}{3}S_1 + 0S_2 = P - 8, \quad\quad \textbf{-2 times row 2 added to row 1}$$

$$x + \tfrac{2}{3}y + \tfrac{1}{3}S_1 + 0S_2 = 4,$$

$$0x + \tfrac{4}{3}y - \tfrac{1}{3}S_1 + \ \ S_2 = 4 \quad\quad \textbf{-1 times row 2 added to row 3}$$

With y and S_1 the zero variables, the values of the remaining variables can be determined.

$$0x - \tfrac{1}{3}(0) - \tfrac{2}{3}(0) + 0S_2 = P - 8 \longrightarrow P = 8,$$

$$x + \tfrac{2}{3}(0) + \tfrac{1}{3}(0) + 0S_2 = 4 \quad\quad \longrightarrow x = 4,$$

$$0x + \tfrac{4}{3}(0) - \tfrac{1}{3}(0) + \ \ S_2 = 4 \quad\quad \longrightarrow S_2 = 4.$$

In other words, profit is \$8.00 with $x = 4$, $y = 0$, $S_1 = 0$, $S_2 = 4$.

Four of the x items and none of the y items are produced with no slack time in constraint (1). Production of four x items does not use all the available time in the second constraint; the slack, or idle, time is four hours.

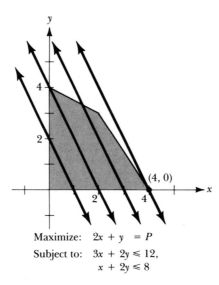

Maximize: $2x + y = P$

Subject to: $3x + 2y \leqslant 12,$

$x + 2y \leqslant 8$

Figure 12.24

Can this solution be improved upon further? No. By examining the "new" revised objective function, we see that no positive coefficient remains. We do not continue the process in this case. The optimal solution of $(x, y) = (4, 0)$ has been reached for a maximum profit of $8.00.

To substantiate that $(x, y) = (4, 0)$ is the optimal solution, we sketch the graph, showing the feasible region, several objective functions, and the optimal point, in Fig. 12.24. □

Exercise Set 12.4

1. To better understand the algebraic method, an outline of the solution to a linear programming problem is provided; however, the solution is only partially completed. By completing the outline below, you will obtain a more thorough grasp of the methodology. The problem is:

Maximize: $4x + 5y = P$
Subject to: $x + 4y \leq 6,$
$7x + 6y \leq 20.$

The system of equations, with slack variables added, is

$4x + 5y + 0S_1 + 0S_2 = P,$
$x + 4y + S_1 + 0S_2 = 6,$

a) _____?_____.

b) Which variables are initially the zero variables?

c) The initial solution is

$$x = \underline{\quad}, \quad y = \underline{\quad}, \quad S_1 = \underline{\quad}, \quad S_2 = \underline{\quad}, \quad P = \underline{\quad}.$$

Next, determine the key variable and the pivotal equation. Choosing the largest positive coefficient in the objective function, and then finding the smallest quotient, we have identified the key variable and pivotal equation.

$$4x + \quad 5y \; + 0S_1 + 0S_2 = \quad P,$$
$$\longrightarrow \quad x + \textcircled{4y} + \quad S_1 + 0S_2 = \quad 6, \quad (6 \div 4 = \tfrac{3}{2})$$
$$7x + \quad 6y \; + 0S_1 + \quad S_2 = 20 \quad (20 \div 6 = \tfrac{10}{3})$$

d) The pivot is transformed into 1 by dividing by _____. The transformed system then becomes the following system:

$$4x + \boxed{5y} + 0S_1 + 0S_2 = P,$$
$$\tfrac{1}{4}x + \left(\, y \,\right) + \tfrac{1}{4}S_1 + 0S_2 = 1.5,$$
$$7x + \boxed{6y} + 0S_1 + \quad S_2 = 20.$$

This "new" pivotal equation is then used to eliminate the $5y$ term from the first equation. The result is written below. Now eliminate the $6y$ term from the third equation.

$$\tfrac{11}{4}x + 0y - \tfrac{5}{4}S_1 + 0S_2 = P - 7.5,$$
$$\tfrac{1}{4}x + \; y + \tfrac{1}{4}S_1 + 0S_2 = 1.5,$$

e) $\underline{\hspace{4cm} ? \hspace{4cm}}.$

f) For the transformed system above, what are now the zero variables?

g) An improved solution is

$$P = \underline{\quad}, \quad \text{with } x = \underline{\quad}, \quad y = \underline{\quad}, \quad S_1 = \underline{\quad}, \quad \text{and } S_2 = \underline{\quad}.$$

Again, a search is made for a new key variable and for a new pivotal equation. Examine the system below.

h) The new key variable is _____.

i) The new pivotal equation is which equation? _____

    i) $\tfrac{11}{4}x + 0y - \tfrac{5}{4}S_1 + 0S_2 = P - 7.5$

    ii) $\tfrac{1}{4}x + \; y + \tfrac{1}{4}S_1 + 0S_2 = 1.5$

    iii) $\tfrac{11}{2}x + 0y - \tfrac{3}{2}S_1 + \; S_2 = 11$

j) The transformed pivotal equation, with pivot equal to 1, is _____

$$\underline{\hspace{9cm}}.$$

(Cover the third transformed system of equations below before continuing reading.) The first two equations, after "$\frac{11}{4}x$" and "$\frac{1}{4}x$," respectively, are eliminated from the x column would be:

k) ───────────────────────────────── ,

l) ───────────────────────────────── ,

$$x + 0y - \tfrac{3}{11}S_1 + \tfrac{2}{11}S_2 = 2.$$

The third transformed system is:

$$0x + 0y - \tfrac{1}{2}S_1 - \tfrac{1}{2}S_2 = P - 13,$$
$$0x + \ y + \tfrac{7}{22}S_1 - \tfrac{1}{22}S_2 = 1,$$
$$x + 0y - \tfrac{3}{11}S_1 + \tfrac{2}{11}S_2 = 2.$$

m) The new solution is now:

$$P = 13, \quad \text{with } x = \underline{\quad}, \quad y = \underline{\quad}, \quad S_1 = \underline{\quad}, \quad \text{and } S_2 = \underline{\quad}.$$

Can we continue? The new objective function has no positive coefficients. Further improvements in the maximum profit are not possible. The original objective function has been maximized with $x = 2$ and $y = 1$, and a profit of $13.00.

Write a system of equations to replace the objective function and constraints in each of the following.

2. Maximize: $4x + 3y = P$
    Subject to: $x + \ y \le \ 60,$
            $3x + 2y \le 150$

3. Maximize: $x + 2y = P$
    Subject to: $x + 5y \le 10,$
            $3x + \ y \le \ 6$

4. Maximize: $x + \ y = P$
    Subject to: $2x + 3y \le 180,$
            $3x + 2y \le 240$

5. Maximize: $30x + 25y = P$
    Subject to: $20x - 10y \le 400,$
            $40x + 15y \le 1200,$
            $10x + 30y \le \ 300$

6. Maximize: $x + 2y = P$
    Subject to: $3x + \ y \le 6,$
            $x + 3y \le 6,$
            $x + \ y \le 2.5$

7. Maximize: $x + 2y + 3z = P$
    Subject to: $4x + 3y + 2z \le 400,$
            $3x + 2y + 4z \le 500,$
            $x + \ y + 2z \le 200$

8. Solve each linear programming problem in Exercises 2–7 using the algebraic method.

9. Solve each of the maximization problems in Exercises 16–27 of Exercise Set 12.3 using the algebraic method.

10. Find x and y such that

$$3x + 2y \le 6,$$
$$x + 4y \le 6,$$
$$x \ge 0, y \ge 0,$$

and $2x + 5y$ is maximized.

11. Find x and y such that

$$2x + \ y \le 18,$$
$$x + 3y \le 15,$$
$$x \ge 0, y \ge 0,$$

and $6x + 5y$ is maximized.

12. Find x and y such that

$$x + 3y \leq 1,$$
$$x + y \leq 2,$$
$$x + 2y \leq 1,$$
$$x \geq 0, y \geq 0,$$

and $10x + 20y$ is maximized.

12.5
Using Matrix Form with the Algebraic Method

As another step toward a better understanding (and eventually a presentation) of the simplex method, we will make some additional refinements in the algebraic method. A very important feature of the simplex method is its reliance on many of the matrix techniques developed for work on systems of equations (see Chapters 10 and 11).

The first step in the algebraic method is to formulate a system of equations to represent the linear programming problem. The matrix form presented in Chapter 11 had the distinct advantage of clearly displaying the variables involved. Use of matrices and matrix forms will also simplify the process of manipulating the rows of coefficients in the constraint equations and the equation representing the objective function.

As an illustration of the use of these matrix notations, we return to Example 10.

■ **EXAMPLE 10 (continued)**

Maximize the profit: $2x + y = P$
          Subject to $3x + 2y \leq 12,$
    these constraints: $x + 2y \leq 8$

Solution

We rewrite the linear programming problem as a system of equations:

$$2x + y + 0S_1 + 0S_2 = P,$$
$$3x + 2y + S_1 + 0S_2 = 12,$$
$$x + 2y + 0S_1 + S_2 = 8.$$

Now, by reviewing the matrix representation of a system of equations, we know that a system such as

$$3x + 2y = 5,$$
$$8x + 6y = 7$$

could be rewritten as

$$\begin{pmatrix} 3 & 2 \\ 8 & 6 \end{pmatrix} \cdot \begin{pmatrix} x \\ y \end{pmatrix} = \begin{pmatrix} 5 \\ 7 \end{pmatrix}.$$

Rewriting the system of equations that represents our linear programming example, we have

$$\begin{pmatrix} 2 & 1 & 0 & 0 \\ 3 & 2 & 1 & 0 \\ 1 & 2 & 0 & 1 \end{pmatrix} \cdot \begin{pmatrix} x \\ y \\ S_1 \\ S_2 \end{pmatrix} = \begin{pmatrix} P \\ 12 \\ 8 \end{pmatrix}.$$

Instead of referring to Eqs. (1), (2), or (3), we shall now refer to rows, or columns, of the coefficient matrix. Changes will be made in the coefficient matrix. And, as changes are made on the left-hand side of each row (equation), the corresponding changes will also be made on the right-hand side of each equation (in the constant matrix). Because no changes will be made in the column vector (x, y, S_1, S_2), the variable matrix, this matrix will be represented by the capital letter X.

Our linear programming system becomes

$$\begin{pmatrix} 2 & 1 & 0 & 0 \\ 3 & 2 & 1 & 0 \\ 1 & 2 & 0 & 1 \end{pmatrix} X = \begin{pmatrix} P \\ 12 \\ 8 \end{pmatrix}.$$

We then proceed with the following steps.

1. The key variable, x, is identified by the largest positive coefficient in row one. This column, which contains the coefficients of the key variable, is called the *key column*.

Key column
↓

$$\text{Pivotal} \rightarrow \begin{pmatrix} 2 & 1 & 0 & 0 \\ 3 & 2 & 1 & 0 \\ 1 & 2 & 0 & 1 \end{pmatrix} \cdot X = \begin{pmatrix} P \\ 12 \\ 8 \end{pmatrix}$$
row

2. The pivotal equation is determined by dividing each constant on the right by the respective coefficient of the key variable. This row, when determined, will be called the *pivotal row*.

3. The intersection of the key column and the pivotal row is called the *pivot*.

4. The pivot is changed to 1 by dividing by the value of the pivot.

$$\begin{pmatrix} 2 & 1 & 0 & 0 \\ ① & \frac{2}{3} & \frac{1}{3} & 0 \\ 1 & 2 & 0 & 1 \end{pmatrix} X = \begin{pmatrix} P \\ 4 \\ 8 \end{pmatrix} \qquad \text{(Row two ÷ by 3)}$$

5. The "new" pivotal row is used to eliminate the other entries in the key column (that is, to change them to 0).

Abbreviated row operations are given on the right to indicate how the new transformed matrix was obtained from the prior matrix (in step 4).

$$\begin{pmatrix} 0 & -\frac{1}{3} & -\frac{2}{3} & 0 \\ 1 & \frac{2}{3} & \frac{1}{3} & 0 \\ 0 & \frac{4}{3} & -\frac{1}{3} & 1 \end{pmatrix} \cdot \mathbf{X} = \begin{pmatrix} P - 8 \\ 4 \\ 4 \end{pmatrix} \qquad \begin{matrix} (-2 \cdot R2 + R1) \\ \\ (-1 \cdot R2 + R3) \end{matrix}$$

Because no positive coefficients are present in row one (representing the transformed objective function), the optimal solution has been reached. No further progress is possible! But where are the solutions? Even multiplying out the matrix forms does not provide a *clear* idea of what the solutions are or what variables are now nonzero!

$$\begin{aligned} 0x - \tfrac{1}{3}y - \tfrac{2}{3}S_1 + 0S_2 &= P - 8, \\ x + \tfrac{2}{3}y + \tfrac{1}{3}S_1 + 0S_2 &= 4, \\ 0x + \tfrac{4}{3}y - \tfrac{1}{3}S_1 + S_2 &= 4 \end{aligned}$$

However, by studying the matrix form above, and simultaneously studying the final solution from working the previous section,

$$x = 4, \quad y = 0, \quad S_1 = 0, \quad S_2 = 4, \quad P = 8,$$

we note that the nonzero variables x and S_2 have only 1's and 0's present in their respective columns. (This is no coincidence!)

In general, the nonzero variables occur in a column containing a single 1 with the other entries in that column being zeros. The remaining columns, then, represent the zero variables; in the case above, the zero variables are y and S_1.

A further illustration is given to show how the nonzero variables attain their values.

Because the second row times the variable vector $\mathbf{X}$ equals the second element in the column vector on the right of the equals sign,

$$\begin{pmatrix} 0 & -\frac{1}{3} & -\frac{2}{3} & 0 \\ 1 & \frac{2}{3} & \frac{1}{3} & 0 \\ 0 & \frac{4}{3} & -\frac{1}{3} & 1 \end{pmatrix} \begin{pmatrix} x \\ y \\ S_1 \\ S_2 \end{pmatrix} = \begin{pmatrix} \\ 4 \\ \end{pmatrix}$$

$1(x) + \tfrac{2}{3}(0) + \tfrac{1}{3}(0) + 0S_2 = 4$ yields $x = 4$. (Remember that y and S_1 are the zero variables.)

The third row times the variable vector $\mathbf{X}$ equals the third element in the column vector on the right side of the equals sign:

$$\begin{pmatrix} 0 & -\frac{1}{3} & -\frac{2}{3} & 0 \\ 1 & \frac{2}{3} & \frac{1}{3} & 0 \\ 0 & \frac{4}{3} & -\frac{1}{3} & 1 \end{pmatrix} \begin{pmatrix} x \\ y \\ S_1 \\ S_2 \end{pmatrix} = \begin{pmatrix} \\ \\ 4 \end{pmatrix}.$$

Therefore, $0(x) + \frac{4}{3}(0) - \frac{1}{3}(0) + 1(S_2) = 4$ yields $S_2 = 4$.

Because row one of the coefficient matrix times the variable matrix equals the first element in the column vector on the right of the equals sign,

$$
\begin{pmatrix} 0 & -\frac{1}{3} & -\frac{2}{3} & 0 \\ 1 & \frac{2}{3} & \frac{1}{3} & 0 \\ 0 & \frac{4}{3} & -\frac{1}{3} & 1 \end{pmatrix}
\begin{pmatrix} x \\ y \\ S_1 \\ S_2 \end{pmatrix} =
\begin{pmatrix} P - 8 \\ \\ \end{pmatrix},
$$

then $0(x) - \frac{1}{3}(y) - \frac{2}{3}(S_1) + 0(S_2) = P - 8$ yields $0 = P - 8$ or that the profit $= 8$.

In much simpler form, the solution can now be read directly from the final matrix in the following manner:

$$
\begin{pmatrix} 0 & -\frac{1}{3} & -\frac{2}{3} & 0 \\ 1 & \frac{2}{3} & \frac{1}{3} & 0 \\ 0 & 1\frac{1}{3} & -\frac{1}{3} & 1 \end{pmatrix}
\quad X = \quad
\begin{pmatrix} P - 8 \\ 4 \\ 4 \end{pmatrix} \quad \text{Profit}
$$

Matrices will greatly aid in the consolidation of a linear programming problem as manipulation and calculations are performed on the coefficient (and simultaneously, the constant) matrix.

As one final illustration (before you begin your own work) we will solve the linear programming problem in Example 9 using matrix methods. A good review of the mechanics of the algebraic method (in simpler form) will be presented, in addition to providing a rather revealing comparison of the amount of work needed for the algebraic method versus the algebraic method in matrix form!

■ **EXAMPLE 11**

Using the algebraic method in matrix form, determine how many gadgets and widgets should be manufactured in order to maximize profit if:

$$P = 20G + 30W \quad \text{and}$$
$$4G + 12W \le 120,$$
$$2G + 2W \le 40.$$

By algebraic method in matrix form.

Solution

Once we have introduced the slack variables, the following system becomes our starting point:

$$\begin{array}{cccc} G & W & S_1 & S_2 \end{array}$$

$$\rightarrow \begin{pmatrix} 20 & 30 & 0 & 0 \\ 4 & 12 & 1 & 0 \\ 2 & 2 & 0 & 1 \end{pmatrix} X = \begin{pmatrix} P \\ 120 \\ 40 \end{pmatrix} \quad \begin{array}{l} (120 \div 12 = 10) \\ (40 \div 2 = 20) \end{array}$$

$$\uparrow$$

1. The key variable is W, or the second column (the largest positive coefficient in the objective function).
2. The pivotal equation is row two (the smaller of the two quotients).
3. We reduce the pivot to 1. (We divide the pivotal row by the value of the pivot.)

$$\begin{pmatrix} 20 & 30 & 0 & 0 \\ \frac{1}{3} & \textcircled{1} & \frac{1}{12} & 0 \\ 2 & 2 & 0 & 1 \end{pmatrix} X = \begin{pmatrix} P \\ 10 \\ 40 \end{pmatrix}$$

4. The "new" pivotal row is used to eliminate all other entries in the key column. (The row operations used are shown in abbreviated form to the left of the transformed system.)

$$\begin{array}{l} (-30)R2 + R1 \\ \\ (-2)R2 + R3 \end{array} \begin{pmatrix} 10 & 0 & -\frac{5}{2} & 0 \\ \frac{1}{3} & 1 & \frac{1}{12} & 0 \\ \frac{4}{3} & 0 & -\frac{1}{6} & 1 \end{pmatrix} \begin{pmatrix} P - 300 \\ 10 \\ 20 \end{pmatrix}$$

$$\begin{array}{cccc} G & W & S_1 & S_2 \end{array}$$

An improved solution can now be determined. Recall that the columns with a 1 and remaining zeros represent the nonzero variables; the other columns represent the zero variables.

$$\begin{array}{ll} 10G + \boxed{0W} - \frac{5}{2}S_1 + \boxed{0S_2} = P - 300 & \bigg| \quad 0 = P - 300 \\ \frac{1}{3}G + \boxed{1W} + \frac{1}{12}S_1 + \boxed{0S_2} = 10 & \bigg| \quad W = 10 \\ \frac{4}{3}G + \boxed{0W} - \frac{1}{6}S_1 + \boxed{1S_2} = 20 & \bigg| \quad S_2 = 20 \end{array}$$

With $G = 0$ and $S_1 = 0$, we get the reduced system on the right above.

A solution is that the profit $= 300$, with $G = 0$, $W = 10$, $S_1 = 0$, and $S_2 = 20$.

Examine the last matrix representation of the system. Determine how the solutions could have been read directly from the matrices.

On to a better solution? Examining the first row (the transformed objective function), we find a positive coefficient.

$$\begin{array}{cccc} G & W & S & S \end{array}$$

$$\rightarrow \begin{pmatrix} 10 & 0 & -\frac{5}{2} & 0 \\ \frac{1}{3} & 1 & \frac{1}{12} & 0 \\ \frac{4}{3} & 0 & -\frac{1}{6} & 1 \end{pmatrix} X = \begin{pmatrix} P - 300 \\ 10 \\ 20 \end{pmatrix} \quad \begin{array}{l} (10 \div \frac{1}{3} = 30) \\ (20 \div \frac{4}{3} = 15) \end{array}$$

1. The new key variable is G.
2. The new pivotal equation is row three; the new pivot is $\frac{4}{3}$.
3. Reduce the pivot to 1 by dividing by $\frac{4}{3}$.

$$\begin{pmatrix} 10 & 0 & -\frac{5}{2} & 0 \\ \frac{1}{3} & 1 & \frac{1}{12} & 0 \\ \textcircled{1} & 0 & -\frac{1}{8} & \frac{3}{4} \end{pmatrix} X = \begin{pmatrix} P - 300 \\ 10 \\ 15 \end{pmatrix}$$

4. The new pivotal row is used to eliminate the coefficients of the key variables in all other rows by forming linear combinations.

Again, the specific linear combinations that are formed are mentioned to the left of each "new" transformed matrix representation below:

$$\begin{array}{c} \\ (-10)R3 + R1 \\ (-\frac{1}{3})R3 + R2 \end{array} \quad \begin{array}{cccc} G & W & S_1 & S_2 \\ \begin{pmatrix} 0 & 0 & -\frac{5}{4} & -\frac{15}{2} \\ 0 & 1 & \frac{1}{8} & -\frac{1}{4} \\ 1 & 0 & -\frac{1}{8} & \frac{3}{4} \end{pmatrix} \end{array} X = \begin{pmatrix} P - 450 \\ 5 \\ 15 \end{pmatrix}$$

Solutions can now be determined! By remembering which columns represent the nonzero variables (a single 1 and remaining zeros), we can read the solutions directly from the matrix form.

$$\begin{array}{cccc} G & W & S_1 & S_2 \\ \downarrow & \downarrow & & \\ \begin{pmatrix} 0 & 0 & -\frac{5}{4} & -\frac{15}{2} \\ \downarrow & \downarrow & & \\ 0 & 1 \rightarrow & \frac{1}{8} \rightarrow & -\frac{1}{4} \rightarrow \\ \downarrow & & & \\ 1 \rightarrow & 0 \rightarrow & -\frac{1}{8} \rightarrow & \frac{3}{4} \end{pmatrix} \end{array} X = \begin{pmatrix} P - 450 \\ \rightarrow 5 \\ \rightarrow 15 \end{pmatrix}$$

We have $G = 15$, $W = 5$, and the zero variables are now S_1 and S_2. Thus an improved solution is that the profit = \$450 with 15 gadgets, 5 widgets, and zero slack time.

Can we improve further? No! Because no positive coefficients exist in the new revised objective function, any further work would cause a decrease in our profit of \$450.00.   □

Matrices are useful and more efficient than the algebraic method— yet they retain all the workings of the algebraic method. However, as you shall see in the next section, matrix forms will be further simplified to yield the method most widely used in solving linear programming problems: the simplex method.

Exercise Set 12.5

Transform each of the following linear programming problems into a system of equations in matrix form.

1. Maximize: $3x + 2y = P$
 Subject to: $4x + y \leq 14,$
 $2x + 3y \leq 22$

2. Maximize: $25x + 40y = P$
 Subject to: $x + 3y \leq 15,$
 $4x + 2y \leq 16$

3. Maximize: $x + y = P$
 Subject to: $20x + 30y \leq 120,$
 $50x + 10y \leq 170$

4. Maximize: $6x + 4y = P$
 Subject to: $x - y \leq 0,$
 $5x + 3y \leq 75$

5. Maximize: $4x + 4y = P$
 Subject to: $2x + 3y \leq 18,$
 $3x + 2y \leq 24$

6. Maximize: $1.3x + 2.6y = P$
 Subject to: $20x + 10y \leq 160,$
 $10x + 30y \leq 150$

For each inequality, a different slack variable must be introduced to transform the constraint inequality into an equation. Most examples thus far involved only two constraints. Try your ability to set up linear programming problems by writing each problem below as a system of equations in matrix form.

7. Maximize: $3x + 2y = P$
 Subject to: $2x + y \leq 10,$
 $x + 3y \leq 9,$
 $4x + 6y \leq 24$

8. Maximize: $x + 2y = P$
 Subject to: $10x + 30y \leq 60,$
 $30x + 10y \leq 60,$
 $10x + 10y \leq 25$

9. Maximize: $10x + 10y = P$
 Subject to: $2x + 3y \leq 90,$
 $5x + 4y \leq 200,$
 $y \leq 20$

10. Solve each of the linear programming problems in Exercises 1–9 using matrix representation.

12.6
The Simplex Method

By visualizing the feasible region of the gadget–widget linear programming problem (Example 9) simultaneously with the listing of the solutions as profit is increased from \$0 to \$450, we can see the "idea" behind not only the algebraic method, but also the simplex method.

 By referring to Fig. 12.25, we can see that, in essence, the optimal solution was found by (1) selecting a corner point of the feasible

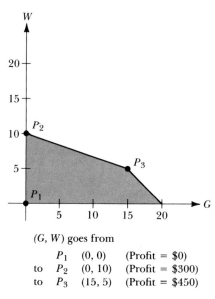

(G, W) goes from

	P_1	(0, 0)	(Profit = $0)
to	P_2	(0, 10)	(Profit = $300)
to	P_3	(15, 5)	(Profit = $450)

Figure 12.25

region (0, 0); (2) choosing a boundary to proceed to the next corner point such that the objective function increased in value; (3) finding an improved solution by evaluating the objective function at the "new" corner point (0, 10); and (4) repeating steps (2) and (3) until an optimal solution was reached.

The simplex method, formulated by G. B. Dantzig, is a systematic procedure to identify the corner points of the feasible region and narrow down the search of these extreme points until an optimum solution has been located. As the number of variables increases, the process of finding the optimal point from the set of all corner points becomes very time-consuming—and, if done by hand, extremely tedious. Because the simplex method is very algorithmic in nature, it is readily adaptable to computers. And, except for all but the smallest of linear programming problems, it is not feasible, nor often possible, to solve these problems without the aid of a digital computer.

A brief description or outline of the simplex method will be given first, and then our familiar gadget–widget example will be reworked, this time to illustrate the simplex method.

The simplex method for solving linear programming problems is based on the workings of the algebraic method, but with even more refinements than the algebraic method in matrix form. It makes use of the elimination procedure for solving systems of simultaneous linear equations discussed in Chapter 10.

An outline of our work will follow this pattern: (1) problem, (2) problem in matrix form, (3) transition to simplex tableaus, (4) introduction of simplex terminology, (5) the algorithmic process, and (6) an example worked!

The Gadget–Widget Problem (Again)

For comparative purposes, we will use the simplex method on the previous gadget–widget problem.

$$\begin{array}{ll} \text{Maximize:} & 20G + 30W = P \\ \text{Subject to:} & 4G + 12W \le 120, \\ & 2G + 2W \le 40 \end{array}$$

When this linear programming problem with the constraints is converted to equations, we can write these coefficients in matrix form:

$$\begin{pmatrix} 20 & 30 & 0 & 0 \\ 4 & 12 & 1 & 0 \\ 2 & 2 & 0 & 1 \end{pmatrix} \mathbf{X} = \begin{pmatrix} P \\ 120 \\ 40 \end{pmatrix}.$$

A slightly different representation, in form, from the matrix form is called a *simplex tableau*. The above matrix form is written as a simplex tableau below:

	G	W	S_1	S_2	
S_1	4	12	1	0	120
S_2	2	2	0	1	40
	-20	-30	0	0	0

The entries in the last row are the *negatives* of the coefficients of the objective function. As you recall, the last step in the solution of this linear programming problem had

$$0G + 0W - \tfrac{5}{4}S_1 - \tfrac{15}{2}S_2 = P - 450$$

as the transformed objective function. Because the left-hand side of the equation was equivalent to 0, $0 = P - 450$ was the objective function. The profit was the "opposite" of -450. By multiplying the objective function by -1 ($20G + 30W = P$ becomes $-20G - 30W = -P$) and also eliminating the variable P, the "changing" values—and the optimal values—of the objective function will appear in the *lower right corner* of the simplex tableau. Presently the entry in this cell is 0 (the value at the beginning of the process). The entries in this last row of the tableau are called the *indicators*.

The entries in each column are the coefficients for the variables named at the top of each column (G, W, S_1, S_2). The nonzero variables (S_1, S_2) are also recorded to the *left* of each tableau; initially the two slack variables are nonzero, as shown in the following tableau:

	G	W	S_1	S_2	
→ S_1	4	⑫	1	0	120
S_2	2	2	0	1	40
	−20	−30	0	0	0

The initial simplex tableau, and all subsequent tableaus, display the same information that can be given in matrix form. The procedure for solving the linear programming problem using the simplex method is almost identical to the algebraic method and the algebraic method in matrix form. Instead of choosing the largest positive coefficient in the objective function, now (because of multiplying by −1) we use the smallest negative coefficient to determine the key variable. The remainder of the procedure is the same!

	G	W	S_1	S_2		
S_1	4	⑫	1	0	120	($120 \div 12 = 10$)
S_2	2	2	0	1	40	($40 \div 2 = 20$)
	−20	−30	0	0		

The indicator −30 determines the key variable, W. (Note that −30 is smaller than −20!) The key variable W will be entering the set of nonzero variables. The smaller nonnegative quotient of $120 \div 12$ and $40 \div 2$ is used to determine the pivotal row. The intersection of the key column and the key row is called the pivot—in this case, 12. The variable S_1, presently a nonzero variable, will be replacing the entering variable W, and S_1 will be departing the set of nonzero variables. The tableau is transformed so that the pivot is 1.

	G	W	S_1	S_2		
W	$\frac{1}{3}$	①	$\frac{1}{12}$	0	10	(Row 1 ÷ 12)
S_2	2	2	0	1	40	
	−20	−30	0	0		

By row elimination techniques, all other numbers in this column (W) become 0. (Row elimination procedures are noted to the right of the tableau for your reference.)

The second simplex tableau is

	G	W	S_1	S_2		
W	$\frac{1}{3}$	1	$\frac{1}{12}$	0	10	
S_2	$\frac{4}{3}$	0	$-\frac{1}{6}$	1	20	$(-2 \times R1 + R2)$
	-10	0	$\frac{5}{2}$	0	300	$(30 \times R1 + R3)$

Reading the solution at each stage is now much simpler. The nonzero variables are to the left of the tableau, with their respective values to the right of the tableau ($W = 10$, $S_2 = 20$). The remaining variables are zero-valued. The present optimal value is in the lower right-hand corner: $P = \$300$ with $G = 0$, $W = 10$, $S_1 = 0$, $S_2 = 20$.

To continue, the same process is repeated.

1. Determine the key variable by the smallest negative number in the last row. This number determines the entering variable, G.

	G	W	S_1	S_2	
W	$\frac{1}{3}$	1	$\frac{1}{12}$	0	10
S_2	$\frac{4}{3}$	0	$-\frac{1}{6}$	1	20
	-10	0	$\frac{5}{2}$	0	300

2. The pivotal row is determined by finding the smallest nonnegative quotient. This also establishes the departing variable. The pivot is then transformed to be 1 by dividing row two by $\frac{4}{3}$.

	G	W	S_1	S_2	
W	$\frac{1}{3}$	1	$\frac{1}{12}$	0	10
G	①	0	$-\frac{1}{8}$	$\frac{3}{4}$	15
	-10	0	$\frac{5}{2}$	0	300

3. Transform the tableau by row-elimination methods until each other entry in the key column is zero.

	G	W	S_1	S_2	
W	0	1	$\frac{1}{8}$	$-\frac{1}{4}$	5
G	1	0	$-\frac{1}{8}$	$\frac{3}{4}$	15
	0	0	$\frac{5}{4}$	$\frac{15}{2}$	450

The third simplex tableau is

	G	W	S_1	S_2	
W	0	1	$\frac{1}{8}$	$-\frac{1}{4}$	5
G	1	0	$-\frac{1}{8}$	$\frac{3}{4}$	15
	0	0	$\frac{5}{4}$	$\frac{15}{2}$	450

Final solution? Steps (1)–(3) are repeated, each time yielding another tableau with an improved solution. We continue doing this until we have obtained a simplex tableau with no negative numbers in the bottom row. (Recall that in the algebraic method, progress continued so long as there were positive entries. After multiplying the objective function by -1, the criteria to continue, now, become negative indicators.) The final solution is a profit of $450 with $G = 15$, $W = 5$, $S_1 = 0$, $S_2 = 0$.

Before introducing the next example, the algorithmic nature of the simplex method is given in outline form to summarize our work.

■

1. Construct the initial simplex tableau.
2. Find the pivot by locating the entering variable (smallest negative entry in the objective function) and the departing variable (smallest nonnegative quotient).
3. Derive a new improved simplex tableau by using the pivotal row and elimination techniques.
4. Repeat steps (2) and (3) until a terminal simplex tableau is obtained (no more negative entries in the objective function).
5. The nonzero variables are to the left of the final tableau, their values to the right, and the optimal value of the objective function in the lower right-hand corner.

■ **EXAMPLE 12**

A practical application. An investor wants to purchase three types of stocks: a potential growth stock (A), a conservative blue chip stock (B), and a high dividend stock (C). A maximum of $8000 is available for investing. No more than $6000 should be invested in the B and C stocks combined; no more than $3000 should be invested in A and C combined. The rates of return on stocks A, B, and C are 10%, 6%, and 14%, respectively. How much should be invested in each stock in order to maximize return?

Solution

We first determine the variables. We let A stand for money invested in growth stock, B for money invested in blue chip, and C for money invested in high dividend stock. Then we derive the objective function,

$$0.10A + 0.06B + 0.14C = \text{Return},$$

and the constraints,

$$A + B + C \le 8000,$$
$$B + C \le 6000,$$
$$A \quad + C \le 3000.$$

We introduce slack variables, one per constraint, to form a system of equations:

$$A + B + C + S_1 = 8000,$$
$$B + C + S_2 = 6000,$$
$$A \quad + C + S_3 = 3000.$$

Now we begin the algorithmic nature of the simplex method.

1. We set up the initial tableau.

	A	B	C	S_1	S_2	S_3	
S_1	1	1	1	1	0	0	8000
S_2	0	1	1	0	1	0	6000
S_3	1	0	1	0	0	1	3000
	-0.1	-0.06	-0.14	0	0	0	0

2. We then find the entering variable, the departing variable, and the pivot.

	A	B	C	S_1	S_2	S_3	
S_1	1	1	1	1	0	0	8000
S_2	0	1	1	0	1	0	6000
$\to S_3$	1	0	①	0	0	1	3000
	-0.1	-0.06	-0.14	0	0	0	0

3. After transforming the pivot to 1 (it already is in this case), we derive a new improved tableau.

	A	B	C	S_1	S_2	S_3	
S_1	0	1	0	1	0	-1	5000
S_2	-1	1	0	0	1	-1	3000
C	1	0	1	0	0	1	3000
	0.04	-0.06	0	0	0	0.14	\$420

A return of $420 is obtained by $A = 0$, $B = 0$, $C = \$3000$, $S_1 = \$5000$, $S_2 = \$3000$, $S_3 = 0$.

4. We repeat steps (2) and (3) because a negative coefficient exists in the objective function (-0.06).

↓

	A	B	C	S_1	S_2	S_3	
S_1	0	1	0	1	0	-1	5000
S_2	-1	①	0	0	1	-1	3000
C	1	0	1	0	0	1	3000
	0.04	-0.06	0	0	0	0.14	$420

2. We find the entering and departing variables and the pivot.
3. After transforming the pivot to 1 (already done in this case), we eliminate all other entries in the key column.

	A	B	C	S_1	S_2	S_3	
S_1	1	0	0	1	-1	0	2000
B	-1	①	0	0	1	-1	3000
C	1	0	1	0	0	1	3000
	-0.02	0	0	0	0.06	0.08	$600

4. Again, we repeat steps (2) and (3), because of the negative coefficient (-0.02).
2. We find the new entering variable (A), the new departing variable (S_1), and the pivot (1).
3. After transforming the pivot to 1 (already done), we again eliminate all other entries in the key column using row elimination.

	A	B	C	S_1	S_2	S_3	
A	1	0	0	1	-1	0	2000
B	0	1	0	1	0	-1	5000
C	0	0	1	-1	1	1	1000
	0	0	0	0.02	0.04	0.08	$640

With no negative entries in the new revised objective function, there can be no more entering variables. Our final simplex tableau has been obtained. The solutions are read: A, B, and C on the left are the nonzero variables. By investing $2000 in growth stocks ($A$), $5000 in blue chip stocks (B), and $1000 in high dividend stocks (C), the investor will realize an optimal return of $640, while not exceeding any of the original constraints. □

A Minimization Problem

Up until now almost all the problems we worked with were maximization problems. Before completing our discussion of the simplex method, let's examine a minimization problem.

■ **EXAMPLE 13**

Minimize: $4x + 2y$
Subject to: $x + 2y \geq 10,$
       $4x + 3y \geq 12$

Solution

We have placed most of our emphasis in this section on maximization problems because of the following fact.

■

> Every minimization problem can be "transformed" into a maximization problem (and when necessary, vice versa).

How? Without introducing the slack variables, we first write the linear programming problem in matrix form:

$$\begin{array}{cc|c} 1 & 2 & 10 \\ 4 & 3 & 12 \\ \hline 4 & 2 & \end{array}$$

For this "matrix," we will find its dual. The *dual* is obtained by using the transpose of the given matrix. (For a discussion of transpose, see Chapter 11.) Briefly, to obtain the transpose of a matrix, we *interchange* the rows and columns; that is, we write row one as column one, row two as column two, and so on. In this case the transpose is

$$\begin{array}{cc|c} 1 & 4 & 4 \\ 2 & 3 & 2 \\ \hline 10 & 12 & \end{array}$$

We now can form a dual problem. For every given minimizing linear programming problem (called the primal problem), there is an associated maximizing problem called its *dual problem,* and vice versa.

For the minimization example, the corresponding dual problem, with the slack variables included, is as follows:

$$x + 4y + S_1 + 0S_2 = 4,$$
$$2x + 3y + 0S_1 + S_2 = 2,$$
$$10x + 12y + 0S_1 + 0S_2 = ?$$

The initial tableau is then

	x	y	S_1	S_2	
S_1	1	4	1	0	4
S_2	2	3	0	1	2
	-10	-12	0	0	0

(Note: The "new" objective function has been multiplied by -1.)

The work from this stage on is identical to that developed earlier. Because we have worked several simplex problems before now, the following work is presented in skeletal form, even though each step is shown.

1	4	1	0	4
$\rightarrow$ 2	3	0	1	2
-10	-12	0	0	0

$\uparrow$

The key variable and pivotal equation are identified.

1	4	1	0	4
$\frac{2}{3}$	①	0	$\frac{1}{3}$	$\frac{2}{3}$
-10	-12	0	0	0

The pivot is changed to 1.

$-\frac{5}{3}$	⓪	1	$-\frac{4}{3}$	$\frac{4}{3}$
$\frac{2}{3}$	1	0	$\frac{1}{3}$	$\frac{2}{3}$
-2	⓪	0	4	8

Numbers other than the pivot in the key column are eliminated by row operations.

$-\frac{5}{3}$	0	1	$-\frac{4}{3}$	$\frac{4}{3}$
$\rightarrow$ $\frac{2}{3}$	1	0	$\frac{1}{3}$	$\frac{2}{3}$
-2	0	0	4	8

$\uparrow$

A new key variable is determined, and the pivotal equation is found.

$-\frac{5}{3}$	0	1	$-\frac{4}{3}$	$\frac{4}{3}$
①	$\frac{3}{2}$	0	$\frac{1}{2}$	1
-2	0	0	4	8

The pivot is changed to 1.

⓪	$\frac{5}{2}$	1	$-\frac{1}{2}$	3
1	$\frac{3}{2}$	0	$\frac{1}{2}$	1
⓪	3	0	5	10

Other numbers in the key column are eliminated using row operations.

Because no negative numbers exist in the bottom line (the revised objective function), further reiterations are not possible. The optimal solutions are read from the following tableau:

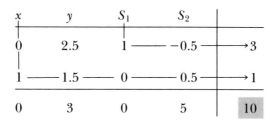

x	y	S_1	S_2	
0	2.5	1 — -0.5		$\rightarrow 3$
1 — 1.5		0 — 0.5		$\rightarrow 1$
0	3	0	5	10

Thus the minimum cost is $10, and the other final values are $x = 1$, $y = 0$, $S_1 = 3$, and $S_2 = 0$.

Exercise Set 12.6

In each of the following tableaus, (a) find the key variable, (b) label the pivotal equation, and (c) circle the pivot.

1.

A	B	S_1	S_2	
3	5	1	0	10
2	4	0	1	12
-5	-8	0	0	0

2.

G	W	S_1	S_2	
2	0.5	1	0	4
0.8	6	0	1	12
2	-4	0	0	0

3.

R	T	L	S_1	S_2	S_3	
4	2	3	1	0	0	6
-2	-1	2	0	1	0	6
4	3	4	0	0	1	6
-1	-2	3	0	0	0	0

4.

A	B	C	S_1	S_2	S_3	
1.2	3	-0.5	1	0	0	4.8
0.6	2	1.3	0	1	0	7.2
1.8	0.2	4.6	0	0	1	7.2
-5	-3	1	0	0	0	0

Transform each of the following tableaus into the final simplex tableau (this may require several steps). (*Note:* The objective function has not been multiplied by -1 yet!)

5.

1	1	1	0	10
1	6	0	1	15
4	7	0	0	

6.

2	3	1	0	39
2	1	0	1	25
3	4	0	0	

7.

1.0	1.0	1	0	0	24
1.2	0.4	0	1	0	24
0.6	1.2	0	0	1	24
4	8	0	0	0	

8.

3	1	1	0	0	28
2	3	0	1	0	39
2	1	0	0	1	25
8	3	0	0	0	

Use the simplex method to solve each of these linear programming problems.

9. Find x and y such that the constraints

$$x + 5y \le 10,$$
$$3x + y \le 6,$$
$$x \ge 0, y \ge 0$$

are satisfied, and $P = x + 2y$ is maximized.

10. Maximize: $2x + 5y = P$
Subject to: $3x + 2y \le 6,$
            $x + 2y \le 7,$
            $x \ge 0,$
            $y \ge 0$

11. Find the maximum value of $2.0G + 3.0W$, where G and W must be nonnegative and satisfy the inequalities

$$G + W \le 6,$$
$$3G + 5W \le 20,$$
$$G - W \le 1,$$
$$G \ge 0, W \ge 0.$$

12.   Minimize: $x + y$
        Subject to
the constraints: $5x + y \ge 10,$
                $x + 4y \ge 8$

(*Hint:* Consider this problem to be the primal problem; form the dual problem.)

13. Minimize: $5A + 4B = \text{Cost}$
Subject to: $A + 2B \ge 50,$
            $5A + 10B \ge 75$

14. Solve Exercises 16–19, 22, and 24–25 of Exercise Set 12.3 using the simplex method.

15. Artificial variables (optional)

    a) In linear programming problems involving "greater than" constraints, artificial variables, as well as slack variables, must be introduced. For example,

$$\text{to maximize:} \quad 5x + 3y$$
$$\text{subject to:} \quad 3x + 2y \le 150,$$
$$x + 2y \ge 40$$

we first add slack variables:

$$\begin{array}{cccc} x & y & S_1 & S_2 \\ \begin{pmatrix} 5 & 3 & 0 & 0 \\ 3 & 2 & 1 & 0 \\ 1 & 2 & 0 & 1 \end{pmatrix} & & & \left. \begin{matrix} 0 \\ 150 \\ 40 \end{matrix} \right) . \end{array}$$

However, when the solution process begins, the initial solution is zero. The slack variables are used to transform the "less than" inequality to equality (that is, take up the slack). While S_2 will take up the slack if the constraint $x + 2y \ge 40$ is less than 40, a variable is needed to account for the amount initially in excess of 40. Because slack variables are not negative, another variable, called an *artificial variable,* is introduced in $x + 2y \ge 40$: $x + 2y - A_1 \ge 40$. If we subtract the excess from $x + 2y$, equality with 40 can be achieved. The problem in matrix form, ready for solving, is

$$\begin{array}{ccccc} x & y & S_1 & S_2 & A_1 \\ \begin{pmatrix} 5 & 3 & 0 & 0 & 0 \\ 3 & 2 & 1 & 0 & 0 \\ 1 & 2 & 0 & 1 & -1 \end{pmatrix} & & & & \left. \begin{matrix} 0 \\ 150 \\ 40 \end{matrix} \right) . \end{array}$$

Rewrite the problem above in a simplex tableau, and then solve by the "usual" procedure explained under the simplex method.

    b) Solve using the simplex method. (Introduce slack and artificial variables where necessary.)

$$\text{Maximize:} \quad x + 3y = P$$
$$\text{Subject to:} \quad x + y \le 5,$$
$$x + y \ge 2,$$
$$x \le 4, y \le 3$$

16. Solve the linear programming problems in Exercises 2–7 and 10–12 of Exercise Set 12.4 using the simplex method.

17. Solve Exercises 7–9 of Exercise Set 12.5 using the simplex method.

Solve each of the following by the simplex method.

18. An ice cream manufacturer blends ingredients A, B, and C in varying proportions to make three types of ice cream: low calorie (L), regular (R), and extra-rich (E). Each gallon of L contains 0.2 gallon of A, 0.3 gallon of B, and 0.2 gallon of C. Each gallon of R contains 0.4 gallon of A, 0.2 gallon of B, and 0.2 gallon of C. Each gallon of E contains 0.5 gallon of A, 0.2 gallon of B, and 0.1 gallon of C. Suppose that the profits on the three ice creams L, R, and E are $0.20, $0.40, and $0.20, respectively. If the manufacturer has 50, 80, and 100 gallons of A, B, and C available, how many gallons of each type of ice cream should be made in order to maximize the profit?

19. An investment banker has up to $12,000 available for investing. She can purchase a type-A bond yielding an 8% return on the amount invested, and she can purchase a type-M bond yielding a 12% return. Her client insists that she invest at least twice as much in A as in M, but no more than $6000 in A and no more than $8000 in M. How much money should be invested in each kind of bond in order to maximize the client's return?

Summary

A relatively recent topic in mathematics, linear programming has found a wide diversity of applications in such areas as agriculture, business management, economics, marketing, and transportation. Many real-life situations require finding an optimal solution to a given problem; however, the task is made more difficult and the complexity is greatly increased as certain constraints or restrictions are imposed on the variables.

For example, in the production of two items, A and B, profit is obtained by the function

$$\$3A + \$4B = P$$

($3.00 for each A and $4.00 for each B). The problem is to find A and B such that the profit will be the greatest, yet A and B will be subject to the following restrictions:

$$7A + 2B \le 140,$$
$$2A + 5B \le 80.$$

Linear programming provides a solution to problems far more complex than this. Linear programming can either maximize or minimize a given linear function in several variables (the objective function), subject to a set of linear constraints, or restrictions, on these same variables. A typical linear programming problem, albeit small, is to optimize

$$c_1x_1 + c_2x_2 + c_3x_3$$

subject to the constraints

$$a_1x_1 + a_2x_2 + a_3x_3 \le K_1,$$
$$b_1x_1 + b_2x_2 + b_3x_3 \le K_2,$$
$$c_1x_1 + c_2x_2 + c_3x_3 \le K_3.$$

Several methods are used for solving linear programming problems. To provide an insight into the linear programming process, the *graphical method* was shown first. From the common intersection of all the constraints, a *feasible region* can be determined. The *optimal point* is then chosen from this area by selectively choosing the corner point of the feasible region with the optimal value from the objective function.

Although the graphing method does provide a clear picture of the processes behind linear programming problems, it is really suitable only for linear programming problems containing two variables. Furthermore, as the number of constraints increases, the number of vertices to be evaluated increases drastically. (With 5 variables and 6 constraints, there would be more than 400 vertices to evaluate to determine the optimal value for the objective function.)

An alternative, very systematic, means of solving linear programming problems is the *simplex method*. Geometrically, the simplex method proceeds systematically from one vertex to another vertex such that the value of the objective function is automatically improved. In essence, the simplex method depends on the use of row operations on matrices to solve systems of linear equations. Toward the goal of understanding the many manipulations required in the simplex method, a format was developed beginning with the *algebraic method*. By transforming the constraints into a system of equations, we built on procedures learned in Chapter 10.

■

The algorithmic nature of the algebraic method is summarized as follows:

1. A key variable is identified in the objective function; this variable is used to determine the pivotal equation.
2. The pivot in the pivotal equation is then used to eliminate the key variable in the other equations.
3. An improved solution is then determined.

These steps are repeated progressing from a zero solution to continued improved solutions until an optimal solution is finally reached.

The algebraic method is then further refined and simplified using *matrix notation*. Several additional refinements are made; then the form of presentation becomes a tableau of the simplex method. From an illustrative graphical method, the progression has been made through the algebraic method, through the algebraic method in matrix form, to the simplex method in tableau form.

The simplex method of solving linear programming problems is summarized as follows:

1. Replace the constraints with a system of equations.
2. Write the simplex tableau.
3. Determine the pivot.
4. Perform pivotal row operations to arrive at an improved solution.
5. Repeat steps (2)–(4) until no further progress is possible.
6. An optimal solution to the linear programming problem has then been reached.

Frequently, practical applications involve 10, 20, and 30 or more variables, and as many restrictions or constraints. (Industrial examples exist involving over 300 variables and constraints!) Since it is a rather involved and intricate technique, the simplex method is tedious and subject to computational errors. However, because the simplex method is very systematic, it can be readily adapted for digital computers. Without the high-speed computers, linear programming solutions for some problems would be practically unattainable. And, without the use of the simplex method, it is not possible to solve any but the smallest linear programming problems.

Review Exercises

For each of the following linear programming problems, circle all points that satisfy all the constraints. From these feasible points, pick the one point that optimizes the given objective function.

1. Maximize: $x + 2y = P$ (10, 10), (20, 10),
   Subject to: $x + y \leq 40$,   (25, 10), (13, 27),
   $\phantom{\text{Subject to:}}$ $x + 4y \leq 80$ (27, 13), (13, 26)

2. Maximize: $2x + 3y$ (2, 3), (3, 2),
    Subject to: $x + 4y \leq 20,$ (4, 1), (2.1, 3.3),
               $x + y \leq 8,$ (5.2, 2.9), (6, 4)
             $2x - 2y \geq -4$

3. Minimize: $4x + 3y$ (1, 1), (2, 1),
    Subject to: $x + y \geq 2,$ (1, 2), (2, 2),
               $x + 3y \geq 3,$ (3, 2), (2, 3),
             $3x + y \geq 3$ (4, 1), (1, 4)

For each of the following systems of constraints, graph the feasible region.

4. $3x + 6y \leq 12,$
$2x + y \leq 8$

5. $4x + y \geq 8,$
$x + y \geq 5,$
$2x + 6y \geq 20$

6. $x + y \leq 100,$
$5x + 6y \leq 600,$
$0 \leq x \leq 70,$
$0 \leq y \leq 70$

7. $x + 2y \leq 6,$
$x + y \leq 4,$
$x - y \leq 2,$
$x \geq 0, y \geq 0$

8. For the feasible region shown in Fig. 12.26, find a point at which:

a) $x + y$ is a maximum;

b) $3x + 2y$ is a maximum;

c) $4x + y$ is a minimum;

d) $x + 4y$ is a maximum.

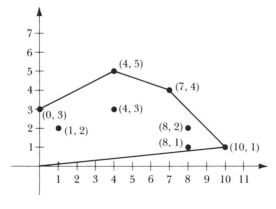

Figure 12.26

Use the graphical method to solve each of these linear programming problems.

9.   Maximize: $7x + 2y$
     Subject to: $2x + y \leq 2,$
     $x - y \leq 1,$
     $x \geq 0, y \geq 0$

10.   Maximize: $40x + 40y$
     Subject to: $3x + 9y \leq 27,$
     $5x + 2y \leq 10,$
     $x \geq 0, y \geq 0$

11. Minimize: $15x + 25y$
     Subject to: $x + 9y \leq 9,$
     $5x + 2y \leq 30,$
     $0 \leq y \leq 7,$
     $0 \leq x,$
     $5x - y \geq 0$

12. Find the minimum and maximum values for $P = 5x + 3y$ subject to the constraints in Exercise 11.

Solve each of the following linear programming problems by (a) the algebraic method and (b) the algebraic method in matrix form.

13. Maximize: $3x + 2y = P$
     Subject to: $2x + y \leq 2,$
     $x + 2y \leq 8,$
     $x \geq 0,$
     $y \geq 0$

14. Maximize: $8x + 10y$
     Subject to: $2x + 3y \leq 180,$
     $3x + 2y \leq 240,$
     $0 \leq y \leq 40,$
     $x \geq 0$

15. Maximize: Profit $= 4x + 3y + 2z$
     Subject to: $x + y + z \leq 10,$
     $3y + z \leq 50,$
     $2x + z \leq 35$

Set up the initial simplex tableau for each of the following.

16. Maximize: $30x + 40y$
     Subject to: $x + y \leq 40,$
     $x + 2y \leq 60,$
     $2x + y \leq 60$

17. Maximize: $40x + 39y + 30z$
     Subject to: $4x + 3y + 2z \leq 90,$
     $x + y + z \leq 32,$
     $3x + 4y + 2z \leq 84$

18. Minimize: $3x - 2y$
     Subject to: $x + y \leq 40,$
     $2x - y \leq 20,$
     $-x + 2y \geq 10$

19. Solve each of Exercises 13–18 using the simplex method.

20. An individual with $4000 to invest considers the following investments at the indicated after-tax rates of return: bonds, 4%; second mortgages, 8%; stocks, 3%; and loans, 5%. A broker advised him to invest no more than $1500 in mortgages and loans, at least $2000 in bonds and stocks, but less than $2500 in stocks and mortgages. Find the most profitable investment combination to maximize the return. (Write the objective function and the constraints; then form a tableau. Solve using the simplex method.)

Statistics
and Probability

CHAPTER 13 OBJECTIVES

After completing this chapter, you should be able to:

1. Differentiate between descriptive statistics and inferential statistics.
2. Select a representative sample from a population using the techniques of random sampling, systematic sampling, stratified sampling, or cluster sampling.
3. Organize data using frequency distributions, relative frequency distributions, and cumulative frequency distributions.
4. Present data graphically using histograms, frequency polygons, bar graphs, circle graphs, or pictographs.
5. Summarize data computing these descriptive statistics: n, mean, median, mode, range, standard deviation, and/or percentiles.
6. Explain why the study of probability is essential for work with statistics.
7. Solve basic probability problems.
8. Identify when probability events are mutually exclusive, independent, or dependent.
9. Use counting techniques such as tree diagrams, the multiplication principle, combinations, and permutations to determine the entire sample space of a statistical experiment.

13

13.1
Statistics (An Overview)

What is statistics? You might want to begin this chapter by briefly answering, in your own words, the following two questions: "What is *statistics?*" and "what is *probability?*"

Descriptive vs. Inferential Statistics

For many people, the word "statistics" conjures up visions of long columns and tables of numbers, or complex and often seemingly incomprehensible formulas. While these views are not completely wrong, statistics is much more than simply charts, tables, and formulas. Statistics and statistical methods are applicable to almost every area of human endeavor; it is virtually impossible to scan a newspaper or magazine (see Fig. 13.1), or turn on a news broadcast, without being confronted with statistical information. In data processing, you may frequently be asked to (a) provide management with statistical information in the form of *descriptive summaries*, or (b) make *predictions* or *inferences* from a set of gathered data.

It is rather difficult to be an educated person without having at least a passing acquaintance with basic statistics—so great is the extent of our dependency on this area of mathematics. Predictions have been made that having some knowledge of statistics may be as important a factor in good citizenship as the ability to read and write.

As a science, statistics is that branch of mathematics concerned with the *collection, organization, analysis, presentation,* and *interpretation* of numerical data. As a field, it is divided into two major areas: descriptive statistics and inferential statistics. **Descriptive statistics** deals with the presentation of numerical facts (data) in the form of charts, graphs, and tables, and with finding or calculating numbers to succinctly summarize the numerical facts.

■ **EXAMPLE 1**   Figure 13.1(g), (h), and (i) are examples of the *descriptive nature* of statistics (statistics used to describe the nature of the data collected).   □

■ **EXAMPLE 2**   If a computer installation were to determine the "average" job cost per week, the total cost per week could be divided by the total run time for jobs in that week. If it were determined that the *average* cost per job is $3.12 per minute, then this number describes approximately how much money should be allocated per minute of run time.   □

Statistical inference is concerned with drawing useful conclusions about a larger group, called the *population*, on the basis of a study of a relatively small portion of that population, called the *sample*.

■ **EXAMPLE 3**   Figure 13.1(a), (b), and (c) are examples of using statistics to make *inferences*. In Fig. 13.1(a), the population is *all* elderly people and the *sample* for the study is that subset of the population for which data were actually gathered—the 200 elderly people.   □

When using statistical inference, we rely heavily on selecting a sample that is "representative" of the population. In a study, it is entirely possible to select elements that are not representative of the population. For example, suppose we are testing light bulbs (35 bulbs will be tested) so we can later make a claim to our consumers that the average life of our light bulbs is 800 hours. Suppose the first 35 bulbs manufactured are selected and tested, and their average life found to be 740 hours. Can we make our claim? If our sample is not representative, our inference will not be valid. (Here we borrow an expression from computer science: GIGO (Garbage In, Garbage Out)!)

Regardless of how the 35 bulbs were selected, the possibility exists, however small, that when a sample of light bulbs is selected and tested, they will all burn out in fewer than 200 hours! By pure chance, the 35 worst light bulbs in the entire lot of 465,000 bulbs manufactured could be chosen. In making a selection, the sample chosen is obtained by some degree of chance; there is some uncertainty in the selection process of how "representative" our sample is of *all* the light bulbs manufactured.

Why Study Probability?

Quite often in business, predictions (or inferences) must be made involving quite a bit of uncertainty. For example, "From studies of market analyses, we are 90% sure (not absolutely certain though) that production levels for the next six months should be double the current month's level," or "The odds are good that the stock prices will increase by at least 12% for the next year." As Wallis and Roberts write in *The Nature of Statistics*, "statistics is a body of methods for making wise decisions in the face of uncertainty" (Free Press, 1965, p. 11).

When using such terms as "likelihood," "unlikely," "chance," "90% sure," and "odds," we are using concepts from the area of mathematics called *probability*. A knowledge of probability makes it possible to interpret statistical results. The basis for statistical inference, probability permits the numerical expression of the inevitable uncertainties in drawing conclusions.

Probability is very much a part of our everyday lives, since there are very few things in life of which we can be absolutely certain. If the weather forecaster predicts rain for tomorrow with a probability of 90%, do you take your umbrella? Is it possible that instead the sun will shine tomorrow? Analyzing the forecast, however, we see that what is really meant is that on the basis of past records and weather conditions, when the weather has been as it is today, rain followed on the next day 90% of the time.

(a)

According to a recent study of 200 elderly people, a four month program of walking improved the people's reaction time, their reasoning power and also their short-term memory.

(b)

□ **A sampling of Italian wines.** Most popular are Lambrusco and Soave, which are staples at most re~taurants and cocktail parties

(c)

According to a study in the Journal of the American Medical Association, hypnosis therapy aided over 70% in staying off cigarettes for over one year. (Before the study the 100 smokers averaged 1.2 packs a day.)

USA SNAPSHOTS
A look at statistics that shape our lives

Eatery beefs

More than 1 in 4 consumers (28%) say they have found restaurant menus to be misleading. When recently questioned in a national survey, consumers raised these five complaints most frequently:

Description of food	34%
Size of portion	23%
Price	13%
Quality of food	10%
Didn't receive some items	9%

Source: The Gallup Monthly Report on Eating Out

By Karren Loeb, USA TODAY

(e)

Slightly up at the gate

Attendance at National Basketball Association games is up this year by 184 fans per game, but only two teams — Boston and Portland — had sold out all their 1983-84 home games through Sunday. The NBA average attendance since the NBA-ABA merger in 1976-77 (through Sunday):

Season	Total Att	Teams	Games	Att/G
1976-77	9,898,521	22	902	10,974
1977-78	9,874,155	22	902	10,947
1978-79	9,761,377	22	902	10,822
1979-80	9,937,575	22	902	11,017
1980-81	9,449,340	23	943	10,021
1981-82	9,989,410	23	943	10,593
1982-83	9,637,614	23	943	10,220
1983-84	2,205,705	23	212	10,404

Source: NBA

(d)

DOW JONES STOCK AVGS.

	Open	High	Low	Close	Chg.
30 Ind.	1244.51	1260.26	1240.24	1254.98	+ 13.01
20 Trn.	591.83	600.71	588.75	596.79	+ 5.70
15 Util.	130.73	131.98	129.87	131.71	+ 0.39
65 Stks	497.45	503.90	495.28	501.31	+ 4.60

(g)

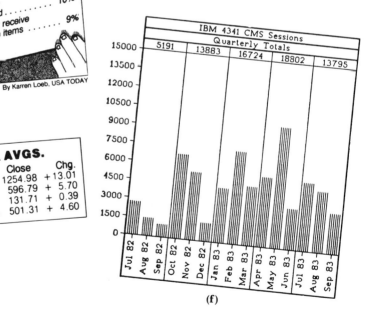

IBM 4341 CMS Sessions
Quarterly Totals

| 5191 | 13883 | 16724 | 18802 | 13795 |

(f)

| Redskins | 14 | 0 | 7 | 10 - 31 |
| Cowboys | 7 | 3 | 0 | 0 - 10 |

ATTENDANCE — 65,074 (8 no-shows)

REDSKINS	STATISTICS	COWBOYS
20	First downs	13
42-166	Rushes-yards	20-33
170	Pass yards	172
124	Return yards	148
17-11	Att/Cmp	35-20
1	Had intercepted	3
3-25	Sacks by	4-33
0-0	Fumbles/lost	2-1
2-25	Punts	4-43
5-56	Penalties/yds	6-109
31:51	Possession	28:09

Missed FGs: Moseley (Redskins) 49 yards.

REDSKINS RUSHING
Riggins 27-89, J. Washington 8-44, Theismann 3-24, Giaquinto 4-9.

COWBOYS RUSHING
Dorsett 14-34, Springs 6-(-1).

REDSKINS PASSING
Theismann 11-17-1 (int.)-203 yards and 2 TDs.

COWBOYS PASSING
White 20-35-3 (int.)-197 yards and 1 TD.

REDSKINS RECEIVING
Monk 6-119, Brown 2-31, Garrett 2-13, Didier 1-40.

COWBOYS RECEIVING
Hill 5-63, Cosbie 4-53, Dorsett 4-30, Springs 3-17, Pearson 2-22, Johnson 2-12.

(h)

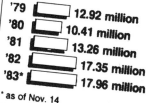

Comex volume

Here's the number of futures contracts and options traded on the New York Commodity Exchange:

'79	12.92 million
'80	10.41 million
'81	13.26 million
'82	17.35 million
'83*	17.96 million

* as of Nov. 14

Source: New York Commodities Exchange Inc.

USA TODAY

(i)

The sexual habits of American women as compiled from several recently published polls

(j)

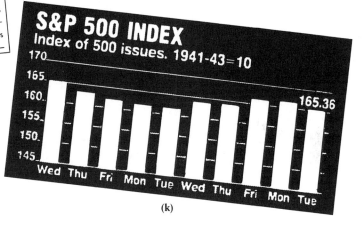

(k)

Tennessee size just average

Special for USA TODAY

WASHINGTON — Where's the USA's average state?
A new analysis says:
■ **Population:** USA average was 4,530,000; Tennessee was closest at 4,591,000.
■ **Median age:** Iowa, 30 years.
■ **Average college graduates:** Illinois, 16.2 percent.
■ **Average women in workforce:** Iowa, 50 percent.
■ **Median home income:** Pennsylvania, $16,880.
■ **Average housing cost:** Massachusetts, at $55,800.

(l)

In a study reported in *American Health*, retirees at age 62 died younger than those retiring at age 65. For this group of 64,000 people, staying active seems to increase longevity.

(m)

Figure 13.1

Probability Defined

Specifically, what is *probability*? In our weather forecasting example, the probability of rain was 90%. This means that 90 times out of 100, or 9 out of every 10 days similar to today's, rain will occur on the next day. This idea of probability is based on *relative frequency*—the ratio of the times an event will occur (9 days of rain) to the number of total possible outcomes (for every 10 days).

■ **DEFINITION**

PROBABILITY

Probability is a measure of the likelihood that some stated event will occur. Often probability can be expressed as the ratio of "favorable outcomes" to the "total possible outcomes."

■ **EXAMPLE 4**

Suppose a "fair" coin were tossed in the air. When the coin lands, the probability of getting a head is 1/2. For any coin there are two sides (total outcomes). For any coin there is 1 head (in this case, the favorable outcome or desired event). Thus the ratio of the *favorable outcomes* to the *total outcomes* is 1 to 2, or 1/2.   □

■ **EXAMPLE 5**

If you are rolling a die, what is the likelihood, or probability, of obtaining an even number (see Fig. 13.2)?

Figure 13.2

Solution

The total possible outcomes for rolling one die are shown in Fig. 13.3; the number of total outcomes is six.

Figure 13.3

The favorable outcomes, or desired outcomes, are shown in Fig. 13.4. Therefore, the probability of rolling an even number is the ratio

Figure 13.4

of the favorable outcomes (3) to the total outcomes (6). The probability of rolling an even number is 3/6, or 1/2. $\square$

$$\text{Probability} = \frac{\text{Number of favorable outcomes}}{\text{Number of total outcomes}}$$

If you are asked to choose a letter of the alphabet, what is the probability that the letter selected will be a vowel? Out of 26 possible choices (total outcomes) only 5 letters are vowels (a, e, i, o, u). The probability of picking a vowel then is 5/26.

In these examples, probability is regarded as the ratio of favorable outcomes to total outcomes. Another term used for the set of all possible outcomes is *sample space*. In statistical work, quite often a portion of the sample space is selected and studied; after a thorough analysis, an inference is made about the entire set based on the findings from the portion studied.

Population and Samples

In the example mentioned earlier, from a total of over 465,000 light bulbs, 35 were selected to be tested. Intended was the making of an inference "from the *testing of a few* to a *prediction about the many*" concerning the length of the average life of all light bulbs. In this case, the 465,000 bulbs representing all the possible bulbs from which to pick is the *population*. The portion of bulbs chosen to test (35) is called the *sample*.

The majority of work in statistical inference is centered on two questions: (1) "How is a representative sample selected from the population?" and (2) "how certain are we that the sample selected is indeed representative of the population?" (Recall that our 35 bulbs may not be representative of all the bulbs.) Methods to select a representative sample will be discussed next. Later, more work will be done with probability, showing how probability helps provide answers to the second question.

How is a representative sample selected from the population?

■ DEFINITION

POPULATION

The *population* is the total well-defined group of individuals, objects, or data about which information is desired.

■ **DEFINITION**

SAMPLE

A *sample* is a finite portion (subset) of the population.

Numerical descriptions of a population are called *parameters*. Numerical descriptions of a sample are called *statistics*. Generally, information is gathered from a sample in order to draw conclusions or make predictions (statistical inferences) about the population. The sample statistics are obtained to make suggestions about the population parameters. Certain industries often conduct a complete checkout called 100% inspection; every single item is checked for defective parts. However, modern statistical quality control has proven that more reliable—and much less expensive—conclusions can be reached by *sampling*.

Often it may not be possible to include every individual item of the population. Some quality control cannot be done by 100% inspection, especially in cases where the items tested are destroyed in the process. The selection of a suitable sample is then required. For example, to determine a light bulb's average life, clearly a sample must be selected; if all light bulbs were to be tested until they burned out, there would be no light bulbs left to sell!

● Question

In medicine, in order to determine the proportion of red blood cells in your blood, would you prefer that the doctor take only a sample, or the entire population?

To predict election results, samples are taken of the entire population, and if these have been chosen representatively, predictions of the population's vote can be accurately made. (Often by good sampling techniques, accurate predictions can be made about populations in excess of 200 million by sampling fewer than 2000!) However, if the selection of the sample is not representative, the inference can be misleading. A good example, from history, is the 1936 Presidential election in which *The Literary Digest* predicted Landon would beat Roosevelt. After the actual election, Roosevelt had swept all but 2 of the 48 states!

"Five out of seven people prefer . . . ." To make an advertising claim such as this, a population was identified, a representative sample chosen, and then statistical inferences made. Market research is designed specifically to discover consensus preferences and usages of products by appropriate sampling. Accounting and auditing procedures do not verify each and every entry; most often, well-chosen representative samples are taken.

Sampling

The term sample comes from the word example; thus the sample selected should be an example of what the population is like. In making a study of voter preferences in a city, if the ratio of Democrats to Republicans is 3 to 1, then our sample should include approximately three Democrats for each Republican. In a market research study to determine the approximate anticipated purchases of a new $380 microcomputer, the population was divided into three categories: income level under $12,000 (65%), income level between $12,000 and $25,000 (30%), and income in excess of $25,000 (5%). When selecting a sample of 100 people, how many should have incomes falling in the $12,000–$25,000 income level?

A sample that is nonrepresentative of the population is said to be *biased*. In the 1936 Presidential election prediction by *The Literary Digest,* the sample selected was chosen from listings of people who owned either a car or a telephone. The sample was biased—prejudiced in the sense that the poll favored the wealthier voters; the lower socioeconomic voters, those without cars or phones, were not represented.

What methods of sampling can be used to select a sample from the population of interest, and ensure a representative sample? To aid the process of statistical inference, we will discuss the following sampling techniques: simple random sampling, systematic sampling, stratified sampling, and cluster sampling.

Regardless of the sampling technique decided on, the prime objective is to eliminate—as much as possible—any bias in the selection of the sample. One technique of selection is *simple random sampling.* In a simple random sample, any item in the population has at any time as much chance of being chosen as any other item in the population. To sample voter preference, would opening the telephone directory and picking every 500th name be a random sample? No, because the 348th person, for example, has no chance of being selected, and many voters may have unlisted numbers.

■ DEFINITION

RANDOM SAMPLE

A *random sample* from a population is one in which each population item has an equal probability of being selected for the sample.

The importance of random sampling is that the probability of selected samples is known or can be determined; then valid inferences are more likely to be made about the parent population.

How do we select a simple random sample? One method is the "goldfish bowl," or "name in the hat," technique. If possible, we number all items in the population and put these numbers into a large container, mixing them thoroughly. Items are then picked—mixing after each choice—until the desired sample size is reached. A more usual procedure is to use a table of random numbers (see the endpapers at the back of the book) to select items from the population. After picking a starting point, at random, we move in any direction—right, left, up, down, or even on a diagonal—continuing to select numbers from the table. Another method for generating random numbers (or "pseudorandom" numbers) is to use a computer, since most models do have a function to generate random numbers.

To choose a sample of 800 voters from all registered voters in the United States, the technique of simple random sampling would not only be very time-consuming, but also quite expensive. Other sampling techniques would prove more useful.

In a *stratified random sample,* the population is first subdivided into strata, and a random sample is then chosen from each stratum. From past records, different subpopulations often show different voting preferences. Thus in our voting example, sampling should be stratified. Suppose, for example, the average salary was to be estimated for all employees of a large business firm. The following strata might be formed: day hourly workers, evening hourly workers, salaried employees, and finally supervisors and/or administrators. Then either a fixed random sample can be chosen from each stratum, or the technique can be modified by *proportional sampling* within each stratum. For example, suppose the ratio of females to males is 3 to 1; then approximately three times as many women should be selected from a stratum as men. (Additional modifications exist but are beyond the scope of this text.) The key to remember, for our purposes, is that the success of stratified sampling depends on getting the data as much alike as possible (homogeneous) within each stratum (for example, all evening hourly employees).

Another easy and frequently used technique for selecting a sample is called *systematic sampling.* As the name implies, some predetermined system is used to choose the sample from the population. A typical example is to select every 100th (or 30th, or 5th, etc.) item from the population. Or, in a quality control environment, every third part could be systematically marked for inspection.

As is the case with all sampling, the technique itself should be well understood and the question asked, "Is this particular sampling technique appropriate for this situation?" The following example illustrates when systematic sampling is not appropriate. Suppose a machine assembles items in lots of ten and makes a recurring error on every ninth item of each lot of ten. If systematic sampling were used starting with the ninth item and selecting every tenth item thereafter, the sample would contain all defective items! Should the machine be shut down? Or suppose in the same situation, sampling began with the eighth item and every tenth item were selected thereafter. Our sample now would contain no defective items, yet in reality, 10% of all items are indeed defective!

The last technique is related to stratified sampling. In *cluster sampling* the population is divided into clusters (often geographical) and then certain clusters are selected at random. From these clusters, individual items are selected at random (sometimes, the entire cluster is sampled). The key to cluster sampling is to have the items in each cluster heterogeneous (as varied as possible). Although this is a relatively inexpensive method, predictions using cluster sampling can sometimes be inaccurate. For example, to determine voter preference, it wouldn't be wise to choose samples (clusters) from one precinct or neighborhood. Family income may have an effect on voting preferences (especially tax issues), and many living in a particular neighborhood (cluster) may have approximately the same income level.

In practice, the different techniques are often used in conjunction with each other. For example, to gather test data on high school students, school districts could be stratified on a statewide basis on the basis of size. Within each stratum, individual schools could be randomly selected. Then within randomly selected schools, clusters of students in given schools could be tested. The variations for sampling techniques are numerous and complex.

Exercise Set 13.1

1. Define the term statistics.

2. What is the difference between *descriptive* statistics and *inferential* statistics?

3. From local newspapers, magazines, radio, and television, tabulate, for a one-week period, all references you notice on statistics or statistical information.

4. In the statistics cited below, which refer to descriptive statistics, and which to inferential statistics?

   a) The monthly telephone bills over the last year ranged from a low of $14.95 to a high of $84.39.

   b) On the basis of a questionnaire completed by every 100th registered freshman, the popular consensus on campus is to do away with curfew.

c) Of all monies collected, 30% came from taxes.

d) Having tested the new medical discovery on over 100 patients, the FDA has concluded that the medicine is now safe to be marketed to the public.

e) To audit the account, 20 entries were picked at random and thoroughly checked.

f) The average income for all people in Faraway County is $18,500.10.

5. Explain what the forecaster means by this prediction: "The probability of snow tomorrow is 80%."

6. If one die is rolled, what is the probability of:

    a) rolling a 4?   **b)** rolling a 2 or a 3?

    c) rolling an odd number?   **d)** rolling a number that is prime?

7. The following table shows the party affiliations by sex for all voters in the local city.

	MALE	FEMALE	
DEMOCRAT	400	200	600
REPUBLICAN	300	100	400
	700	300	

If a voter is selected, at random, what is the probability that the voter is:

    a) a Democrat?   **b)** a female?

    c) a male?   **d)** a Republican?

    e) a female Republican?   **f)** a male Democrat?

8. A man buys four lottery tickets; the numbers on those tickets are 137, 295, 764, and 731. The winning number will be a three-digit number. What is the man's chance of winning?

9. **a)** If three coins are tossed, show all the possible outcomes. (For example, HTH is one of the outcomes.)

    b) Find the probability of getting two tails when tossing three coins.

In the following statements, does the italicized portion refer to the population or a sample?

10. When visiting my sister at college, *all six times that we ate in the dining hall we had fish.* That's all the food they ever seem to serve there!

11. Go to Washington for your summer vacation! We had *two solid weeks* of beautiful sunny weather (no rain) when we vacationed there in June.

12. To attract customers, the *costs of all flights have been reduced by 35%*.

13. A publisher surveyed the *heads of fifty households* in a larger urban center to determine the amount of time and money spent on pleasure reading.

14. To ensure quality control, *every 20th part* was carefully inspected.

15. To determine the type of virus, the medical team wants to take *some of your blood*.

In Exercises 16–20, identify the type of sampling used in the proposed studies.

16. To determine the average family income in a suburban area, every 1000th name was selected from the phone book and a telephone survey was conducted.

17. To select two people from the six volunteers, each person was assigned a number: 1, 2, 3, 4, 5, 6. A die was rolled. A person was selected. A die was rolled again. A second person was selected. (Assume that the same number was not rolled twice!)

18. To obtain a sample of 20 accounts receivable from a list of over 200, each account was numbered. All numbers were placed in an envelope. Twenty numbers were selected.

19. To accurately determine entertainment preferences, questionnaires were sent to 50 people in each of these age categories: (a) under 16, (b) 16 to 24, and (c) over 24.

20. To assess attitude toward an upcoming tax issue, the federal government researcher randomly selected 100 households from rural counties and 300 households from urban counties.

21. A person is born every 9.1 seconds in the United States. Explain how this figure might have been reached. Could sampling have been used? If so, how?

22. In a table of random numbers, how often does 7 appear? Test this empirically by selecting perhaps 100 random numbers and computing the proportion of 7's.

23. You are given the task of determining the favorite fast food of all students at your college. Explain, in detail, how you would design such a statistical study. (Define the population, explain how you would sample the population, etc.)

24. For the temperatures listed below:

20	31	43	26	27	51	38	45	61	34
95	71	64	83	71	35	45	71	72	70
61	11	71	85	35	10	92	58	81	67
43	14	27	91	12	25	46	59	74	74
50	27	67	54	49	37	38	45	61	82

a) Find the average temperature (add all the temperatures and divide by 50).

b) Estimate the average temperature by randomly selecting ten temperatures and computing the sample's average. (This is one estimate of all the temperatures.)

c) Treat each row of temperatures as a stratum and select two temperatures from each stratum. Use this average of ten temperatures as an estimate.

d) Repeat steps (b) and (c) several more times. Should one method, (b) or (c), approximate the population of 50 temperatures better than the other? Why or why not?

25. Do you agree or disagree?

"Smoking may be hazardous to your health."

a) Assume that you are a representative from a national medical association. Prepare a statistical argument to show your agreement with this claim.

b) Assume that you are the president of the nation's largest tobacco producer. Prepare a statistical argument to counter this claim.

26. Read several examples from Darrell Huff's book *How to Lie with Statistics* (New York: Norton, 1954). Then try to find several "misuses" of statistics in daily newspapers, magazines, radio, and/or television.

13.2
Descriptive Statistics

Statistics, in general, is concerned with the collection, organization, presentation, and analysis of data, with the purpose often being to draw inferences. Whether the data gathered are from a sample or from a population, one of the objectives of statistics is to summarize the data and make their interpretation and analysis more manageable.

■ DEFINITION

DESCRIPTIVE STATISTICS

Descriptive statistics is collecting data, tabulating results, summarizing the data, and recording these results in a succinct yet meaningful way.

The presentation of the data and/or results can be numerical, in the form of numbers that summarize the data, or graphical, in the form of tables, graphs, and charts.

Table 13.1

Data:	138	145	156	146	179	200	210	220	140	143
	285	230	219	199	176	145	104	130	132	154
	288	187	204	256	185	167	186	194	203	205
	202	215	165	185	188	164	177	184	198	201
	189	180	174	276	300	187	108	148	174	175

Most often the purpose of descriptive statistics is to facilitate making more meaningful inferences or decisions. The data should be *organized,* perhaps in the form of a frequency distribution and *presented graphically,* perhaps in the form of a histogram. Then the *appropriate descriptive summaries* should be given, such as the number in the study, measures of central tendency, and measures of variation.

At first glance, it would be very difficult to analyze the scores shown in Table 13.1. However, if these same data were presented in a more sophisticated tabular form and organized as a frequency distribution, patterns and trends could be noticed, and overall summaries could be made more readily.

The term "statistics" when used by the general public generally refers to the descriptive branch of statistics, that is, classified "facts or graphs" about a particular class of objects:

"Statistics for the past month indicate the employment rate down 7.3%!"

The Organization of Data

After the collection of data by appropriate sampling techniques, the raw data (data that have not been numerically organized) must be organized and summarized in a meaningful fashion. (See the articles in Figs. 13.1 and 13.5.)

Frequency Distribution

A convenient way of grouping data in order to find a meaningful pattern is to classify the data using a *frequency distribution.* (Forming a frequency distribution is also the first step before graphical presentations of the data can be made.)

On a data processing aptitude test, the scores for 40 applicants are as shown in Table 13.2. A score of 20 represents a perfect test score. If a score of 14 or higher were needed in order to allow the applicant to be considered for admission, what percent of the applicants would not be considered? This and many other questions could be answered more easily if the data were organized in a more meaningful manner; this is even more important as the number of figures increases.

Table 13.2

19	20	18	17	16	16	16	16	14	13
19	18	17	16	15	14	15	12	16	17
14	16	17	18	16	17	17	15	16	12
18	15	16	17	18	16	15	13	17	15

BEST PERFORMANCES
TEAM HIGHS-LOWS

(Through Dec. 7)
Longest winning streak:
Longest losing streak:
Most points (game):
Fewest points (game):
Most points (half):
Fewest points (half):
points (quarter):

9, Boston (10/29-11/15)
7, Cleveland (11/5-11/2
156, Port. vs. Denv
83, Indiana at M
81, San Die
31, Clev

(a)

Weather outlook
East Coast: Normal temperatures will prevail in August, wi
wet spots in northern Ver
and New York and all of
ropical-storm activi
ida at th

(c)

A look at statistics that shape your finances

USA's net worth

The USA's net worth — the value
of all tangible assets owned by
the government, business
and individuals — has
increased 39 percent
since 1978. Here is the
USA's net worth for the
past five years:

Year	Net worth (current dollars)	Per capita
1978	$9.0 trillion	$40,436
1979	$10.2 trillion	$45,322
1980	$11.5 trillion	$50,516
1981	$12.3 trillion	$53,524
1982	$12.5 trillion	$53,880

Source: U.S. Department of Commerce

By Marcy Eckroth Mullins, USA TODAY

(b)

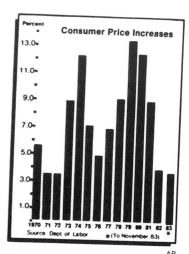

Retail prices so far this year
are up only 3.8%, prompting
economic experts to say it could
be the country's best perform-
ance since the 3.4% figure in
1971 and 1972.

(d)

Figure 13.5

Table 13.3

COLUMN 1 SCORE	COLUMN 2 TALLY	COLUMN 3 FREQUENCY	COLUMN 4 CUMULATIVE FREQUENCY	COLUMN 5 RELATIVE FREQUENCY		COLUMN 6 RELATIVE CUMULATIVE FREQUENCY	
20	I	1	40	1/40	2.5%	40/40	100 %
19	II	2	39	2/40	5 %	39/40	97.5%
18	ʘʘ	5	37	5/40	12.5%	37/40	92.5%
17	ʘʘ III	8	32	8/40	20 %	32/40	80 %
16	ʘʘ ʘʘ I	11	24	11/40	27.5%	24/40	60 %
15	ʘʘ I	6	13	6/40	15 %	13/40	32.5%
14	III	3	7	3/40	7.5%	7/40	17.5%
13	II	2	4	2/40	5 %	4/40	10 %
12	I	1	2	1/40	2.5%	2/40	5 %
11		0	1	0	0	1/40	2.5%
10	I	1	1	1/40	2.5%	1/40	2.5%

In order to construct a frequency distribution, we first list the possible numbers from highest to lowest, in this case starting at 20 and ending at 10. (See Table 13.3.) In column 2, a tally mark is made next to each possibility for each score in the list of raw scores. Finally, in column 3, the frequency (that is, how many times each score appears in the raw data) is recorded. Columns 1 through 3 make up a basic frequency distribution.

To answer our question concerning the number of applicants not considered for admission (see Table 13.2), three additional columns might provide even more information. Column 4 represents the *cumulative frequency,* which indicates the number of applicants making a score of less than or equal to a given score. Column 5 represents the *relative frequency* of a given score—a comparison of the number of applicants obtaining a given score to the total number of applicants. For example, the relative frequency of a score of 17 is 8/40, or 1/5; 20% of all applicants obtained a score of 17. (Often the relative frequency is expressed in terms of percent.) Finally, column 6 represents the *relative cumulative frequency distribution.* Combining the ideas of both a relative and a cumulative frequency distribution, the relative cumulative frequency distribution shows the percentage of scores less than or equal to a given score.

The original question "What percent of the applicants would not be considered?" can now be answered. All those applicants with scores of 13 or less are *not* considered for admission. Checking the relative cumulative frequency distribution column, we see that 4/40 of all applicants had scores less than or equal to 13. Thus 10% will not be admitted and 90% will be.

Before we discuss other examples, some points should be mentioned concerning the frequency distribution in Table 13.3. The sum of the frequencies (column 3) should equal n, the number in the study. The top cumulative frequency distribution (column 4) should also equal n, and the top entry in the relative cumulative frequency column should be 100%.

Frequency distributions can be either ungrouped or grouped. The preceding example listed every possible raw score from 20 to 10; thus the data in the table were *ungrouped*. What would happen if we tried to list the average weekly temperatures (see Table 13.4) that were recorded for a city over the past year? Indeed, the list would be rather long!

A more convenient way to summarize and describe the data in Table 13.4 would be to group the temperatures into classes. Although any number of classes could be used, a general rule of thumb is to use between 10 and 20 classes. For example, if 52 classes were used, the data would be too spread out, and no gain would be realized in summarizing the average temperatures. On the other hand, if only 1 or 2 classes were used, the temperatures would be too "bunched together" for us to see any patterns in the data. Statistical information would be lost. A happy medium must be reached in grouping the data—not too many classes, not too few. In addition, classes should be of the same width, and every data item must fit into exactly one of the classes.

To form a *grouped frequency distribution,* we first find the range of the data. The *range* is computed as the highest number minus the lowest number in the given data.

$$\boxed{\text{Range} = \text{High} - \text{Low}}$$

The temperatures have a range of $92 - 12$, or 80. This range must be equally divided between 10 and 20 classes. If 10 classes were to be used, $80 \div 10$ would indicate that each *class width* will be 8. If 20 classes were used, the *class width* would be 4 ($80 \div 20$). Selection of the number of classes and determination of the class width is somewhat arbitrary; it depends on the purposes and setting of the particular statistical investigation. For illustrative purposes, we will opt for more,

Table 13.4
Average weekly
temperatures

12	17	20	80	90	84	81	78	89	24	74	76	36
15	50	82	53	84	83	82	26	87	56	70	30	43
72	78	32	76	78	65	80	68	88	69	64	70	56
80	36	74	40	85	59	90	45	92	48	62	48	64

rather than fewer classes. Further, we will choose our class width to be 5 for more easily definable class limits. Beginning with 10 as the lower limit of class number 1, the individual classes are formed as shown in Table 13.5. Each of the temperatures is then tallied in the appropriate class interval, and the grouped frequency distribution is then completed. The *class mark,* to be used later, indicates the midpoint of the class interval. Again, a relative frequency distribution, a cumulative frequency distribution, and a relative cumulative frequency distribution could be used to further organize and present the data.

Note: While the apparent class limits for class number 1 are 10 through 14, technically the true class limits (called *class boundaries*) are from 9.5 to 14.5. The class boundaries for class number 2 are then from 14.5 to 19.5, and so on. This will be helpful in our later work on graphing.

For comparison purposes, Table 13.6 shows the same temperatures, but grouped in classes with class widths of 10. Here, a fewer number of classes adequately summarizes the 52 average weekly temperatures, and yet still portrays the dispersion of the temperatures over the entire range.

Some helpful notes in completing a table such as Table 13.5:

1. The class mark, 92, can be obtained by "averaging" the class limits ((90 + 94)/2).

Table 13.5

CLASS NUMBER	CLASS INTERVAL	CLASS MARK	TALLY	FRE-QUENCY
17	90–94	92	|||	3
16	85–89	87	|||	3
15	80–84	82	ЖЖ	10
14	75–79	77	Ж	5
13	70–74	72	Ж	5
12	65–69	67	|||	3
11	60–64	62	|||	3
10	55–59	57	|||	3
9	50–54	52	||	2
8	45–49	47	|||	3
7	40–44	42	||	2
6	35–39	37	||	2
5	30–34	32	||	2
4	25–29	27			1
3	20–24	22	||	2
2	15–19	17	||	2
1	10–14	12			1

Table 13.6

CLASS INTERVAL	TALLY	FREQUENCY
90–99	|||	3
80–89	ЖЖ |||	13
70–79	ЖЖ	10
60–69	Ж		6
50–59	||||	5
40–49	||||	5
30–39	||||	4
20–29	|||	3
10–19	|||	3

2. You may have noticed that the class marks differ by five. Once one class mark is found, subsequent ones can be obtained by subtracting five (or the class width), as in Table 13.5.

3. If the class width selected is an odd number, the class mark leads to a whole number.

The Presentation of Data

For most people, data are much more meaningful if presented graphically rather than in tabular form. (Here the adage, "A picture is worth a thousand words" definitely applies!) Through graphs, several objectives can be achieved: simplification, clarity, summarization, reinforcement, and explanation.

Frequency Histogram

To pictorially demonstrate the characteristics of any group of data, the frequency distribution can be presented as a graph, called a **frequency histogram.**

Let's construct a frequency histogram for the data given in Table 13.6. The first step is to construct a frequency distribution (Fig. 13.6). The class boundaries are then marked along the *horizontal* axis; in this case, the lower boundary for each class interval is shown. The frequencies are labeled along the *vertical* axis.

Rectangles are then constructed over each interval, with the height of the rectangle equal to the class frequency. It is interesting to note that a similar pattern of temperatures is present in the frequency distribution itself, if the tallies are examined. (You may wish to turn the page with Table 13.6 sideways to better see this!)

A similar histogram could be made for a relative frequency distribution, the only difference being that the vertical axis would be labeled with relative frequencies rather than frequencies. This fact is mentioned because the relative frequency histogram is one of the most important graphical descriptive methods in making statistical inferences.

The histogram is an excellent means of describing a set of measurements because the areas under the histogram tell the fraction of the total number of measurements falling in each given interval. At a glance, we can see the range, with both low and high data items. At a glance, it is possible to see whether the data are clustered, and if so, where and to what extent.

In our example, the temperatures, shown in Fig. 13.6, tend to cluster around the higher temperatures; the lower portion of the distribution tails off to the left, in a "negative" direction. Such distributions are called *skewed distributions* (specifically, here, a *negatively skewed distribution*). On the other hand, if the scores were clustered to the left and

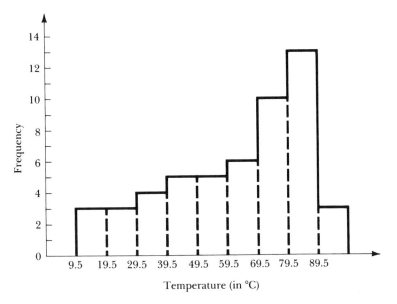

Figure 13.6 Average weekly temperatures

the distribution tailed off to the right (as in Fig. 13.7), the distribution would be *positively skewed,* or skewed to the right.

If the scores, or data, are clustered in the middle with almost equal tails on the right and the left, as in Fig. 13.8, the distribution is said to be *normal.* A normal distribution is *symmetrical:* If it is folded vertically in the middle, the left half will be a mirror image of the right half.

Normal distributions are most useful in that a good majority of statistical operations are based on the assumption that the data are normally distributed. Often before proceeding with specific statistical analyses, one should construct a frequency histogram or distribution to check the "normality" of the data.

Other types of pictorial presentations of data include *frequency polygons, bar graphs,* and *circle* or *pie diagrams.*

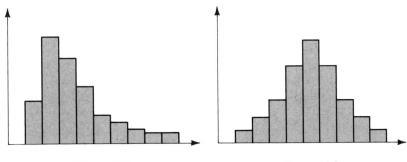

Figure 13.7   **Figure 13.8**

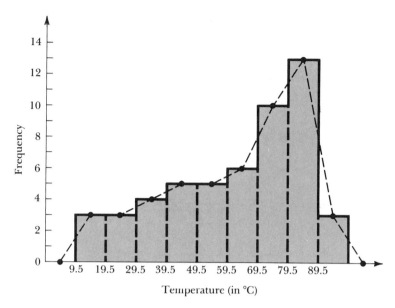

Figure 13.9   Frequency polygon for weekly average temperatures

Frequency Polygon

A **frequency polygon** is constructed by connecting the midpoints, or class marks, of each class interval with straight-line segments. Figure 13.9 shows a frequency polygon superimposed on the frequency histogram for the weekly average temperatures in our example. The bars would not be shown in a frequency polygon. If it makes sense, a zero frequency is assumed at each end to bring the polygon down to the horizontal axis. Generally, frequency polygons are used to emphasize changes, increases, or decreases.

Bar Graph

Closely related to a histogram is the *bar graph*, or bar chart. This form is used extensively in business, especially to display economic data. Several examples taken from newspapers and magazines are shown in Fig. 13.10.

Bar graphs are used to display data that can be categorized. If the categories represent distinct different classes, the bars are generally separated from each other, unlike histograms. But like histograms, both axes are labeled. Usually frequencies, or proportions or percents, are on the vertical axis with the height of each bar being proportional to the number of members in that category. For the horizontal axis, generally, bar graphs use intervals, histograms use boundaries, and frequency polygons use class marks.

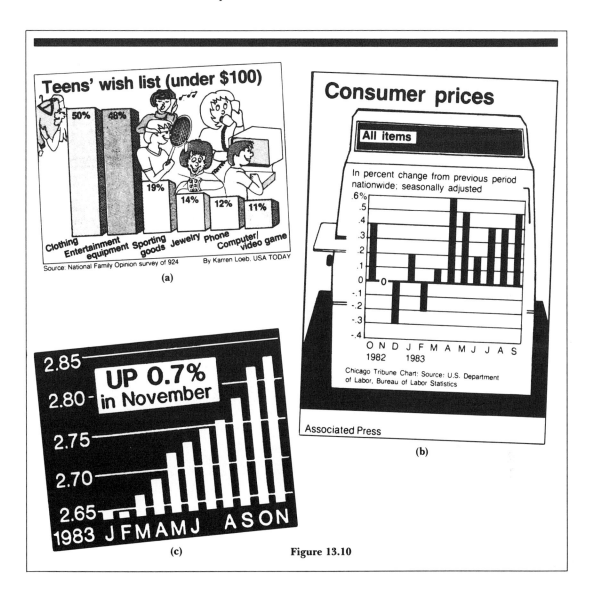

Figure 13.10

To illustrate, the five businesses with the greatest number of company-owned franchises are shown in Fig. 13.11. Which method of data presentation do you prefer to show the relationship between the franchises?

Pictograph
A variation of the bar graph uses objects (coins, pictures, or symbols) to replace the bars. This type of graph is called a *pictograph;* examples are shown in Fig. 13.12.

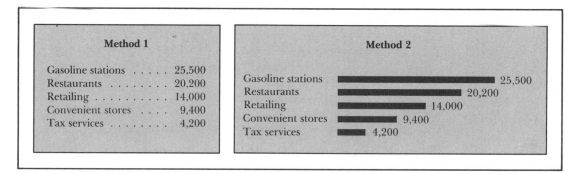

Method 1

Gasoline stations 25,500
Restaurants 20,200
Retailing 14,000
Convenient stores 9,400
Tax services 4,200

Method 2

Gasoline stations   25,500
Restaurants   20,200
Retailing   14,000
Convenient stores   9,400
Tax services   4,200

Figure 13.11

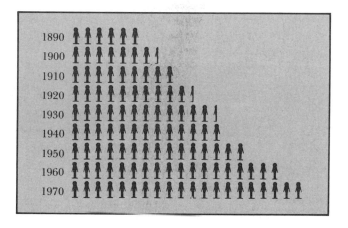

1890
1900
1910
1920
1930
1940
1950
1960
1970

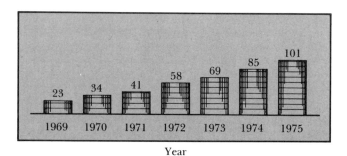

101
85
69
58
41
34
23

1969   1970   1971   1972   1973   1974   1975

Year

(b)

Figure 13.12

Circle Graph

A *circle graph,* or pie chart, is commonly used to illustrate *the relationship of component parts not only to each other,* but *also to the whole.* In this case, relative frequencies or percentages are used. Before constructing a circle graph, we must first construct a frequency distribution for the data. Some general guidelines in constructing a circle graph are (1) limit the number of categories used to summarize the data, and (2) display the pieces of the graph in either ascending order or descending order. Several circle graphs are shown in Fig. 13.13.

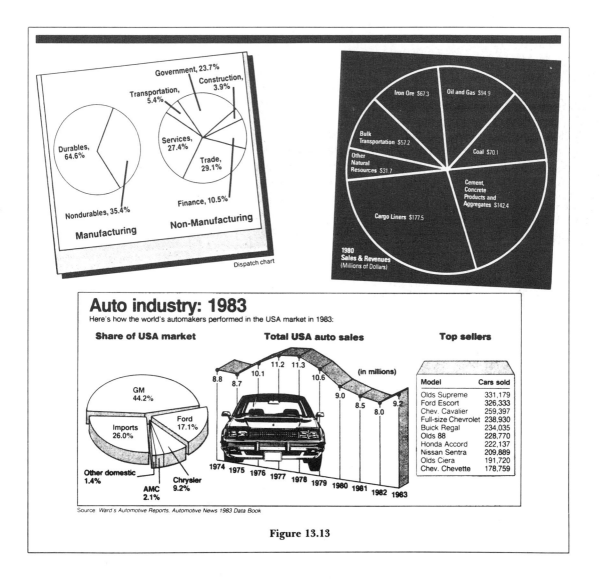

Figure 13.13

The Summarization of Data

Frequency distributions, histograms, pie charts, and other forms of presenting "a picture" of data also summarize the data more succinctly than a listing or table can. Another important way of describing a set of items or data—whether a sample or a population—is to use *numerical descriptive measurements*. The two most important types of measures are those that locate the *center of the distribution* (measures of central tendency) and those that describe the *spread* or *variability of a distribution* (measures of dispersion). These two measures can be used either as a good description of the data, which may be the desired goal in and of itself, or as a first step in phrasing inferences about populations on the basis of measurements obtained from a sample (or samples).

Measures of Central Tendency

The Arithmetic Mean

You have probably used the concept "average" for several years already. For example, "To obtain my *average* test score, I add up all 5 of my test scores (70, 60, 80, 90, 50), and divide the sum by the number of tests I have taken (5). My average is 70!"

Actually, there are several averages: the arithmetic mean, the median, and the mode. The average that we computed above really has a more sophisticated name; it is called the *arithmetic mean*.

■ DEFINITION

ARITHMETIC MEAN

The *arithmetic mean* is the sum of a set of measurements divided by the number of measurements, and is denoted by $\overline{X}$ for a sample and by the Greek letter μ for a population.

■ EXAMPLE 6

If four test scores are denoted by

$$X_1 = 70, \quad X_2 = 60, \quad X_3 = 80, \quad X_4 = 90, \quad \text{and } X_5 = 50,$$

the mean can be obtained by the equation

$$X = \frac{X_1 + X_2 + X_3 + X_4 + X_5}{5}$$

$$70 = \frac{70 + 60 + 80 + 90 + 50}{5},$$

or, in more general form,

$$\bar{X} = \frac{\Sigma X}{N},$$

where the symbol ΣX means "the sum of all the X-values, or raw scores that are recorded," and N is the number of values being added.

Conceptually, the mean can be represented as the "balance point" of the distribution of raw scores. For example, if we arranged our five test scores on a "beam," as shown in Fig. 13.14, the balancing point (or fulcrum) would be 70.

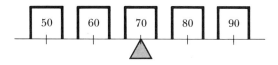

Figure 13.14

The Median

The term **median** refers to "the middle," or an intermediate position. For example, a median strip on a highway is the strip that divides the highway into two equal halves.

■ **DEFINITION**

MEDIAN

The *median* of a distribution of values is the middle value after the values have been arranged from lowest to highest. The symbol for median is Md.

For our five test scores in Example 6—70, 60, 80, 90, and 50—the middle score is 70. Thus 70 divides the distribution into two equal halves.

● Question

What if we had ten stock prices: 16, 13, 31, 14, 14, 15, 14, 13, 15, and 15? What would be the median price?

Answer

Arranging the stock prices in ascending order,

13, 13, 14, 14, 14, 15, 15, 15, 16, 31,

we count in from each end and find that we have two middle scores. For an even number of values, the median is the mean of the two middle scores. Thus in this case, the median stock price would be 14.5 ((14 + 15)/2).

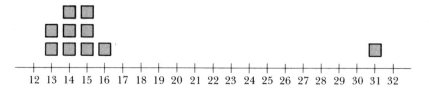

Figure 13.15 Stock prices

What would be the *mean* for the ten stock prices? Adding the ten prices and dividing by 10, we find that the arithmetic mean is 16. Now the question is, which measure of central tendency best reflects, or describes, the central location of most of the prices?

Appropriate Applications

Remember that in most cases, a statistic is a descriptive measurement of a sample, but it is also often used as an *estimate* of the population. To best reflect the central tendency of the skewed distribution of stock prices in our example, the median would be used; the majority of stock prices cluster around 14.5 (the median) rather than 16.0 (the mean). (See Fig. 13.15.)

Here, a sketch of a distribution is most helpful as the first step in analyzing a set of data. In general, the mean is preferred if the distribution is symmetrical (bell-shaped or normal). The median is used when the distribution is not symmetrical (skewed), or when a quicker measurement than the mean is needed. For large volumes of data and/or for sociological or business data, the median is frequently used (median income, median age, median price index, etc.).

If the distribution is normal, or very nearly normal, are the mean and median different? (*Hint:* Try computing the mean and median for hours worked for the eleven employees shown in Fig. 13.16.)

The Mode

This measure of central tendency is the easiest to use. The term *mode* refers to the *most frequent,* or most common, data item.

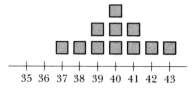

Figure 13.16 Hours worked for eleven employees

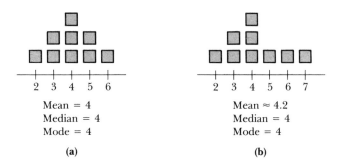

Figure 13.17

■ **DEFINITION**

MODE

The *mode* of a set of measurements is that value that occurs most frequently. The symbol for mode is Mo.

For the distribution of employee hours shown in Fig. 13.16, the number of hours worked most frequently is 40: Three workers worked 40 hours. The mode of this distribution is 40.

In our example on stock prices (refer to Fig. 13.15), three stocks are listed at $14 per share and three stocks are listed at $15 per share. In this case, the distribution has two modes, or is said to be *bimodal*.

Appropriate Applications

For distributions that are normal, or approximately normal, the mean, median, and mode are equal (or approximately equal). See Fig. 13.17.

The more skewed the distribution, the greater the difference between the mean, median, and mode. Conversely, if the mean, median, and mode are computed and great differences exist, we know that the distribution is not normal, but skewed. The placement of the three measures of central tendency are predictable, as shown in Fig. 13.18.

From the figure, we can obtain additional insights on the appropriate applications of these measures of central tendency. The more skewed the distribution, the more affected the mean becomes. The mean is "pulled" toward the extremes. For skewed distributions with extreme scores, consideration should be given to the median or mode as measures of central tendency. For distributions that approximate a normal distribution, the mean is the most representative measure. In addition, the mean is more stable than the median and the mode; with continued sampling of a population, the computed means will probably have less fluctuation from sample to sample than will the median

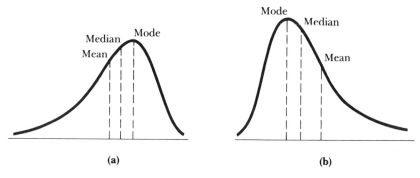

Figure 13.18   (a) Negative or left skew; (b) positive or right skew

or the mode. The mean is more reliable, and as such it is also the prerequisite statistic for the computation of many other statistics that we will discuss later.

Measures of Dispersion

By studying the two distributions shown in Fig. 13.19, we can see that merely knowing the average monthly temperature is not sufficient statistical information to describe living in these two cities.

In addition to locating measures of central tendency, statistical information should include some measure of the "spreadability" or "variability" of the measures. Would you be able to use the same wardrobe of clothing in both the cities in the figure? (They both have a comfortable average monthly temperature of 70°!)

Some statistical measures of data variation are the *range, percentiles, variance,* and *standard deviation.*

The Range

The simplest and probably most overused measure of variability is the *range.*

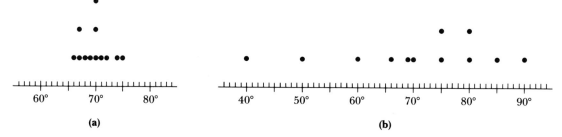

Figure 13.19   (a) Mean monthly temperature is 70 in Warmville; (b) mean monthly temperature is 70 in Littletown.

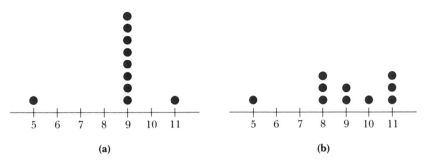

Figure 13.20

■ **DEFINITION**

RANGE

The *range* is the difference between the largest and the smallest measurements.

For our two cities, the range of temperatures is 20° in Warmville (80° − 60°) and 50° (90° − 40°) in Littletown. The temperatures in the latter city display a wider range because they fluctuate more widely.

Because the range is calculated by using only two measurements, it is very easy to compute. But this fact also creates a disadvantage. No description is given of the remaining temperatures. Take a look at the two distributions shown in Fig. 13.20. Both have the same mean (9) and both have the same range (6). Must the two distributions necessarily be the same?

Percentiles

Percentiles are used to provide a statistical description of variation for a large set of measurements. For example, "Of all test applicants on the IOWA Test, Reb scored at the 95th percentile!" This describes Reb's score in relationship to all other scores. The 95th percentile represents the value that 95% of all measurements are less than (see Fig. 13.21).

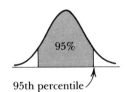

95th percentile

Figure 13.21

■ **DEFINITION**

PERCENTILE

In general, the *p*th *percentile* is the value such that *p*% of all measurements in the distribution are less than and (100 − *p*)% are greater than that value.

Special values are given to the 25th percentile, 50th percentile, and 75th percentile; they are, respectively, *first quartile* (Q_1), *second quartile* or median (Q_2), and *third quartile* (Q_3). The first quartile, median, and third quartile, together with the minimum and maximum measurements, provide a description of the distribution.

Knowledge of the quartiles provides a means of calculating a measure of dispersion that is used most frequently when the median is the measure of central tendency. (Ideal for skewed distributions!) The *semi-interquartile range* (sometimes called simply the *quartile deviation*) is one half the distance from the first quartile to the third quartile.

■ DEFINITION

$$\text{Semi-interquartile range} = \frac{Q_3 - Q_1}{2}$$

The Variance and Standard Deviation

The statistic of dispersion used most often in conjunction with the mean is the *standard deviation*. Although it is more complex to compute than either the range or the quartile deviation, it does give us more complete statistical information concerning the distribution. Unlike the range, the standard deviation utilizes every measurement in the distribution for its calculation.

To illustrate the computation of the standard deviation, we shall proceed in a step-by-step manner. When we are finished, we will look at the formula and provide a formal definition.

Suppose a quiz were given and the following scores were obtained: 3, 7, 7, 8, and 10. We will use Table 13.7 to explain, column by column, how to compute the standard deviation, after which we will learn of its meaning and appropriate application.

Step 1. List the scores from highest to lowest. (This is done in column I.)

Step 2. Compute the mean and place that value in column II.

Table 13.7
The computation of standard deviation

	I	II	III	IV
	X_i	μ	$(X_i - \mu)$	$(X_i - \mu)^2$
X_1	10	7	+3	9
X_2	8	7	+1	1
X_3	7	7	0	0
X_4	7	7	0	0
X_5	3	7	−4	16

Step 3. Find how far each quiz score deviates from the mean by subtracting the mean from each quiz score. (This has been done and recorded in column III.)

Keeping in mind that we want a measure of the "average variability" of the scores, why couldn't we simply add the deviations in column III, and in this case, divide by 5? Won't the sum of column III always be zero? Why?

Note: An alternative method is to ignore the signs in column III. Then adding the values in column III (absolute values, 3 + 1 + 0 + 0 + 4) and dividing by 5 yields a measure called the *average deviation*. In this case, the average deviation is 1.6. However, the standard deviation is the measure more applicable and hence more frequently encountered, so we continue onward with its computation.

Step 4. Square the results of column III and place these "squared deviations" in column IV.

Step 5. Now find the mean of the values in column IV:

$$9 + 1 + 0 + 0 + 16 = 26 \quad \text{and} \quad 26 \div 5 = 5.2.$$

This value 5.2 is called the **variance.**

■ DEFINITION

VARIANCE

The *variance* is the average of the squares of the differences of the measurements from their mean.

A good portion of later statistical work involves the calculation and then the manipulation of the variance. However, for our purposes, variance does not lend itself directly to a meaningful interpretation for our single set of quiz scores. *What does 5.2 mean?* (For example, if data were measured in pounds, then the "label" for variance would be square pounds!) Indirectly, though, variance is useful. To obtain a measure of variation that is more easily interpreted, we compute the standard deviation from the variance.

■ DEFINITION

STANDARD DEVIATION

The *standard deviation* is the square root of the variance.

In our example, the variance of the quiz scores is 5.2, and the standard deviation is $\sqrt{5.2}$, or approximately 2.28. Standard deviation is a type of average of the individual deviations from the mean. (*Note:* Did you notice that when computing variance we are dealing with square units? Thus to get back to our original units, we should perform the opposite operation of squaring used in step 4, and take the square root!)

In summary, to compute the standard deviation for a set of measurements:

1. Arrange the numbers from highest to lowest.   Column I X_i
2. Compute the mean.   Column II μ
3. Compute the deviation scores.   Column III $X_i - \mu$
4. Square the deviation scores.   Column IV $(X_i - \mu)^2$
5. Add column IV. $\Sigma(X_i - \mu)^2$
6. Find the average of column IV. $\Sigma(X_i - \mu)^2 \div N$
7. Take the square root of the variance. $\sqrt{\Sigma(X_i - \mu)^2 \div N}$

Thus the formula for the standard deviation is as follows.

$$\text{Standard deviation} = \sqrt{\frac{\Sigma(X_i - \mu)^2}{N}}$$

By studying our example carefully, do you see that this formula is merely a summary of the step-by-step work that we did in Table 13.7?

With one small modification, we will be ready to use standard deviation in our analyses of future sets of data. Remembering that often a sample is taken to make inferences about a population, we must keep in mind that the sample will not be "exactly" the same as the population. The sample will be biased one way or another. To correct for this, different formulas for standard deviation and variance are used, depending on whether you are dealing with a sample or with a population.

For a population:

$$\text{Variance } \sigma^2 = \frac{\Sigma(X_i - \mu)^2}{N} \qquad \text{Standard deviation } \sigma = \sqrt{\frac{\Sigma(X_i - \mu)^2}{N}}$$

For a sample:

$$\text{Variance } S^2 = \frac{\Sigma(X_i - \overline{X})^2}{N - 1} \qquad \text{Standard deviation } \sigma = \sqrt{\frac{\Sigma(X_i - \overline{X})^2}{N - 1}}$$

An explanation for dividing by $N - 1$ can be found in more advanced statistics texts; suffice it to say that this mathematically derived divisor "adjusts" for bias in sampling.

Interpretation of standard deviation

As the variability of a set of measurements increases, both the variance and the standard deviation become larger. On the other hand, as the variability decreases, the variance and standard deviation get smaller. If there were no variability in the scores in our example, what would the variance, or the standard deviation, equal?

The standard deviation is most useful in describing a distribution of measurements, especially if the distribution is approximately normal or bell-shaped. A rule of thumb called the *empirical rule* states that given a normal distribution:

a) If 1 standard deviation is measured off on each side of the mean, then this interval will contain approximately 68% of all the measurements.

b) If 2 standard deviations are measured off on each side of the mean, then this interval will contain approximately 95% of all the measurements.

c) If 3 standard deviations are measured off on each side of the mean, then this interval will contain approximately 99.7% of all the measurements.

The empirical rule is summarized in Fig. 13.22.

A typical application of the empirical rule is in the area of quality control. Let's say that for excellent quality a manufacturer decides to eliminate all parts that are not within 3 standard deviations of the mean. If the mean weight is 18.5 grams and the standard deviation is 0.5 gram, the limits for acceptance would be $18.5 - (3)(0.5)$ and $18.5 + (3)(0.5)$, or between 17.0 grams and 20.0 grams. Any parts not having a weight within this "confidence" region would be rejected!

In summary, measures of dispersion include the range, percentiles, semi-interquartile range, and standard deviation. Although the easiest to compute, the *range* is also the crudest in that reliance is on only the minimum and the maximum measurements in the data set. The *semi-interquartile range* makes use of the first and third quartiles and is commonly used with the median in the analysis of skewed distributions. The measure of dispersion that uses all the measurements in the

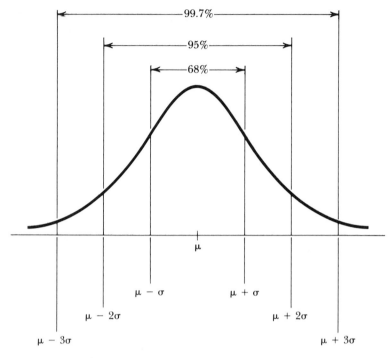

Figure 13.22   The empirical rule

data set is the *standard deviation,* and is, in essence, the square root of the average of the squares of the deviation scores. Variance and standard deviation allow good statistical descriptions of normal distributions, as well as comparability of variation between different sets of measurements.

Exercise Set 13.2

1. Define the term descriptive statistics.

2. For the set of data in Table 13.8, construct:

   a) a frequency distribution,

   b) a relative frequency distribution,

   c) a cumulative frequency distribution,

   d) a cumulative relative frequency distribution.

3. For the data given in Table 13.8, construct:

   a) a frequency histogram,

Table 13.8

3	7	6	5	4	7	8	4	2	10	11	3	2	3	1	4
3	2	3	5	8	9	6	4	4	1	2	5	2	3	4	3

b) a relative frequency histogram,

c) a cumulative frequency histogram (replace frequencies on the vertical axis with cumulative frequencies),

d) a cumulative relative frequency histogram (called an *ogive*).

4. For the set of data given in Table 13.8, compute and list the following descriptive statistics.

a) n **b)** Range   **c)** High   **d)** Low

e) Mean   **f)** Median   **g)** Mode

5. Using the information from Exercises 2–4, answer the following question: The data given in Table 13.8 is:

a) very skewed left,   **b)** skewed left,

c) normal,   **d)** skewed right, or

e) very skewed right?

6. Monthly net income after taxes for 32 computer programmers is shown below.

1800	2100	3570	1650	2140	1500	1450	1940	1320	1850
1580	2310	1750	1950	1850	2025	2400	2110	2200	1500
2810	1840	3450	2220	1840	2060	1760	1690	1950	1640
1820	1910								

Construct a frequency distribution and the corresponding histogram to graphically present these data. Use class widths of 200 and let the lower class limit be 1300 to 1499.

7. Regional sales offices accounted for the following company profits (given in hundred thousands):

Albuquerque	120
Topeka	180
Chicago	240
Seattle	60

Erie	80
Orlando	150
Denver	170
Boston	200

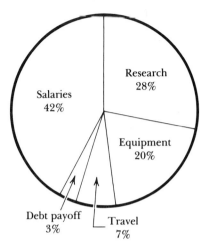

Figure 13.23

Display the regional sales using each of these graphical techniques.

a) A bar chart or bar graph

b) A circle graph or pie chart

8. If the total budget of the data shown in Fig. 13.23 is $140,650, how much money is allocated for each of the following?

a) Salaries   **b)** Research

c) Equipment   **d)** Debt payoff

9. Criticize the graphical presentation of each of the following.

a) To demonstrate the dramatic decline in the inflation rate, the graph in Fig. 13.24 was used.

b) To demonstrate the significant increase in taxes the graph in Fig. 13.25 was displayed in a candidate's TV presentation (the candidate was against higher taxes).

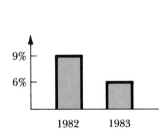

Figure 13.24

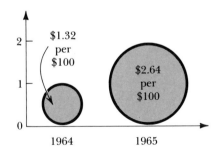

Figure 13.25

10. Find the mean, median, and mode for each of the following sets of data.

   a) 1, 3, 4, 2, 2, 2, 5, 6, 9

   b) 10, 20, 30, 40, 50, 60, 70

   c) 22, 25, 25, 47, 4, 22, 36, 22, 13, 29, 32, 41

11. Earnings per share for a common stock over the last ten years were $3.05, $4.12, $6.05, $2.10, $3.05, $3.00, $3.85, $4.60, $9.80, and $5.20.

   a) Compute the following measures of central tendency:

   i) Mean = _____

   ii) Median = _____

   iii) Mode = _____

   b) Which measure best reflects the "center of the distribution"?

12. Two "thinkers" for you.

   a) If the annual mean salary for 40 employees is $24,500, what is the total annual salary to be budgeted for the group?

   b) Find 6 scores whose mean is 19 and whose mode is 21.

13. For each of the following sets of data, compute the mean and the standard deviation.

   a) 2, 4, 6, 8, 10

   b) 6, 6, 6, 6, 6

   c) 12, 14, 16, 18, 20

   d) 20, 40, 60, 80, 100

   e) 200, 400, 600, 800, 1000

   After studying the statistics for the sets of data, what observations did you make?

14. The sample of batting averages below were randomly chosen from a list of All-American League batting averages.

0.320	0.285	0.310	0.210	0.276
0.245	0.345	0.315	0.294	0.287

   a) Compute the mean batting average.

   b) Compute the standard deviation for the batting averages.

15. For the monthly salaries given in Exercise 6, give a complete descriptive summary, as follows.

   a) n **b)** Range

c) High   d) Low

e) Mean   f) Median

g) Mode   h) Variance

i) Standard deviation   j) First quartile

k) Third quartile   l) Semi-interquartile range

13.3
Inferential Statistics and Probability

In addition to being a descriptive tool, statistics is used to make statistical inferences. On the basis of the data gathered from a sample, conclusions can be drawn about the population from which the sample was drawn. Probability is the "backbone" of statistical inference—in estimation and in hypothesis testing, in all walks of life: medicine, sports, business, the military, agriculture, and so on. An understanding of the concepts of probability is basic to an appreciation of statistical inference. In making estimates about the population you may wish to be a little more definite than "very likely," or "not so possible."

While anyone can use sample data to estimate what percent of the television viewers favor a particular product, the real question is, how far is the estimate from the true population parameters? How much of a deviation exists? How confident can we be in the estimate? To answer questions such as these, we must be able to determine the probabilities associated with the values of the random variables that appear in the sample. We need to know the distribution of the sample's probabilities. Probability can be estimated by the sampling method. Probability distributions can be, and are, derived mathematically and serve as models for the population's relative frequency histogram. These distributions can be described by the numerical descriptive statistics mentioned earlier, such as the mean, standard deviation, and so on. With this knowledge about the sample, populations can be described (or inferences can be made) by estimating and/or making decisions based on the representative sample.

What Is Probability?

As we defined earlier in the chapter:

■ **DEFINITION**

Probability is the likelihood that some stated event will occur. Probability is the ratio of the favorable outcomes to the total possible outcomes.

Table 13.9

TOSSES	HEADS	PROPORTION OF HEADS		
10	4	4/10	or	0.4000
100	53	53/100	or	0.5300
1,000	512	512/1,000	or	0.5120
10,000	4,986	4,986/10,000	or	0.4986
100,000	50,147	50,147/100,000	or	0.50147

Example 4, used earlier, involved tossing a fair coin. The probability of obtaining a head is 1/2. Does this mean that if we flipped 20 coins we would obtain 10 heads? Not necessarily! There is a chance (not very likely, however) that even with a fair coin we could obtain no heads (all tails) with 20 tosses.

■ **DEFINITION**

PROBABILITY OF AN EVENT

The *probability of an event* is the proportion of times that the event can be expected to occur "in the long run."

Listed in Table 13.9 are values obtained by tossing a fair coin several times using a computer simulation. Note that as the number of tosses increases, the proportion of heads approaches 1/2.

Theoretical vs. Empirical Probability

At this stage, a distinction should be made between theoretical probability and empirical probability. If possible to calculate, *theoretical probability* is the ratio of favorable outcomes to total possible outcomes. The theoretical probability of obtaining a head by tossing a fair coin is 1/2. In practice, however, we may not obtain a 1-to-2 ratio of heads to coins flipped (as seen in Table 13.9). For example, by actually flipping 10 coins, 4 heads were obtained. This ratio is referred to as *empirical probability*. Theoretical probability is what should happen (ideally), whereas empirical probability is what actually does happen (realistically).

For example, find the probability of obtaining a prime number when one die is rolled. The ratio of favorable outcomes (2, 3, 5) to total possible outcomes (1, 2, 3, 4, 5, 6) is 1 to 2; thus the theoretical probability is 1/2. However, if you actually rolled a die 6 times, could you be certain that half the outcomes would be prime numbers? Perhaps in the long run, say with 6,000,000 tosses or so, we could expect close to half of the outcomes to be prime numbers. (You might try that . . . let a computer help you!)

Table 13.10
Possible outcomes for
the sum of two dice

2	3	4	5	6	7	8	9	10	11	12	
$\frac{}{36}$	$\frac{}{36}$	$\frac{}{36}$	$\frac{}{36}$	$\frac{}{36}$	$\frac{}{36}$	$\frac{}{36}$	$\frac{}{36}$	$\frac{}{36}$	$\frac{}{36}$	$\frac{}{36}$	Your actual results here as empirical probability
$\frac{1}{36}$	$\frac{2}{36}$	$\frac{3}{36}$	$\frac{4}{36}$	$\frac{5}{36}$	$\frac{6}{36}$	$\frac{5}{36}$	$\frac{4}{36}$	$\frac{3}{36}$	$\frac{2}{36}$	$\frac{1}{36}$	Theoretical probabilities

An Experiment

Roll two dice, add the spots on the "up" faces of the dice. Record your answer. Repeat this 36 times. Compare your empirical results to the theoretical results listed in Table 13.10. Were you close? If you repeated the process 360 times, would your results be closer?

Could you determine how the theoretical probabilities in Table 13.10 were computed? To obtain a sum of 5, for example, we would need the following rolls: (1, 4), (2, 3), (3, 2), (4, 1), for four favorable outcomes. The total possible outcomes can be easily determined from Table 13.11. (You may recall from earlier in the chapter that a listing of all possible outcomes is called the *sample space*.)

The probability, then, is the ratio of favorable outcomes to total outcomes, or 4/36. In the long run, *one "sum of 5"* should show up for every nine tosses of the two dice.

■

The probability of an event A occurring is denoted by $\Pr(A)$.

■ **EXAMPLE 7**   If A is the simple event of obtaining a head when tossing a fair coin, then $\Pr(A) = 1/2$.   □

■ **EXAMPLE 8**   If A is the event of obtaining a sum of 5 when rolling two dice, then $\Pr(A) = 4/36$ or $1/9$.   □

Table 13.11
All possible tosses of
two dice

		DIE 1					
		1	2	3	4	5	6
	1	(1, 1)	(1, 2)	(1, 3)	(1, 4)	(1, 5)	(1, 6)
	2	(2, 1)	(2, 2)	(2, 3)	(2, 4)	(2, 5)	(2, 6)
	3	(3, 1)	(3, 2)	(3, 3)	(3, 4)	(3, 5)	(3, 6)
DIE 2	4	(4, 1)	(4, 2)	(4, 3)	(4, 4)	(4, 5)	(4, 6)
	5	(5, 1)	(5, 2)	(5, 3)	(5, 4)	(5, 5)	(5, 6)
	6	(6, 1)	(6, 2)	(6, 3)	(6, 4)	(6, 5)	(6, 6)

■ EXAMPLE 9

If B is the simple event of picking a Democrat from the people listed below (by party affiliation),

$$\{D, R, D, D, D, R, D, D, R, D, D, D\},$$

then $\Pr(B) = 9/12$ or 0.75.   □

Probability Laws, or Axioms

Associated with probability are certain laws.

■

If A is a desired outcome of an experiment, then:

a) $\Pr(A) = 1$ if A always occurs,
b) $\Pr(A) = 0$ if A never occurs,
c) $0 \leq \Pr(A) \leq 1$,
d) $\Pr(\text{not } A) = 1 - \Pr(A)$.

The following examples will help to clarify these axioms.

■ EXAMPLE 10

What is the probability of selecting a Democrat from the following list of people?

$$\{D, D, D, D, D, D, D, D, D\}$$

Solution

Because *all* elements in the sample space are Democrat, you are absolutely certain (100% sure) to pick a Democrat. This is an example of a *certain event,* or a *definite event;* thus $\Pr(A) = 1$.   □

■ EXAMPLE 11

What is the probability of picking a Republican from a list of all Democrats?

$$\{D, D, D, D, D, D, D, D, D\}$$

Solution

Because there are no favorable outcomes, the ratio of 0 to 9 is 0/9, or 0. This is an example of an *impossible event;* the probability is zero.   □

■ EXAMPLE 12

Consider the theoretical probabilities given to the sums obtained when rolling two dice (see Table 13.9). The probabilities ranged from 1/36 to 6/36, all in the range of 0 to 1. The probability of an event happening ranges somewhere between "definitely not occurring" (zero) to "occurring with absolute certainty" (one).   □

If the probability of rain is 0.87, or 87%, do you consider taking your umbrella? If the chances of your passing the entrance exam without studying are 0.13, do you study?

■ **EXAMPLE 13**   What is the probability of picking an ace when drawing one card from a deck of 52?

Solution   Because there are 4 aces

$$A \spadesuit, \quad A \clubsuit, \quad A \spadesuit, \quad \text{and} \quad A \heartsuit$$

the probability of favorable outcomes to total outcomes is 4/52, or 1/13. □

What is the probability of *not* picking an ace? If 4 cards are aces, then the remaining 48 cards are not aces; hence prob(not an ace) = 48/52. *Note:*

$$\Pr(\text{ace}) + \Pr(\text{not an ace}) = 4/52 + 48/52 = 52/52 = 1.$$

If A represents an event, then not A, or $\overline{A}$, or A' is called the *complement* of A. Thus

$$\Pr(A) + \Pr(\overline{A}) = 1, \quad \text{or } \Pr(A) = 1 - \Pr(\overline{A}), \quad \text{or } \Pr(\overline{A}) = 1 - \Pr(A).$$

Mutually Exclusive Events

The list of all possible outcomes of an experiment is called the sample space, as you may recall. An *event* is a subset of the sample space. For example, when tossing a die, the sample space is

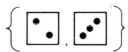

If event A is rolling a number between 1 and 4,

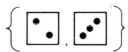

is a *subset* of the original sample space.

Two events A and B are said to be **mutually exclusive** if the sets have an empty intersection ($A \cap B = 0$). In this case, A and B are *disjoint sets.* Events A and B cannot occur simultaneously!

■ **EXAMPLE 14**   If A is the event of drawing an ace and K is the event of drawing a king from a deck of 52 cards, are events A and K mutually exclusive? Can a card drawn be simultaneously both a king and an ace? Events A and K are *alternative outcomes.*

■

For mutually exclusive events A and B,
$$\Pr(A \text{ or } B) = \Pr(A) + \Pr(B).$$

The probability of drawing an ace or a king is equal to the *sum* of their respective probabilities:

$$Pr(ace\ or\ king) = Pr(ace) + Pr(king)$$
$$= 4/52 + 4/52 = 8/52.$$

This makes sense because if we are hoping for a king or an ace, there actually are 8 favorable cards to pick from the deck of 52. □

• Question

What is the probability of picking either a jack or a heart?

Answer

$$Pr(jack\ or\ heart) = Pr(jack) + Pr(heart)$$
$$= 4/52 + 13/52 = 17/52$$

But wait! Actually count all the favorable outcomes!

There are only 16 distinct different choices. Why doesn't our formula work?

In this case, the events "drawing a jack" and "drawing a heart" are not mutually exclusive. The set of hearts intersects the set of jacks:

$$\{\ Jacks\} \cap \{Hearts\} = \{\ Jack\ of\ hearts\}.$$

To correct for this common intersection—actually counting the jack of hearts twice—the common intersection is subtracted from our previous answer.

■

> In summary, for events A and B that are not mutually exclusive,
>
> $$Pr(A\ or\ B) = Pr(A) + Pr(B) - Pr(A \cap B),$$
> $$\text{or } Pr(A) + Pr(B) - Pr(A\ and\ B).$$

Independent and Dependent Events

Two events A and B are said to be *independent* if one event, say B, is not influenced by whether A has, or has not, occurred.

■ **EXAMPLE 15**

Suppose we have a container with 5 red, 2 green, and 3 blue balls. If we draw two balls in succession and *replace* each ball after it is drawn, what is the probability of obtaining a blue ball followed by a red ball? Thus

$$Pr(blue\ and\ red) = ?$$

Solution

The probability of obtaining a blue ball is 3/10. After replacing the blue ball, the probability of obtaining a red ball is 5/10. One drawing does not influence the drawing of the other ball. These events are said to be independent.

■

> For independent events,
>
> $$\Pr(A \text{ and } B) = \Pr(A) \cdot \Pr(B).$$

So, for our problem,

$$\Pr(\text{blue and red}) = \Pr(\text{blue}) \cdot \Pr(\text{red})$$
$$= (3/10)(5/10) = 15/100. \quad \square$$

An Example of Dependent Events

Suppose we wished to draw from a deck of cards an ace followed by a king. However, after selecting a card, we will not replace the first card. Thus $\Pr(\text{ace}) = 4/52$. However, assuming we did draw an ace and did not replace that card, the probability of getting a king is no longer $4/52$, but $4/51$! Event A has "influenced" the probability of event K. These are examples of *dependent events*. To compute the probability of (ace and king) without replacement, our earlier formula could be used; however, the second probability (of getting a king) must be "adjusted" for the decreased sample space:

$$\Pr(\text{ace and king}) = \Pr(\text{ace}) \cdot \Pr(\text{king (adjusted)})$$
$$= 4/52 \cdot 4/51 = 16/2652.$$

The formula for independent events can be rewritten to accommodate dependent events as follows:

■

> For dependent events,
>
> $$\Pr(A \text{ and } B) = \Pr(A) \cdot \Pr(B|A),$$
>
> where $\Pr(B|A)$ represents the conditional probability of B, provided that A has already occurred. The sample space has been appropriately adjusted. (Note that $\Pr(B|A)$ does not represent B divided by A!)

Counting Techniques

In concluding this section on probability, we should mention some methods of counting all possible outcomes for an experiment. Quite

often in probability problems, the easiest way to determine the probability is to list all the possible outcomes (sample space) and then determine how many of the outcomes are favorable.

The Multiplication Principle

● Question

Suppose that in an election for the offices of president, secretary, and
treasurer Jan and Ed are presidential candidates; Mary, Frank, and
Beth are vying for the position of secretary; and Laura and Al are
candidates for the position of treasurer. In how many possible ways
can these offices be filled?

Answer

An attempt to list some possibilities becomes futile (we have used
initials for names):

PRESIDENT	SECRETARY	TREASURER
J	M	L
J	M	A
E	F	L
J	F	L
	etc.	

A more convenient way to "count" and/or "show" all possibilities is to
use a *tree diagram*. For the position of president, there are two possibilities, as shown in Fig. 13.26(a). For the position of secretary, there are
3 possibilities. Thus for each of the two presidential candidates, there
are now 3 branches added, as in Fig. 13.26(b).

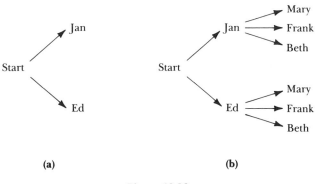

(a)   (b)

Figure 13.26

Pausing momentarily in this problem, we can trace 6 different "branches" of the tree thus far. Find and trace the branches in Fig. 13.26(b) for these possibilities for the offices of president and secretary, respectively:

(Jan, Mary),   (Jan, Frank),   (Jan, Beth),

(Ed, Mary),   (Ed, Frank),   and   (Ed, Beth).

Continuing on with the problem, we find that for each of these 6 possibilities, there exist 2 choices for treasurer. The completed tree diagram is shown in Fig. 13.27, together with all 12 possibilities.

Summarizing the work of the tree diagram, we arrive at what is called the *multiplication principle*. If there are 2 treasurer choices for each of 3 secretary choices for each of 2 president choices, then there are $2 \times 3 \times 2 = 12$ total possible outcomes.

MULTIPLICATION PRINCIPLE

If one event can be done in m ways and another event can be done in n ways, then there are $m \times n$ total possible ways of combining these two events. (This can, of course, be generalized beyond just two events.)

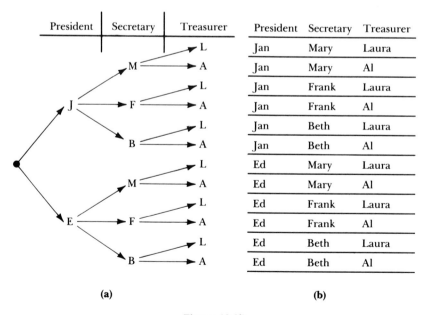

President	Secretary	Treasurer
Jan	Mary	Laura
Jan	Mary	Al
Jan	Frank	Laura
Jan	Frank	Al
Jan	Beth	Laura
Jan	Beth	Al
Ed	Mary	Laura
Ed	Mary	Al
Ed	Frank	Laura
Ed	Frank	Al
Ed	Beth	Laura
Ed	Beth	Al

(a)   (b)

Figure 13.27

Permutations

The first of two other methods used not only for counting, but also for several other areas in statistics, is the **permutation.**

■ **DEFINITION**

PERMUTATION

A *permutation* is an arrangement of objects in which the order of the arrangement is relevant.

■ **EXAMPLE 16**

To answer the question, "How many two-digit codes can be formed from the letters R, T, and S?" permutations are used. The code RS is different from the code SR. If we list all the possible codes, we find a total of six possibilities:

RS, RT, SR, ST, TS, TR. □

■

A formula to find the permutation of n objects selected r at a time is

$$_nP_r = \frac{n!}{(n - r)!}.$$

In Example 16, n is 3 and r is 2. If we substitute this into the formula, we get

$$_3P_2 = \frac{3!}{(3 - 2)!}.$$

Now, what does 3! mean?

The symbol ! represents *factorial notation* and indicates a multiplication process. For example, 6! means that 6 is multiplied by all integers between itself and 1:

$$6! = 6 \cdot 5 \cdot 4 \cdot 3 \cdot 2 \cdot 1 = 720.$$

Now, completing our problem, we have

$$_3P_2 = \frac{3!}{1!} = \frac{3 \cdot 2 \cdot 1}{1} = 6.$$

There are 6 ordered arrangements when selecting 2 items from a group of 3 items!

Combinations

If the order of RS and SR is not relevant, then instead of permutations, we are interested in combinations.

■ **DEFINITION**

COMBINATION

A *combination* is a selection of items in which the order of the items is irrelevant. The number of combinations of n things taken r at a time is the number of sets of r items, with no two sets containing the same r items.

■ **EXAMPLE 17**

Suppose we wish to select two accounts from a total of six to audit. The accounts are labeled A, B, C, D, E, and F. In this case, the order of A and B is irrelevant. How many combinations of accounts are possible?

■

The formula for the combination of n items selected r at a time is

$$_nC_r = \frac{n!}{(n-r)!\, r!}.$$

Thus, since we have $n = 6$ and $r = 2$,

$$_6C_2 = \frac{6!}{(6-2)!\, 2!} = \frac{6!}{4!\, 2!}$$
$$= \frac{6 \cdot 5 \cdot 4 \cdot 3 \cdot 2 \cdot 1}{4 \cdot 3 \cdot 2 \cdot 1 \cdot 2 \cdot 1} = 15.$$

There are 15 possible combinations of accounts, taken two at a time, from an original set of six accounts. □

Answers to other probability questions can now be found more easily using these counting techniques: the multiplication principle, permutations, and combinations.

● Question

If you are an officer in charge of account A in Example 17, what is the probability that your account will be audited? (*Hint:* Don't you already know the size of the sample space?)

Exercise Set 13.3

1. What is meant by the term statistical inference?

2. Define the term probability.

3. How will the study of probability aid in work with statistical inference?

4. Find ten newspaper articles dealing with or mentioning probability concepts.

5. Toss a die 60 times. Make a bar graph showing how many 1's, 2's, 3's, 4's, 5's, and 6's were obtained.

6. Explain the difference between theoretical probability and empirical probability.

7. If one die is rolled, find the probability of:

   a) obtaining a 3.

   b) obtaining a 1 or a 3.

   c) obtaining a 7.

   d) obtaining a 1, 2, 3, or 5.

   e) obtaining a 1, 2, 3, 4, 5, or 6.

8. If a letter were selected at random from the word C O M P U T E R, find the probability that:

   a) the letter will be an R.

   b) the letter will be an O.

   c) the letter will be an A.

   d) the letter will be a vowel.

   e) the letter will be a consonant.

9. Roll two dice, a red one and a green one.

   a) Find Pr(sum of 7).

   b) Find Pr(sum > 7).

   c) Find Pr(sum < 7).

   d) Find Pr(sum = 1).

   e) Find Pr(sum = 12).

   f) Find Pr(sum is an even number).

   g) Find Pr(sum is an odd number).

   h) Find Pr(sum is a prime number).

   i) Find Pr(4 < sum and sum < 8).

   j) Find Pr(3 < sum or sum > 9).

10. Ten accounts are to be audited by sampling two from the following: 3 accounts from Baybridge, Inc., 5 accounts from Chapman Assoc., and 2 accounts from Merriman.

   a) Find the probability that the two accounts will both be from Baybridge, Inc.

b) Find the probability that the first account will be from Chapman Assoc., and the second from Merriman.

c) Find the probability that one account will be from Chapman Assoc., and the second from Baybridge, Inc.

d) Find the probability that both will be from Merriman.

11. A survey was taken to determine the level of education of employees of a business firm in relation to the employees' seniority with the firm. The results are shown below.

YEARS WITH FIRM / HIGHEST DEGREE	1–5	6–10	OVER 10
High school diploma	5	2	1
Bachelor's degree	3	12	5
Master's degree	2	9	21

Using the results from this table and assuming that a person is picked at random, find each of the following.

a) The probability that the person has been with the firm for more than 10 years

b) The probability that the person has been with the firm for 10 years or less

c) The probability that the person has only a high school diploma

d) The probability that the person has either a bachelor's or a master's degree

e) The probability that the person has a master's degree and has been with the company for more than 10 years

f) The probability that the person has a master's degree or has been with the company for more than 10 years

12. For the following questions, A, B, and C refer to events.

a) If the probability of A happening is 0.15, what is the probability of A not happening?

b) If $\Pr(B) = 0.80$, find $\Pr(B')$.

c) $\Pr(C) + \Pr(C') = ?$

d) If A and B are mutually exclusive events, find $\Pr(A \text{ or } B)$.

e) If B and C are independent events, find $\Pr(A \text{ and } B)$.

f) If $\Pr(A) = 0.60$ and $\Pr(B|A) = 0.2$, find $\Pr(A \text{ and } B)$.

g) If $\Pr(A) = 0.60$ and $\Pr(A|B) = 0.60$, are A and B independent or dependent events?

13. State two events that represent:

    a) mutually exclusive events;

    b) independent events;

    c) dependent events;

    d) not mutually exclusive events.

14. Four offices—president, vice-president, secretary, and treasurer—are to be filled with the following people:

Jim or Mary	for president,
Bill, Karen, or Nancy	for vice-president,
Ann, Rob, Cora, or Dick	for secretary,
Ed or George	for treasurer.

    a) In how many possible ways can these offices be filled?

    b) Show the entire sample space.

15. How many different 4-digit codes can be formed from the digits {1, 3, 5, 6, 7, 8, 9}?

16. How many 3-digit account codes can be made if the first digit is a vowel, the second digit either a 1, 3, or 5, and the last digit either an N or an E?

17. Three coins are tossed. In how many possible ways can the three coins land?

18. Four coins are tossed.

    a) In how many possible ways can the coins land?

    b) Find Pr(obtaining 3 heads).

    c) Find Pr(obtaining no heads).

    d) Find Pr(obtaining 2 heads).

    e) Find Pr(obtaining 1 head).

19. Compute each of the following combinations.

    a) $_3C_2$ **b)** $_5C_2$

    c) $_7C_2$ **d)** $_{10}C_4$

20. Compute each of the following permutations.

    a) $_3P_2$ **b)** $_5P_2$

    c) $_7P_2$ **d)** $_{10}P_4$

21. a) How many different 4-digit codes can be formed with the symbols B, C, F, and J?

b) How many of the codes in part (a) would begin with the letter J?

c) How many of the codes in part (a) would have B as the second letter?

22. To check on the accuracy of computer printouts, 5 of a group of 21 printouts will be carefully inspected. In how many different ways could 5 printouts be selected from the 21?

23. Five playing cards are to be picked one at a time and then replaced from a deck of 52.

a) Find the probability that all 5 cards are hearts.

b) Find the probability that all 5 cards are red.

c) Find the probability that the 5 cards are all face cards.

d) Find the probability that 4 of the 5 cards selected are kings.

24. Assume in Exercise 23 that the cards will not be replaced. What would the probabilities then be for parts (a)–(d)?

Summary

Statistics and statistical methods are present in almost every part of our daily lives. By scanning news articles and being attentive to the media, you will notice averages, bar graphs, and terminology such as "median income." To be an "average" educated citizen requires some statistical literacy. To be involved with data processing requires having more than a mere cursory overview of statistics and statistical concepts. Frequently, personnel in data processing are asked to either (a) write programs that will provide statistical information, or (b) analyze already generated statistics from given data.

Data for the question being investigated are collected from either the entire group (a *population*) or a subset of the group (a *sample*). Usually the population is very large; thus collecting data from the population can be extremely costly, time-consuming, or even impossible! Methods of collecting the data from a sample of the population include *simple random sampling, systematic sampling, stratified sampling,* or *cluster sampling.* Regardless of the techniques used, the important criterion to emphasize, in sampling, is that the sample be *representative* of the population.

After the data have been collected, the statistics computed generally fall into one of two areas: either descriptive statistics or inferential statistics. *Descriptive statistics*—as the adjective implies—are statistics that "describe." The data can be organized by using a frequency distri-

bution for a more succinct and meaningful interpretation than a mere
listing of all items. The data may be presented graphically in several
different ways, depending on the purpose for the statistical investigation and/or the level of sophistication of the intended audience. *Frequency histograms, frequency polygons, bar graphs, pie charts,* and *pictographs* are several ways of describing vast quantities of data in a clear
and efficient manner.

In addition to graphical representations, descriptive statistics can
also be presented in the form of *numerical measures.* The "usual" measures to describe a set of data include *n* (how many items?), measures
of central tendency (*mean, median,* and *mode*), and measures of dispersion (*range, standard deviation,* and occasionally *quartiles.*) A good
understanding of each of these descriptive statistics is absolutely necessary, because while each one may be used quite appropriately, each
can be misused rather easily.

The goal of *inferential statistics* is to infer, or make inferences from, a
representative sample to the entire population. The question of ultimate concern in work with statistical inference is, "How far (or close) is
the statistical estimate obtained from the sample in relation to the true
population parameter?"

To aid in finding the answer to this question, a working knowledge
of *probability concepts* is of great value. An understanding of what probability is, and the difference between *mutually exclusive events* and *independent* and *dependent events* is required. A mastery of such counting
techniques as the *multiplication principle, permutations,* and *combinations*
is also most helpful. With these fundamental probability concepts, a
much greater "degree of confidence" can be assigned to the statistical
inferences made from our representative sample to the population of
interest. Probability allows us to make wiser decisions from our gathered data, even in the face of uncertainty!

Review Exercises

Determine whether each of the following statements (1–4) is true or false.

1. Statistical inference is the study of a sample to allow the making of predictions or estimates about the population from which the sample was taken.

2. The name for a measure of some characteristic of a sample is a parameter.

3. Probability is a tool to aid in more accurately making statistical inferences.

4. The arithmetic mean is a measure of central tendency used in descriptive
   statistics.

5. A business executive would like to know the average net worth of all

businesses in a large city. Identify each of the following as either a parameter or a statistic.

a) All businesses were listed alphabetically and numbered. Every 25th business was selected and its net worth computed. For the companies selected, the average net worth was $145,000.

b) The average of 20 randomly selected businesses was computed to be $128,000.

c) The net worth of all 137 companies averaged was $131,000.

6. Provide a "statistically literate" description or definition of each of these sampling techniques.

a) Systematic sampling

b) Simple random sampling

c) Proportional stratified sampling

7. Define probability in a much more detailed manner than simply "probability is the chance of something happening."

8. A container has 7 green chips, 3 blue chips, 2 white chips, and 8 yellow chips. A chip is drawn at random. What is the probability that the chip you select is:

a) a white chip?   b) a blue chip?

c) a black chip?   d) a green chip?

e) not a green chip?   f) not a white chip?

g) either a blue or white or yellow chip?

9. In a batch of 80 computer runs, 4 had errors in the development of the algorithm.

a) If a computer run were selected at random, what is the probability that there is an algorithmic error?

b) If 200 additional runs were processed, how many runs would be "expected" to have algorithmic errors?

10. For the following data, compute these descriptive statistics.

28	26	25	18	46	28
23	17	36	14	35	12
22	28	29	28	25	45
33	39	28	42	18	30
25	29	27	29	28	48

a) n b) Range

c) High   d) Low

e) Mean   f) Median

g) Mode   h) Variance

i) Standard deviation

11. Using the same data given for Exercise 10, estimate the mean by each of the following.

 a) Selecting a simple random sample.

 b) Selecting a sample systematically.

 c) Selecting a sample by randomly picking two numbers from each row.

 d) Selecting 8 random samples, each of size 10. Compute each sample mean. Add all 8 sample means and divide by 8. (This is really the arithmetic mean of the sample means!)

 e) Which method in parts (a)–(d) provided the best estimate of the actual population mean as computed in Exercise 10?

12. For the data in Exercise 10, prepare each of the following by grouping the data with the limits of the first class being 10 to 13.

 a) A frequency distribution and its histogram

 b) A relative frequency distribution and its histogram

 c) A cumulative frequency distribution and its histogram

 d) A cumulative relative frequency distribution and its histogram

 e) A frequency polygon

13. Prepare a bar chart showing the annual average yield over the last five years for the different investments shown below.

U.S. Treasury bonds (5 yr)	11.48%
U.S. Treasury bonds (20 yr)	11.29%
U.S. Government long-term bonds	7.84%
Municipal bonds	8.59%
Corporate AAA bonds	11.94%
Preferred stocks	10.64%
Common stocks	5.26%
Home mortgages	13.27%

14. Using the following data, prepare a bar chart showing how the percent yield on state and local government bonds rated Aaa have changed over the ten-year span from 1960 to 1969.

1960	3.26%	1965	2.51%
1961	3.16%	1966	5.71%
1962	6.12%	1967	3.57%
1963	4.84%	1968	4.68%
1964	3.54%	1969	7.25%

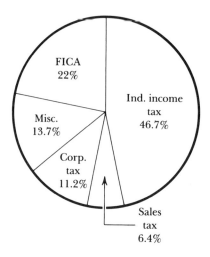

Figure 13.28

15. Using Fig. 13.28, determine how much money was received by the federal government from each of the following sources if total receipts were $475,085,652.

    a) Corporate taxes

    b) Individual income taxes

    c) FICA taxes and contributions

16. Do the data in Exercise 10 represent a normal distribution, a positively skewed distribution, or a negatively skewed distribution?

17. If the spinner shown in Fig. 13.29 were spun once:

    a) What is the probability that the arrow will land on a vowel?

    b) What is the probability that the arrow will land on F?

    If the spinner were spun twice:

    c) List the entire sample space.

    d) What is the probability of getting either AB, BC, CD, or EF?

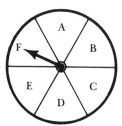

Figure 13.29

18. In how many different ways can five objects be chosen from eight objects:

    a) if order is not important?

    b) if order is important?

19. How many different microcomputer systems (a terminal, a CRT, a disc drive, and a printer) could be formed if each system can be selected from the following: 3 terminals, 4 CRT's, 2 disc drives, and 5 printers?

Appendix A

Hexadecimal and Decimal Conversion

To use this table to convert from *hexadecimal to decimal*:

1. Locate each hexadecimal digit in the column corresponding to its position in the hexadecimal number.
2. Read the decimal value equivalent to each hexadecimal digit.
3. Add these decimal values to obtain the decimal number equivalent to the original hexadecimal number.

Hexadecimal Columns

6		5		4		3		2		1	
HEX	= DEC	HEX	= DEC	HEX	= DEC	HEX	= DEC	HEX	= DEC	HEX	= DEC
0	0	0	0	0	0	0	0	0	0	0	0
1	1,048,576	1	65,536	1	4,096	1	256	1	16	1	1
2	2,097,152	2	131,072	2	8,192	2	512	2	32	2	2
3	3,145,728	3	196,608	3	12,288	3	768	3	48	3	3
4	4,194,304	4	262,144	4	16,384	4	1,024	4	64	4	4
5	5,242,880	5	327,680	5	20,480	5	1,280	5	80	5	5
6	6,291,456	6	393,216	6	24,576	6	1,536	6	96	6	6
7	7,340,032	7	458,752	7	28,672	7	1,792	7	112	7	7
8	8,388,608	8	524,288	8	32,768	8	2,048	8	128	8	8
9	9,437,184	9	589,824	9	36,864	9	2,304	9	144	9	9
A	10,485,760	A	655,360	A	40,960	A	2,560	A	160	A	10
B	11,534,336	B	720,896	B	45,056	B	2,816	B	176	B	11
C	12,582,912	C	786,432	C	49,152	C	3,072	C	192	C	12
D	13,631,488	D	851,968	D	53,248	D	3,328	D	208	D	13
E	14,680,064	E	917,504	E	57,344	E	3,584	E	224	E	14
F	15,728,640	F	983,040	F	61,440	F	3,840	F	240	F	15

To use this table to convert from *decimal to hexadecimal*:

1. Find the largest *decimal* value in the table that will divide into the decimal number you wish to convert.
2. Note its hexadecimal equivalent *and* column position.
3. Divide and determine the decimal remainder.
4. Repeat steps 1–3 on this remainder and any subsequent remainders.

■ **EXAMPLE 1**   Convert $4BD_{16}$ into base ten.

1. From column 3, 4_{16} = 1024.
2. From column 2, B_{16} =   176.
3. From column 1, D_{16} =   <u>13.</u>
                       1213

Therefore, $4BD_{16}$ = 1213. □

■ **EXAMPLE 2**   Convert 1213 into base sixteen.

1. The largest decimal value that will divide into 1213 is in column 3, row 4, that is, 1024:

$$1024 \overline{)\begin{array}{r} 1 \\ 1213 \\ -1024 \\ \hline 189 \end{array}}$$

2. The largest decimal value that will divide into 189 is in column 2, row B, that is, 176:

$$176 \overline{)\begin{array}{r} 1 \\ 189 \\ -176 \\ \hline 13 \end{array}}$$

3. The largest decimal value that will divide into 13 is in column 1, row D, that is, 13:

$$13 \overline{)\begin{array}{r} 1 \\ 13 \\ -13 \\ \hline 0 \end{array}}$$

The hexadecimal equivalent is constructed by noting the row and column numbers.

Column 3, row 4 yields 4.
Column 2, row B yields B.
Column 1, row D yields D.

Therefore, 1213 = $4BD_{16}$. □

Appendix
B

Octal and Decimal Conversion

To use this table to convert from *octal to decimal*:

1. Locate each octal digit in the column corresponding to its position in the octal number.
2. Read the decimal value equivalent to each octal digit.
3. Add these decimal values to obtain the decimal number equivalent to the original octal number.

To use this table to convert from *decimal to octal*:

1. Find the largest *decimal* value in the table that will divide into the decimal number you wish to convert.
2. Note its octal equivalent *and* column position.
3. Divide and determine the decimal remainder.
4. Repeat steps 1–3 on this remainder and any subsequent remainders.

Octal Columns

6		5		4		3		2		1	
OCT	= DEC	OCT	= DEC	OCT	= DEC	OCT	= DEC	OCT	= DEC	OCT	= DEC
0	0	0	0	0	0	0	0	0	0	0	0
1	32,768	1	4,096	1	512	1	64	1	8	1	1
2	65,536	2	8,192	2	1,024	2	128	2	16	2	2
3	98,304	3	12,288	3	1,536	3	192	3	24	3	3
4	131,072	4	16,384	4	2,048	4	256	4	32	4	4
5	163,840	5	20,480	5	2,560	5	320	5	40	5	5
6	196,608	6	24,576	6	3,072	6	384	6	48	6	6
7	229,376	7	28,672	7	3,584	7	448	7	56	7	7

■ **EXAMPLE 1**   Convert 743_8 into base ten.

1. From column 3, 7_8 = 448.
2. From column 2, 4_8 =   32.
3. From column 1, 3_8 = $\underline{\quad 3.}$
                                             483

Therefore, 743_8 = 483. □

■ **EXAMPLE 2**   Convert 483 into base eight.

1. The largest decimal value that will divide into 483 is in column 3, row 7, that is, 448:

$$
\begin{array}{r}
1 \\
448{\overline{)\,483}} \\
-448 \\
\hline
35
\end{array}
$$

2. The largest decimal value that will divide into 35 is in column 2, row 4, that is, 32:

$$
\begin{array}{r}
1 \\
32{\overline{)\,35}} \\
-32 \\
\hline
3
\end{array}
$$

3. The largest decimal value that will divide into 3 is in column 1, row 3, that is, 3:

$$
\begin{array}{r}
1 \\
3{\overline{)\,3}} \\
-3 \\
\hline
0
\end{array}
$$

The octal equivalent is constructed by noting the row and column numbers.

Column 3, row 7 yields 7.
Column 2, row 4 yields 4.
Column 1, row 3 yields 3.

Therefore, 483 = 743_8. □

Glossary

Absolute value. The quantitative value of a number representing the distance between the number and zero on the number line.

Algorithm. A step-by-step list of instructions for accomplishing some specific task.

ANSI. American National Standards Institute. An organization that acts as a national clearinghouse and coordinator for voluntary standards in the United States.

Array. A rectangular arrangement of elements in which the elements are displayed in rows and columns.

Associative. If the result of combining three or more elements of a set, in groups of two, is independent of the manner in which the elements are grouped, the set is said to be associative with respect to the method by which they are combined.

BASIC. *B*eginners *A*llpurpose *S*ymbolic *I*nstruction *C*ode, a high-level computer language.

Biconditional (equivalence). The compound statement formed by joining two simple statements by the IF AND ONLY IF connective.

Binary numeration system. A numeration system consisting of only the symbols 0 and 1.

Boole, George (1815–1864). An English mathematician who helped develop modern symbolic logic.

Boolean algebra. A mathematical structure, named after George Boole, that is composed of a set of elements (usually 0 and 1), two binary operations (+ and ·), and a unary operation (').

Cantor, Georg (1845–1918). The German mathematician who developed the theory of sets.

Cardinal number. The number of elements in a set.

COBOL. *CO*mmon *B*usiness *O*riented *L*anguage. A high-level computer language used in business.

Coefficient matrix. The matrix formed from the coefficients of a system of linear equations.

Cofactor. The minor of an element of a matrix preceded by the appropriate sign (+ or −) for that element.

Combination.   An arrangement of objects in counting theory, in which the order of the arrangement is unimportant, that is, m, a, t is the same as t, m, a.

$$_nC_r = \frac{n!}{r!(n-r)!}$$

Commutative.   If the result of combining two or more elements of a set is independent of the order in which they are combined, the set is said to be commutative with respect to the method used to combine them. The real numbers are commutative with respect to addition and multiplication, but not to subtraction and division. In Boolean algebra, $A + B = B + A$ and $A \cdot B = B \cdot A$.

Complement.   In set theory, the complement of a given set is the set of all the elements from the universal set that are not in the given set. In Boolean algebra, the complement of a switch A is a switch that is in the opposite state of A, that is, if A is closed, the complement of A, written A', is open.

Conditional.   The IF–THEN logic statement.

Conformable.   If the number of rows of a matrix A is equal to the number of columns of another matrix B, then matrices A and B may be multiplied.

Conjunction.   A compound statement in which the connective is AND and which is true only when both component statements are true.

Contradiction.   A compound statement that is false for all possible cases.

Cramer's rule.   A process of solving systems of linear equations using determinants.

Decision table.   A chart that shows a list of actions that can be taken for various combinations of conditions.

DeMorgan's laws.   In Boolean algebra, $(A + B)' = A'B'$ and $(AB)' = A' + B'$.

Dependent equations.   A system of equations with infinitely many solutions. Graphically, a linear dependent system represents lines that coincide.

Descriptive statistics.   The collecting, tabulating, and summarizing of data collected from a population and reported in a succinct, yet meaningful way.

Determinant.   The value associated with a square matrix.

Dimension (order).   The size of the matrix as given by the number of rows and the number of columns. An $m \times n$ matrix has m rows and n columns.

Discriminant.   The radicand in the quadratic formula: $b^2 - 4ac$.

Disjoint sets.   Sets that have no elements in common.

Disjunction.   A compound statement in which the connective is OR and which is false only when both components are false.

Distributive.   If the result of performing an operation on a group of elements is equivalent to performing the operation on the individual elements and then combining the results, the operation is said to be distributive. With real numbers, $a(b + c) = ab + ac$. In Boolean algebra there are two distributive properties: (1) $A \cdot (B + C) = (A \cdot B) + (A \cdot C)$, and (2) $A + AB = (A + B)(A + B)$.

Domain.   The set of all possible input values for the independent variable in a function.

Duality.   In Boolean algebra, if the operations of $+$ and $\cdot$ are interchanged in a property, a second property is obtained and is called the dual of the original property.

Elementary transformations. Operations that are performed on a system of linear equations to produce other equivalent linear systems. The operations include (1) interchanging two equations, (2) multiplying an equation by a nonzero constant, and (3) adding one equation to another.

Equal matrices. Two matrices whose elements in the corresponding positions are equal.

Equation. A statement that two algebraic expressions are equal.

Euler, Leonard (1707–1783). The Swiss mathematician who first attempted to draw pictures of sets.

Exclusive or (EOR). A compound statement that is true when one or the other component statements is true, but is false when both components are true and when both components are false.

Exponential function. A function of the form $y = b^x$, where b is a fixed constant greater than zero.

Extraneous root. A number obtained in the process of solving an equation that appears to be a solution to the equation but is not.

Feasible region. The area bounded by the lines of constraint in the graphical solution of a linear programming problem.

Flowchart. A pictorial representation of an algorithm.

FORTRAN. *FOR*mula *TRAN*slation. A high-level computer language used in science and engineering.

Frequency distribution. A tabular display of the number of items (frequency) in each class for a given collection of data.

Frequency histogram. A type of bar graph in which class intervals are marked on the horizontal axis and the frequencies are shown on the vertical axis.

Frequency polygon. A geometric figure formed by connecting the midpoints, or class marks, of each class interval with straight-line segments.

Function. A special type of relation between two sets of numbers, input values x and output values y, in which for each value of x there is one and only one value of y.

Gaussian elimination. A process for solving a system of linear equations in which the system is transformed into an equivalent, but simpler, system by forming linear combinations.

Hexadecimal numeration system. The numeration system consisting of the symbols 0, 1, 2, 3, 4, 5, 6, 7, 8, 9, A, B, C, D, E, F.

Idempotent properties. In Boolean algebra, $A + A = A$ and $A \cdot A = A$.

Identity matrix. A square matrix with 1's in the main diagonal and 0's in each of the other positions.

Inclusive or (OR). A compound statement that is true when one or the other of the component statements is true or when both components are true, and is false when both components are false.

Inconsistent equations. A system of equations in which no common solution exists. Graphically, a linear inconsistent system represents parallel lines.

Indcpendent equations. A system of equations that has exactly one solution. Graphically, a linear independent system represents lines that intersect in one unique point.

Independent events. Two events are independent if the occurrence or nonoccurrence of one event has no effect on the occurrence of the other event.

Inequality. A statement that one algebraic expression is less than or equal to or greater than or equal to another algebraic statement.

Integers. The natural numbers together with their negatives, plus zero.

Intercepts. The point(s) at which a graph crosses an axis: the point on the x-axis is the x-intercept and the point on the y-axis is the y-intercept.

Intersection. The set operation that picks out the elements common to two sets.

Inverse of a function. The relation obtained by interchanging the x- and y-values of the original function. If (x, y) is an ordered pair from the original function, then (y, x) is the corresponding ordered pair of the inverse. The inverse relation may or may not be a function.

Invertible. A matrix that has an inverse. It is invertible if and only if its determinant is not zero.

Irrational numbers. Numbers that are not rational.

Key variable. In linear programming, the variable in the objective function that has the largest positive coefficient.

Linear function. An equation of the form $ax + b = y$, where a and b are real numbers and $a \neq 0$ in which the degree of any term is one or less. The graph of a linear function is a straight line.

Linear programming. The mathematical method of determining the maximization or minimization of a linear function subject to certain constraints.

Literal. A nonnumerical symbol used to represent certain elements from a set of numbers.

Literal equation. An equation that contains more than one literal symbol.

Logarithm. From the exponential function, $y = b^x$, the logarithm is the value of x to which b must be raised in order to obtain y. Generally, the logarithm is written $x = \log_b y$ or $y = \log_b x$.

Logic circuits. A circuit in Boolean algebra that contains one or more input devices but has exactly one output device.

Logical product. In Boolean algebra $A \cdot B$, called the logical product, is interpreted as A AND B and represents a series circuit.

Logical sum. In Boolean algebra $A + B$, called the logical sum, is interpreted as A OR B and represents a parallel circuit.

Matrix. A rectangular array of elements.

Mean. A measure of the central tendency of a set of data. It is found by finding the sum of the set of data and dividing that sum by the total number of items.

Median. The value that falls in the middle in a set of ranked data.

Minor. The determinant that remains after the row and column containing a given element of a matrix have been deleted.

Mode. The value in a set of data that occurs most frequently.

Modular arithmetic. Arithmetic performed in a finite mathematical system, also called *clock arithmetic*.

Modulus.   The numeral representing the number of symbols in a modular system.

Mutually exclusive.   Two events that have no common outcomes.

Natural numbers.   The numbers used in counting: 1, 2, 3, 4, 5, ...

Negation.   The statement formed by reversing the truth value of a given statement.

Nonsingular.   The name given to a matrix that does have an inverse. The determinant of a nonsingular matrix is not zero. A matrix for which an inverse does not exist is said to be singular.

Null set.   A set that contains no elements; denoted by $\emptyset$ or { }.

Objective function.   The function that is to be maximized or minimized in a linear programming problem.

Octal numeration system.   The numeration system consisting of the symbols 0, 1, 2, 3, 4, 5, 6, 7.

One-to-one correspondence.   A correspondence between the set of x-values and the set of y-values of a function, $y = f(x)$, in which no two x-values have the same y-values.

Parabola.   The graph of a quadratic function.

Parallel circuit.   A switching circuit that will allow current to flow if either one switch or the other switch is closed (ON).

Parameters.   Numerical descriptions of a population.

Pascal.   A high-level computer language named after the French mathematician Blaise Pascal (1623–1662).

Permutation.   An arrangement of objects in counting theory in which the order of the arrangement is important, that is, m, a, t is different from m, t, a. The formula is

$$_nP_r = \frac{n!}{(n - r)!}$$

Pivot.   The name given to the coefficient on the key variable in the pivotal equation.

Pivotal equation.   In linear programming, this equation is determined by the smallest quotient resulting from the division of the constant term in each constraint by the nonzero coefficient of the key variable.

Polynomial function.   A function defined by an equation of the form

$$a_nx^n + a_{n-1}x^{n-1} + a_{n-2}x^{n-2} + \cdots + a_1x + a_0,$$

where n is a positive integer or zero and the coefficients of each term are real numbers.

Population.   The total well-defined group of individuals, objects, or data about which information is desired.

Principal diagonal.   The diagonal of a matrix that stretches from the upper left corner to the lower right corner.

Probability.   A measure of the likelihood that some stated event will occur; it is the ratio of favorable outcomes to total possible outcomes.

Pseudocode.   The short simple commands of an algorithm written in English.

***p*th percentile.**   The value in a distribution that is greater than $p\%$ of all the values in that distribution. Also, $100 - p\%$ of the values are larger than the value.

Quadratic equation.   Any equation of the form $ax^2 + bx + c = 0$, where $a \neq 0$.

Quadratic formula.   A formula for solving quadratic equations:

$$x = \frac{-b \pm \sqrt{b^2 - 4ac}}{2a}.$$

Quartiles.   The values in a distribution that divide the data into four equal segments. The 25th, 50th, and 75th percentiles are the first, second, and third quartiles, respectively.

Radical equation.   Any equation that contains a variable in the radicand of a radical.

Random sample.   A sample from a population in which each item of the population has an equally likely chance of being selected.

Range.   The set of all possible output values for the dependent variable in a function. In statistics, the range is the difference between the largest value and smallest value for a set of data and is a measure of the dispersion of the data.

Rational number.   A number of the form a/b, where a and b are integers and $b \neq 0$.

Relation.   A mathematical relationship between two sets of numbers represented by the ordered pairs (x, y).

Sample.   A finite subset of a population.

Series circuit.   A switching circuit that will allow current to flow only when both switches are closed (ON).

Set.   A collection of similar objects.

Simplex method.   An iterative process for solving linear programming problems in which successive, more optimal, feasible solutions are determined until the maximum or minimum solution has been reached.

Simultaneous equations.   Two or more equations that impose conditions simultaneously on the same variables.

Slope.   The ratio of the vertical change to the horizontal change of a line.

Slope–intercept form.   A linear equation that has been explicitly solved for y in which the coefficient of x is the value of the slope of the line and the constant is the y-intercept ($y = mx + b$).

Standard deviation.   The square root of the variance.

Statement.   A declarative sentence, or mathematical equation, that is either true or false, but not both.

Statistical inference.   The process of making predictions or inferences about a population on the basis of data gathered and studied in a random sample.

Statistics.   The collection, organization, analysis, presentation, and interpretation of numerical data.

Tautology.   A compound statement that is true for all possible cases.

Union.   The set operation that joins together the elements of two sets.

Universal set.   The set that contains all elements in a discussion.

Variance.   The average of the squares of the differences of the values in a distribution from the mean of the distribution.

Vector.   A matrix that consists of only one row or only one column.

Venn, John (1834–1923). The British mathematician who extended Leonard Euler's work and for whom Venn diagrams are named.

Vertex.   The highest point in the graph of a parabola that opens downward and the lowest point in the graph of a parabola that opens upward.

Vertical-line test.   A test to determine whether a relation is a function. If all possible vertical lines intersect the graph of the relation in only one point, the relation is a function.

Zero of a function.   The value of the independent variable that makes the value of the dependent variable zero. The real-number zeros can be determined as the points at which the graph of the function crosses the x-axis.

Solutions to Odd-Numbered Exercises

Chapter 0

Exercise Set 0.1, pp. 6–7

1. No; yes **3.** No; yes **5.** No; yes **7.** The associative property of multiplication
9. The commutative property of addition **11.** Yes
13. (a) Either $a < 0$ and $b > 0$ or $a > 0$ and $b < 0$; (b) either $a = 0$ or $b = 0$ or $a = b = 0$;
(c) either $a < 0$ and $b < 0$ or $a > 0$ and $b > 0$.

Exercise Set 0.3, p. 9

1. 0 **3.** 0 **5.** Undefined **7.** y^4 **9.** $3^5 = 243$ **11.** $9b^4$ **13.** x^{-2} **15.** x^8
17. b^{-3} **19.** $16x^6$ **21.** x^4 **23.** $x^6/27y^6$ **25.** b^6/a^4 **27.** $b^2/9a^4$ **29.** $-2y/3$

Exercise Set 0.4, p. 11

1. -5 **3.** -3 **5.** 9 **7.** 11 **9.** 1 **11.** -3 **13.** 12 **15.** 8 **17.** -24

Exercise Set 0.5, pp. 15–17

1. $xy - 5xz$ **3.** $-6x^2$ **5.** $7ab^2 + 3abc$ **7.** $5x^3y^2 + 2xy^2$ **9.** $-3x - 5$ **11.** $-3x - 8$
13. $-x + 1$ **15.** -37 **17.** $-8a^2bc^2$ **19.** $24x^3y^4$ **21.** $a^2x^4y^2z^2$ **23.** $-6a^3bx - 10a^2b^2x$
25. $-12cx^2y + 21ac^3x^2$ **27.** $30a^4bc - 42a^3b^3c - 63a^2b^5c^2$ **29.** $x^2 + 7x + 10$ **31.** $2x^2 + 3x - 9$
33. $x^2 - y^2$ **35.** $a^2 + 2ab + b^2$ **37.** $a^2 - 2ab + b^2$ **39.** $x^4 - y^4$ **41.** $x^3 - x^2 - 9x + 9$
43. $2x^3 - 7x^2 - 3x + 18$ **45.** $x^2 - y^2 - z^2 + 2yz$ **47.** $9x^2$ **49.** $8xy^3/a$ **51.** $8z/x$

53. $-2y + 3x$ **55.** $5y^2 - 2x/y + 3x$ **57.** $\dfrac{5b^2}{3} + \dfrac{5bc}{2a} - \dfrac{5ac}{6}$

Exercise Set 0.6, pp. 24–25

1. $x = 2$ **3.** $x = 5$ **5.** $x = 3$ **7.** $x = -3$ **9.** $x = 0$ **11.** $x = 2$ **13.** $x = 2$
15. $x = 4$ **17.** $x = 24$ **19.** $x = 8$ **21.** $x = 1/4$ **23.** $x = 3$ **25.** $x = -6$
27. $y = -1/3$ **29.** $x = 20/3$
31. **33.** **35.**

37. $x \le -13/3$ **39.** $x > 10/27$ **41.** $x < 25/11$

13. $3/2 < x \le 6$

45. $1 \le x \le 6$

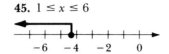

47.

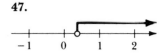

49.

51.

53.

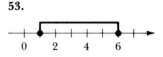

Exercise Set 0.7, pp. 30–31

1. $3(4x - 5y)$ **3.** $5ab(4a - 9a^2c)$ **5.** $ab(ab + a + b)$ **7.** $3x(8x - 4y + 6y^2)$ **9.** $x^n(x + 1)$
11. $(a + b)(a - b)$ **13.** $(x + 6)(x - 6)$ **15.** $(x + 1)(x - 1)$ **17.** $2(a + 4b)(a - 4b)$
19. $(x^2 + y^2)(x + y)(x - y)$ **21.** 8   **23.** 30   **25.** 25   **27.** $(a + 3)^2$ **29.** $(y - 11z)^2$
31. $(3x - y)^2$ **33.** $(x - 3)(x - 4)$ **35.** $(y + 5)(y - 4)$ **37.** $(a - 12b)(a + 3b)$
39. $(y + 7)(y + 5)$ **41.** $(2x + 1)(x + 1)$ **43.** $(4b + 3)(2b + 3)$ **45.** $4x(x - 4)(x + 2)$

Exercise Set 0.8, p. 34

1. $\dfrac{2a - 3b}{2(2a + 3b)}$ **3.** $\dfrac{2}{(x - 2y)}$ **5.** $\dfrac{-3}{2}$ **7.** $\dfrac{(x + y)}{4}$ **9.** $\dfrac{x - 4}{x - 3}$ **11.** $\dfrac{1}{x}$ **13.** $\dfrac{5}{3y}$

15. $\dfrac{y(x + y)}{x}$ **17.** $\dfrac{a - 4}{a + 3}$ **19.** $\dfrac{6x - 5m}{30}$ **21.** $\dfrac{8c^2 + 3a^2 - 30b^2}{36abc}$ **23.** $\dfrac{b - c}{b(b + c)}$

25. $\dfrac{x - 5y}{(x^2 - 4y^2)(x + y)}$

Exercise Set 0.9, pp. 37–38

1. $x = 1$ **3.** $x = 1$ **5.** $x = 33/2$ **7.** $x = -1$ **9.** $x = 1/6$ **11.** $x = 10$ **13.** $x = -5$
15. No solution   **17.** $x = 5/3$ **19.** $x = -2$

Review Exercises, pp. 38–39

1. 0   **3.** 5   **5.** 7   **7.** $3x^2y + 6xy^2$ **9.** $3 - 2x - y$ **11.** $-15x + 66$ **13.** $12x^3y^2$
15. $6ayx - 8ay^2$ **17.** $2x^2 - 7xy + 6y^2$ **19.** $8xy$ **21.** $3y - x + 1$ **23.** $x = 3$ **25.** $x = -1$
27. $x > 4$ **29.** $x > 5$ **31.** $x = 2$ **33.** $8x(3y - 1)$ **35.** $(2x + 3y)(2x - 3y)$ **37.** $(x + 10)^2$

39. $(x + 3)(x - 5)$ **41.** $(y + 4)(y - 7)$ **43.** $\dfrac{(x - 3)}{4}$ **45.** $\dfrac{a + 6}{(a - 1)(x + 3y)}$ **47.** x

49. $\dfrac{(5a + 6b)}{9}$ **51.** $\dfrac{5x - y}{(x + y)(x - y)^2}$

Chapter 2

Exercise Set 2.1, pp. 54–55

1. 3   **3.** 3   **5.** 4   **7.** 2   **9.** 5   **11.** 3   **13.** 4   **15.** 2   **17.** 3.14159   **19.** 4.5402
21. 3.14   **23.** 0.736   **25.** 24.3   **27.** 3.142   **29.** 0.005   **31.** 24.372   **33.** 14.295

Exercise Set 2.2, pp. 56–57

1. 7.406×10^3 **3.** 9.2743×10^3 **5.** 9.3704×10 **7.** 6.34E4 **9.** 8.321E − 2
11. 7.6E7 **13.** 7.42E − 6 **15.** 3.32E8 **17.** 34,200 **19.** 0.0000004735 **21.** 7260
23. 0.00023754 **25.** 432,700

Exercise Set 2.3, pp. 61–62

1. 23 **3.** 11 **5.** 72 **7.** 40 **9.** 6 **11.** 2 **13.** 4 **15.** 4 **17.** 0 **19.** 2
21. 1 **23.** 0.327×10^2 **25.** 0.845×10 **27.** 0.745×10^{-3} **29.** 0.724E2 **31.** 0.743E6
33. 0.643E2 **35.** 0.4172E − 3 **37.** 0.7983×10^2 **39.** 0.8893×10^3 **41.** 0.3077×10^3
43. 0.1587278×10^4 **45.** 0.398266×10^5 **47.** 0.4×10^5 **49.** 0.5×10^5

Review Exercises, p. 63

1. 3 **3.** 4 **5.** 4 **7.** 2 **9.** 3 **11.** 0.4325 **13.** 0.4752 **15.** 2.718 **17.** 0.2047
19. 2.7183 **21.** 24.7152 **23.** 0.006745 **25.** 3,752,400 **27.** 2,710,000,000
29. 9.3×10^7 **31.** 6 **33.** 2 **35.** 1 **37.** 0.834×10^3 **39.** 0.622×10^2
41. 0.366×10^4 **43.** 0.25×10^3

Chapter 3

Exercise Set 3.1, pp. 69–70

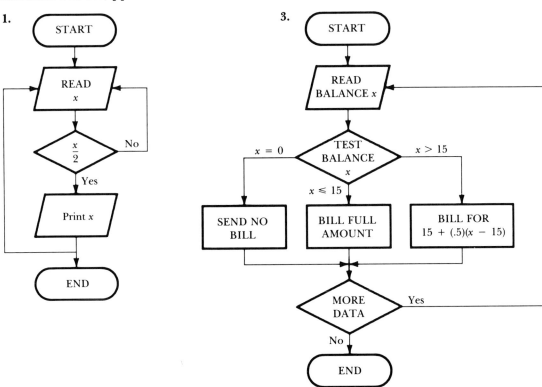

5.

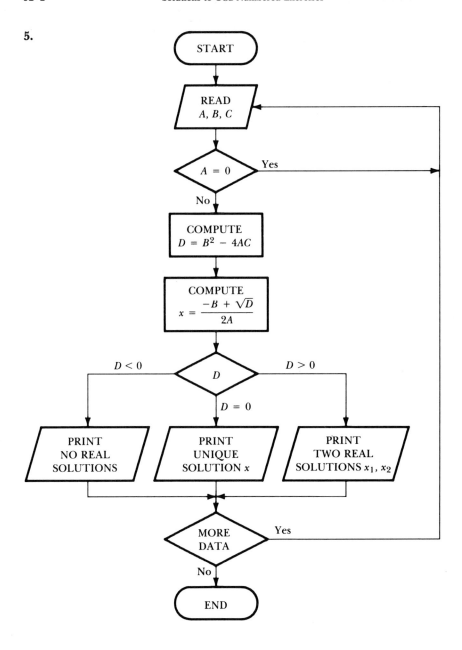

Exercise Set 3.2, p. 75

1. (a) 6 loops: N = 1
  N = 2
  N = 3
  N = 4
  N = 5
  N = 6

(b) 6 loops: N = 1
  N = 3
  N = 5
  N = 7
  N = 9
  N = 11

(c) 7 loops: N = 2
  N = 2.5
  N = 3
  N = 3.5
  N = 4
  N = 4.5
  N = 5

(d) 5 loops: N = −3
  N = −1
  N = 1
  N = 3
  N = 5

3. READ total sales
IF sales ≤ $5000
  THEN salary = .03 ∗ sales + $100
ELSE
  salary = .05 ∗ sales
ENDIF
WRITE salary
END

5. DO $x = 1$ to 5
  $y = x^2 - 3x$
  WRITE y
ENDDO
END

Exercise Set 3.3, p. 78

1. *Conditions:* $x < 0$ *Actions:* $y = x + 1$
  $x \geq 0$ $y = -(x + 1)$

$x < 0$	Y	N
$x \geq 0$	N	Y

$y = x + 1$	×	—
$y = -(x + 1)$	—	×

3. Let C = cost.
  Conditions: $1 \leq C \leq 10$ *Actions:* Tax = $0.05 \ast C$
    $10.01 \leq C \leq 50$ Tax = $0.055 \ast C$
    $C > 50$ Tax = $0.06 \ast C$

$1 \leq C \leq 10$	Y	N	N
$10.01 \leq C \leq 50$	N	Y	N
$C > 50$	N	N	Y

$0.05 \ast C$	×	—	—
$0.055 \ast C$	—	×	—
$0.06 \ast C$	—	—	×

5. Let P = total pay.
  Conditions: sales ≤ $5000 *Actions:* $P = \$250 + 0.04 \ast$ sales
    sales > $5000 $P = \$175 + 0.05 \ast$ sales

Sales ≤ $5000	Y	N
Sales > $5000	N	Y

$\$250 + 0.04 \ast$ sales	×	—
$\$175 + 0.05 \ast$ sales	—	×

Review Exercises, p. 79

1.
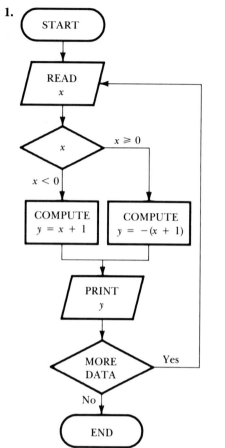

3.
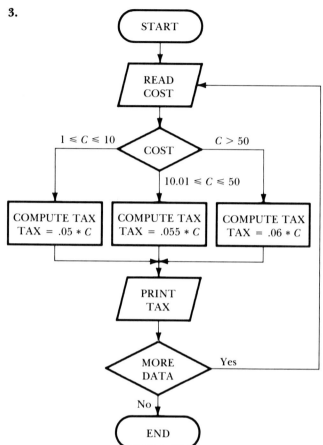

5. READ x
IF $x < 0$
    THEN $y = x + 1$
ELSE
    $y = -(x + 1)$
ENDIF
WRITE y
END

7. DO $x = 1$ to 10 STEP 0.5
    $y = 2x - 3$
        WRITE y
ENDDO
END
    19 loops: $y = -1, 0, 1, 2, 3, 4, 5, 6, 7, 8,$
        $9, 10, 11, 12, 13, 14, 15,$
        $16, 17$

9. $x = 1$
    DOUNTIL $x > 10$
        $y = 2x - 3$
            WRITE x, y
        $x = x + 0.5$
    ENDDO
    END

11. *Conditions:* $x < 0$ *Actions:* $y = -x$
            $x = 0$ $y = x^2$
            $x > 0$ $y = -x^2$

$x < 0$	Y	N	N
$x = 0$	N	Y	N
$x > 0$	N	N	Y
$y = -x$	X	—	—
$y = x^2$	—	X	—
$y = -x^2$	—	—	X

Chapter 4

Exercise Set 4.1, pp. 85–86

1. 3   **3.** 10   **5.** 4   **7.** 2

9.

$\oplus$	0	1	2
0	0	1	2
1	1	2	0
2	2	0	1

$\otimes$	0	1	2
0	0	0	0
1	0	1	2
2	0	2	1

11.

$\oplus$	0	1	2	3	4	5	6
0	0	1	2	3	4	5	6
1	1	2	3	4	5	6	0
2	2	3	4	5	6	0	1
3	3	4	5	6	0	1	2
4	4	5	6	0	1	2	3
5	5	6	0	1	2	3	4
6	6	0	1	2	3	4	5

$\otimes$	0	1	2	3	4	5	6
0	0	0	0	0	0	0	0
1	0	1	2	3	4	5	6
2	0	2	4	6	1	3	5
3	0	3	6	2	5	1	4
4	0	4	1	5	2	6	3
5	0	5	3	1	6	4	2
6	0	6	5	4	3	2	1

13.

$\oplus$	0	1	2	3	4	5	6	7	8
0	0	1	2	3	4	5	6	7	8
1	1	2	3	4	5	6	7	8	0
2	2	3	4	5	6	7	8	0	1
3	3	4	5	6	7	8	0	1	2
4	4	5	6	7	8	0	1	2	3
5	5	6	7	8	0	1	2	3	4
6	6	7	8	0	1	2	3	4	5
7	7	8	0	1	2	3	4	5	6
8	8	0	1	2	3	4	5	6	7

$\otimes$	0	1	2	3	4	5	6	7	8
0	0	0	0	0	0	0	0	0	0
1	0	1	2	3	4	5	6	7	8
2	0	2	4	6	8	1	3	4	7
3	0	3	6	0	3	6	0	3	6
4	0	4	8	3	7	2	6	1	5
5	0	5	1	6	2	7	3	8	4
6	0	6	3	0	6	3	0	6	3
7	0	7	5	3	1	8	6	4	2
8	0	8	7	6	5	4	3	2	1

15. **(a)** Yes, yes; **(b)** a, b; **(c)** $b:e, c:d, d:c, e:b$; **(d)** $b:b, c:d, d:c, e:e$

Exercise Set 4.2, p. 87

1. $(7 \times 10^3) + (4 \times 10^2) + (0 \times 10) + (6 \times 10^0)$
3. $(9 \times 10^3) + (2 \times 10^2) + (7 \times 10) + (4 \times 10^0) + (3 \times 10^{-1})$
5. $(9 \times 10) + (3 \times 10^0) + (7 \times 10^{-1}) + (0 \times 10^{-2}) + (4 \times 10^{-3})$
7. $(3 \times 10^5) + (2 \times 10^4) + (4 \times 10^3) + (5 \times 10^2) + (0 \times 10) + (7 \times 10^0)$
9. $(1 \times 10) + (3 \times 10^{-2}) + (7 \times 10^{-3}) + (8 \times 10^{-4})$ **11.** 0,2617   **13.** 345.07   **15.** 4.3039
17. $(3 \times 8^2) + (4 \times 8) + (7 \times 8^0)$

Exercise Set 4.3, p. 94

1. 29   **3.** 51.25   **5.** 14.375   **7.** 11.6875   **9.** 27.6875   **11.** 1111_2 **13.** 10000111_2
15. 1100.11_2 **17.** 101000.0101_2 **19.** 101.0100110011_2 **21.** 0.1001100110011_2

Exercise Set 4.4, p. 99

1. 114   **3.** 2638   **5.** 83.21875   **7.** 170_8 **9.** 3103_8 **11.** 173.54_8 **13.** $0.46\overline{31463}1_8$
15. 0.231463146_8 **17.** 156_8 **19.** 126.1_8 **21.** 100101111_2 **23.** 111110000010_2
25. 101100011001.011010_2

Exercise Set 4.5, p. 104

1. 441   **3.** 8656   **5.** 15789.125   **7.** 78_{16} **9.** $10C8_{16}$ **11.** $CB1_{16}$ **13.** 0.54_{16}
15. $0.19\overline{9}_{16}$ **17.** $6E_{16}$ **19.** $2B.A_{16}$ **21.** 11110111101100_2 **23.** 10010101110.11_2
25. $A27_{16}$ **27.** 35174_8

Exercise Set 4.6, pp. 109–110

1. 101001_2 **3.** 110101110_2 **5.** 1001010_2 **7.** 1511_8 **9.** 13345_8 **11.** $E7B1_{16}$
13. 1221_{16} **15.** 11_2 **17.** 10011_2 **19.** 216_8 **21.** 2017_8 **23.** $6CE_{16}$ **25.** $68C71_{16}$
27. 101010_2 **29.** 100100111_2 **31.** 14727_8 **33.** 534002_8 **35.** $1363E_{16}$
37. $17DAD11_{16}$ **39.** 1010_2 **41.** 11001.1_2 **43.** 1011.01001_2

Exercise Set 4.7, pp. 114–115

1. 1001   **3.** 110110   **5.** 111111   **7.** 00110100   **9.** 01000000   **11.** 11100011
13. 11000000   **15.** 01001000   **17.** 00100000 overflow   **19.** 01010001   **21.** 11111000
23.   8 bit: largest + 01111111; smallest + 00000001
          largest − 11111111; smallest − 10000001
       16 bit: largest + 0111111111111111; smallest + 0000000000000001
          largest − 1111111111111111; smallest − 1000000000000001
25. 0000000001111101   **27.** 1111111110000000

Review Exercises, pp. 115–117

1. $(3 \times 10^3) + (2 \times 10^2) + (4 \times 10)$ **3.** $(1 \times 2^5) + (1 \times 2^3) + (1 \times 2^2) + (1 \times 2^0)$
5. $(2 \times 10^2) + (1 \times 10) + (2 \times 10^{-1}) + (4 \times 10^{-2})$ **7.** $(1 \times 2^2) + (1 \times 2^0) + (1 \times 2^{-1}) + (1 \times 2^{-3})$
9. 21   **11.** 3864   **13.** 3131   **15.** 101011_2 **17.** 164_8 **19.** 70_{16} **21.** 11.0001100011_2
23. $0.3146\overline{3146}_8$ **25.** $0.66\overline{6}_{16}$ **27.** 56_8 **29.** 111100101011_2 **31.** $E1E_{16}$ **33.** 10000_2
35. 10331_8 **37.** $C85E_{16}$ **39.** 1000_2 **41.** 733_8 **43.** $12EE_{16}$ **45.** 11111101_2
47. 37166_8 **49.** $5DAE_{16}$ **51.** 01001011; 0000000001001011
53. 11100000; 1111111111100000   **55.** 00111000   **57.** 01001000

Chapter 5

Exercise Set 5.1, pp. 122–123

1. $\{1, 2\}$, $\{a, b, c\}$, $\{1, 2, 3, 4, \ldots\}$, $\{\ldots, -3, -2, -1\}$
3. 8; $\{a\}$, $\{b\}$, $\{c\}$, $\{a, b\}$, $\{a, c\}$, $\{b, c\}$, $\{a, b, c\}$, $\emptyset$
5. **(a)** $\{4, -5, 17, \sqrt{9}\}$; **(b)** $\{4, \frac{2}{3}, -\frac{1}{2}, -5, 17, \sqrt{9}\}$; **(c)** $\{\sqrt{3}, \pi\}$; **(d)** $\{4, 17, \sqrt{9}\}$

Exercise Set 5.3, pp. 128–130

1. (a) U; **(b)** {2}; **(c)** {3, 5}; **(d)** $\emptyset$; **(e)** $\emptyset$; **(f)** U; **(g)** $\emptyset$; **(h)** U; **(i)** {4, 6}; **(j)** {1, 4, 6}

3. (a) {all male employees in home office}; **(b)** {all female employees in home office and all male employees between the ages of 21 and 30}; **(c)** {all male employees between the ages of 21 and 30}; **(d)** {all employees between 21 and 50}; **(e)** {all employees over 50 years of age}; **(f)** {all employees under 21 or over 50}; **(g)** {all employees of the DP department 50 years of age and under}; **(h)** {all female employees of the DP department 50 years of age and under}; **(i)** {all male employees between 21 and 30 together with all DP department employees}; **(j)** {all employees of the DP department under 21 or over 50}

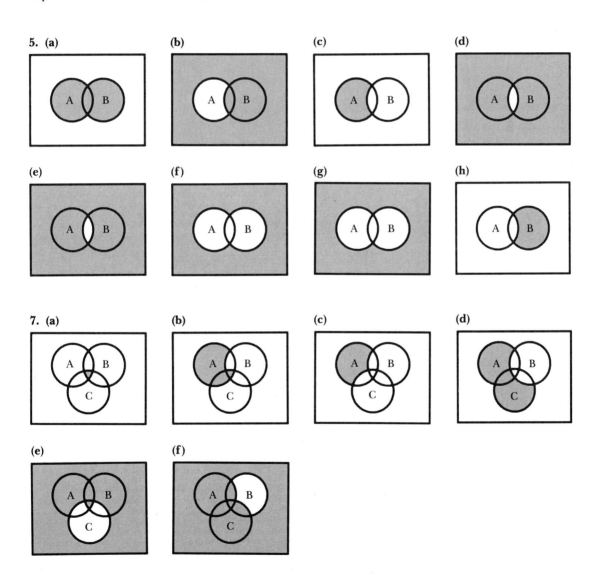

5. (a) **(b)** **(c)** **(d)**

(e) **(f)** **(g)** **(h)**

7. (a) **(b)** **(c)** **(d)**

(e) **(f)**

Exercise Set 5.4, pp. 137–139

1. (a)   (b)   (c)

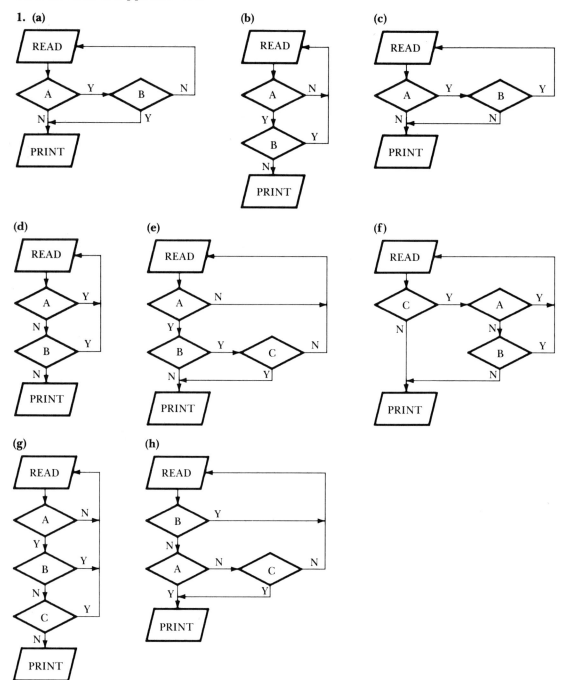

(d)   (e)   (f)

(g)   (h)

3. (a)

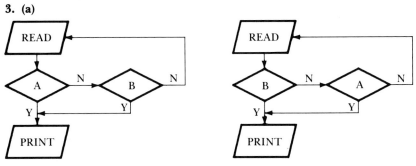

(b) 160; **(c)** B ∪ A; **(d)** A ∩ B: 345, B ∩ A: 370; **(e)** 35; **(f)** A ∩ B

5. (a) Checks: 408, list: 8; **(b)** checks: 684, list: 202

(c) checks: 666, list: 112   **(d)** checks: 666, list: 42   **(e)** checks: 400, list: 230

(f) checks: 684, list: 122   **(g)** checks: 444, list: 262   **(h)** checks: 586, list: 12

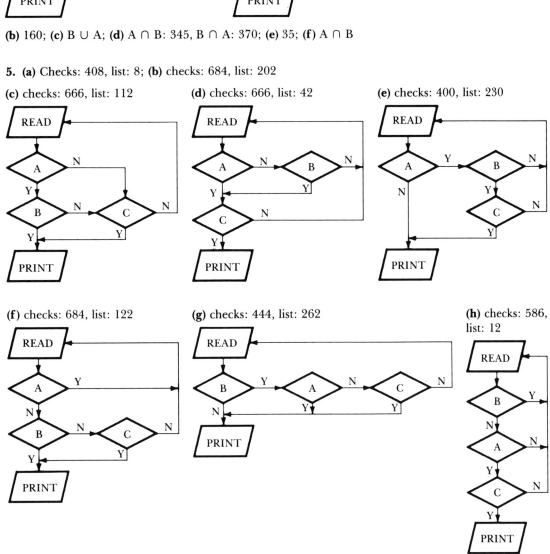

Exercise Set 5.5, pp. 140–141

1. (a) {a, b, d} ∪ ({c, d, f, g} ∩ {b, c}) = ({a, b, d} ∪ {c, d, f, g}) ∩ ({a, b, d} ∪ {b, c})
{a, b, d} ∪ ({c}) = {a, b, c, d, f, g} ∩ {a, b, c, d}
{a, b, c, d} = {a, b, c, d}

These are all "verify"-type problems.

Review Exercises, pp. 142–143

1. (a) {5, 6, 7, 8, 9, 10, 11, 12, 13, 14}; **(b)** {11, 12, 13, 14, . . .}; **(c)** {1, 2, 3, 4, 5, 6, 7, 8, 9};
(d) {. . . , −5, −4, −3, −2, −1}
3. (a) {6, 8}; **(b)** {1, 2, 3, 4, 5, 6, 7, 8, 9}; **(c)** {3, 7, 9, 10}; **(d)** {1, 2, 3, 4, 5, 6 ,7, 8, 9};
(e) {1, 2, 3, 4, 5, 7 ,9}; **(f)** {1, 5}; **(g)** {10}; **(h)** ∅
5. (a) 95; **(b)** 45; **(c)** 55
7. (a) 180, 95; **(b)** 165, 10; **(c)** 180, 45; **(d)** 170, 95; **(e)** 265, 35; **(f)** 200, 15; **(g)** 200, 110; **(h)** 180, 5

Chapter 6

Exercise Set 6.1, pp. 147–148

1. 5   **3.** 3   **5.** −3   **7.** 3   **9.** 7   **11.** $\frac{1}{5}$ **13.** −32   **15.** $\frac{1}{2}$ **17.** $\frac{1}{16}$ **19.** $\frac{1}{25}$
21. $x^{7/12}$ **23.** x^2 **25.** $x^{7/6}$ **27.** $x^{1/5}$ **29.** $x^{3/5}y^{1/4}$

Exercise Set 6.2, pp. 149–150

1. $3x$ **3.** $-3x$ **5.** $2\sqrt[3]{5}$ **7.** $2xy\sqrt[3]{3xy^2}$ **9.** 4   **11.** $\dfrac{\sqrt[3]{5x^2}}{2x}$ **13.** 5   **15.** $5\sqrt{2}$
17. $5x\sqrt[3]{y}$ **19.** y **21.** $10x^2y$ **23.** 18

Exercise Set 6.3, p. 156

1. $-3, 5$ **3.** $4, 0$ **5.** $7, -7$ **7.** $2, -6$ **9.** $-3, 1$ **11.** $-2 \pm \sqrt{-1}$ **13.** $\dfrac{3 \pm \sqrt{29}}{2}$
15. $2 \pm \sqrt{-5}$ **17.** Two equal real roots   **19.** Two unequal complex roots
21. Two unequal complex roots   **23.** $2, -1$ **25.** $\dfrac{3 \pm \sqrt{-23}}{2}$ **27.** $\dfrac{-1 \pm \sqrt{22}}{2}$ **29.** $\frac{1}{2}, \; -4$

Exercise Set 6.4, pp. 158–159

1. 32   **3.** 40   **5.** 45   **7.** No solution   **9.** 5   **11.** 5   **13.** 4   **15.** 2, 6

Exercise Set 6.5, pp. 162–163

1. $x < 3$ or $x > 5$ **3.** $x \leq -7$ or $x \geq 3$ **5.** $-4 < x < -1$

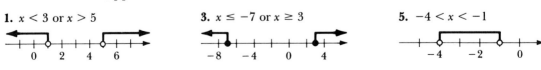

7. $-1 \le x \le 8$

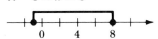

9. $0 \le x \le 3$

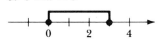

11. $x \le -2$ or $x \ge 0$

13. $x < -4$ or $x > 4$

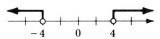

15. $-3 < x < 3$

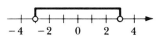

17. $x < -5$ or $x > 2$

19. $-8 \le x \le -5$

21. $x < \frac{1}{2}$ or $x > 3$

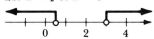

Review Exercises, pp. 164–165

1. 11 **3.** $2x$ **5.** $2x$ **7.** 2 **9.** $2x$ **11.** $2x\sqrt[4]{2x}$ **13.** $2x^2y\sqrt[3]{3y}$ **15.** $-\frac{1}{2}$ **17.** $x^{16/15}$

19. $-4, 0$ **21.** $-3, 3$ **23.** $\pm\sqrt{5}$ **25.** $-5, 3$ **27.** $5 \pm \sqrt{7}$ **29.** $-5, -3$ **31.** $3 \pm 2\sqrt{5}$

33. $\dfrac{3 \pm \sqrt{-39}}{4}$ **35.** 24 **37.** No solution **39.** 3 **41.** No solution **43.** $\dfrac{169}{16}$

45. $\dfrac{183 \pm 12\sqrt{129}}{9}$

47. $x < -3$ or $x > 0$

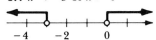

49. $x \le -7$ or $x \ge 7$

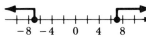

51. $-6 \le x \le -2$

53. $x \le -3$ or $x \ge \frac{5}{2}$

Chapter 7

Exercise Set 7.1, pp. 172–173

1.

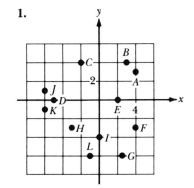

3. (a) $(2, 3), (0, 4), (8, 0)$; (b) $(6, 0), (3, 2), (-3, 6), (9, -2)$; (c) $(6, 0), (0, -2)$; (d) $(1, 1), (4, 10), (0, -2)$

5.

Graph	Crosses x-axis	Crosses y-axis
4(a)	$(\frac{3}{2}, 0)$	$(0, -3)$
4(b)	$(-2, 0), (2, 0)$	$(0, -4)$
4(c)	Doesn't cross	$(0, 1)$
4(d)	Doesn't cross	$(0, 1)$
4(e)	$(-2, 0), (2, 0)$	$(0, 4)$
4(f)	$(-2, 0), (2, 0)$	$(0, 2)$
4(g)	Doesn't cross	Doesn't cross

Exercise Set 7.2, pp. 182–184

1. $(-3, -9)$, $(-2, -7)$, $(-1, -5)$, $(0, -3)$, $(1, -1)$, $(2, 1)$, $(3, 3)$; a function

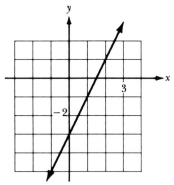

3. $(-3, -5)$, $(-2, 0)$, $(-1, 3)$, $(0, 4)$, $(1, 3)$, $(2, 0)$, $(3, -5)$; a function

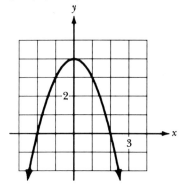

5. $(-3, 2)$, $(-2, 2)$, $(-1, 2)$, $(0, 2)$, $(1, 2)$, $(2, 2)$, $(3, 2)$; a function

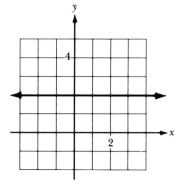

7. $(-3, -\frac{1}{3})$, $(-2, -\frac{1}{2})$, $(-1, -1)$, $(0, \text{undefined})$, $(1, 1)$, $(2, \frac{1}{2})$, $(3, \frac{1}{3})$; a function

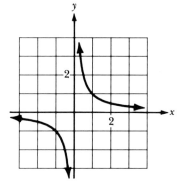

9. $(-3, \sqrt{-3})$, $(-3, -\sqrt{-3})$, $(-2, \sqrt{-2})$, $(-2, -\sqrt{-2})$, $(-1, \sqrt{-1})$, $(-1, -\sqrt{-1})$, $(0, 0)$, $(1, 1)$, $(1, -1)$, $(2, \sqrt{2})$, $(2, -\sqrt{2})$, $(3, \sqrt{3})$, $(3, -\sqrt{3})$; not a function

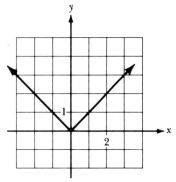

11. The function takes a value for x, then squares the value, and then adds 3.
13. The function takes a value for x, takes the square root, and then divides by 4.
15. A function   **17.** Not a function   **19.** A function   **21.** A function   **23.** Not a function
25. $f(2) = 0, f(-2) = 8, f(0) = 4, f(w) = 4 - 2w$
27. $f(2) = 0, f(-2) = 0, f(0) = -4, f(w) = w^2 - 4$
29. $f(2) = -4, f(-2) = 12, f(0) = 0, f(w) = w^2 - 4w$ **31.** $f(2) = 2, f(-2) = 2, f(0) = 0, f(w) = |w|$
33. Domain $\mathbb{R}$, range $y \geq 1$ **35.** Domain $\mathbb{R}$, range $y \leq 0$
37. Domain $x < 0$ or $x > 0$, range $y > -1$ **39.** Domain $\mathbb{R}$, range $y = 7.1$ **41.** $2x^2 + 2x - 4$
43. $2x^2 - 2x - 12$ **45.** 2   **47.** 192   **49.** 64
51. **(a)** \$1166.40; **(b)** \$2163.20; **(c)** \$8160.80; **(d)** \$10,475.35

Exercise Set 7.3, pp. 189–191

1.

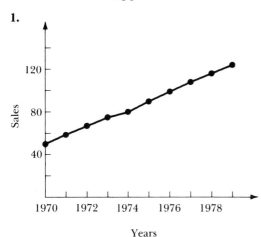

Years

3. (a)

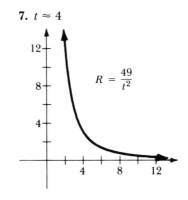

Items

(b) $i = 25$; **(c)** $R = 2500$

5. The cost will be the smallest at the production rate of zero.

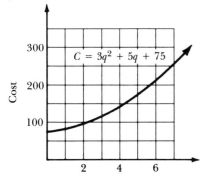

Units produced

7. $t \approx 4$

9. (a) The graph is shifted up 3 units.

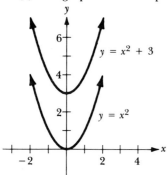

(b) The graph increases in value twice as fast.

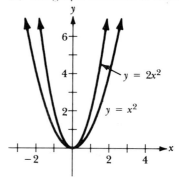

(c) Reflection of the graph around the x-axis.

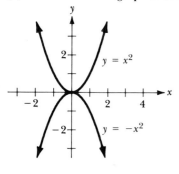

(d) The graph is shifted to the left 2 units.

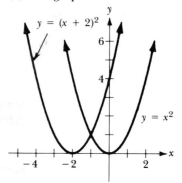

Exercise Set 7.4, pp. 201–202

1. $m = 1$ **3.** $m = 5$ **5.** $m = \frac{2}{3}$ **7.** $m = \frac{5}{4}$ **9.** $m = 1$ **11.** $m = 1$

13. (a) $y = \frac{2}{3}x - 2$; **(b)** $m = \frac{2}{3}$; **(c)** $(0, -2)$;
(d)

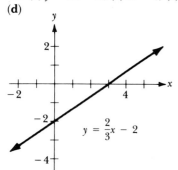

15. (a) $y = -\frac{2}{3}x + 2$; **(b)** $m = -\frac{2}{3}$; **(c)** $(0, 2)$;
(d)

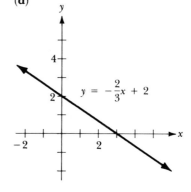

17. (a) $y = \frac{3}{2}x - 2$; **(b)** $m = \frac{3}{2}$; **(c)** $(0, -2)$;
(d)

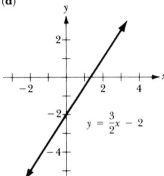

$y = \dfrac{3}{2}x - 2$

19. (a) $y = \frac{1}{3}x$; **(b)** $m = \frac{1}{3}$; **(c)** $(0, 0)$;
(d)

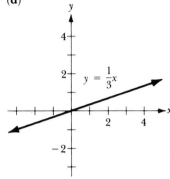

$y = \dfrac{1}{3}x$

21. (a) $y = \frac{5}{2}x$; **(b)** $m = \frac{5}{2}$; **(c)** $(0, 0)$;
(d)

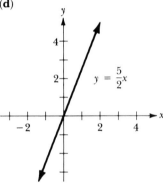

$y = \dfrac{5}{2}x$

23.

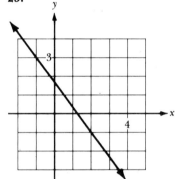

25.

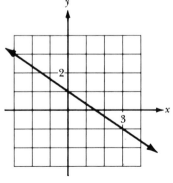

27.

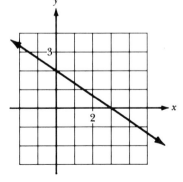

29. $3x - y = 12$ **31.** $3x - 4y = -4$ **33.** $3x - 2y = 4$ **35.** (a) $b = 5$; (b) $m = \frac{2}{3}$; (c) $(2, 4)$

37. (a) \$8400; (b) 1200; (c) 0.3 per unit; (d) \$1200; (e) slope $= \frac{3}{10}$, u-intercept $= (-4000, 0)$, TC-intercept $= (0, 1200)$

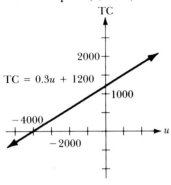

Exercise Set 7.5, p. 206

1.

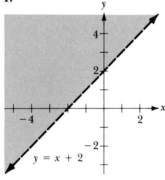

3.

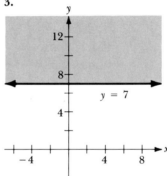

5.

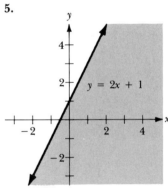

7.

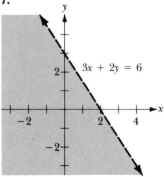

9.

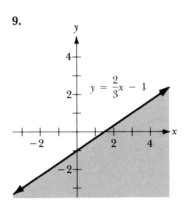

11.

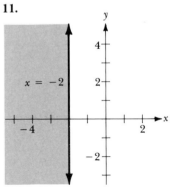

13.

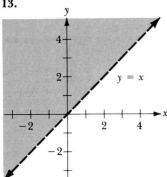

15.

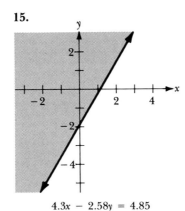

$$4.3x - 2.58y = 4.85$$

17.

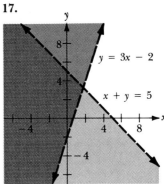

19.

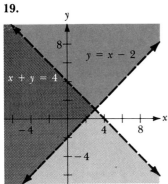

21. Each element in the domain has more than one element in the range.

Exercise Set 7.6, pp. 214–216

1. $(-3, 7), (-2, 0), (1, -9),$
$(0, -8), (4, 0), (5, 7), (-1, -5)$

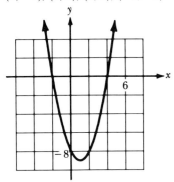

$$y = x^2 - 2x - 8$$

3. (a) $(5, 0)$ and $(-1, 0)$

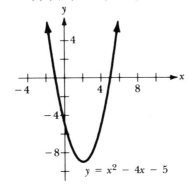

(b) $x = 5$ and $x = -1$;
(c) The x-coordinate of a zero is a root of the function.

5. x-intercepts are
$(-2, 0), (2, 0)$; vertex is $(0, -4)$

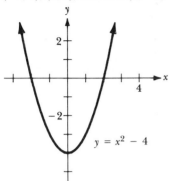

$y = x^2 - 4$

7. x-intercepts are
$(-2, 0), (6, 0)$; vertex is $(2, -16)$

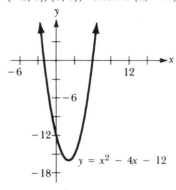

$y = x^2 - 4x - 12$

9. x-intercepts are
$(-6, 0), (1, 0)$; vertex is $(\frac{5}{2}, \frac{51}{4})$

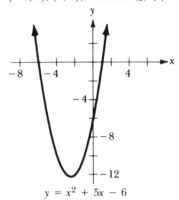

$y = x^2 + 5x - 6$

11. x-intercepts are
$(0, 0), (4, 0)$; vertex is $(2, -12)$

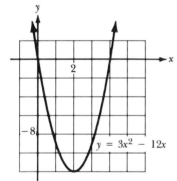

$y = 3x^2 - 12x$

13. x-intercepts are
$(-2, 0), (-1, 0)$; vertex is $(-\frac{3}{2}, \frac{35}{4})$

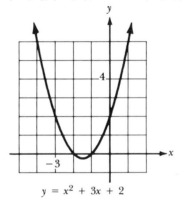

$y = x^2 + 3x + 2$

15. x-intercepts are
$(0, 0), (30, 0)$; vertex is $(15, 900)$

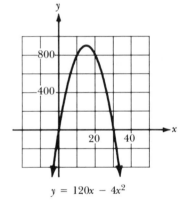

$y = 120x - 4x^2$

17. x-intercepts are $(-4 - \sqrt{21}, 0)$,
$(-4 + \sqrt{21}, 0)$; vertex is $(-4, 21)$

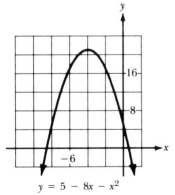

$y = 5 - 8x - x^2$

19. If $b^2 - 4ac > 0$, then the intercepts are real and distinct.
If $b^2 - 4ac = 0$, then there is one real intercept.
If $b^2 - 4ac < 0$, then the intercepts are nonreal;
$y = ax^2 + bx + c$ will then not touch the x-axis.

21. $K = 2$, tangent to the x-axis;
$K < 2$, crosses the x-axis at two distinct points;
$K > 2$, does not intersect the x-axis.

23. (a)

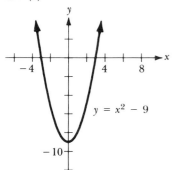

(b)

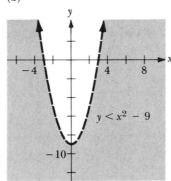

(c)

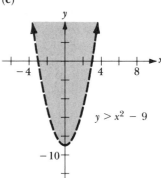

25. $C(x) = x^2 - 6x + 10$

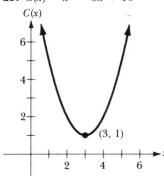

27. (a)

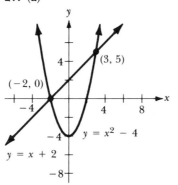

(b)

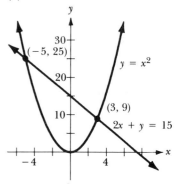

Review Exercises, pp. 216–219

1. II   **3.** III

5. III

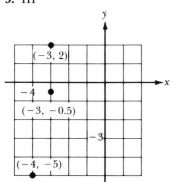

7. $(1, 1), (0, -2), (2, 4)$
9. $(-1, 2), (1, 2), (0, 1)$

11. $y = 3x - 2$

x	y
-2	-8
-1	-5
0	-2
1	1
2	4

Domain = $\mathbb{R}$ (all real numbers)
Range = $\mathbb{R}$ (all real numbers)

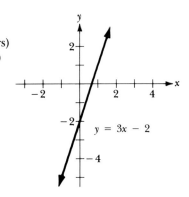

13. $y = x^2 - 9$

x	y
-3	0
-2	-5
-1	-8
0	-9
1	-8
2	-5
3	0

Domain = $\mathbb{R}$ (all real numbers)
Range = $\{y \mid y \geq -9\}$

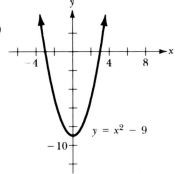

15. $y = \sqrt{4 - x^2}$

x	y
-2	0
-1	$\sqrt{3}$
0	2
1	$\sqrt{3}$
2	0

Domain = $\{x \mid -2 \leq x \leq 2\}$
Range = $\{y \mid 0 \leq y \leq 2\}$

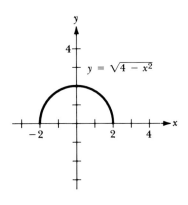

17. Points do represent a function.   **19.** A function   **21.** Not a function   **23.** A function
25. A function is a relationship such that for every element in the domain there is one unique element in the range.
27. $g(3) = 7$; $g(-3) = -11$; $g(0) = -2$ **29.** $f(3) + g(3) = 14$; $f(3) - g(3) = 0$
31. $f(g(1)) = -1$; $g(f(1)) = -5$

33. $m = 2$, x-intercept $(-\frac{1}{2}, 0)$,
y-intercept $(0, 1)$

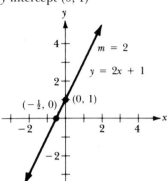

35. $m = \frac{2}{3}$, x-intercept $(6, 0)$,
y-intercept $(0, -4)$

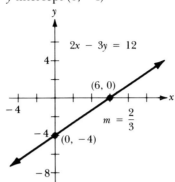

37. $m = \frac{3}{2}$, x-intercept $(1, 0)$,
y-intercept $(0, -\frac{3}{2})$

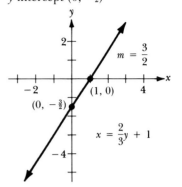

39.

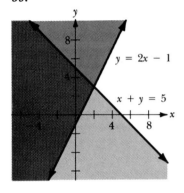

41.

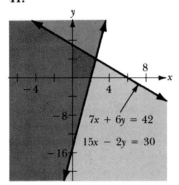

43. intercepts $(3, 0)$, $(-3, 0)$,
$(0, 9)$; vertex $(0, 9)$, concave
down

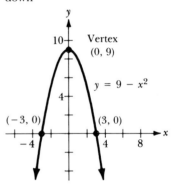

45. intercepts $((7 + \sqrt{41})/4, 0)$,
$((7 - \sqrt{41})/4, 0)$, $(0, 1)$; vertex
$(7/4, -41/8)$, concave up

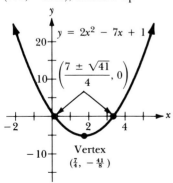

47. intercepts $((5.1 - \sqrt{27.95})/.8, 0)$, $((5.1 + \sqrt{27.95})/.8, 0)$,
$(0, -4.85)$; vertex $(51/8, -559/8)$, concave up

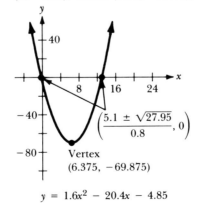

49. $x - 3$

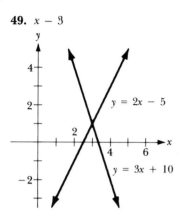

51. At $x = 45 + 5\sqrt{41}$ and $45 - 5\sqrt{41}$, the profit will be zero. The profit is greatest at $x = 45$. At $x = 45 + \sqrt{1005}$ and $x = 45 - \sqrt{1005}$, a profit of \$20 results.

Chapter 8

Exercise Set 8.1, pp. 227–228

1. Statement   **3.** Statement   **5.** Not a statement   **7.** Statement   **9.** Not a statement
11. I like to study mathematics and I study at least 20 hours per week.
13. I do not like to study mathematics or I study at least 20 hours per week.
15. I do not like to study mathematics and I study at least 20 hours per week.
17. I like to study mathematics.
19. P AND Q **21.** P OR Q **23.** NOT P AND NOT Q

25.

P	Q	$\sim P$	$\sim P \wedge Q$
T	T	F	F
T	F	F	F
F	T	T	T
F	F	T	F

27.

P	Q	NOT Q	P AND NOT Q
T	T	F	F
T	F	T	T
F	T	F	F
F	F	T	F

29.

P	Q	NOT P	NOT P EOR Q
T	T	F	T
T	F	F	F
F	T	T	F
F	F	T	T

31.

P	Q	$P \vee Q$	$\sim(P \vee Q)$
T	T	T	F
T	F	T	F
F	T	T	F
F	F	F	T

33.

P	Q	$P \wedge Q$	$\sim(P \wedge Q)$
T	T	T	F
T	F	F	T
F	T	F	T
F	F	F	T

35. $\sim(P \wedge \sim Q) \equiv \sim P \vee Q$ **37.** NOT(NOT P OR Q) $\equiv P$ AND NOT Q

Exercise Set 8.2, pp. 231–232

1. $P \bigvee_{3} \underset{2}{\sim}(\underset{1}{Q \wedge R})$ **3.** $P \underset{2}{\wedge} \underset{1}{\sim}Q \underset{3}{\vee} R$ **5.** P OR Q OR NOT R
                                          2   3   1

7. $(\underset{1}{P \text{ AND } Q}) \underset{4}{\text{ OR }} (\underset{3}{R \text{ AND }} \underset{2}{\text{NOT } S})$

9.

P	Q	R	$Q \wedge R$	$P \vee (Q \wedge R)$
T	T	T	T	T
T	T	F	F	T
T	F	T	F	T
T	F	F	F	T
F	T	T	T	T
F	T	F	F	F
F	F	T	F	F
F	F	F	F	F

11.

P	Q	R	$\sim P$	$\sim P \wedge Q$	$(\sim P \wedge Q) \vee R$
T	T	T	F	F	T
T	T	F	F	F	F
T	F	T	F	F	T
T	F	F	F	F	F
F	T	T	T	T	T
F	T	F	T	T	T
F	F	T	T	F	T
F	F	F	T	F	F

13.

P	Q	R	Q AND R	NOT (Q AND R)	P OR NOT (Q AND R)
T	T	T	T	F	T
T	T	F	F	T	T
T	F	T	F	T	T
T	F	F	F	T	T
F	T	T	T	F	F
F	T	F	F	T	T
F	F	T	F	T	T
F	F	F	F	T	T

15.

P	Q	R	NOT Q	NOT R	P AND NOT Q	P AND NOT Q OR NOT R
T	T	T	F	F	F	F
T	T	F	F	T	F	T
T	F	T	T	F	T	T
T	F	F	T	T	T	T
F	T	T	F	F	F	F
F	T	F	F	T	F	T
F	F	T	T	F	F	F
F	F	F	T	T	F	T

17.

P	Q	R	$\sim Q$	$\sim Q \overline{\vee} R$	$P \wedge (\sim Q \overline{\vee} R)$
T	T	T	F	T	T
T	T	F	F	F	F
T	F	T	T	F	F
T	F	F	T	T	T
F	T	T	F	T	F
F	T	F	F	F	F
F	F	T	T	F	F
F	F	F	T	T	F

19.

P	Q	R	$Q \vee R$	$P \vee (Q \vee R)$	$\sim[P \vee (Q \vee R)]$
T	T	T	T	T	F
T	T	F	T	T	F
T	F	T	T	T	F
T	F	F	F	T	F
F	T	T	T	T	F
F	T	F	T	T	F
F	F	T	T	T	F
F	F	F	F	F	T

21.

23.

25.

27.

29.

Exercise Set 8.3, pp. 235–236

1. (I work hard) → (I will succeed)   **3.** (I get the job) → (I will buy a new stereo)
5. (A is an odd number) ↔ (A is not divisible by 2)

7.

P	Q	~Q	P → ~Q
T	T	F	F
T	F	T	T
F	T	F	T
F	F	T	T

9.

P	Q	NOT Q	NOT Q → P
T	T	F	T
T	F	T	T
F	T	F	T
F	F	T	F

11.

P	Q	P ∨ Q	(P ∨ Q) → P
T	T	T	T
T	F	T	T
F	T	T	F
F	F	F	T

13.

P	Q	P AND Q	NOT Q	(P AND Q) → NOT Q
T	T	T	F	F
T	F	F	T	T
F	T	F	F	T
F	F	F	T	T

15.

P	Q	NOT Q	P ↔ NOT Q
T	T	F	F
T	F	T	T
F	T	F	T
F	F	T	F

17.

P	Q	~Q	P ↔ ~Q	~(P ↔ ~Q)
T	T	F	F	T
T	F	T	T	F
F	T	F	T	F
F	F	T	F	T

19.

P	Q	~P	~Q	P ∨ ~Q	~P ↔ (P ∨ ~Q)
T	T	F	F	T	F
T	F	F	T	T	F
F	T	T	F	F	F
F	F	T	T	T	T

21.

P	Q	R	Q ∧ R	P ∨ (Q ∧ R)	P ∨ Q	P ∨ R	(P ∨ Q) ∧ (P ∨ R)	↔
T	T	T	T	T	T	T	T	T
T	T	F	F	T	T	T	T	T
T	F	T	F	T	T	T	T	T
T	F	F	F	T	T	T	T	T
F	T	T	T	T	T	T	T	T
F	T	F	F	F	T	F	F	T
F	F	T	F	F	F	T	F	T
F	F	F	F	F	F	F	F	T

23. Converse: If NOT R, then (P AND Q).
Inverse: If NOT (P AND Q), then R.
Contrapositive: If R, then NOT (P AND Q).

25. Converse: If $x = 1$, then $x(x - 1) = 0$.
Inverse: If $x(x - 1) \neq 0$, then $x \neq 1$.
Contrapositive: If $x \neq 1$, then $x(x - 1) \neq 0$.

27. (b): contrapositive

Exercise Set 8.4, pp. 237–238

1. Tautology   **3.** Tautology   **5.** Contradiction   **7.** Contradiction   **9.** Tautology
11. Logically equivalent   **13.** Logically equivalent   **15.** Not logically equivalent
17. Logically equivalent

Exercise Set 8.5, p. 241

1. Logical expression: A >= 30
   Relational operator: >=
   Dependent statement: 70

3. Logical expression: PRICE > 50
   Relational operator: >
   Dependent statement: MULTIPLY PRICE BY 0.06 GIVING TAX

5. Logical expression: B ** 2 − 4 * A * C
   Relational operator: <, =, >
   Dependent statement: 40, 60, 80

7. Logical expression: A <= 10
   Relational operator: <=
   Dependent statement: 3 + A (True branch)
                        ELSE 3 * A (False branch)

9. Logical expression: PROFIT > (0.20 * REVENUE)
   Relational operator: >
   Dependent statement: TAX: = 0.15 * PROFIT

11. Logical expression: A .EQ. 90
    Relational operator: .EQ.
    Dependent statement: GO TO 100

13. Logical expression: (B .GT. A)
    Relational operator: .GT.
    Dependent statement: X = B (True branch)
                         X = A (False branch)

Review Exercises, pp. 242–243

Only the last column is given.

1. FTFF   3. TFTT   5. TTFTFTFT   7. FFFFFTFF   9. TFFFTTTT   11. TTTF
13. TTFT   15. TFTTTFFF   17. TFTTFTFF   19. TTFF

21.   **23.**   **25.**

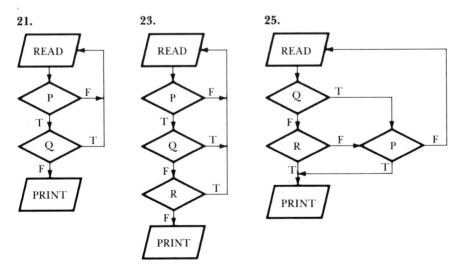

27.

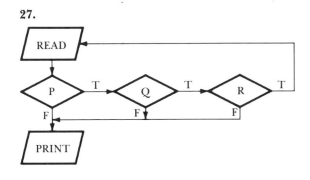

29.

31. Converse: If NOT R, then (P OR Q).
   Inverse: If NOT (P OR Q), then R.
   Contrapositive: If R, then NOT (P OR Q).
33. Converse: $(\sim Q \vee R) \to P$
   Inverse: $\sim P \to \sim(\sim Q \vee R)$ or $\sim P \to (Q \vee \sim R)$
   Contrapositive: $\sim(\sim Q \vee R) \to \sim P$ or $(Q \vee \sim R) \to \sim P$
35. Converse: If a WIM is not a WIDGET, then a BOBE is a GADGET.
   Inverse: If a BOBE is not a GADGET, then a WIM is a WIDGET.
   Contrapositive: If a WIM is a WIDGET, then a BOBE is not a GADGET.
37. Neither   **39.** Tautology   **41.** Neither   **43.** Contradiction   **45.** Not logically equivalent
47. Logically equivalent   **49.** Logically equivalent

Chapter 9

Exercise Set 9.1, pp. 247–248

1. False   **3.** True   **5.** False   **7.** OR is $A \vee B$; NOT is A'; AND is $A \wedge B$.

Exercise Set 9.2, pp. 252–255

1. No, no   **3.** No, no   **5.** No, no   **7.** Yes, yes
9. (a)

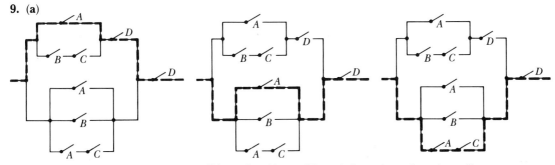

(b) none, D is always open; (c) 5 possible paths, 16 possible switch settings, 6 settings allow
current flow.

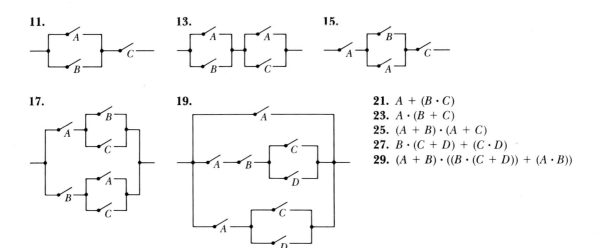

11.   **13.**   **15.**

17.   **19.**

21. $A + (B \cdot C)$
23. $A \cdot (B + C)$
25. $(A + B) \cdot (A + C)$
27. $B \cdot (C + D) + (C \cdot D)$
29. $(A + B) \cdot ((B \cdot (C + D)) + (A \cdot B))$

31. $A \cdot ((B \cdot (C + D) \cdot C) + (C \cdot (C + D))) \cdot D$ **33.** $A \cup B, A \text{ OR } B, A + B$

Exercise Set 9.3, pp. 263–265

1. Commutative law for addition   **3.** First distributive law   **5.** Associative law for addition
7. Idempotency   **9.** $A + (B + C)$ or $(B + A) + C$ **11.** A **13.** $A + B$ **15.** $A + B \cdot C$
17. $(A + B) \cdot (A + C)$ **19.** 1   **21.** False   **23.** False   **25.** False
27. $C + C \cdot D = C \cdot 1 + C \cdot D$
$\qquad\qquad\qquad = C \cdot (1 + D)$
$\qquad\qquad\qquad = C \cdot (1)$
$\qquad\qquad\qquad = C$
Here the second distributve property $C + C \cdot D = (C + C) \cdot (C + D)$ is *not* directly useful.

29.

A	B	$A + A' \cdot B$	$A + B$
0	0	0	0
0	1	1	1
1	0	1	1
1	1	1	1

31. (a) $A \cdot B + A \cdot C = A \cdot (B + C)$ (b) $(A + B) \cdot (A + C) \cdot (A + D) = A + B \cdot C \cdot D$ by the second
distributive property

33. $A \cdot (B + C)$ **35.** $A \cdot (B + B' + 1) = A$ **37.** $(A' \cdot B') + (A + B) = (A + B)' + (A + B) = 1$
39. $A \cdot (B + B')$ **41.** $A \cdot B + C'$
43. $A \cdot B + A' \cdot B + A \cdot B' + A' \cdot B' = (A + A') \cdot B + (A + A') \cdot B' = 1 \cdot B + 1 \cdot B' = 1$
45. $A \cdot B \cdot C + A \cdot B' \cdot C + A \cdot B \cdot C' + A \cdot B' \cdot C' = A(B \cdot C + B' \cdot C + B \cdot C' + B' \cdot C') = A \cdot 1 = A$

47. 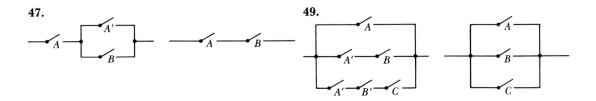 **49.**

51. $(A \cdot B)'$ **53.** $(A \cdot (B + C))'$

Exercise Set 9.4, pp. 266–267

1. $A \cdot B = B \cdot A$ **3.** $A + (B \cdot C) = (A + B) \cdot (A + C)$ **5.** $A \cdot (A + B) = A$
7. $(A + B + C)' = A' \cdot B' \cdot C'$ **9.** $(A \cdot C) + (B \cdot A) = B + C \cdot A$ **11.** $(A \cdot B)'$ or $A' + B'$

Exercise Set 9.5, pp. 268–269

1. No   **3.** Yes   **5.** No   **7.** No   **9.** Yes   **11.** No
13. $(A \cup B) \cap (A \cup C) = A \cup (B \cap C)$ **15.** $A \cdot (B + C)' = A \cdot (B' \cdot C')$
17. $(C' \cdot B) + (A \cdot C') + (B \cdot C); B + (A \cdot C')$; Anyone receiving financial support or anyone that has no living relatives and is not over 65.
19. Sets would form a Boolean algebra with the empty set replacing 0 and the universal set replacing 1.

Exercise Set 9.6, pp. 276–279

1. $A \cdot B$ **3.** $A \cdot B + A' \cdot B'$

5.

A	B	C	D	ABCD
1	1	1	1	1
1	1	1	0	0
1	1	0	1	0
1	1	0	0	0
1	0	1	1	0
1	0	1	0	0
1	0	0	1	0
1	0	0	0	0
0	1	1	1	0
0	1	1	0	0
0	1	0	1	0
0	1	0	0	0
0	0	1	1	0
0	0	1	0	0
0	0	0	1	0
0	0	0	0	0

7. $ABC'D$

0
0
1
0
0
0
0
0
0
0
0
0
0
0
0
0

9. $AB'C'D$

0
0
0
0
0
0
1
0
0
0
0
0
0
0
0
0

11. $AB'C'D'$

0
0
0
0
0
0
0
1
0
0
0
0
0
0
0
0

13. $A \cdot B' \cdot C \cdot D + A' \cdot B \cdot C \cdot D$ or $(AB' + A'B)CD$ **15.** $A \cdot B'$
17. $A' \cdot B' \cdot C + A \cdot B + A \cdot C + B' \cdot C$ **19.** $A \cdot B$ **21.** $A + B' + A' \cdot B' + A \cdot B$

23. For Figure 9.17(a):
(a) $A \cdot (A \cdot B + A' \cdot B) + A' \cdot (A \cdot C + A' \cdot B)$;
(b) $A \cdot B + A' \cdot B$; (c) B;
(d)

For Figure 9.17(b):
(a) $(A' + B \cdot A' + C \cdot A') \cdot (A' \cdot C' + A \cdot B + A \cdot C)$;
(b) $A' \cdot C' + A' \cdot B \cdot C'$; (c) $A' \cdot C'$;
(d)

25. $A' \cdot C$

Exercise Set 9.7, pp. 281–284

1.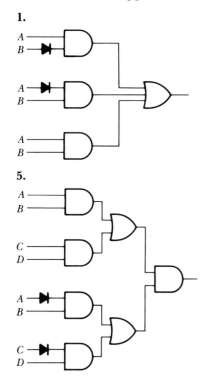

3.

5.

7. $A \cdot C + A'B$
9. $A' \cdot B \cdot C + A' + A \cdot B'$
11. $A \cdot B' \cdot C + A' \cdot B' \cdot C$
13. Yes

15. $A \cap B \cup B \cap C$ becomes $A \cdot B + B \cdot C$

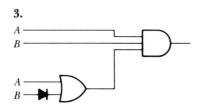

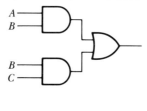

Review Exercises, pp. 285–289

1. Boolean algebra is a mathematical structure composed of (a) a set of elements containing at least 0 and 1; (b) two binary operations, usually denoted by + and ·; (c) a unary operation, denoted by '.

3.

A	B	C	(a)	(b)
0	0	0	no	no
0	0	1	no	no
0	1	0	no	no
0	1	1	yes	yes
1	0	0	yes	yes
1	0	1	yes	yes
1	1	0	yes	yes
1	1	1	yes	yes

5. For 2 variables, 4 results; for 3 variables, 8 results; for n variables, 2^n results

7.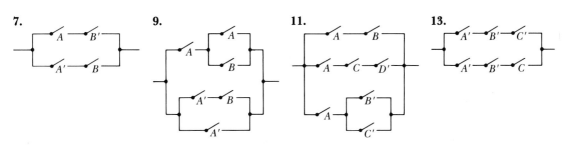

9.

11.

13.

15.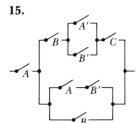

17. $ABC + AB'C + AB'C'$
19. $((A + B)A) + (A' + A'B)$
21. $(A + A') \cdot (ABA' + (A + B))$
23. B OR A AND P AND C; $B + APC$

25. (a) $(A \cup B) \cap (A \cup C)$; (b) $(A$ OR $B)$ AND $(A$ OR $C)$;
(c)

(d)

27. Second distributive law
29. Involution
31. Commutative law of multiplication
33. AB
35. 1
37. $A + B$

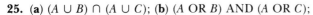

39. 1

41. $A(C + B')$

43. $A'B'C + A'BC' + AB'C + ABC'$, which simplifies to $B'C + BC'$
45. $A' \cdot B + A \cdot B' + A \cdot B + A \cdot C$; $A + B$
47. $AB'C + AB'C' + ABC$; $A \cdot B' + A \cdot C$
49. $B + A$
51. $AB + A'C'$

53.

55.

57.

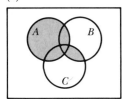

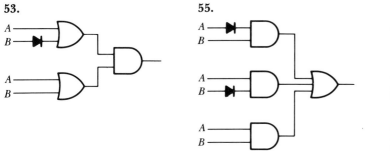

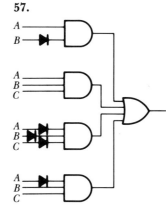

59. $AB'C + A'BC' + A'B'C$ **61.** $ABC + AB'C' + A'BC'$

63. Part (1)

EQUATION	SWITCHING DIAGRAM	VENN DIAGRAM	TRUTH TABLE	LOGIC CIRCUIT
$K = A \cdot B$	$-\!\!\!\!\sphericalangle A -\!\!\!\!\sphericalangle B - K$		$\begin{array}{ccc} A & B & K \\ 0 & 0 & 0 \\ 0 & 1 & 0 \\ 1 & 0 & 0 \\ 1 & 1 & 1 \end{array}$	
$K = A' \cdot B$	$-\!\!\!\!\sphericalangle A' -\!\!\!\!\sphericalangle B - K$		$\begin{array}{ccc} A & B & K \\ 0 & 0 & 0 \\ 0 & 1 & 1 \\ 1 & 0 & 0 \\ 1 & 1 & 0 \end{array}$	
$K = A \cdot B'$	$-\!\!\!\!\sphericalangle A -\!\!\!\!\sphericalangle B' - K$		$\begin{array}{ccc} A & B & K \\ 0 & 0 & 0 \\ 0 & 1 & 0 \\ 1 & 0 & 1 \\ 1 & 1 & 0 \end{array}$	
$K = A' \cdot B'$	$-\!\!\!\!\sphericalangle A' -\!\!\!\!\sphericalangle B' - K$		$\begin{array}{ccc} A & B & K \\ 0 & 0 & 1 \\ 0 & 1 & 0 \\ 1 & 0 & 0 \\ 1 & 1 & 0 \end{array}$	

63. Part (2)

EQUATION	SWITCHING DIAGRAM	VENN DIAGRAM	TRUTH TABLE	LOGIC CIRCUIT
$K = A + B$			A B K 0 0 0 0 1 1 1 0 1 1 1 1	
$K = A' + B$			A B K 0 0 1 0 1 1 1 0 0 1 1 1	
$K = A + B'$			A B K 0 0 1 0 1 0 1 0 1 1 1 1	
$K = A' + B'$			A B K 0 0 1 0 1 1 1 0 1 1 1 0	

Chapter 10

Exercise Set 10.1, pp. 296–297

1. $3x - 1y = -6$; $a = 3, b = -1, c = -6$ **3.** $2x - 1y = 8$; $a = 2, b = -1, c = 8$
5. $0x + 1y = 7$; $a = 0, b = 1, c = 7$

7. (a)

(b)

(c)

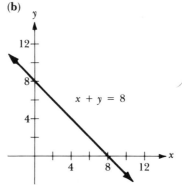

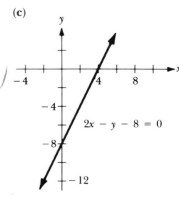

(d)

(e)

(f)

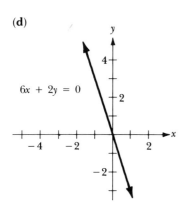

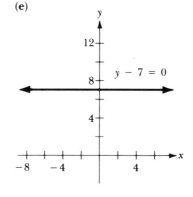

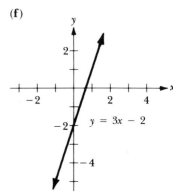

9. $(x, y) = (10, 1)$ **11.** $(x, y) = (4, 1)$ **13.** $x = 2, y = 0$; $x = 0, y = -4$

Exercise Set 10.2, pp. 303–304

	Slope	x-intercept	y-intercept	Additional points
1.	$\frac{3}{2}$	$(2, 0)$	$(0, -3)$	$(1, -\frac{3}{2}), (5, \frac{9}{2})$
3.	-1	$(12, 0)$	$(0, 12)$	$(1, 11), (5, 7)$
5.	$\frac{1}{3}$	$(6, 0)$	$(0, -2)$	$(1, -\frac{5}{3}), (5, -\frac{1}{3})$

Substitute the x- and y-coordinates into the original equation.

7.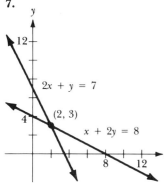

9.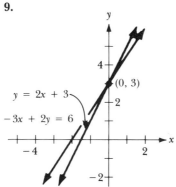

11.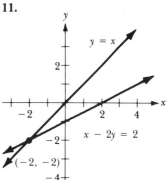

13.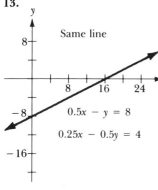

15.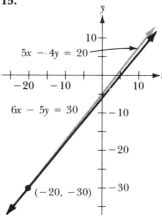

17.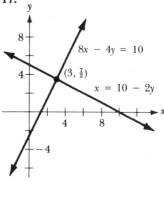

19. Inconsistent   **21.** Dependent   **23.** If $ab \neq 0$, then dependent.

Exercise Set 10.3, pp. 309–310

1. $x = 1, y = 2$ **3.** $x = \frac{1}{2}, y = -2$ **5.** $x = \frac{20}{13}, y = \frac{31}{13}$ **7.** $x = 1.1, y = 3.02$
9. $x = -3, y = -2, z = -1$ **11.** $x = 1, y = 2, z = 3$ **13.** $x = 2, y = 6$; or $x = 1, y = 3$
15. $x = 2, y = 0$; or $x = -1, y = -3$ **17.** $x = 0, y = 0$; or $x = 1, y = 1$
19. (a) When solved by substitution the step is reached where $16 = 5$. This is not possible.
Therefore, the original system is not possible to solve; hence parallel lines or an inconsistent system.
(b) When attempting to solve this system a step is reached where $16 = 16$. This is always true—for
any value of x and any value of y. The two equations represent the same line. Whatever point
"works" in one equation works in the other.
21. $U > 30$

Exercise Set 10.4, pp. 316–318

1. $x = 8, y = 1$ **3.** $x = 1, y = \frac{1}{2}$ **5.** $x = 8, y = -2$
7. $x = (ce - bf)/(ae - bd), y = (cd - af)/(bd - ae)$ **9.** $R = 21.780, S = -1.424$

11. $x = 4, y = 1, z = 1$ **13.** $A = 2, B = 1, C = 5$ **15.** Inconsistent
17. $R = 100, S = 50, T = 10$ **19.** $A = 0, B = 50, C = 50$
21. (a–i) Inconsistent; (a–ii) independent; (a–iii) dependent; (b–i) $10 = 2$; (b–ii) $x = 1, y = -\frac{1}{2}$;
(b–iii), $0 = 0$; (c) two numbers equal to each other, such as $0 = 0$; (d) two numbers not equal to
each other, such as $10 = 2$
23. $m = -1, b = 5$

Exercise Set 10.5, pp. 328–330

1. $x = 2, y = 3$ **3.** $x = 4, y = -5, z = 2$ **5.** $w = -4, x = -2, y = 5, z = 6$
7. $x = 73/408, y = 11/6, z = 1/8$ **9.** $x = 1, y = 5, z = 3$ **11.** $x = 25, y = 3$
13. $x = \frac{7}{9}, y = -\frac{7}{9}, z = \frac{40}{9}$ **15.** $x = -1, y = -1, z = 1$ **17.** $x = 30, y = 10, z = -10$
19. $r = 10, s = 5, t = 100$
21. As the value of each variable is substituted, all of the other variables are eliminated from the
problem and the step is reached where $0 = 0$
23. $x = \frac{23}{17}, y = \frac{44}{51}, z = \frac{131}{51}, w = \frac{8}{3}$
25. $R = 5, S = -2, T = 3, U = 4, V = 1, W = 10$

Exercise Set 10.6, pp. 334–335

1. The two numbers are 116 and 16.   **3.** The two numbers are 3 and 129.
5. (a) 40 units; (b) at the breakeven point there is no profit or loss, $0; (c) $1400 loss, $4200 profit;
(d) 70 units
7. $13428.57 was loaned at 8.5%, $3357.14 was loaned at 9%, and $1214.29 was loaned at 10%
9. $a = -0.029, b = 2.167, c = 1.250$
11. (15, 260) Algebraically, p could also equal -20; however, a negative price is not valid in this
situation.

Review Exercises, pp. 336–337

1. Systems (b) and (d)   **3.** Slope $= -\frac{2}{3}$; y-intercept $= (0, -2)$; x-intercept $= (-3, 0)$
5. Slope $= \frac{4}{3}$; y-intercept $= (0, 8)$; x-intercept $= (-6, 0)$

7. (5, 1)

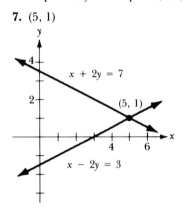

9. (4, 5)

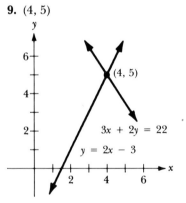

11. Dependent   **13.** Independent; $(x, y) = (3, -1.5)$ **15.** $x = 7, y = 1$ **17.** $x = 3, y = -2$
19. $x = -171/28, y = -331/28, z = 9/28$ **21.** No unique solution; dependent system
23. $x = 2, y = 8, z = 7, w = 0$ **25.** $6000 at 10% and $2000 at 5%

Chapter 11

Exercise Set 11.1, pp. 344–345

1. (a) 6; (b) 12; (c) 5   3. False   5. False   7. 1   9. 5   11. 3   13. 7   15. 1.5
17. Square matrix   19. Row vector   21. Square matrix   23. $\begin{bmatrix} 7.2 & 1 & 3 & 5 \\ 0 & 4 & -2 & -6 \end{bmatrix}$

25. $\begin{bmatrix} 2 & 3 & 4 & 5 & 6 \\ 3 & 4 & 5 & 6 & 7 \\ 4 & 5 & 6 & 7 & 8 \\ 5 & 6 & 7 & 8 & 9 \\ 6 & 7 & 8 & 9 & 10 \end{bmatrix}$ For example, $m_{11} = 1 + 1 = 2$;
$m_{32} = 3 + 2 = 5$

Exercise Set 11.2, pp. 353–355

1. 13   3. -13 5. 13   7. 44.18   9. 0   11. $6K - (-4K) = 0; K = 0$
13. $-2K^2 - 4K = 0; K = 0, K = -2$ 15. 0   17. -4 19. 40   21. 69
23. $r^2t + rt^2 - r^3 - t^3$ 25. $x = -3$ 27. $x = 2, x = 3$ 29. -24

Exercise Set 11.3, pp. 362–364

1. $x = \dfrac{17}{6}, y = \dfrac{-13}{18}$ 3. $x = -1, y = -2$ 5. $x = -4, y = 0$ 7. $x = -5, y = -2$
9. $x = 1.1, y = -1.8$
11. If $h = 2$, an infinite number of solutions (same line). If $h = 1$, one solution $(x, y) = (2, -1)$.
13. $x = 1.08, y = 0.68, z = 0.52$ 15. $x = 2.12, y = -2.48, z = 4.28$ 17. $a = 7, b = -1, c = 3$
19. $x = 1, y = 1, z = 1$ 21. $r = 1, s = 2, t = 3$ 23. $x = \dfrac{7}{3}, y = \dfrac{4}{3}, z = \dfrac{13}{3}, w = \dfrac{10}{3}$

Exercise Set 11.4, pp. 373–377

1. (b) and (g) are equal   3. $\begin{bmatrix} 1 & 0 \\ 0 & 1 \end{bmatrix}$ 5. $\begin{bmatrix} 9 & 6 & 0 & 6 \\ 0 & 13 & 6 & 9 \end{bmatrix}$ 7. $\begin{vmatrix} 5 & 0.5 & 10 \\ 0 & 3 & 6 \end{vmatrix}$

9. $a = 2, b = 5, c = 1$ 11. $\begin{bmatrix} 2.3 & -3.1 \\ -4.1 & -5.2 \\ -1.2 & 7.3 \end{bmatrix}$ 13. $\begin{bmatrix} 3 & 9 & 6 \\ 12 & 3 & 15 \end{bmatrix}$ 15. $\begin{bmatrix} 5 & -5 & 14 \\ -16 & 13 & -3 \end{bmatrix}$

17. $\begin{bmatrix} -2 & -6 & -4 \\ -8 & -2 & -10 \end{bmatrix}$ 19. $\begin{bmatrix} 10 & 0 & 26 \\ -14 & 22 & 8 \end{bmatrix}$ 21. $x = 3, y = 3, z = 4, w = -8$

23. $x = 2, y = 1$ 25. $\begin{bmatrix} -2 & 3 & -4 \\ -1 & -5 & -6 \\ -4 & 0 & 1 \end{bmatrix}$ 27. $\begin{bmatrix} 15 & 14 \\ 5 & 30 \end{bmatrix}$ 29. $[51 \quad 53 \quad 62]$ 31. $\begin{bmatrix} r & s & t \\ u & v & w \\ x & y & z \end{bmatrix}$

33. $\begin{bmatrix} -21 & 13 & -16 \\ 8 & 64 & -22 \\ -28 & 36 & -40 \end{bmatrix}$ 35. $\begin{bmatrix} r - 2s + 4t + 5u \\ 3r + 2u \\ s + 7t \\ 5s + 2u \end{bmatrix}$

37. (a) $\begin{bmatrix} 1 & 0 \\ 0 & 1 \end{bmatrix}$; (b) $\begin{bmatrix} 1 & 0 \\ 0 & 1 \end{bmatrix}$; (c) multiplication is commutative

39. $\begin{bmatrix} r \\ s \end{bmatrix}$; a 2-by-1 matrix identical to the original.

41. $\begin{bmatrix} x \\ y \\ z \end{bmatrix}$; a 3-by-1 matrix identical to the original.

43. $\begin{bmatrix} x \\ 3y \\ z \end{bmatrix}$; row 2 of original matrix multiplied by 3.

45. $\begin{bmatrix} y \\ x+y \\ z \end{bmatrix}$; new first row was former second row; new second row is sum of former first two rows.

47. $\begin{bmatrix} z \\ y \\ z \end{bmatrix}$; rows one and three of original interchanged.

49. (a) $\begin{bmatrix} 4 & 2 & 1 & 1 \\ 1 & 0 & 0 & 1 \\ 6 & 1 & 0 & 1 \end{bmatrix}$; (b) $\begin{bmatrix} 10 & 12 & 13 & 13 \\ 8 & 9 & 9 & 8 \\ 0 & 5 & 6 & 5 \end{bmatrix}$

Exercise Set 11.5, pp. 389–391

1. (a) $\begin{pmatrix} 1 & 0 \\ 0 & 1 \end{pmatrix}$; (b) $\begin{pmatrix} 1 & 0 & 0 \\ 0 & 1 & 0 \\ 0 & 0 & 1 \end{pmatrix}$; (c) $\begin{pmatrix} 1 & 0 & 0 & 0 \\ 0 & 1 & 0 & 0 \\ 0 & 0 & 1 & 0 \\ 0 & 0 & 0 & 1 \end{pmatrix}$;

(d) $\begin{pmatrix} a_{11} & a_{12} & a_{3} & \cdots & a_{1n} \\ a_{21} & a_{22} & a_{23} & \cdots & a_{2n} \\ \vdots & & & & \\ a_{n1} & a_{n2} & a_{n3} & \cdots & a_{nn} \end{pmatrix}$ For any a_{ij}, if $i = j$, then $a_{ij} = 1$. Otherwise, $a_{ij} = 0$.

3. $\begin{bmatrix} 3 & -2 \\ 10 & 7 \end{bmatrix}\begin{bmatrix} x \\ y \end{bmatrix} = \begin{bmatrix} 13 \\ 43 \end{bmatrix}$ **5.** $\begin{bmatrix} 14 & 1 \\ 8 & -2 \end{bmatrix}\begin{bmatrix} x \\ y \end{bmatrix} = \begin{bmatrix} 0 \\ 18 \end{bmatrix}$ **7.** $\begin{bmatrix} 1.2 & 1.7 \\ 0.8 & 0.3 \end{bmatrix}\begin{bmatrix} x \\ y \end{bmatrix} = \begin{bmatrix} -11.6 \\ -4.4 \end{bmatrix}$

9. $\begin{bmatrix} a & b \\ c & d \end{bmatrix}\begin{bmatrix} x \\ y \end{bmatrix} = \begin{bmatrix} e \\ f \end{bmatrix}$ **11.** $\begin{bmatrix} 1 & 5 \\ 3 & 7 \end{bmatrix}$ **13.** $\begin{bmatrix} 1 & -2 & 8 \\ 3 & 6 & 7 \\ 4 & 5 & 2 \end{bmatrix}$

15. (a) $\begin{vmatrix} 1 & -1 & 1 \\ 4 & 5 & 6 \\ 3 & 2 & -7 \end{vmatrix} = -100$; (b) $m_{11} = -47, m_{12} = 46, m_{13} = -7$ (c) $\begin{bmatrix} -47 & -5 & -11 \\ 46 & -10 & -2 \\ -7 & -5 & 9 \end{bmatrix}$
$m_{21} = -5, m_{22} = -10, m_{14} = -5;$
$m_{31} = -11, m_{32} = -2, m_{15} = 9$

(d)
$$M^{-1} = \begin{bmatrix} 0.47 & 0.05 & 0.11 \\ -0.46 & 0.10 & 0.02 \\ 0.07 & 0.05 & -0.09 \end{bmatrix}$$

17. $\begin{bmatrix} 1 & -3 \\ -2 & 7 \end{bmatrix}$ **19.** $\begin{bmatrix} -0.8 & 0.7 \\ 0.6 & -0.4 \end{bmatrix}$ **21.** Singular **23.** $\begin{bmatrix} 0.15625 & -0.3125 & -0.72916\overline{6} \\ -0.375 & -0.125 & 0.375 \\ -0.15625 & 0.3125 & 0.40625 \end{bmatrix}$

25. $\begin{bmatrix} 2 & 1 & 0 \\ -1 & 3 & 1 \\ -5 & 4 & 2 \end{bmatrix}$ **27.** Singular **29.** $x = 4, y = 0.5, z = -3$; see Exercise 25.

Review Exercises, pp. 397–401

1. (a) $a_{13} = 7$; (b) $a_{24} = 1$; (c) $a_{23} = 8$; (d) $a_{21} = 0$; (e) $a_{32} = 6$; (f) $a_{31} = 4.2$
3. (a) 1 by 4; (b) row vector   **5.** (a) 2 by 4; (b) none of these

7. (a) $\begin{pmatrix} a & b \\ c & d \\ e & f \\ g & h \end{pmatrix}$; **(b)** 4 different matrices could be formed with eight elements {a 1×8, a 2×4, a 4×2, and an 8×1}

9. $r = 9, s = 3, t = -2, u = 8, v = 7$ **11.** $\begin{bmatrix} 1 & 0 \\ 0 & 1 \end{bmatrix}$ **13.** $\begin{bmatrix} -6 & 7 & 4 \\ -11 & 39 & -24 \end{bmatrix}$

15. (a) $\begin{bmatrix} 13 & -2 \\ 13 & -13 \end{bmatrix}$; **(b)** $\begin{bmatrix} 13 & -2 \\ 13 & -13 \end{bmatrix}$; **(c)** $\begin{bmatrix} 13 & -2 \\ 13 & -13 \end{bmatrix}$; **(d)** yes   **17.** $\begin{bmatrix} -6 \\ 18 \end{bmatrix}$ **19.** $\begin{bmatrix} 43 & 61 \\ 40 & 13 \\ 29 & -5 \end{bmatrix}$

21. $\begin{bmatrix} x \\ 2y \\ 3z \end{bmatrix}$ **23.** 70   **25.** $\begin{bmatrix} y \\ x \end{bmatrix}$ **27.** $\begin{bmatrix} x + 2y - 3z + w \\ 2x - 4y - z - 3w \\ -x - y - z - w \\ x - 2y + 3z + 2w \end{bmatrix}$

29. 4   **31.** 24   **33.** -69 **35.** 69   **37.** $3x - 2y = 10, 7x + 6y = 22$
39. $r - s + 2t = 7, r - 3s + 3t = -1, r + 2s - 3t = 5$

41. (a) -2; **(b)** $\begin{bmatrix} 7 & -11 & 1 \\ -5 & 7 & -1 \\ -3 & 5 & -1 \end{bmatrix}$; **(c)** $\begin{bmatrix} -3.5 & 5.5 & -0.5 \\ 2.5 & -3.5 & 0.5 \\ 1.5 & -2.5 & 0.5 \end{bmatrix}$ **43.** $\begin{bmatrix} -3.5 & 5.5 & -0.5 \\ 2.5 & -3.5 & 0.5 \\ 1.5 & -2.5 & 0.5 \end{bmatrix}$

45. $\begin{bmatrix} 3/38 & 7/38 \\ 1/19 & -4/19 \end{bmatrix}$ **47.** Coefficient matrix $\begin{bmatrix} 23 & 9 \\ 5 & 2 \end{bmatrix}$; inverse $\begin{bmatrix} 2 & -9 \\ -5 & 23 \end{bmatrix}$; $x = -10, y = 30$

49. Coefficient matrix $\begin{bmatrix} 1 & -1 & 2 \\ 3 & 1 & -1 \\ 1 & -2 & -2 \end{bmatrix}$; inverse $\begin{bmatrix} 0.174 & 0.261 & 0.043 \\ -0.217 & 0.174 & -0.304 \\ 0.304 & -0.043 & -0.174 \end{bmatrix}$ or

$\begin{bmatrix} 4/23 & 6/23 & 1/23 \\ -5/23 & 4/23 & -7/23 \\ 7/23 & -1/23 & -4/23 \end{bmatrix}$; $x = -1/23, y = 7/23, z = 27/23$

51. For inverse, see Exercise 43; $x = 1, y = 1, z = 1$

Chapter 12

Exercise Set 12.1, pp. 410–412

1.
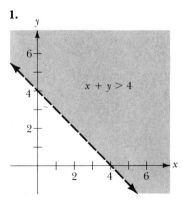

3.
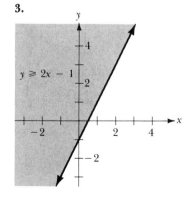

5.
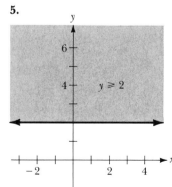

7.

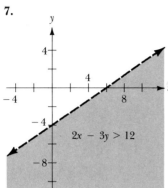

$2x - 3y > 12$

9.

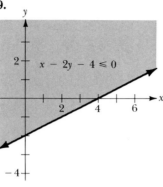

$x - 2y - 4 \leq 0$

11.

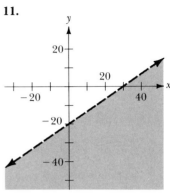

$0.16x - 0.24y > 4.8$

13.

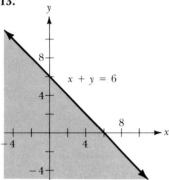

$x + y = 6$

15.

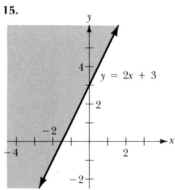

$y = 2x + 3$

17.

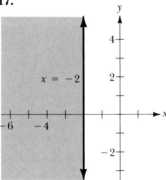

$x = -2$

19.

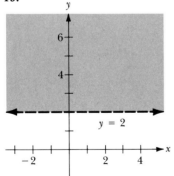

$y = 2$

21.

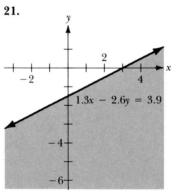

$1.3x - 2.6y = 3.9$

23.

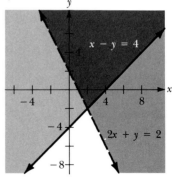

$x - y = 4$

$2x + y = 2$

25.

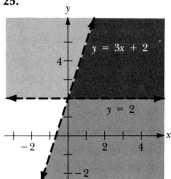

Note: $x \geqslant 0$

27.

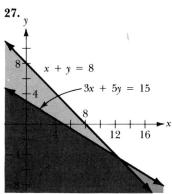

29.

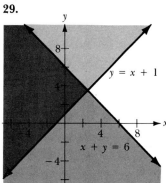

31.

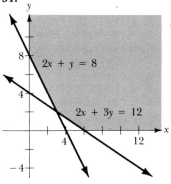

Note: $x \geqslant 0, y \geqslant 0$

33.

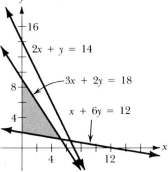

Note: $x \geqslant 0, y \geqslant 0$

35.

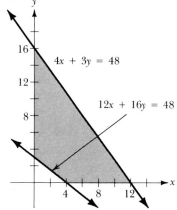

Note: $x \geqslant 0, y \geqslant 0$

37.

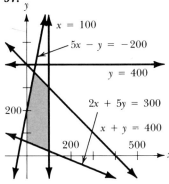

Note: $x \geqslant 0, y \geqslant 0$

39.

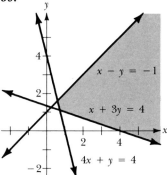

41.

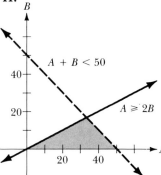

Note: $A \geqslant 0, B \geqslant 0$

Exercise Set 12.2, pp. 420–421

1. (a) $p = 130$; all constraints are satisfied; **(b)** $p = 230$, all constraints are satisfied; **(c)** $p = 180$, all constraints are satisfied; **(d)** $p = 270$, the second constraint is not satisfied; **(e)** $p = 390$, none of the constraints is satisfied; **(f)** $p = 240$, the second constraint is not satisfied; **(g)** $p = 398$, the first constraint is not satisfied; **(h)** $p = 405$, the first constraint is not satisfied; **(i)** $p = 400$, the first constraint is not satisfied; **(j)** $p = 407$, the first constraint is not satisfied.
3. (a) $p = 12$, one constraint is not satisfied; **(b)** $p = 13$, one constraint is not satisfied; **(c)** $p = 14$, two constraints are not satisfied; **(d)** $p = 15$, two constraints are not satisfied; **(e)** $p = 16$, two constraints are not satisfied; **(f)** $p = 13$, one constraint is not satisfied; **(g)** $p = 10$, one constraint is not satisfied; **(h)** $p = 7$, all constraints are satisfied; **(i)** $p = 4$, all constraints are satisfied; **(j)** $p = 9$, all constraints are satisfied; **(k)** $p = 10$, all constraints are satisfied; **(l)** $p = 11$, all constraints are satisfied; **(m)** $p = 12$, one constraint is not satisfied; **(n)** $p = 9$, all constraints are satisfied; **(o)** $p = 6$, all constraints are satisfied; **(p)** $p = 3$, all constraints are satisfied.
5. (19, 29) yields 115 for cost.
7. $0.14c + 0.18d = p$; $c + d < 15{,}000$, $c - d \geq 1000$, $0 < c \leq 7500$, $0 < d \leq 5000$

Exercise Set 12.3, pp. 430–433

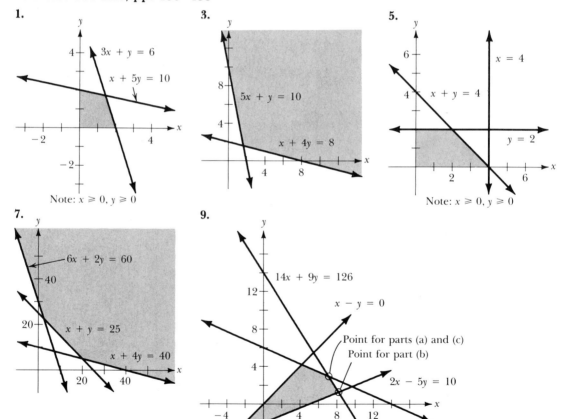

11. Maximum is (5, 3); minimum is (1, 2).   **13.** Maximum is (5, 3); minimum is (1, 2).
15. Maximum is (6, 3); minimum is (1, 2).

17. $(x, y) = (50, 0); P = 250$ **19.** $(x, y) = (12, 1); P = 129$ **21.** $(x, y) = (0, 2);$
$C =$ unbounded, no minimal solution.

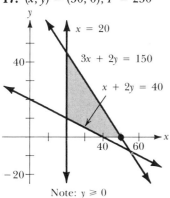 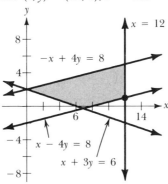 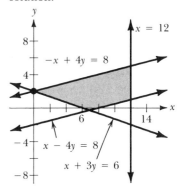

23. $(x, y) = (2.5, 22.5);$
$C = 78.75$

25. $(x, y) = (1.5, 2); P = 6.5$

27. $(x, y) = (2.29, 5.14);$
$C = 19.43$

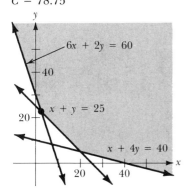 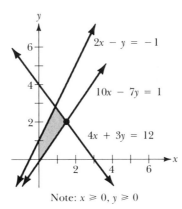 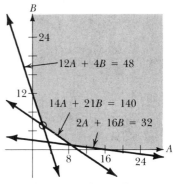

29. The objective function is $4A + 3B = P$. The constraints are:
$$30A + 50B \le 200,$$
$$40A + 20B \le 160,$$
$$60A + 20B \le 240.$$
$A = 3, B = 2$, and the profit is \$18.00.

Exercise Set 12.4, pp. 444–447

1. **(a)** $7x + 6y + 0S_1 + S_2 = 20$; **(b)** x and y; **(c)** $x = 0, y = 0, S_1 = 6, S_2 = 20, P = 0$; **(d)** 4;
(e) $5.5x + 0y - 1.5S_1 + S_2 = 11$; **(f)** x and S_1; **(g)** $P = 7.5, x = 0, y = 1.5, S_1 = 0, S_2 = 11$; **(h)** x;

(i) iii **(j)** $x + 0y - \dfrac{3}{11} S_1 + \dfrac{2}{11} S_2 = 2$; **(k)** $0x + 0y - 0.5S_1 - 0.5S_2 = P - 13$;

(l) $0x + y + \dfrac{7}{22} S_1 - \dfrac{1}{22} S_2 = 1$; **(m)** $x = 2, y = 1, S_1 = 0, S_2 = 0$

3. $x + 2y + 0S_1 + 0S_2 = P$ **5.** $30x + 25y + 0S_1 + 0S_2 + 0S_3 = P$
 $\quad x + 5y + S_1 + 0S_2 = 10$ $\quad 20x - 10y + S_1 + 0S_2 + 0S_3 = 400$
 $\quad 3x + y + 0S_1 + S_2 = 6$ $\quad 40x + 15y + 0S_1 + S_2 + 0S_3 = 1200$
 $\qquad\qquad\qquad\qquad\qquad\qquad 10x + 30y + 0S_1 + 0S_2 + S_3 = 300$

7. $x + 2y + 3z + 0S_1 + 0S_2 + 0S_3 = P$
 $\quad 4x + 3y + 2z + S_1 + 0S_2 + 0S_3 = 400$
 $\quad 3x + 2y + 4z + 0S_1 + S_2 + 0S_3 = 500$
 $\quadx + y + 2z + 0S_1 + 0S_2 + S_3 = 200$

9. (16) $x \approx 1.429,\ y \approx 1.714,\ P \approx 4.857$ (answers rounded to 3 decimal places)
 (17) $x = 50,\qquad y = 0,\qquad P = 250$
 (18) $x = 0,\qquad y = 2,\qquad P = -22$
 (19) $x = 12,\qquad y = 1,\qquad P = 129$
 (20) $x = 0,\qquad y = 2,\qquad c = -22$
 (21) $x = 0,\qquad y = 2,\qquad c$ is unbounded
 (22) $x = 12,\qquad y = 5,\qquad P = 27$
 (23) $x = 2.5,\qquad y = 22.5,\quad c = 98.75$
 (24) $x \approx 0.231,\ y \approx 3.692,\ P \approx 11.308$ (answers rounded to thousandths)
 (25) $x = 1.5,\qquad y = 2,\qquad P = 6.5$
 (26) $A = 3,\qquad B = 2.5,\qquad C = 110$
 (27) $x = 2.29,\quad y = 5.14,\quad c = 19.43$

11. $x = 7.8, y = 2.4$

Exercise Set 12.5, p. 453

1. $\begin{pmatrix} 3 & 2 & 0 & 0 \\ 4 & 1 & 1 & 0 \\ 2 & 3 & 0 & 1 \end{pmatrix} X = \begin{pmatrix} P \\ 14 \\ 22 \end{pmatrix}$ **3.** $\begin{pmatrix} 1 & 1 & 0 & 0 \\ 20 & 30 & 1 & 0 \\ 50 & 10 & 0 & 1 \end{pmatrix} X = \begin{pmatrix} P \\ 120 \\ 170 \end{pmatrix}$ **5.** $\begin{pmatrix} 4 & 4 & 0 & 0 \\ 2 & 3 & 1 & 0 \\ 3 & 2 & 0 & 1 \end{pmatrix} X = \begin{pmatrix} P \\ 18 \\ 24 \end{pmatrix}$

7. $\begin{pmatrix} 3 & 2 & 0 & 0 & 0 \\ 2 & 1 & 1 & 0 & 0 \\ 1 & 3 & 0 & 1 & 0 \\ 4 & 6 & 0 & 0 & 1 \end{pmatrix} X = \begin{pmatrix} P \\ 10 \\ 9 \\ 24 \end{pmatrix}$ **9.** $\begin{pmatrix} 10 & 10 & 0 & 0 & 0 \\ 2 & 3 & 1 & 0 & 0 \\ 5 & 4 & 0 & 1 & 0 \\ 0 & 1 & 0 & 0 & 1 \end{pmatrix} X = \begin{pmatrix} P \\ 90 \\ 200 \\ 20 \end{pmatrix}$

Exercise Set 12.6, pp. 463–466

1.

	a	b	s_1	s_2	
$P \rightarrow$	3	⑤	1	0	10
	2	4	0	1	12
	-5	-8	0	0	0
		↑			
		k			

3.

	r	t	l	s_1	s_2	s_3	
	4	2	3	1	0	0	6
$P \rightarrow$	-2	⑴	2	0	1	0	6
	4	3	4	0	0	1	6
	-1	-2	3	0	0	0	0
		↑					
		k					

5.

1	0	1.2	-0.2	9
0	1	-0.2	0.2	1
0	0	3.4	0.6	43

7.

0.5	0	1	0	-0.83	4
1	0	0	1	-0.33	16
0.5	1	0	0	0.83	20
0	0	0	0	6.67	160

rounded to hundredths

9. $x = 1.429, y = 1.714$ **11.** 12.625   **13.** $a = 0, b = 25$, cost $= 100$
15. (a) $x = 50, y = 0, p = 250$; (b) $x = 2, y = 3, p = 11$

17. (7) $x = 4.5, y = 1, p = 15.5$; (8) $x = 0.75, y = 1.75, p = 4.25$; (9) $x = 34.286, y = 7.143$, $p = 414.286$

19. $6000 invested in bond A and $3000 invested in bond M with a return of $840

Review Exercises, pp. 468–471

1. (27, 13) maximum; other points are (10, 10), (20, 10), and (25, 10)
3. (1, 1) minimum; all these points satisfy the constraints

5.
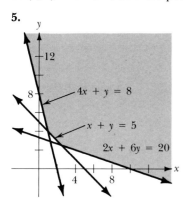

7.
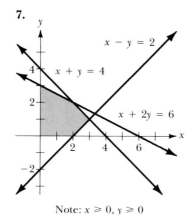

Note: $x \geq 0, y \geq 0$

9. $x = 1, y = 0$; maximum value is 7
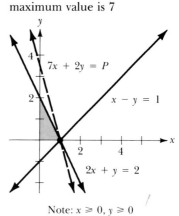

Note: $x \geq 0, y \geq 0$

11. $x \approx 5.86, y \approx 0.35$; maximum value is 96.63 (values rounded to hundredths)
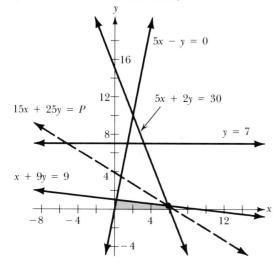

Note: $x \geq 0, y \geq 0$

13. $x = 8, y = 0, p = 24$
15. $x = 10, y = 0, z = 0, p = 40$
17.

4	3	2	1	0	0	90
1	1	1	0	1	0	32
3	4	2	0	0	1	84
40	39	30	0	0	0	

Objective function has not been multiplied by -1 yet.

19. (13) $x = 8, y = 0, p = 24$; (14) $x = 72, y = 12, p = 696$; (15) $x = 10, y = 0, z = 0, p = 40$; (16) $x = 20, y = 20, p = 1400$; (17) $x = 10\frac{2}{3}, y = 4\frac{2}{3}, z = 16\frac{2}{3}, p = 1108\frac{2}{3}$; (18) $x = 0, y = 40, c = -80$

Chapter 13

Exercise Set 13.1, pp. 483–486

1. Statistics is the branch of mathematics concerned with the collection, organization, analysis, presentation, and interpretation of numerical data.

5. If all conditions were similar to the present ones, 80 times out of 100 snow would fall.

7. (a) $\dfrac{600}{1000} = 0.6$; (b) $\dfrac{300}{1000} = 0.3$; (c) $\dfrac{700}{1000} = 0.7$; (d) $\dfrac{400}{1000} = 0.4$; (e) $\dfrac{100}{1000} = 0.10$; (f) $\dfrac{400}{1000} = 0.4$

9. (a)

	1	2	3	4	5	6	7	8
Coin 1	H	H	H	H	T	T	T	T
Coin 2	H	T	H	T	H	H	T	T
Coin 3	H	H	T	T	H	T	H	T

8 possible outcomes

(b) From the 8 columns (possible outcomes), columns 4, 6, and 7 contain 2 tails. Therefore, prob(2 tails) = $\frac{3}{8}$.

11. The two weeks is a sample   **13.** A sample   **15.** Some of your blood would be a sample.

17. Simple random sampling   **19.** Stratified sampling

21. One method could have been to simply divide the total number of births for a given time period by the number of seconds in that time period (i.e., total number of births in a year divided by the number of seconds in a year). Sampling could have been used whereby selected hospitals were studied for a given time period on a certain date. (Many other answers are possible.)

Exercise Set 13.2, pp. 508–512

1. Descriptive statistics is that branch of statistics dealing with the collection, tabulation, and summarization of a given set of data.

3. (a)

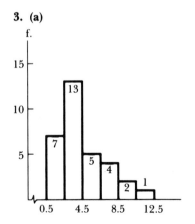

Frequency histogram

(b)

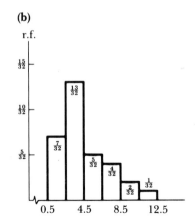

Relative frequency histogram

(c)

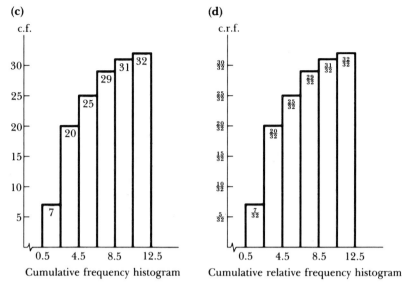

Cumulative frequency histogram

(d)

Cumulative relative frequency histogram

5. The data is skewed right (the sharp tail of the distribution is to the right); in addition, the positioning of the mode, median, and mean (respectively, 3, 4, and 4.5) indicate a distribution skewed right.

7. (a)   **(b)**

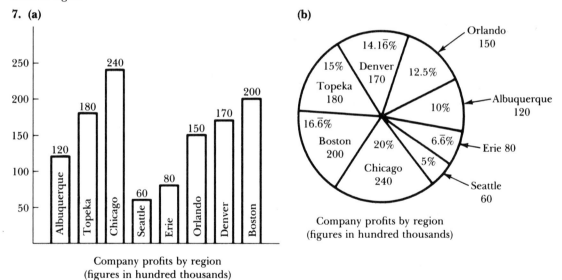

Company profits by region
(figures in hundred thousands)

Company profits by region
(figures in hundred thousands)

9. (a) The graph in Fig. 13.24 "seems to indicate" that the inflation rate was cut in half (according to the relative heights of the bars). However, by examining the vertical axis, we see that the distance from 0% to 6% is the same as the distance from 6% to 9%. Either the graph should contain a break, or the vertical axis should be extended to include 9%, 6%, and 3%.
(b) While the actual increase in taxes doubled ($1.32 to $2.64), the size of the circles is misleading. As the radius or diameter is doubled, the area of a circle is increased by four.

11. (a) The mean is $4.48, the median is $3.99, and the mode is $3.05 (rounded to the nearest cent).
(b) Either the median or the mode would better reflect the "center of the distribution"; the mean is influenced by the extreme price of $9.80.
13. (a) Mean is 6; standard deviation is 3.162. **(b)** Mean is 6; standard deviation is 0.
(c) Mean is 16; standard deviation is 3.162. If some constant number is added (or subtracted) to all the data items, the mean is also increased (or decreased) by that constant—compare the results of parts (a) and (c). The standard deviation remains unaffected.
(d) Mean is 60; standard deviation is 31.62. If all the data items are multiplied by some constant number, the mean and the standard deviation of the original data are also multiplied by that same constant—compare parts (a) and (d). **(e)** Mean is 600; standard deviation is 316.2—compare to part (a).
15. $n = 32$; range = 2250; high = 3570; low = 1320; mean = 1999.53; median = 1880; each of these salaries has a frequency of 2: 1550, 1840, 1850, and 1950; first quartile = 1720; third quartile = 2125; semi-interquartile range = 202.5.

Exercise Set 13.3, pp. 522–526

1. Statistical inference is making generalizations (or inferences) to a population from data taken from a representative sample of that population.
3. Probability aids in determining how close the estimate of the population is to that actual population parameter. The study of probability assists in determining the degree of confidence that can be placed in the estimates of the population parameters.
5. Assuming a "fair" die, the bar graph should be almost uniformly distributed; there should be approximately an equal number of 1's, 2's, . . . , and 6's.
7. (a) 1/6; **(b)** 2/6 = 1/3; **(c)** 0/6 (no 7's on one die); **(d)** 4/6;
(e) 6/6 (definitely has to be one of these outcomes).
9. (a) 6/36 or 1/6; **(b)** 15/36; **(c)** 15/36; **(d)** 0/36; **(e)** 1/36; **(f)** 18/36 or 1/2; **(g)** 18/36 or 1/2;
(h) 13/36; **(i)** 15/36; **(j)** 7/36.
11. All answers could be reduced. **(a)** 27/60; **(b)** 33/60; **(c)** 8/60; **(d)** 52/60; **(e)** 21/60;
(f) 32/60 + 27/60 − 21/60 = 38/60
13. (a) Rolling two dice and (i) obtaining a sum of 7 and (ii) obtaining a sum of 11;
(b) picking two cards from a deck of cards and (i) getting an ace and (ii) getting an ace on the second pick, provided the first card is not replaced;
(c) picking two cards from a deck of cards and (i) getting an ace and (ii) getting an ace on the second pick, provided the first card is replaced;
(d) (i) studying for a mathematics exam and (ii) passing the exam.
15. Assuming the digits are not repeated in a single code (3533), $P = 840$.
17. Using a tree diagram for an example shows there are $2^3 = 8$ possible ways.
19. (a) 3; **(b)** 10; **(c)** 21; **(d)** 210
21. (a) 24 (assuming no letter is repeated in the same code; no CCBC); **(b)** 1/4 of all the codes, or 6;
(c) 1/4 of all the codes, or 6
23. (a) $(\frac{1}{4})^5$; **(b)** $(\frac{1}{2})^5$; **(c)** $(\frac{12}{52})^5$; **(d)** $5 \times (4/52) \times (4/52) \times (4/52) \times (4/52) \times (48/52)$

Review Exercises, pp. 527–531

1. True   **3.** True   **5. (a)** Statistic (from a sample); **(b)** statistic; **(c)** parameter (all businesses)
7. Probability is the likelihood of some stated event occurring; the ratio of favorable outcomes to all possible outcomes.

9. (a) $\frac{4}{80}$ or 0.05; **(b)** For the 200 additional runs, 200×0.05 or 10 errors.

11. Individual answers will vary for parts (a)–(d). For part (e): Taking several samples and "averaging" the means of the different samples *should* provide the best estimate.

13.

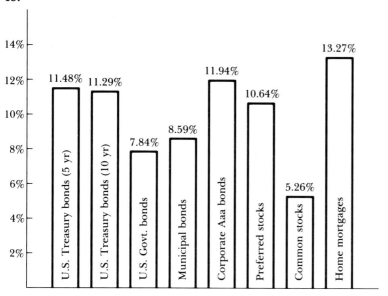

Annual average investment yields

15. (a) \$5,320,959; **(b)** \$22,186,500; **(c)** \$10,451,884 (answers to nearest dollar)

17. (a) 2/6 or 1/3; **(b)** 1/6;

(c)	A	B	C	D	E	F
A	AA	AB	AC	AD	AE	AF
B	BA	BB	BC	BD	BE	BF
C	CA	CB	CC	CD	CE	CF
D	DA	DB	DC	DD	DE	DF
E	EA	EB	EC	ED	EE	EF
F	FA	FB	FC	FD	FE	FF

19. $3 \times 4 \times 2 \times 5$ by the multiplication principle, or 120 different systems

Index

L.-Brault

✓

DATE DE RETOUR

2 8 JAN 2007